LAWRENCE
AND HIS
LABORATORY

The publisher gratefully acknowledges
the generous contribution provided
by the Director's Circle of the
Associates of the University of California Press
whose members are

Edmund Corvelli, Jr.
Leslie and Herbert Fingarette
Diane and Charles L. Frankel
Florence and Leo Helzel
Sandra and Charles Hobson
Valerie and Joel Katz
Robert Marshall
Ruth and David Mellinkoff
Joan Palevsky

A CENTENNIAL BOOK

One hundred books
published between 1990 and 1995
bear this special imprint of
the University of California Press.
We have chosen each Centennial Book
as an example of the Press's finest
publishing and bookmaking traditions
as we celebrate the beginning of
our second century.

UNIVERSITY OF CALIFORNIA PRESS

Founded in 1893

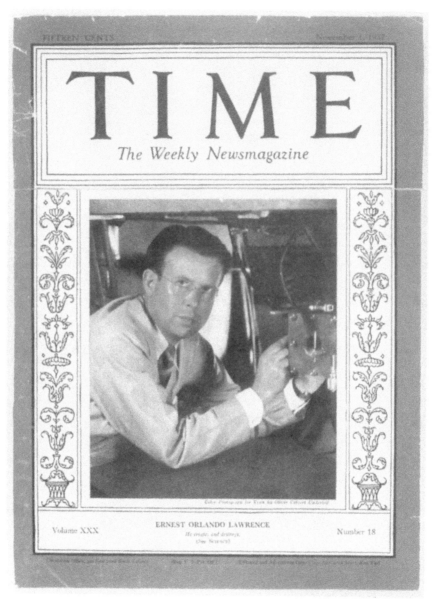

LAWRENCE
AND HIS
LABORATORY

A HISTORY
of the LAWRENCE
BERKELEY
LABORATORY

VOLUME I

J. L. Heilbron
and Robert W. Seidel

University of California Press

Berkeley   Los Angeles   Oxford

CALIFORNIA STUDIES IN THE HISTORY OF SCIENCE
J. L. Heilbron, Editor

The Galileo Affair: A Documentary History, edited and translated by Maruice A Finocchiaro

The New World, 1939–1946 (A History of the United States Atomic Energy Commission, volume 1) by Richard G. Hewlett and Oscar E. Anderson, Jr.

Atomic Shield, 1947–1952 (A History of the United States Atomic Energy Commission, volume 2) by Richard G. Hewlett and Francis Duncan

Atoms for Peace and War, 1953–1961: Eisenhower and the Atomic Energy Commission (A History of the United States Atomic Energy Commission, volume 3) by Richard G. Hewlett and Jack M. Holl

Lawrence and His Laboratory: A History of the Lawrence Berkeley Laboratory, Volume 1 by J. L. Heilbron and Robert W. Seidel

University of California Press
Berkeley and Los Angeles, California

University of California Press, Ltd.
Oxford, England
© 1989 by
The Regents of the University of California

Library of Congress Cataloging-in-Publication Data

Heilbron, J. L.
    Lawrence and his laboratory: a history of the Lawrence Berkeley Laboratory/ J.L. Heilbron and Robert W. Seidel.
    p. cm. — (California studies in the history of science)
        Bibliography: v. 1, p.
        Includes index.
        ISBN 0-520-06426-7 (v. 1: alk. paper)
        1. Lawrence Berkeley Laboratory—History.   2. Lawrence, Ernest Orlando. 1901–1958.   3.   Physicists—United   States—Biography.   I.   Seidel,   Robert W.   II. Title.   III. Series.
QC789.2.U62L384   1989   89–4820
539.7'0720794'67—dc20   CIP

Printed in the United States of America
1   2   3   4   5   6   7   8   9

Contents

Plates

Figures

Tables

Introduction

The Lawrence Berkeley Laboratory is a scientific institution of the first importance. It was the forerunner of the modern multipurpose national research laboratory, the direct parent of Livermore and Los Alamos, an essential contributor to the wartime work of Oak Ridge and Hanford, the inspiration for the founders of Brookhaven. Its achievements have long been recognized through awards of Nobel prizes, memberships in the National Academy of Sciences, and other high scientific honors. The hundreds of accelerator laboratories throughout the world give ample testimony to the Laboratory's contribution to modern big science. Many of the leaders of these institutions began their careers at the Laboratory, and most make or made use of its technology.

As with most novel technologies, the art of accelerator building had to be learned via apprenticeship. Berkeley was the center of the art. Former apprentices trained there opened up new technologies that fed back for further development. The modern linear accelerator, electron and proton synchrotrons, heavy-ion accelerators, bubble chambers, and computers to analyze accelerator-produced data all owe their inspiration or success to the work of the Radiation Laboratory. The achievement of these technologies alone would be enough to lend the Laboratory great historical interest.

The motives and mechanisms that shaped the growth of the Laboratory helped to force deep changes in the scientific estate and in the wider society. In the entrepreneurship of its founder, Ernest Orlando Lawrence, these motives, mechanisms, and changes came together in a tight focus. He mobilized great and small philanthropies, state and local governments, corporations and plutocrats, volunteers and virtuosos. The work they supported, from astrophysics to atomic bombs, from radio chemistry to nuclear medicine, shaped the way we observe, control, and manipulate our environment. To do justice to the Laboratory's history requires a global perspective because its influence was global.

The scale of our history and the quantity of available documentation gave us an opportunity to study systematically questions of concern to historians, sociologists, and philosophers of science, and also to makers of science policy. The social construction of scientific knowledge should manifest itself, if ever, in the organized labor of thousands of scientists over half a century. The interaction of the individual with the organization, and of the intended with the unintended products of scientific activity, are present with exemplary clarity in the rich historical record of the Lawrence Berkeley Laboratory.

The results of our study will be presented in several volumes, of which this, which explains the growth of the Laboratory up to its mobilization for war in 1940/41, is the first. The second volume will carry the story forward through World War II, the early years of the Cold War, and the Korean War. The third will treat the evolution of the modern national laboratory. This periodization reflects the strong interaction between the development of the Laboratory and the main forces of recent history, and suggests our weighting of the social, economic, technological, and scientific factors at play.

The first two chapters of this volume set the stage—time, place, and intellectual milieu—for the invention of the Laboratory. We begin with the local conditions that led to the growth of physical science research in California, as epitomized by the Panama-Pacific Exposition of 1915, and opportunities in the state's academic institutions opened by the national transformation of science effected by the first world war. The recruitment and retention of Ernest Lawrence by the University of California could not have occurred apart from these broader conditions and opportunities.

The second chapter adopts a perspective localized in the world of knowledge, that of nuclear and high-energy physics. The problems of achieving high voltages for scientific investigations occupied minds and hands in many countries. The solutions ranged from the fanciful to the fatal. Most contributed something to Lawrence's thinking. Berkeley's first efforts at accelerator technology and institution-building, and the revolution in the understanding of nuclear physics that gave Lawrence a lever to work further on both, occupy our attention in chapter 3. Here we apply the lessons learned from our study of the temporal and geographic context to understand why Lawrence succeeded where others failed.

The account of Lawrence's early program of research in the fourth chapter introduces the reader, as the program did Lawrence, to the international world of nuclear physics. His first attempts to change the world were flawed by overenthusiasm and under preparation; they brought him forcibly and permanently to the attention of Ernest Rutherford, Werner Heisenberg, Niels Bohr, and other leaders of international physics. By following up discoveries made in Europe, Lawrence soon established himself as a prolific producer of new radioactive substances. The Laboratory earned a reputation for generosity by giving away the products of the cyclotron as well as information about the machine itself. Chapter 5 analyzes the social and financial underpinnings of the Laboratory. They included not only the givers—the private and public patrons Lawrence cultivated so effectively—but also the receivers, the disciples who submitted to the discipline of scientific and technical production to have the opportunity to work on a unique machine. Later, as missionaries from what they called their "Mecca," the disciples spread the word and the machine throughout America and the world.

In the sixth chapter we examine this missionary activity in the United States, and in the seventh in Europe and Japan. The examination brings out the general interests and concerns that conditioned the development of accelerator physics and related disciplines. It also shows how local conditions shaped the reception of the new tools and techniques disseminated from Berkeley.

The productivity and capacity of the cyclotron, as well as the need to raise money for its support and multiplication, brought cyclotroneers into other disciplines than physics and electrical engineering. Chapter 8 describes the Laboratory's contributions to radiochemistry and its pioneering in nuclear medicine. During the late 1930s, physicists tested cyclotron beams for applications to biology and medicine as assiduously as their predecessors had applied x rays half a century earlier.

Chapter 9 continues with research in nuclear physics and chemistry. The Laboratory discovered the true transuranic elements, which had eluded the nuclear physicists of Europe, and it confirmed and extended the European discovery of nuclear fission. Both lines of work presaged a new application of cyclotron technology and a decisive influence of the Laboratory on the course of human history.

The transition from peacetime to wartime occupies the final chapter, in which we examine the origins of what was to have been the ultimate cyclotron; the consequences of the recognition of Lawrence's achievements by the Nobel prize; and the first, tentative applications of accelerator personnel and principles to the instruments of war. This period of transition, during which the Laboratory only partially mobilized, came to an end with Pearl Harbor.

Because we treat both scientific and general aspects of the Laboratory's growth, our book divides into segments that make unequal technical demands upon the reader. People unacquainted with physics at the university level may find chapters 2, 4, 8, and 9 challenging. There is no help for it. An understanding of the modern world of big science demands coming to grips with its scientific and technological, as well as with its social, economic, and political, imperatives. We have tried to make the technical material accessible to the general reader.

A few words about sources are in order. We have chosen the primary source, the contemporaneous record, whenever possible, relied upon secondary sources when these primaries failed us, and had recourse to oral histories and interviews only when absolutely necessary. The primary materials were sufficient to keep us occupied. Our study of them has been eased by the kind help of many individuals. First among them are Vicki Davis, archivist of the Lawrence Berkeley Laboratory, and Robin Rider, head of collections in history of science and technology at the Bancroft Library of the University of California. We are also much obliged to archivists at other national laboratories, at the Historian's Office of the Department of Energy, and at many other repositories at home and abroad. We thank Bruce R. Wheaton for useful discussions on technical points. Our debt to Alice Walters and Diana Wear is incalculable, like the time and good humor they put into preparing the final copy for the press.

Our deepest obligations are to Edward Lofgren, builder of the Berkeley Bevatron, and also the prime and persistent mover in this project, and to James Clark, director of the University of California Press, who has supported it from its inception. To Lofgren, and to the other veterans of the Laboratory—Luis Alvarez, Jackson Laslett, Glenn Seaborg, Emilio Segrè, and Robert Wilson—who read and commented on the penultimate draft of the manuscript, both we, and our readers, owe a great many thanks, which we here express on behalf of us all.

Partial support for this work came from the Department of Energy through the Lawrence Berkeley Laboratory. We thank its director, David Shirley, for procuring this support, and its public affairs staff, especially Judy Goldhaber, for their kindness and promptness in furnishing photographs. It is important to state that our arrangements with the Department of Energy left us with complete editorial freedom.

Note on the Notes

Sources are cited in the notes in a short form, which should identify most of them to anyone familiar with the literature and the abbreviations that follow. Full bibliographical citations to all articles and books—though not to editorials in journals or to interventions at scientific meetings—appear in the Bibliography. All references to unpublished materials include the archive code except for the Ernest O. Lawrence Papers, which are identified by the form (x/y), where x refers to box and y to folder.

Abbreviations used in the notes:

AHQP	Archive for the History of Quantum Physics
AIP	American Institute of Physics
AJP	*American journal of physics*
Alvarez P	Luis Alvarez Papers, Lawrence Berkeley Laboratory
APhilS	American Philosophical Society
APS	American Physical Society
BAAS	British Association for the Advancement of Science
BAS	*Bulletin of the atomic scientists*
Barrett P	E.C. Barrett Papers, California Institute of Technology
BSC	Niels Bohr Scientific Correspondence, in AHQP
Birge	Birge, R.T. *History of the physics department.* Typescript. 5 vols. Berkeley, 1966–.
Birge P	R.T. Birge Papers, TBL
Bretscher P	Egon Bretscher Papers, Churchill College, Cambridge
CF	The Chemical Foundation
CIW	Carnegie Institution, Washington

CKFT   John Cockcroft Papers, Churchill College, Cambridge

CP   *Collected papers*, identified by author

CR   Académie des Sciences, Paris, *Comptes rendus*

Crowther   Crowther, J.G. *The Cavendish Laboratory, 1874–1974.* New York: Science History Publications, 1974.

CW   *Collected works*, identified by author

ER   Ernest Rutherford Papers, AHQP

Eve   Eve, A.S. *Rutherford.* New York: Macmillan, 1939.

FAL   F.A. Lindemann (Lord Cherwell) Papers, Nuffield College, Oxford

Fermi P   Enrico Fermi Papers, University of Chicago Library, Chicago

Flexner P   Simon Flexner Papers, American Philosophical Society

Frisch P   O.R. Frisch Papers, UA

Geiger   Geiger, Roger L. *To advance knowledge. The growth of American research universities, 1900–1940* Oxford: Oxford University Press, 1986.

GPT   G.P. Thomson Papers, Trinity College, Cambridge, courtesy of the Master and Fellows

HAB   H.A. Bethe Papers, Cornell University Library

Hale P   G.E. Hale Papers, California Institute of Technology

Hartcup &
Allibone   Hartcup, Guy, and T.E. Allibone. *Cockcroft and the atom.* Bristol: Adam Hilger, 1984.

Harteck P   Paul Harteck Papers, Rensselaer Polytechnic Institute, Archives and Special Collections, Troy, New York

Hendry   Hendry, John, ed. *Cambridge physics in the thirties.* Bristol: Adam Hilger, 1984.

HSPS   *Historical studies in the physical and biological sciences*

ICHS   International Congress of History of Science

IEEE   Institute of Electrical and Electronics Engineers

IRE   Institute of Radio Engineers

JACS	American Chemical Society, *Journal*
JFI	Franklin Institute, *Journal*
JP	Frédéric Joliot Papers, Institut du Radium, Paris
JP	*Journal de physique*
Kamen	Kamen, Martin. *Radiant science, dark politics.* Berkeley: University of California Press, 1985.
Kevles	Kevles, Daniel J. *The physicists. The history of a scientific community in modern America.* New York: Knopf, 1978.
KTC	MIT, Office of the President, 1930–1958 (Compton-Killian), (AC4) Institute Archives and Special Collections, MIT Libraries, Cambridge, Massachusettes.
Langevin P	Paul Langevin Papers, Ecole municipale de chimie et physique industrielles, Paris
Lauritsen P	C.C. Lauritsen Papers, California Institute of Technology
Lewis P	G.N. Lewis Papers, TBL
Livingston	Livingston, M.S., ed. *The development of high-energy accelerators.* New York: Dover, 1966.
Livingston & Blewett	Livingston, M.S., and J.P. Blewett. *Particle accelerators.* New York: McGraw-Hill, 1962.
Loeb P	Leonard Loeb Papers, TBL
MAT	M.A. Tuve Papers, Library of Congress
Mendelssohn P	Kurt Mendelssohn Papers, Bodleian Library, Oxford
McMillan P	E.M. McMillan Papers, Lawrence Berkeley Laboratory
Millikan P	R.A. Millikan Papers, California Institute of Technology
MPG	Archives, Max-Planck-Gesellschaft, Berlin
Nahmias	Nahmias, Maurice E. *Machines atomiques.* Paris: Editions de la Revue d'Optique, 1950. Mrs. Bretscher Papers kept by Mrs Egon Bretscher, Cambridge, England.
NAS	National Academy of Sciences
Neylan P	John Francis Neylan Papers, TBL

Nobel	Protokoll ang. Nobelärenden, Royal Swedish Academy of Sciences, Stockholm
NRC	National Research Council
Nwn	*Die Naturwissenschaften*
Pegram P	George B. Pegram Papers, Rare Book and Manuscript Library, Columbia University, New York City
PR	*Physical review*
PRS	Royal Society of London, *Proceedings*
PT	*Physics today*
RC	Research Corporation Archives, New York City
Reingold	Reingold, Nathan, ed. *The sciences in the American context.* Washington, D.C.: Smithsonian Institution, 1979.
RF	Rockefeller Foundation Archives, North Tarrytown, New York
Richardson P	O.W. Richardson Papers, Harry Ransom Humanities Reseach Center, The University of Texas at Austin
RMP	*Review of modern physics*
RSI	*Review of scientific instruments*
SEBM	Society for Experimental biology and medicine
Seidel	*Seidel, Robert W. Physics research in California: The rise of a leading sector in American physics.* Ph.D. thesis, University of California, Berkeley, 1978. (*DAI*, 7904599.)
Swann P	W.F.G. Swann Papers, APS
SzP	Leo Szilard Papers, University of California, San Diego
TBL	The Bancroft Library, University of California, Berkeley
UA	Cambridge University Archives
UAV	Harvard University Archives
UCA	Archives, University of California, Berkeley
UCPF	President's files, University of California
Urey P	Harold Urey Papers, University of California, San Diego

Weart   Weart, Spencer *Scientists in power.* Cambridge, MA: Harvard University Press, 1979.

Weart &
Phillips   Weart, Spencer, and Melba Phillips, eds. *History of physics. Readings from* Physics today. New York: American Institute of Physics, 1985.

Weart &
Szilard   Weart, Spencer, and Gertrud Weiss Szilard, eds. *Leo Szilard: His version of the facts. Selected recollections and correspondence.* Cambridge, MA: MIT Press, 1978.

El Dorado

Ernest Orlando Lawrence, the protagonist of our story, had much in common with the first European to explore the coasts of Northern California. Sir Francis Drake was an adventurer, a master of navigation and other contemporary high technologies, and withal a scientist, a collector and student of the flora and fauna of the new worlds he discovered. And, like Lawrence, Drake missed spectacular discoveries under his nose: he overlooked or overran the Golden Gate, the entrance to San Francisco Bay, which lies a day's sail south of his probable anchorage.[1]

Today's sailor who enters the Gate, or today's tourist who surveys the panorama from its northern shore, sees two great steel bridges, each an engineering wonder when built, which define a body of water hemmed in by factories and harbors, and much reduced from its size in Drake's time by the operations of later entrepreneurs. The south shore of the Gate lies in San Francisco, which spreads over a peninsula washed by the Pacific Ocean on the west and the Bay on the east, and carrying, to the south, Silicon Valley, the home of electronic high technology. Across the Bay, on its eastern shore, the conurbation continues in Oakland and Berkeley. Above Berkeley, on hills facing the Gate, rises a great Laboratory that commands a magnificent view of the Bay Area. The people who work there do not study the view, however, or much else that they can see unaided. Most of their exploring concerns the behavior of molecules, atoms, nuclei, and radiation.

The Lawrence Berkeley Laboratory is now functionally a multipurpose research facility of the Department of Energy, admini-

1. For Drake's science see Allen and Parkinson, *Examination.*

stered by the University of California. Its bureaucratization, growth, diversification, and nationalization have reduced the significance of local color and influence on its operations. But in the period covered by this volume, the Radiation Laboratory of the University of California, or "Rad Lab," which preceded the complex on the hill, grew in a manner and at a pace that are unintelligible without reference to the character of the region and the state that supported it.

1. THE ASCENT OF THE WEST

Half a century after the Gold Rush of 1849, California showed the marks of an urbanized, commercialized society: although agriculturally rich, it began to rely heavily upon electricity, petroleum, and steel. The technological requirements rose sharply during the first twenty years of the twentieth century, when the state's population more than doubled, its output in agriculture, mining, petroleum, lumbering, and fishing increased severalfold, and hydroelectric power, motion pictures, and the cult of the automobile gave it a distinctive character.[2]

The culture derived from these elements mixed the pioneering of the Gold Rush, the boosterism of rapid expansion, the make-believe of Hollywood, and the confidence of technological expertise. A precocious expression of the California spirit was the foundation of an Academy of Sciences in San Francisco in 1853. Its immediate stimulus was a survey of coasts and harbors undertaken to safeguard navigation. We hear the booster and the entrepreneur, the student and the dreamer, in the founders' declaration, "It is due to science, it is due to California, to her sister states, and to the scientific world, that early means be adopted for a thorough survey of every portion of the State and the collection of a cabinet of her rare and rich productions." And we are sensible of the rawness and violence of the society that produced the academy in its adjourning to a discussion of the trees of North America after attending the hanging of an outlaw who had killed its president. Two decades later, the great captains and pirates of

2. Cleland, *California*, 105.

California's first commercialization, men like Leland Stanford, David Colton, Charles Crocker, and James Lick, gave the academy the wherewithal to build a museum of natural history, sponsor a lecture series, and represent culture at the edge of the civilized world.[3]

The rawness and vigor, the culture and pretentiousness, the hucksterism and fantasy, were celebrated in the grand Panama-Pacific Exposition held on newly filled waterfront in San Francisco in 1915. The official poster of the exposition (plate 1.1) might have been designed as frontispiece for Lawrence's first Radiation Laboratory: a monumental Hercules cracking open the resistant earth (he is making the Panama Canal, but could be smashing atoms) to disclose a dream city in fairyland (it is the exposition itself, but could be the future powered by atomic energy). Within the exposition, "the only place...in the United States where romance seems pervasive and inevitable," stood halls filled with California's natural plenty and with the machinery that harvested, transformed, and transported it.[4] The Palace of Mines and Metallurgy proffered two big balls of gold, and a hydraulic mine and gold dredger, as proof of California's mastery of the precincts of Pluto. Further to the theme, it presented "ample evidence of the great figure which steel now makes in the world, and of the vast extent of the petroleum industry." And—an unreadable indicator of a theme that will preoccupy us—there was a tiny sample of radium. "Being so little of it in the world," a children's guide to the fair explained, "it is tremendously expensive."[5] Lawrence would change that.

The Palace of Machinery also pointed directly to California's future. It emphasized, in the words of a contemporary journalist, "the increasing displacement of coal by hydroelectric plants and liquid fuels." To this proposition the prime mover of the exposition gave silent testimony. Power came from the Sierra via high-tension electrical lines; there was nothing in San Francisco resembling the mountainous Corliss engine, some thirty-nine feet tall

3. Lewis, *George Davidson*, 60–1.
4. Dobkin in Benedict, *World's fairs*, 67; Brechin, ibid., 94, 101 (quote from Edmund Wilson); Anderson, ibid., 114–5.
5. Quotes from Macomber, *Jewel city*, 150, and Gordon, *What we saw*, 45.

and weighing 1.4 million pounds, whose steam had moved the Centennial Exposition in Philadelphia in 1876. It had delivered some 1,400 horsepower. The exposition of 1915 showed the Corliss's latter-day competitor, one of those instruments for replacing dirty coal with clean liquid fuels, a Diesel engine that also developed 1,400 horsepower. This engine, called to life by a radio signal from President Wilson, stood seven feet tall and weighed 44,000 pounds, an item of domestic furniture in comparison with steam engines. And so it was housed, in a booth trimmed with oriental rugs and finished in fine woods, with attendants in white uniforms who "looked like guests aboard a yacht."[6]

A more direct symbol of transportation was the gasoline engine, exhibited in every size and form, since, as the official historian of the exposition observed, San Francisco, so far from coal and so close to oil, had naturally become a center for its development. A special exhibit by the Ford Motor Company showed how to put a new gas engine into use 144 times a day. It was the latest wonder of technology: the mass production of automobiles, "in which a continuous stream of Fords is assembled and driven away, one every ten minutes."[7] These and many other cars would soon be clogging California's urban centers, requiring more and more highways, freeways, and gasoline, and fixing the contours of cities. A frequent plaint of European visitors to California laboratories was (and is!) that life could not be lived without an automobile.[8]

The exposition not only displayed, but also, like its Hercules, forced technology. The Panama Canal had connected the waters of the Pacific and the Atlantic; AT&T would link the telephones of California with exchanges in New York. In 1911 officers of the company promised the management of the exposition that the link would exist by 1915, although they had not yet proceeded beyond Denver. The Denver line used loading coils spaced eight miles apart to "repeat" the message. To reach San Francisco, which it did on time, AT&T suspended 2,500 tons of wire from 130,000 poles. Those are the sorts of numbers that announce progress.

6. Todd, *Story*, 4, 158–9, 163–4.
7. Macomber, *Jewel city*, 149–50.
8. E.g., Brasch to Szilard, 27 Jan 1939 (Sz P, 29/306); Todd, *Story*, 4, 164–5. Capron, "Report," 4.

Another quantity to conjure with—one that Lawrence did conjure with—was a million volts. With financial support of the exposition, an engineer from Chicago named C.H. Thordarson exhibited an experimental transformer made of 400 miles of paper, aluminum, and copper. A wire net hung from the transformer's high-voltage secondary and suspended some feet above the ground made one of the most popular demonstrations at the fair. Parading under this canopy, while "the ends of the fingers glowed," one had, among other feelings, the sensation of participating in an experiment significant for California's future.[9]

White Gold and Black

The determined harnessing of water in California began with the Forty-niners, who washed away the hills of gold and silted up the rivers with the help of 5,000 miles of sluices and aqueducts. The prolific waters of the Sierra also powered mining operations and machinery and stimulated the imaginations of engineers and entrepreneurs who looked upon the rapidly growing cities on the coast. How to bring the power from the hills, where it was wasted, to the population centers, where it was wanted?

The first of the many technologies that affected the transmission and efficient exploitation of the strength of the Sierra streams was a waterwheel developed by a California metallurgist named Lester Allen Pelton. Pelton found that he could increase the speed and efficiency of conversion of the power of falling water by directing it against the edges of the buckets on the standard wheels in use in the mines. A Pelton wheel weighing only 220 pounds could supply 125 horsepower. Its basic design may still be discerned in modern turbines.[10] As this persistence suggests, Pelton did not chance on his invention. When he began his experiments in 1878, he procured the latest instruments and instructions for hydraulic investigation; he was no ordinary millwright, but an accomplished, although uncertified, engineer.[11] The Pelton Water Wheel

9. Brooks, *Telephone*, 262; Reich, *American industrial research*, 166–70; Macomber, *Jewel city*, 152; Todd, *Story*, 4, 184–6 (quote).

10. Durand, *Mech. eng.*, 61 (1939), 447–51; Wilson in Williams, *History of techn.*, 6, 207; Smith, *Sci. Am.*, 242:1 (1980), 138–48.

11. Constant, *Techn. and cult.*, 24 (1983), 194, and *Soc. stud. sci.*, 8 (1978), 183–210.

Company of San Francisco, established in 1888, became a world leader in the design and manufacture of hydraulic machinery. Its exhibit at the Panama-Pacific Exposition so impressed officials from Columbia University that they bought it outright for their School of Mines. All, that is, except a turbine turned by a wheel twenty-six feet in diameter (plate 1.2), rated at 20,000 horsepower and made for a PG&E project on the Yuba River. "No single unit greater than half the big turbine had been used in Europe....The place [the Pelton exhibit] was haunted every time the Palace [of Machinery] was open by civil engineers from all quarters of the globe."[12] The Pelton Company developed a general expertise in metal working on a large scale. Lawrence was to commission it to machine the magnet of his first big cyclotron.

At the time the Pelton Company was founded, the first steps had been taken, in Europe, toward bringing power from the hills to the plain. In 1886 Italians transmitted electricity from Tivoli to Rome, a distance of 17 miles; five years later the Germans managed 100 miles; and in 1892 a plant in Southern California joined the game, providing power at long distance—some 28 miles—for the first time in the state. This success was noticed. According to the *San Francisco Call*, "the air of California, and the whole Pacific Coast for that matter, has all at once become filled with talk about setting up water wheels in lonely mountain places and making them give light and cheaply turn other wheels in towns miles away." California soon outdistanced Europe. In 1903 the San Francisco Bay Area derived electricity from the Electra powerhouse on the Mokelumne River, a distance of 142 miles. The power pressed at 60,000 volts, twice the maximum potential that General Electric and Westinghouse thought feasible. The line ran through forests and valleys before leaping the Carquinez Strait—4,427 feet at the mouth of the Sacramento River—and so was a triumph of civil as well as electrical engineering. By World War I, California had more high-voltage transmission systems than any other region in the world.[13]

The men who designed the system's lines, tunnels, dams, and stations were engineers from California's new universities. The

12. Coleman, *PG&E*, 114–5; Todd, *Story, 4*, 177–80 (quote).
13. Coleman, *PG&E*, 102, 107, 143–8; Hughes, *Networks*, 263–6.

great Electra project of 1903 employed several electrical engineers trained at Stanford and two graduates in civil engineering from Berkeley. One of the Stanford men became chief of hydroelectric and transmission engineering of the Pacific Gas and Electric Company, which consolidated the Electra operation and most other hydroelectric power systems in Northern California. The men who supplied the capital for the plants that captured the power of the Sierra streams understood the importance of high technology in their enterprises. In the disposition of their wealth they did not neglect the institutions that taught and advanced electrical engineering and the physical sciences on which it grew.

This patronage intensified with the discovery of a new source of energy. Unlike the yellow gold that enriched the Forty-niners, black gold could not be pulled from the ground, washed off, and traded. The successful extraction, refining, and marketing of petroleum depended on continuing advances in geology, chemistry, physics, and their applications. The Union Oil Company, which pioneered in the oil fields of Southern California, set up a geological department under an engineer from Stanford in 1895. He made a success of his assignment and diminished the guesswork in petroleum geology. Union Oil also had a chemical laboratory, which worked, also successfully, to separate fuel oils and asphalt from California crude. The business proceeded so well that in 1917 California petroleum provided a third of the oil and gasoline used in the United States, and between a quarter and a fifth of world consumption.[14] In California oil mixed with water, science with technology, private greed with public service, to create a power system that, by the early 1920s, was serving over 80 percent of the state's population. That number far exceeded the fractions served elsewhere in the country, "especially the extreme East."[15]

A standard form of the patronage of science and engineering by the power brokers of California was service as trustees (a preliminary step to generosity as donors) of the state's educational institutions. An obvious geographical alignment developed: oil barons

14. Taylor and Welty, *Black bonanza*, 82, 197, 231–2; Norberg, *Chemistry*; Goodspeed, NRC, *Bull.*, 5:6 (1923), 30.
15. Forbes, *Men*, 135; *Electrical world, 81:4* (1924), 203 (quote).

of Los Angeles and officers of the Southern California Edison Company helped to create Caltech from the carcass of a trade school during the 1920s; PG&E gave early and sustained support to the University of California.[16] The link brought the power companies more than good press and trained recruits. PG&E gave the University money to fit out railroad cars with an exhibition of new techniques for farming and husbandry; the company coupled on its own cars, filled with electric agricultural equipment, washing machines, and vacuum cleaners. A decade later the University exposed itself more fully by opposing a bill that would have enabled municipalities to combine to develop water power for their own use. Many of the state's farm bureaus—in contrast to the agricultural lobby—unaccountably declared that the proposed legislation went against the public interest. "The mystery was traced down, and in every case it was found that the treacherous resolution had come from the [bureau's] 'experts'—university men, appointed by university regents in the interest of their privately owned power companies."[17] No fewer than ten regents were implicated. It is dangerous to play with electricity.

Spoils of War

The universities that attracted and served the commercial interests of the state had no trouble mobilizing science for military purposes. World War I mingled the interests of science, industry, and government throughout the nation; but nowhere were the consequences of the mix more enduring and efficacious than in California. This privileged position derived in large measure from what no longer exists—the clean, clear air of the Los Angeles basin. The air had inspired George Ellery Hale, a physical astronomer often nominated for the Nobel prize and an accomplished autodidact in the promotional arts, to establish the Mount Wilson Observatory, with the world's largest reflecting telescope, in the hills above Pasadena. In 1916 Hale invented the National

16. Forbes, *Men*, 97–120, 278–300; Wilson, *California Yankee*, 54, 101–2.
17. Sinclair, *Goose-step*, 127–8, 136 (quote). Among the implicated regents were the vice presidents of PG&E and Southern California Edison, a director of the East Bay Water Company, and the president of the California Electrical Generating Company.

Research Council, which, after defeating several federal agencies, took charge of making allies of American science and technology and bringing them to war on the same side. The council came into existence under the auspices of the somnolent National Academy of Sciences and with the approval of President Wilson.[18]

In administering the council, Hale was seconded by A.A. Noyes, professor of chemistry at MIT and Caltech, and Robert A. Millikan, then suspended between Chicago and Pasadena, whose great powers of organization, oratory, and salesmanship would go to Southern California after the war. Caltech turned itself into a training camp when the United States joined the fighting in 1917 and boasted that it had "exceeded all other civil colleges in devotion to war work." Hale tied his stronghold, Mount Wilson, to the NRC; it contributed importantly to the provision of good optical glass for binoculars, range finders, and field telescopes, which previously had come almost exclusively from Germany.[19] California organized a State Council of Defense whose work, in the judgment of the NRC, "[stood] out conspicuously among the contributions of scientific men of the country during the war." California could act so quickly and effectively because isolation had already prompted its scientists to integrate their local societies and to form West Coast branches of national societies; the Pacific Division of the American Association for the Advancement of Science, founded in connection with the Panama-Pacific Exposition, advised the State Council of Defense on behalf of all these organizations.[20]

Under the State Council and the NRC, the University's engineering shops became "veritable war laboratories;" its Chemistry Department agreed, almost to a man, to "work on any problem assigned to them," and to give any patents that might result to the University's regents. They received commissions to hunt for sources of chemicals to replace German imports, to make California petroleum into TNT, and to counter poison gas. Their most notable contribution was an absorbent for carbon monoxide

18. Kevles, *Physicists*, 109–16; Geiger, *Knowledge*, 95–8; Crawford, Heilbron, and Ullrich, *Nobel population*, s.v. "Hale."

19. Scherer, *Nation at war*, 42–3.

20. Henry S. Graves, chairman, Division of State Relations, NRC, in Goodspeed, NRC, *Bull.*, 5:6 (1923), 2, and Goodspeed, ibid., 4–7.

and other noxious gases. "Had nothing additional been accomplished," says the official report of the State Committee on Scientific Research, "this result alone justifies the total expenditure made for chemical investigation by the State of California." Chemists who could outgas the Germans enjoyed something of the reputation at the Armistice that atomic physicists did on V-J Day. One of the best gas chemists was Gilbert Newton Lewis, already an influential member of the Berkeley faculty before he went to war; he came back as the dominant force for building up the University's research capacity in physical science.[21]

California electrotechnology contributed to the war effort in ways decisive for our story. Before the war the Federal Telegraph Company of Menlo Park, California, which had close ties to Stanford, acquired exclusive rights to market a system of wireless telegraphy invented by a Danish engineer, Vladimir Poulsen, in the United States and the Pacific. In 1912 Federal had won a contract from the U.S. Navy for a 100-kW installation at the Panama Canal, although the company had not then managed to get more than 30 kW from its largest Poulsen machine. A crash program under the direction of Leonard Fuller, an electrical engineer who had studied at Stanford and later taught at Berkeley, discovered and removed the impediment. The navy, now getting up steam for war, commissioned Federal to set up 200-kW plants in Puerto Rico and San Diego; the specification jumped to 350 kW for the Philippines and Pearl Harbor, to 500 kW for Annapolis, and, when the United States entered the European War, to 1,000 kW, to link Washington with American forces in France.[22]

The navy simultaneously supported development of another device, also improved at Federal, which soon subverted the Poulsen system. This was the vacuum tube oscillator, accidentally invented by Lee de Forest when, as an employee of Federal, he worked to perfect his audiotron for use as an amplifier on transcontinental telegraph lines. De Forest's audiotron enabled AT&T

21. Quotes from, resp., U.C., *Pres. rep.*, 1917/18, 116, 93, and Goodspeed, NRC, *Bull.*, 5:6 (1923), 13; ibid., 10–2, 32–3, and U.C., *Pres. rep.*, 1916/17, 25, 120; O'Neill, *U.C. Chronicle*, 20:1 (1918), 82–92.

22. Howeth, *Communications*, 143–7, 182–5, 253; Fuller, interview, 50, 100, and letter to Haradan Pratt, 19 May 1965 (TBL); Crenshaw, IRE, *Proc.*, 4 (1916), 35–40; Aitken, *Continuous wave*, 131–61.

to fulfill its promise to the Panama-Pacific Exposition; his oscillator made possible commercial radio broadcasting, which Federal inaugurated in San Jose in 1919; and it became the heart of long-distance wireless after the war. We shall return to Fuller and Federal, to West Coast electronics, and to the big Poulsen machines, several of which were cannibalized for cyclotrons.[23]

In founding the NRC, Hale was no doubt inspired by what he took to be a civic duty. He also saw, as he put it, "the greatest chance we ever had to advance research in America." The success of the mobilization of science for the production of optical glass, for the location of artillery and submarines, for defense and offense in chemical warfare, for improvements in airplanes, radio, gunnery, and a thousand other things—all this helped to vanquish not only the enemy, but also the idea that academic physical scientists were impractical dreamers or incompetent pedagogues. As Millikan and Hale liked to put the point in a neat double entendre, the war had "forced science to the front."[24] The physicist found himself in the unusual, but portentous, condition of enjoying the attention of the press. "That he 'made good' from the beginning is one of the commonplaces of the history of our war. He took hold of a situation as unacademic as the most sceptical of his critics could have imagined, and proceeded as if the war were nothing more baffling than a particularly unruly set of sophomores." "It was a revelation to the country."[25]

The practical payoff of this demonstration of practicality was the enlistment in the support of academic physics of some foundations, businesses, and industries that previously had not thought it worth their notice. An official of the National Bureau of Standards, which had passed through precarious times, now could augur that never again "would anybody question the...economic value of scientific investigation," or, as the *Saturday Evening Post* added, its place in the "bedrock of business." And of labor. Forgetting in the general enthusiasm the menace of unemployment

23. Aitken, *Continous wave*, 233–9; Morgan, *Electronics in the West*; infra, § 3.3.

24. Hale, quoted by Kevles, *Physicists*, 112; Millikan, ibid., 138, and Hale, NRC, *Bull., 1* (1919), 2.

25. Resp., *New York Evening Post*, as quoted in *Science, 51* (1920), 415, and St. John, *PR, 16* (1920), 372–4.

forced by technological advance, the American Federation of Labor declared in 1920 that "a broad program of scientific and technical research is of major importance to the national welfare and should be fostered in every way by the federal government."[26] With all this implied support, Millikan said, American physicists would soon outdistance the Europeans, whom they had trailed for too long. "In a very few years we shall be in a new place as a scientific nation and shall see men coming from the ends of the earth to catch the inspiration of our leaders and to share in the results which have come from our developments in science."[27]

The NRC took as its postwar purposes the strengthening of science in the universities and technical schools and the mingling of scientists with engineers and businessmen in industrial research laboratories. Its watchwords were cooperation and organization, the lessons of the war. Had not the Germans managed to struggle so long, and the allies finally to beat them, because both sides learned to mobilize their scientific manpower? As the chairman of the Carnegie Institution of Washington put it in the NRC's first *Bulletin*, "competency for defense against military aggression requires highly organized scientific preparedness." The same would be true of the coming commercial struggle: the fruits of industry would go to the nation that "organizes its forces most effectively."[28] The NRC divided itself into interdisciplinary task forces and laid siege simultaneously to nature and the foundations under the cry of "the national essentiality of science." Nor was the home front forgotten. The council and the American Association for the Advancement of Science saw to it that accounts of their activities and of all American "discoveries" appeared in newspapers and magazines.[29] The entrepreneurs of California physical science—Hale, Millikan, and their successor, Lawrence—discovered the uses of the popular press, and, when it became available, the radio.

26. E.B. Rosa, 1921, and *Saturday Evening Post*, 1922, quoted in Kevles, *Physicists*, 148, 173; AFL resolution reported in Herrick, *Science, 52* (1920), 94.

27. Millikan, *Science, 50*, (1919), 297.

28. Root, NRC, *Bull., 1* (1919), 10.

29. Tobey, *American ideology*, 62–71; Kevles, *Physicists*, 174; Carter, *Am. scholar, 45* (1975/6), 786–7. The cry is that of A.W. Mellon, president of the Mellon National Bank, in NRC, *Bull., 1* (1919), 17.

Perhaps the most important instrument of the NRC for the improvement of university physics after the war was a fellowship program for postdoctoral training paid for by the Rockefeller Foundation. The first installment, half a million dollars, was repeated, until by 1940 no fewer than 883 National Research Fellows had had their horizons broadened at a cost of $2.7 million. The first physics fellows took their awards in Europe, whence they imported and domesticated quantum physics; their students tended to take their fellowships in the United States; by 1928 National Research Fellows in physics preferred Caltech to any other American institution.[30] In the 1930s Berkeley became a center of choice. The Radiation Laboratory could not have grown so quickly or so well without the contributions of such National Research Fellows as Franz Kurie and Edwin McMillan.

In keeping with its fellowship program, the NRC urged, and the foundations followed, an elitist policy in their awards: the good American research universities should be made even better, while the rest slid deeper into mediocrity.[31] Regional, ethnic, and financial balance, the preoccupations of the era of federal funding, then scarcely came into consideration. The twenty schools that housed one-third of the professional physicists of the nation produced three-quarters of the American doctorates in physics earned between the wars and three-quarters of the papers published in the leading American journal, the *Physical Review*. The same schools received almost all the NRC's physics fellows.[32] To sustain the "surcharged atmosphere of the great university graduate school," and to enhance its capacity to fulfill its duty to multiply research and researchers, the NRC agitated to remove impediments to uninterrupted investigation and graduate instruction by the most able professors. The greatest of these obstacles was the plague of undergraduates indifferent to science who infested the universities in the great swelling of student bodies after the war. "[We] feel

30. St. John, *PR, 16* (1920), 372–4; NRC, *National research fellowships* (1939); reports by F.K. Richtmeyer, NRC secretary, 9 Mar 1931, and W. Weaver, 13 Nov 1939 (RF, 200/171/2079); Rand, *Sci. monthly, 73:8* (1951), 71–80; Seidel, *Physics research*, 82; Geiger, *Knowledge*, 188, 203.

31. Kevles, *Physicists*, 150–1, 197–8, 219–20; Geiger, *Knowledge*, 99–100, 107, 161–2.

32. Weart in Reingold, *Context*, 322–5; Kevles, *Physicists*, 197.

sick," said the NRC, "when we face the facts." Undergraduates, or the teaching thrown away on them, set "the most serious limitation of research productivity."[33] The NRC strove to bring the administrations of major universities to understand that the main business of their professors was research, not teaching. With its help and foundation support, faculty pressure groups did succeed in shifting the balance of professorial obligation toward research. By 1930 physicists led the academic pack in research time: 36 percent of their total academic effort, as against 26 percent for humanists and 30 percent for biologists.[34] Lawrence was a beneficiary of this secondary consequence of the Great War.

2. THE ROARING TWENTIES

The increased reputation of physical science after the war combined with the requirements and fruits of California's high technology to give physics at both Caltech and Berkeley a more rapid acceleration during the 1920s than it experienced at any other research university in the country. The most prominent single theme in this California physics was its concern with high voltages and radio technologies. The theme sounded first at Caltech. In 1920 Southern California Edison provided the new institution's new president, Millikan of the NRC, with a high-tension testing laboratory. The utility's interest is easily stated: according to a rule of the art, to transmit power economically, at least 1,000 volts should be provided for each mile to be traversed.[35] Southern California Edison, then becoming the world's largest producer of hydroelectric power at a cost that exceeded the price of the Panama Canal, aimed to work at over 200,000 volts. The laboratory at Caltech did improve the technique of high-voltage transmission. Later it supplied the power to run a powerful x-ray

33. Quotes from, resp., Weld, *Science, 52* (1920), 46, and Carty, NRC, *Reprint,* no. 8 (1919), 8; Hale likewise lamented the "avalanche of new students," "these swelling throngs," in *Harper's, 156* (1928), 247. Cf. Kellogg, *Science, 54* (1921), 19, 22; Angell, NRC, *Reprint,* no. 6 (1919), 2; H. Hoover, ibid., no. 6 (1925), 4.

34. National Resources Committee, *Research, 1* (1938), 177; Weart in Reingold, *Context,* 298–9; Geiger, *Knowledge,* 262.

35. Compton, *Science, 78* (1933), 21.

plant for cancer therapy designed by Charles Christian Lauritsen, an engineer and physicist who for a time competed successfully with Lawrence in smashing atoms.[36] At Berkeley the high men in voltage were Fuller, who had become professor there without cutting his ties to Federal, and his protégé Ernest Lawrence.

The work of such men and the students they attracted paid dividends on the capital the state had invested in physical science. The president of the American Physical Society alerted its members in 1933 to the distortion of the old ring circle of Eastern institutions worked by the newly scientific West. "In the past decade the State of California has risen from obscurity to become a center in the physical circle, or more precisely one of the two foci in the academic ellipse representing American physics."[37] Europeans also noticed. "The development of science in California is very impressive, even if theoretical physics is not so strongly represented." Hendrik Kramers, professor of physics at the University of Utrecht, teaching in Berkeley in 1931, thus informed the dean of the world's quantum theorists, Niels Bohr of Copenhagen, who would visit the state twice himself in the 1930s to see its physics for himself.[38]

The MIT of the South

Bohr's visits fit a pattern established in the 1920s by the new administration of Caltech, Hale and its "executive head," Millikan, who had the power, but not the duties, of a college president. With plenty of money from local sources and the Rockefeller and Carnegie foundations, Caltech brought in the leaders of international physics for seminars and lecture courses: Max Born, Paul Ehrenfest, Albert Einstein, H.A. Lorentz, A.A. Michelson, Wolfgang Pauli, C.V.R. Raman, Arnold Sommerfeld. With the help of the prestige thus conferred, the former Throop College of Technology recruited first-rate scientists at home and abroad: Ira Bowen, Paul Epstein, William Houston, J.R. Oppenheimer, Linus Pauling, Richard Tolman, Fritz Zwicky. The great luminaries and Nobel

36. Sorenson, *Jl. elect.*, *53* (1924), 242–5; Lauritsen and Bennett, *PR, 32* (1928), 850–7; Holbrouw, in Weart and Phillips, *History*, 86–93.

37. Foote, *RSI, 5* (1934), 65.

38. Kramers to Bohr, 5 Oct 1931 (BSC).

prize winners, among whom Millikan, already widely nominated, would soon be enrolled, attracted hundreds to pay thousands to belong to the Caltech Associates, a fund-raising initiative worthy of Wall Street.[39] By 1930, by most measures, Caltech overshadowed MIT as a technical school and rivaled the University of California in most fields not related to the social sciences.

Caltech's trustees cared less for science, natural or social, than for the practical and electrical potential of the school. Delays in construction occasioned by the first world war had caused an acute shortage of power in the fast-growing Los Angeles basin. In 1920 the creation of the Federal Power Commission and the revocation of an ancient prohibition against damming navigable waterways opened the way to new hydroelectric developments. Southern California Edison, which dreamed of a dam on the Colorado River to generate hydroelectric power for Los Angeles, planned in the meantime $100 million worth of construction, including conversion of its high-voltage lines to 220,000 volts.[40] These undertakings required research. A cheap and reliable way to get it was to build a high-voltage laboratory at Caltech, supply it with electricity, lease it to the school at a nominal fee, and obtain from its high-powered, low-paid staff help with "all problems of the company arising in connection with [its] experimental and research work."[41]

Millikan understood how to couple a scientific program to the applied work of the laboratory. In a successful approach to the Carnegie Corporation, he wrote that "the most promising field of science today is the field of the behavior of matter under enormously high potentials....The physicist wishes to enter it for purely scientific reasons." Millikan proposed studying spark spectra in high vacua, high-potential x rays, and corona discharges at high potentials. Ultimately the knowledge gained might help in the understanding of nature's play at extremes of electric force, temperature, and pressure, in atoms and in the stars.[42]

39. Seidel, *Physics research*, 66–83; Kargon, *Rise*, 101–3.

40. Rendy, *Jl. elect.*, *44* (1919), 479–81; Breckenridge, ibid., 501–5, and *45* (1920), 347; Barre, ibid., 566–7.

41. Arthur Fleming to Hale, 17 Mar 1911, enclosing lease, quote (Hale P, micro edn, 14.72–4). The company pledged $100,000 for the building and 10,000 kWh a month to run it.

42. "Memorandum...to the Carnegie Corporation," 1921 (Hale P, micro edn,

The early work of the high-voltage laboratory centered on immediate application. Royal Sorenson, a Caltech electrical engineer and consultant to Southern California Edison and the Metropolitan Water District of Los Angeles, built a four-stage cascade transformer that gave a million volts. It was used to test vacuum switches that Sorenson and Millikan designed to overcome the arcing in oil circuit breakers. A high-voltage switch, patented per agreement as the joint property of Southern California Edison and Caltech, was sold to GE in 1930 for $100,000, which repaid the cost of the laboratory. Other patented products of the laboratory included high-voltage fuses and discharge tubes.[43] The physics on which they rested was worked out by Millikan and Lauritsen, a graduate student and former employee of the Federal Telegraph Company, and by Oppenheimer, who came to Caltech as a National Research Fellow in 1928.[44]

Lauritsen carried Caltech's high-voltage technology into new fields with the help of R.D. Bennett, likewise a National Research Fellow, who had sought to extend work on the quantum theory of x-ray scattering to high voltages. The internal workings of the very sturdy instrument Lauritsen devised relate to particle accelerators and will be discussed later. Its external workings were more general and immediate. Millikan had asked Seeley G. Mudd, a Pasadena cardiologist, for money to develop Lauritsen's tube. Mudd consulted Francis Carter Wood of the Crocker Institute for Cancer Research at Columbia, who advised that the high-voltage x-ray tube was "the exact step forward that we need for the treatment of cancer." Others disagreed, for example, Fred Stewart of New York's Memorial Hospital, who preferred "small quantities of energy over a long period of time." Millikan accepted the advice that favored the tube.

Preliminary therapy began in 1930 at 600,000 volts, which Wood judged to be insufficient. Higher potential meant more money. Millikan got it from W.K. Kellogg, the Corn Flakes King, who desired only "credit of some sort."[45] While Caltech built the

6/406–7); Seidel, *Physics research*, 321; Kargon, *Rise*, 122–35.

43. Sorenson, AIEE, *Trans.*, *45* (1926), 1102–5; Caltech, Board of Trustees, minutes, 25 Feb 1930 (E.C. Barrett P, 1/8); U.S. patents 1,757,397 and 1,764,273.

44. Millikan and Lauritsen, NAS, *Proc.*, *14* (1928), 45–9; Oppenheimer, ibid., 363–5.

45. Wood to Mudd, 22 Sep 1930, and Stewart to Mudd, 1 Jan 1931 (Millikan

Kellogg Radiation Laboratory, Los Angeles County Hospital used its 600-kV tube to treat 200 patients. They improved. Albert Soiland, the radiologist in charge, ascribed the efficacy of the rays to their energy, or perhaps to their homogeneity.[46] It does not appear that many of the 746 people treated at the Kellogg Laboratory during the 1940s improved dramatically. The sanguine and bold attributed the fault to the voltages. Still higher energies were sought at Berkeley, Stanford, and Caltech. Supervoltage x rays proved no better therapeutically than high-voltage ones. But the attempt to devise equipment to make them and the funding their uncertain promise provided were very good for physics.

The Harvard of the West

Wartime service also advanced research at Caltech's northern rival. An immediate outcome of Berkeley's contribution to national defense, and a harbinger of much greater reform, was the transmutation of a parochial editorial committee charged with publishing their colleagues' contributions to knowledge into an outward-looking Board of Research. In the early 1920s the new board brought the experience it had gained in supervising war work to promoting the cause of research, particularly scientific research, to the University's administration and well-wishers. The board's pushiness fit in well with a general agitation, led by G.N. Lewis, John C. Merriam (chairman of the wartime State Committee on Scientific Research and a high official of the NRC), and Armin O. Leuschner (dean of the Graduate School and head of the NRC's Physical Sciences Section in 1918), to improve salaries and working conditions. The professors won important concessions in self-governance and a new president representative of their new self-image. He was William Wallace Campbell, a research astronomer, formerly head of Mount Wilson's northern rival, the University's Lick Observatory.[47] Under its version of George Ellery Hale, Berkeley rapidly expanded its facilities and staff for physical research.

P); Holbrouw, in Weart and Phillips, *History*, 86–93.
  46. Soiland, *Radiology, 20* (1933), 99–104.
  47. U.C., *Pres. rep.,* 1919/20, 156–8; Seidel, *Physics research,* 85–105.

First things first. The throng of students entering the University after the war included many more would-be engineers and physicists than could be taught in the old headquarters of physics in South Hall. In 1924 Le Conte Hall, the greatest achievement of the chairman of the Physics Department, E. Percival Lewis, opened for business (plate 1.3). It was among the largest physics buildings in the world, and second in size only to those at Cornell, Illinois, and Princeton (all built under uninflated prewar conditions) in the United States. It exceeded its larger counterparts in the proportion of space dedicated to individual research—forty rooms in Berkeley in contrast with twenty-eight rooms at Princeton.[48] The Board of Research willingly helped to furnish Le Conte. The physicists there had the good sense, in the board's opinion, to spend their money on equipment and supplies, "in marked contrast to...other departments, which expend the funds granted almost exclusively for personal research assistance."[49] The board's grant to the Department in 1923, in anticipation of the completion of Le Conte, was $5,000. The amount rose to $10,000 in 1925/26 and $12,250 in 1928/29 before sinking to an average of about $8,000 during the Depression. These grants and outside monies built up an inventory of research equipment valued at $226,500 in 1934, exclusive of the special apparatus in Lawrence's Radiation Laboratory.[50] The Department's operating expenses also increased steeply, reflecting the enlargement of the staff by the capture of several young physicists: from $78,000 in 1922/23, just before Le Conte opened, to $100,000 in 1923/24 and 1924/25, to $130,000 in 1929/30.[51]

Berkeley tried first to fill its physics building with men of established reputation: Niels Bohr, Arthur Compton, Paul Epstein,

48. U.C. *Pres. rep.*, 1918/19, 43, 81; Forman, Heilbron, and Weart, *HSPS*, 5 (1975), table D.1. Le Conte's floor space covered about 5,200 square meters; Simpson, *California monthly*, 16:5 (1924), 251–3, and "Le Conte Hall" (TBL, CU/68).

49. L. Loeb to R.G. Sproul, 8 Nov 1927 (Loeb P); Board of Research, "Minutes," 1928 (TBL).

50. Ibid., 9 Sep, 4, 15, 20 Oct 1923, 4 Apr and 14 Oct 1925 (TBL, CU/9.1); "Research budgets, 1924–1927," and R.T. Birge to Sproul, 4 Aug 1934 (TBL, CU/68).

51. U.C., *Pres. rep.*, 1925/26, 559; 1926/27, 597; 1927/28, 671; 1928/30, 866; Birge, *History*, 2, vii, 15; viii, 46.

W.F.G. Swann, none of whom valued the charms of Le Conte above the attractions of his position elsewhere. When this premature strategy, which was urged by G.N. Lewis and the Board of Research, failed, two ambitious recent recruits to the Department, Raymond T. Birge, who became its chairman in 1932, and Leonard B. Loeb, took over the head-hunting. Their strategy, as Birge later summarized it, was to follow the guidance of the NRC: "In order to guarantee a prospective candidate's scientific standing and research ability, we have chosen from among the most successful of the National Research Fellows." In this way they chose, among others, Samuel Allison, Frederick Brackett, Robert Brode, Francis Jenkins, E.O. Lawrence, J. Robert Oppenheimer, and Harvey White.[52]

Recruitment is one thing, retention another. Campbell understood the danger: "Tempting invitations come every year to half a dozen or more members of our faculties—always to men in the front rank of attainment and promise—to leave our employ and cast in their lot with other universities, at higher salaries than they are here receiving....The competition for able professors is already keen; it is going to be more keen in the near future; and we must find ways and means of retaining the services of our able men." Campbell tried to make good his word. Birge wrote a colleague at Harvard in 1927: "When our own men, at least the younger men, have had definite offers elsewhere, the administration has at least done something to hold them, and we have succeeded in holding them so far."[53] Allison and Brackett left before 1930, Oppenheimer in 1946; the others remained, despite flattering invitations to emigrate. Loeb refused the chairmanship of Brown University's physics department in 1928, and Brode, Lawrence, and (until 1946) Oppenheimer frequently turned down raiders. In consequence their salaries rose, Oppenheimer's marginally (he was a theorist, a wealthy man, and a part-time professor at Caltech), Brode's by 50 percent in six years (from $2,700 in 1927/28 to $4,160 in 1933/34), Lawrence's by almost 150 percent in eight years (from $3,300 in 1928/29 to $8,000 in 1936/37).[54] As

52. Birge, "Budget request," 1935, in U.C. Budget Requests (TBL). For recruitment strategies, failures, and successes, see Seidel, *Physics research*, 85–105.

53. U.C., *Pres. rep.,* 1926/28, 5–6; Birge to Kemble, 27 Oct 1927 (AHQP).

54. U.C. Budget Request, 1934/35 (TBL); Sproul to Lawrence, 11 Apr 1928, 19

the figures suggest, Lawrence was hard to get and expensive to keep.

Lion Hunting

Loeb began stalking Lawrence in 1926. The game was then a National Research Fellow at Yale, where he had arrived in the intricate wake of his teacher W.F.G. Swann, whom Berkeley wooed in vain. Lawrence, a native of South Dakota and a graduate of its state university, had sought out Swann in 1922, on the recommendation of Merle Tuve, likewise from South Dakota, whom Lawrence had known from childhood. (It is singular that Tuve and Lawrence came from similar middle-class backgrounds of Scandinavian origin, had similar training in physics, and later went at atom smashing with instruments similar in cost and size.) Lawrence joined Swann and Tuve at the University of Minnesota to work on a master's degree. In 1923 Swann left Minnesota for the University of Chicago, where Lawrence followed, to begin work on the photoeffect in potassium vapor. He made good progress in designing equipment before Swann wandered to Yale. There Lawrence finished his dissertation and took up another investigation, which also posed delicate instrumental problems, for example, the creation of a monochromatic beam of slow electrons.[55] That project acquainted him with the magnetic analysis of particle streams, which may later have assisted his design of the cyclotron.

As a National Research Fellow at Yale, Lawrence extended his work on the photoeffect to other alkali vapors and used his monochromatic electrons to demonstrate that the excitation function for ionization of an alkali vapor—the dependence of the probability of ionization on the energy of the ionizing agent—was the same for electrons as for x rays. This demonstration attracted Loeb, headhunting at the American Physical Society's meeting of May 1926. He spoke with the demonstrator, and wrote his chairman: "I felt out one of the most brilliant experimental young men in the East—a lad whose name is on everyone's lips on account of his recent papers on Ionizing Potentials....He is personally one of the

Sep 1930 (full professor at $4,500), and 4 Mar 1936 (16/42).
55. Childs, *Genius*, 63–75; Lawrence, NAS, *Proc., 12* (1926), 29–31.

most charming men I have met....When asked whether he would consider an Assistant Prof. at Berkeley following the termination of his fellowship, he was quite enthusiastic." Birge too praised Lawrence's "really splendid experimental work" on the excitation functions.[56]

Two projects were never enough for Lawrence. A fellow midwesterner, Jesse Beams, who had settled at the University of Virginia, had tried to measure the interval between the absorption of a quantum and the ejection of a photoelectron. While failing, he had devised a very fast electric switch incorporating two Kerr cells. (A Kerr cell is a parallel-plate condenser with a gaseous or liquid dielectric.) Beams and Lawrence teamed up to try to study the speed of the onset of birefringence and to incorporate Beams's idea into a practical device for creating very short bursts of light.[57] Lawrence was to stick to the Kerr effect, as he did to photoionization, until the cyclotron turned his attention to bigger things; and he was to stay in productive contact with Beams while the idea of the cyclotron developed.

Berkeley made its first offer to the young electro-optician in the spring of 1927; Yale did the same; Lawrence stayed East (plate 1.4). His rejection of Berkeley would have discouraged men of weaker will than Birge and Loeb: "I like Yale, the personnel, the laboratory and the facilities for research perhaps even as much as I like the friends I have acquired in New Haven." Loeb and Birge persisted. They extolled Berkeley's "democratic spirit," research ethic, and, what the NRC campaigned for and every ambitious academic desired, light teaching load. Lawrence allowed that if Yale made him work too hard, he might go West: "I am more interested in finding some more of Mother Nature's secrets than telling to someone else things I already know about her." The announcement by his flighty mentor, Swann, of a new nest, the Bartol Foundation in Philadelphia, "a hard blow to Yale and to me," increased Lawrence's mobility.[58]

56. Lawrence, *Science, 64* (6 Aug 1926), 142; Loeb to Hall, in Childs, *Genius*, 99; Birge to Lawrence, 9 June 1926 (Birge P).

57. Beams and Lawrence, NAS, *Proc., 13* (Jul 1927), 505–10, and *JFI, 206* (1928), 169–79; Szlvessy, *Handbuch der Physik, 21* (1929), 742–3.

58. Quotes from, resp., Lawrence to Loeb, 13 Apr 1927 (Loeb P); Birge to Lawrence, 23 Feb 1928 (2/32); and Lawrence to Loeb, 7 May 1927 (Loeb P). Cf. Lawrence to Tuve, 30 Sep 1926 (MAT, 4): "Merle, for the nth time, I hope you

Loeb now hinted that Berkeley might offer an associate professorship with unusually few teaching responsibilities. Yale countered by improving Lawrence's laboratory and reducing his
courses.[59] Friends of California were commissioned to enlighten
Lawrence. They reported that he believed that at Berkeley only
full professors could supervise graduate students (a regime to
which he could not submit) and that he had no conception of the
treasures of Le Conte Hall. Loeb and Birge set him right in
February 1928, when Berkeley formally offered Lawrence an associate professorship. They itemized the research staff, budget, and
facilities, pointed to auxiliaries in the Chemistry Department,
praised the liberality of the Physics Department in assigning graduate students to junior faculty, and fired off their biggest gun:
"The teaching schedules are as light as at any place in the country,
with the exception of Harvard."[60] There remained only salary.
The very experienced Swann advised Lawrence to ignore the few
hundred dollars difference between Berkeley's offer and Yale's,
which the lower cost of living in California would cancel, and to
concentrate on the research opportunities. Birge reassured the
captured lion that promotion came rapidly to vigorous young
research men. It was as if the University had been preparing itself
ever since the war for the reception of Ernest Lawrence:
"Younger men are now being appointed and advanced on an
entirely different plane from that of the older men....I doubt if
any man has ever been offered the permanent position of associate
professor at this University with as short a period of teaching and
research experience as in your case....The conduct of this University now is really in the hands of the exact scientists."[61]

Lawrence arrived in Berkeley in the summer of 1928. By September he was writing with such fervor about his new surroundings that Beams rated him "a 'Native Son' of California already."

will run up to New Haven...This is a wonderful lab."

59. Loeb to Lawrence, 20 May 1927, and Lawrence to Loeb, 27 Sep 1927 (Loeb
P).

60. E.U. Condon, "Comments on Birge's History" (UCA); Muriel Ashley to
Loeb, 18 Feb 1928 (Loeb P); E. Hall to Lawrence, 21 Feb 1928 (the offer) (8/8);
Loeb to Lawrence, with passages to Birge, 21 Feb 1928 (Loeb P).

61. Birge to Lawrence, 23 Feb 1928, and reply, 2 Mar 1928, in Birge, *History,*
3, ix, 5–9; Swann to Lawrence, 28 Feb 1928 (17/3); Birge to Lawrence, 5 Mar 1928
(2/32), quote.

The enthusiast took up residence at the Faculty Club, where Gilbert Lewis brokered influence and crossed disciplines nightly at the dinner table.[62] Lewis became a strong supporter of Lawrence's and, after the invention of the cyclotron, a transient, but most influential, collaborator. The antecedents of the grand invention are not to be sought, however, in the research work that Lawrence did between dinners with Lewis. He continued his study of the photoeffect in alkali vapors.[63] What counted more for his future were his sojourns at General Electric's research laboratories in Schenectady during the summers of 1929 and 1930.

In arranging for his visit of 1929, Lawrence wrote A.W. Hull, an expert on x rays and vacuum tubes who would be his host, that he wanted to study the photoeffect and the Kerr cell. Nothing new there. But Hull was then working with Beams on a lightning arrestor and on the development of sparks under high voltages, and Lawrence was drawn into the investigation.[64] Here he faced for the first time the practical difficulty of holding a potential of more than half a million volts. If we credit Beams's unlikely recollection, this experience inspired a line of thought remarkably, indeed astonishingly, close to the reasoning behind Lawrence's great invention. "There's just no use trying to build this [voltage] up," Lawrence argued. "You may get a few million volts. That's limited. What we've got to do is to devise some method of accelerating through a small voltage, repeating it over and over. Multiple acceleration."[65]

In the event, Lawrence did not labor with Beams or Kerr cells or photoeffects. He wandered about GE's well-equipped laboratories, familiarized himself with Hull's state-of-the-art vacuum tubes, with high-voltage equipment, and with so much of the firm's applied research that he was offered (but declined) a consultancy. Lawrence brought this lore back to Berkeley, along with two other items of importance: a glass blower, E.H. Guyon, a specialist in x-ray and vacuum tubes; and a genius at electronics,

62. Swann to Lawrence, 14 Sep 1928 and 7 Jan 1929, and Beams to Swann, 18 Oct 1928 (Swann P); Lachman, *Borderland*, 143.
63. Lawrence to Loeb, 13 Mar, 16 and 25 May 1928 (Loeb P); Swann to Lawrence, 3 and 4 Oct 1928 (Swann P).
64. Lawrence to Hull, 28 Mar 1928 (7/28).
65. Beams, quoted by Childs, *Genius*, 142.

Hull's assistant David Sloan, whom Lawrence persuaded to improve his B.S. in physics from Washington State College with graduate work at the University of California. Sloan came in 1930, on a Coffin Fellowship (established in 1922 in memory of GE's first president, Charles A. Coffin) from GE.[66] He remained throughout the 1930s, kept from his degree by lack of interest in his studies and injury to his back. The early successes of Lawrence's Radiation Laboratory owed much to David Sloan, who would not have come to Berkeley but for Lawrence's sojurns at GE.

3. DEPRESSION AND ITS CURE

Sloan may have been made the readier to leave GE by observation of the havoc the Depression was wreaking on its research staff. Eventually GE cut back to 50 percent of its strength in 1929; the corresponding figure for AT&T was 40 percent. They suffered, but not as much as the bellwether of economic activity, the steel industry, which was running at a jot over one-quarter of its capacity two years after the stock market crashed. The first nine months of 1931 had been "one of the most unsettled, depressed periods ever known," and the balance of the year looked worse. In the first half of September, tax receipts fell lower than at any time in the preceding decade, and the production of electric power, the agent of industrial vitality and domestic comfort, was diminishing at the rate of 2 percent a week. The vital signs of capitalism slowed to just above moribund; in the opinion of *Time*, the country "had reached the lowest possible level of consumption."[67] Even the mortality rate declined, probably, in the opinion of the Health Organization of the League of Nations, because idleness preserved the population from the dangers of fast living and tuberculosis.[68]

66. Sloan, interview by A. Norberg, 10 Dec 1974 (TBL); Birge, *History, 3*, ix, 13–4; Twentieth Cent. Fund, *Am. found., 3* (1934), 25, 30.
67. Kevles, *Physicists*, 250; *Time, 18* (28 Sep 1931), 45, and ibid. (19 Oct 1931), 51, quote.
68. Murlin, *Science, 80* (27 Jul 1934), 82.

The Depression did not hurt most of the nation's physicists as quickly as it did the steel industry. To be sure, industrial research laboratories cut their staffs, and government agencies, particularly the National Bureau of Standards, had to furlough many of their scientific workers without pay and sustain a devastating amputation from their research budgets.[69] Most American physicists— some 75 percent of the membership of the American Physical Society in 1930—did not work for the federal government or for industry, however, but for universities and colleges. And the universities, especially the large research universities like Berkeley and Caltech, had their best years ever in 1929/30 and 1930/31. By all outward signs—size of faculties and student bodies, receipts, expenditures—the research universities were the ivory towers of their legend, cut off from the hard practical life around them. The average annual gifts from private sources to these universities between 1929 and 1931 totaled the average in the mid 1920s.[70]

The towers were not made of ivory, but of steel, and, as the recent war had made sufficiently clear, they were strongly coupled to the welfare of the nation. Beginning in 1932, decreases in giving, in interest on endowment, and in state support, forced the universities to economize. Senior faculty took a reduction in salary, perhaps 15 percent, across the country; junior faculty suffered most, as is their wont, many being cut out altogether; and research funds from endowments, state appropriations, and external grants fell in proportion.[71] The general method of coping during the bad years, 1932/33 through 1934/35, appears from the steps taken by Berkeley's physics department to meet its severest test, the apportionment of a 20 percent cut in its state allocation for 1933/34. After senior faculty had agreed to a "voluntary" reduction in salary that averaged about 7 percent, they righteously slashed the wages of temporary instructors, axed six of twenty-four

69. Breckwedde to Urey, 10 Nov 1932 (Urey P, 1): "We [at the NBS] are all taking a month's furlough without pay in addition to the month's legislative furlough without pay....Every cent we spend for apparatus means that much more time off without pay."

70. Weart in Reingold, *Context*, 297, 317; AAUP, Committee Y, *Depression*; Geiger, *Knowledge*, 246–8.

71. Rider, *HSPS, 15:1* (1984), 125–6; Geiger, *Knowledge*, 249–50; H. Washburn to Lawrence, 12 Sep 1933 (18/10).

teaching assistants, cut the chairman's contingency fund, and knocked down their research provision by a third.[72] It was in this friendly environment that Lawrence went about raising money for his cyclotrons.

Lawrence managed to squeeze money from a university that claimed to have none in ways we shall delight to chronicle. He also teased out money from foundations, a great support of university research during the 1920s, which had of course suffered in the general financial decline. In 1931, their last good year, 122 foundations gave a total of $52 million, of which $19 million went to medicine and public health and $5 million to research in natural sciences; in 1934 the same number could manage only $34 million, of which $9 million went to medicine and public health and $2 million to the natural sciences. Even in 1940, with a third again as many foundations, giving fell considerably short of the levels of 1930/31. Throughout, however, one thing stayed constant: medical science received four times as much as all the natural sciences combined, and the amount given directly for research in physics (as opposed to fellowships) was a small fraction of the amount available for the biological and physical sciences. In 1937, for example, cancer research received ten times the money (some $17,000!) classified by the foundations as their total direct contribution to research in physics.[73] Lawrence eventually looked to medicine to cure his chronic budgetary ills, and to the Rockefeller Foundation, the largest philanthropy during the 1930s in both outlay and book value.[74] His first patron, the Research Corporation, small and not medical, suffered grievously during the depths of the Depression. Its income sank in 1932, and, as its president wrote Lawrence that August urging the expenditure of more ingenuity and fewer dollars, the outlook was "quite

72. Birge to Sproul, 8 Oct 1932 (5 percent salary cut), and to Lawrence, 6 Jul 1933 (2 percent of total plus 10 percent of excess over $2,000), in Birge P, 33. Research expenses dropped from $11,500 in 1932/33 to $8,000 in 1933/34; Birge to H.O. Kneser, 4 June 1932, and to R. Brode, 9 May 1933 (Birge P, 33).

73. Twentieth Cent. Fund, *Am. found.*, *3*, 5, 19, 55–6, and *4*, 32–3, 37–8, 189, 197; Geiger, *Knowledge*, 253.

74. In 1934 the Rockefeller Foundation had about 28 percent of the declared assets and gave 35 percent of the grants of the 122 foundations covered in Twentieth Cent. Fund, *Am. found.*, *3*, 13–7; the dollar value of its grants declined 25 percent between 1930 and 1934. Cf. ibid., *4*, 22, 30.

discouraging." By the following June, with the University and the Research Corporation crying poor, Lawrence had to acknowledge the chief current bar to research initiatives: "Money is considerably scarcer now than it was three years ago."[75]

The drop in available capital was offset, from the standpoint of the academic research entrepreneur, by a pool of cheap trained labor. "The main result of the depression," Birge wrote in the fall of 1932, "has been that we have more graduate students than ever before and about a dozen visiting fellows....We are suffering, as it were, from a superfluity of riches." By then there were no jobs in universities, industries, and government research agencies. Fresh Ph.D.'s remained at their home institutions, continuing their work and "living on I know not what."[76] They came begging to work, to stay in their fields, to keep up to date until jobs opened again. The situation at the very depth of the Depression, when researchers were less than a dime a dozen, emerges from an exchange in the autumn of 1932 between Lawrence and his then recent Ph.D. Laurence Loveridge, who had worked on alkali spectra. Ill, depressed, with no prospects, Loveridge had fished in all the agencies and got not a single nibble. "It looks as if my only chance is to go to one of the larger universities [to] do some research and try to get enough tutoring or reading on the side to live on." Could he return to Berkeley as an unpaid research associate? Lawrence would not advise it: he already had a half-dozen Ph.D.'s working gratis in the Radiation Laboratory and trying to peck a living from the barren academic landscape.[77] Other research universities experienced a like glut of volunteers.[78]

Berkeley was particularly well placed to profit from this grim rush. Its low cost of living compared with most other major university centers sank even lower as the consumer price index fell to three-fourths of its value in 1929/30. Since in 1930 a graduate

75. H.A. Poillon to Lawrence, 10 Aug 1932 (46/25R), and Lawrence to Poillon, 3 June 1933 (15/16A).

76. Birge to H.L. Johnston, 26 Oct 1932, and to M. Deutsch, 23 Nov 1932 (Birge P, 33).

77. Loveridge to Lawrence, 23 June, 16 Aug 1932 (quote), and reply, 8 Sep 1932 (12/18). A similar request came in Charles T. Zahn to Lawrence, 16 Nov 1933 (18/37).

78. Weart in Reingold, *Context*, 309–10; Rider, *HSPS, 15:1* (1984), 127.

student could obtain a room at $12 to $15 a month, and board at a dollar a day, his minimal needs in 1933/34 could be met for under $40 a month. Anyone lucky enough to find a half-time job at the going rate of 50 cents an hour could just make do; a fellowship holder or teaching assistant with $600 a year could afford to treat his less fortunate fellow students to lunch.[79] The sight of the brigade of volunteer Ph.D.'s did not deter graduate students from finishing their programs or recruits from entering; and, with an almost imperceptible stutter, the number of new doctors of physics in the country kept rising through the 1930s at the same rate as it had during the 1920s, doubling each decade to almost 200 in 1940.

By 1935 many of these fledglings were finding jobs as soon as they found their wings. In 1936 MIT had more calls from industry for its graduates in physics than it could hope to fill. The cycle, in which, as A.W. Hull explained in a talk at the University of Pittsburgh, the industrial physicist is "a luxury indulged in as a speculation during prosperity and dropped when adversity threatens," had turned quickly. It is doubtful that many more than a dozen Ph.D.'s in physics created in the 1930s were permanently lost to the profession because of the Depression. The improvement in the prospects of the graduates reflected a general recovery of the research universities, which by 1935/36 had faculties larger than ever before and rising research budgets. Here the great state universities had a better time than the private ones, since state revenues went up faster than return on endowments.[80]

More Science or Less?

"The prevalent feeling is that of imminent perdition and extinction." These are the words not of a depressed observer of the Great Depression but of Max Nordau, a physician and literary critic, who thus characterized European civilization in the 1890s. He and his fellow doctors of degeneracy traced many of the social ills

79. Geiger, *Knowledge*, 248–9; Lawrence to Akeley, 21 Mar 1930 (1/12), and to Wood, 10 Dec 1932 (9/21); Birge to Aebersold, 10 Mar 1932, and to Brode, 9 May 1933 (Birge P, 33).

80. Kevles, *Physicists*, 273; Hull, *RSI, 6* (1935), 379; Weart in Reingold, *Context*, 296, 315; Geiger, *Knowledge*, 251–2; Tuve to Rabi, 13 Aug 1935 (MAT, 14/"lab letters 1935").

of the fin de siècle to rampant science and uncontrolled technology: to science for removing mystery, spirit, poetry, and choice from the world, for undermining religion and the family and breeding socialism; to technology for encouraging the growth of cities, with their bad air, poor public hygiene, adulterated food, and, worst of all, their frantic pace, driven by the factory and aggravated by the daily press. It occurred to some that society would do well to clamp down on its scientists and engineers, to enact a moratorium on research: "Progress...may be too fast for endurance."[81] This mood evaporated with the multiplication of comforts by electricity and the inauguration of the arms race in Europe preceding the Great War. As we know, the war so raised the stock of physical science and its applications that, in the opinion of lobbyists like the NRC, a progressive country could not have enough of it.

Sed contra, nothing could be plainer than that the uncontrolled exploitation of scientific knowledge did not always net an increase in human happiness. Had not science enriched the slaughter on the battlefield? And was it not still claiming victims in the 1920s, like the telephone operators thrown out of work by the geniuses who invented direct dialing?[82] Since destruction and dislocation attended the deployment of science-based technology, said the bishop of Ripon, setting out some themes from the fin de siècle before an unpromising audience at a meeting of the British Association for the Advancement of Science in 1927, "the sum of human happiness, outside of scientific circles, would not necessarily be reduced if for, say ten years, every physical and chemical laboratory were closed and the patient and resourceful energy displayed in them transferred to recovering the lost art of getting together and finding a formula for making the ends meet in the scale of human life." The bishop's half-serious remark excited only the slightest flurry in the United States, where the eagles of science repeated, with George Ellery Hale, that "our place in the intellectual world, the advance of our industries and our commerce, the health of our people, the production of our farms..., and the

81. Respectively, Nordau (1895) and W. Crookes (1891), quoted in Heilbron in Bernhard et al., *Science, technology, and society*, 57–8.
82. Kevles, *Physicists*, 243–5.

prosperity and security of the nation depend upon our cultivation of pure science," and where the general public was preparing to choose a mining engineer as its president. Millikan inveighed (his usual mode of debate) against a moratorium as an "unpardonable sin." The *New York Times* answered the bishop in a metaphor that summed up the experience of the century: "A ten-year truce on the battle fields of science...is absolutely unthinkable."[83]

It became thinkable in the Great Depression. In 1934 the *Times* recalled that "even before the depression the world was puffing in its efforts to keep pace with science" and challenged scientists to explain why the industrial machine had stopped. Many influential people confused science with its applications and blamed both for the collapse. Raymond Fosdick, trustee and future president of the Rockefeller Foundation: "Science has exposed the paleolithic savage, masquerading in modern dress, to a sudden shift of environment which theatens to unbalance his brain." Hoover's Committee on Recent Social Trends: "Unless there is a speeding up of social invention or a slowing down of mechanical invention, grave maladjustments are certain to result." Robert M. Hutchins, president of the University of Chicago: "Science and the free intelligence of men...have failed us." Henry A. Wallace, Roosevelt's secretary of agriculture: "[Scientists and engineers] have turned loose upon the world new productive power without regard to the social implications."[84] The brutal curtailment of research at the National Bureau of Standards and other federal agencies was a matter of mood as well as of money. Similar considerations affected the states. As W.W. Campbell summed it up from his new eminence as president of the National Academy of Sciences: "The attitude of many, perhaps nearly all, of the legislatures toward research at public expense may fairly be

83. Quotes from E.A. Burroughs, bishop of Ripon (1927), from Hale, *Harper's, 156* (Jan 1928), 244, and from the *New York Times*, 7 Sep 1927, in Pursell, *Lex et scientia, 10* (1974), 146, 151, 152; and from Millikan, *Scribner's, 87* (Feb 1930), 119.

84. *New York Times*, 24 Feb 1934, in Pursell, *Lex et scientia, 10* (1974), 156; Fosdick, *The old savage in the new civilization* (1929), quoted in Kevles, *Physicists*, 249, and attacked in Millikan, *Scribner's, 87* (1930), 121–7; Hoover's Committee (2 Jan 1933), in Kevles, *Physicists*, 237; Hutchins (1933), ibid., 239; Wallace, *Science, 79* (5 Jan 1934), 2.

described as unsympathetic and, in some cases..., as severely hostile."[85]

This time prominent scientists tried to listen. Frank Jewett, vicepresident of AT&T and president of Bell Telephone Laboratories, allowed in 1932, at the dedication of the Hall of Science at the Century of Progress Exposition, that science and technology had had adverse effects. He proposed as cure, in addition to more science, the education of scientists to take into account the social problems their work might create. In 1934 the same sour note sounded at the Nobel prize ceremonies, and, in a lesser venue, at the American Physical Society, whose president told its members that "a thorough investigation of the sociological aspects of physics" was one of the two most important matters before them. The other was "organized propaganda for physics."[86]

The members of Roosevelt's Science Advisory Board, set up in 1933, also acknowledged the need for a scientific analysis of social and economic problems, although they could enlist no one to undertake it. Henry Wallace, speaking for the new administration, urged attention to social engineering and to the humanizing of the engineer by courses in philosophy and poetry. It was a desperate remedy, to be sure, since literature might sap the vigor of an engineer, but then the situation was desperate: "I would be tempted to solve [the difficulty] by saying that probably no great harm would be done if a certain amount of technical efficiency in engineering were traded for a somewhat broader base in general culture."[87] Roosevelt himself called attention to declarations by the British Association for the Advancement of Science in favor of serious study of the social relations of science and asked whether American engineering schools had introduced economics and social science into their curricula. He addressed this question in the fall of 1936 to the former head of his defunct Science Advisory Board, Karl T. Compton, president of MIT, the eastern and better behaved counterpart of Caltech's Millikan. Compton

85. Campbell (1933), in Kevles, *Physicists*, 250; Pursell, APS, *Proc., 109* (1965), 342–51, *Techn. soc., 9* (1968), 145–64, and *Agr. hist., 42* (1968), 231–40.

86. Jewett, *Science, 76* (8 Jul 1932), 23–6 and *Mech. eng., 54* (June 1932), 395; Palmaer, in presenting the Nobel chemistry prize to Urey, in *Les prix Nobel en 1934,* 19; Foote, *RSI, 5* (1934), 63, quotes..

87. Wallace, *Science, 79* (5 Jan 1934), 3; Weiner, *PT, 23:10* (1970), 31, 36; Kuznick, *Beyond the laboratory,* 14–64.

returned the opinion shared by the leadership of the NAS and the NRC: the country does not need engineers and scientists distracted by literature and sociology, but more, better, and costlier science.[88]

This hard line was the theme of a symposium held in February 1934 under the auspices of the American Institute of Physics and the New York Electrical Society. Karl Compton led off: "The idea that science takes away jobs, or in general is at the root of our economic and social ills, is contrary to fact, is based on ignorance or misconception, [and] is vicious in its possible social consequences." And popular. "The spread of this idea is threatening to reduce public support of scientific work..., to stifle further technical improvements..., [to bring] economic disadvantage in respect to foreign countries..., [to precipitate] a national calamity." At this Compton, as head of the Science Advisory Board, hoped to draw $5 million annually from the U.S. Treasury for support of scientific research outside of government agencies. In Compton's "best science" vision, the NAS and NRC would supervise the spending: only they could direct fire at the right targets (as he later said, in the common metaphor), and, without political or regional considerations, extract silk from wood, rubber from weeds, gasohol from corn, and, into the bargain, complete the electrification of the country.[89] Roosevelt's social engineers would have nothing of the scheme and postponed eager federal support of physical science to the next shooting war. The president threw a small bone to the hunting dogs of science: in 1935 he qualified scientific research for support by the Works Progress Administration. Since the WPA could only assist people who had lost their jobs, and since, in science, these were often the least qualified, its program detractors sometimes labelled it "worst science" support.[90]

88. The exchange, in October 1936, was published in *Science, 84* (1936), 393–4. The social concerns of the BAAS are detailed in McGucken, APS, *Proc., 123* (1979), 239–49, and *Scientists, society, and state*, 27–46, 95–115; Kuznick, *Beyond the laboratory*, 64–70.

89. Compton, *Sci. monthly, 38* (1934), 297, and *Techn. rev., 37* (1935), 152, 154, 158. Cf. Dietz, *RSI, 7* (1936), 1–5.

90. Geiger, *Knowledge*, 257–60, 263–4; Kevles, *Physicists*, 254–8, 264–5.

Where then to find the money for scientific research that the Comptons and Millikans and Jewetts thought necessary to national recovery? A startling answer was given by Hull, Lawrence's sometime patron at GE's laboratories. Hull took it for granted that physics would lose the privileged position that break-throughs in electronics had given it since the war. He did not expect much help from government, and nothing in the near term from foundations or industry. "We should face the problem of carrying forward the torch of physical science with not only un-abated, but accelerated speed, without additional facilities." How then? By enlisting high-school teachers in research, certainly not a "best science" approach, and, as Hull acknowledged, not an easy one either, since most teachers were already overworked. But, on reflection, that might be an inducement: "If happiness is propor-tional to accomplishment..., [and accomplishment to effort,] then more, not less, overwork should be our goal."[91]

A more practical goal was to prepare the physicists that indus-try might absorb when business improved. It was agreed that the United States had not progressed so rapidly in industrial as in academic physics.[92] It was further agreed that fresh Ph.D.'s did not enter industry with the tools needed to succeed there: an abil-ity to work in groups and to attack problems en masse, the imagi-nation to disregard "the imaginary boundaries between different branches of science and technology," and, above all, a knowledge of chemistry and "the realization that there is a field in physics outside of atomic structure and wave mechanics."[93]

The American Way

American physicists worked and spent their way out of the Depression. There were close approximations to Sinclair Lewis's Doctor Arrowsmith among them—men eager to maximize over-work, rough around the edges (Arrowsmith did not know "a sym-phony from a savory"), thoroughly dedicated to their science,

91. Hull, *RSI, 6* (1935), 377–8.

92. Barton, reporting on the "Conference on Applied Physics," 14 Dec 1934, under the auspices of AIP and NRC, in *RSI, 6* (1935), 30; Hull, ibid., 383.

93. Quotes from, resp., Barton, *RSI, 5* (Aug 1934), 263, and *RSI, 6* (1935), 32, the last reporting the views of O.E. Buckeley, director of research at Bell Labs, and Saul Dushman of GE; Hull, *RSI, 6* (1935), 378–80.

their careers, and, at second remove, their neighbors. Harold C. Urey, a student of G.N. Lewis's, was one of the most successful of these men. His great discovery of heavy water, made in 1932 after much hard work by himself and selfless colleagues, immediately found application in physics and chemistry, and, what had greater social value, biology; it also lifted the spirits of beleaguered scientists and won Urey a Nobel prize. In 1937, in an address at the dedication of a new building at the Mellon Institute for Technological Research, Urey voiced Arrowsmith's creed: "We wish to abolish drudgery, discomfort and want from the lives of men and bring them pleasure, comfort, leisure and beauty....The results of our work completely outdistance our dreams....You may bury our bodies where you will, our epitaphs are written in our scientific journals, our monuments are the industries which we build, which without our magic touch would never be."[94]

One of Lawrence's most brilliant students of the Depression years, Robert Wilson, a midwesterner like Arrowsmith (and Lawrence), fed his fancy with the heroics of Lewis's doctor when riding the range in Wyoming. Arrived at Berkeley, he dismounted to find his ideal in charge of a radiation laboratory.[95] During his early years at Berkeley, Lawrence did have many of Arrowsmith's qualities. Like the doctor, he had two passions, one science, the other the daughter of a physician on the faculty at Yale. "I have two consuming loves," he wrote his great friend Donald Cooksey in the summer of 1931, "Molly and research!" "I am so badly (or goodly) in love that at times it is positively painful." And, like Arrowsmith, Lawrence was boyish and unsophisticated, open in his enthusiasms, in a word the word was Oppenheimer's— "unspoiled." He believed that to start work was to begin to improve, and that more science, not less, would liberate from psychological as well as economic depression. He labored hard on these principles, so hard that he often fell a victim to severe colds, which increased in frequency in step with the growth of his laboratory.[96] As Arrowsmith discovered, fulfilling one's scientific

94. S. Lewis, *Arrowsmith*, quote from chap. 39; Urey, *RSI, 8* (1937), 226–7. On Arrowsmith as the American scientific hero see Rosenberg, *No other gods*, 128–31.

95. Wilson in Holton, *Twentieth cent.*, 468–9, 471.

96. Lawrence to Cooksey, 17 Jul [1931] (4/19), and to Tuve, 27 Aug 1931 (MAT, 8); Lawrence to Beams, 21 Nov 1931 (2/26) and to Poillon, 20 May 1940, and Cooksey to Poillon, 6 June 1940 (15/18), on colds; Oppenheimer to Francis

ambitions according to the highest standards while running a large research institution constantly in need of money may not be possible. Lewis's doctor hero cleared his conscience by an unrealistic escape to a small workplace in the wilderness; Berkeley's Lawrence cut corners, lost innocence, and built the largest laboratory for nuclear science in the world.

Arrowsmith's ferocious pace in pursuit of truth, his almost athletic performances in the laboratory, were characteristically American. The "feverish exploration for the secrets of matter's composition" (as *Science Service* described research in nuclear physics in 1934) impressed European observers of science in the United States. Would you care to think more about it, Rudolf Peierls wrote Hans Bethe about a joint paper, "or do you insist on publishing in American tempo?" Haste makes waste. Bethe had become almost a legend for his error-free calculations. But in a recent paper he had made two numerical mistakes in one single table. "Is that America," Peierls asked, "or the automobile?"[97]

As American as the fast pace was the big machine. Already before the war, the size and variety of equipment in American physics institutes "made [a European's] mouth water."[98] It appeared to Franz Simon, who surveyed facilities for low-temperature research in the United States in 1932, that "Americans seem to work very well, only they obviously insist on making everything as big as possible." A few years later, Paul Capron of the University of Louvain expressed perfectly the standard impression made on Continentals by the research facilities in the United States: "In Princeton as in Columbia University, I was most amazed by the richness of the laboratory....[At MIT] I saw the most extraordinary technics [i.e., instrumentation]." He returned to Belgium, inspired by the American spirit, the "constructive civilization of 'go ahead.'"[99] From machine worship

---

Ferguson, 14 Nov 1926, in Smith and Weiner, *Oppenheimer*, 100; Lawrence to Hedrick, 2 June 1938 (19/35), re moratorium.

97. *Science service*, ca. March 1934, in Cockburn and Ellyard, *Oliphant*, 55; Peierls to Bethe, 25 Aug 1936 (HAB/3).

98. W.E. Ayrton (1904), quoted in Forman, Heilbron, and Weart, 82; Manegold, *Universität*, 116–47.

99. K. Mendelssohn to Silva Critescu, 11 Nov 1932, reporting Simon's observations (Mendelssohn P); Capron, "Report," 2, 5.

there is but a step to materialism, the last and heaviest ingredient in the European depreciation of American culture. A good American answer to this stale charge came from an immigrant inventor, Michael Pupin, who accepted it, played with it, gloried in it. What we do in America, he said, results from close study of nature, which enhances the spirit; and nature happens to be a machine. "The [artificial] machine is the visible evidence of the close union between man and the spirit of the eternal truth which guides the subtle hand of nature."[100]

The American physicist projected his image at home with the willing help of the press. The collaboration, which was first struck just after the war with *Science Service* and with the coverage of the meeting of the American Association for the Advancement of Science in 1922, entered a new stage in April 1934, with the formation of the National Association of Science Writers and the call from the podium of the American Physical Society for propaganda for physics. One of the leaders of the writers' association was David Dietz, the science editor for the Scripps-Howard newspapers. Dietz volunteered the help of his organization in "selling physics to the public," when, at a conference on applied science held late in 1934, Saul Dushman of GE expressed the hope that the selling "would be done without impairing the dignity of science." In 1935, at a conference on industrial physics in Pittsburgh, Dietz observed that the only way to get $50 million a year from the government—$50 million being the loss in research money owing to the Depression, according to the American Institute of Physics—was to work on public opinion. "Your best allies in creating public support for science are the newspapers," Dietz said. "There is a new understanding today between the world of science and the newspaper world."[101] The director of *Science Service*, Watson Davis, took the same line. Nowadays, he said, in a speech in February 1936, science writers follow science with the same attentiveness and understanding that sports reporters lavish on football. Their relation is symbiotic: the scientist reveals, the

100. Pupin, *Scribner's, 87* (1930), 136.
101. Foote, *RSI, 5* (1934), 63; Barton, *RSI, 6* (1935), 35, quoting Dietz and Dushman; Dietz, *Science, 85* (1937), 108, on the history of relations between science and the press.

reporter "detechnicalizes." "These essentially changed attitudes on the part of the press and the world of science are among the most encouraging signs of our times." No fewer than sixteen reporters showed up at the AAAS meeting in December 1935. That suited the science lobby perfectly. Austin Clark, the press director of the AAAS echoed Dietz: "We must all work together in order that the press may have an abundance of suitable material to present to the public."[102] It remained only to take the step, which would have been anathema to Arrowsmith, of fusing business with research. This Maurice Holland, the director of the Division of Engineering and Industrial Research of the NRC, did not disdain to do. "There seems to be some connection between selling and science—I, for one, believe they are brothers under the skin."[103]

At first Arrowsmith-Lawrence drew back from selling, in public at least. "We are not interested in publicity," he wrote a would-be reporter in 1934. In this policy he had been encouraged by a newspaper report that he was trying to transmute base metal into gold and by Molly, who thought it "unfortunate that the Research Council [i.e., Corporation] etc. have demanded so much publicity on your work."[104] But Lawrence could not long affect this other-wordly attitude. When the president of the University asked him to talk to the Rotary Club of Berkeley, and to furnish a copy "in order that we may use it for publicity in the newspapers of the state," Lawrence could only reply that he would be "more than glad to do so." He became expert in dealing with Dietz and company; and his benefactors came to request that he write press releases to satisfy the curiosity of the newspapers. An example of his handiwork, anent a grant from the National Advisory Cancer Council, which dispensed federal money: "[The] strikingly rapid

102. Resp., Dietz, *RSI*, 7 (1936), 5; Davis, *Vital speeches, 2* (1936), 361; and Clark, *Science, 81* (20 Mar 1935), 316. Cf. Dietz, *Science, 85* (1937), 112.

103. Holland, in *RSI*, 6 (1935), 36. Cf. Carter, *Am. schol., 45* (1975/76), 780.

104. Lawrence to Tuve, 10 Nov 1931 (MAT, 3), and to Hugh Kitchen, 1 Feb 1934 (10/6); Molly Blumer to Lawrence, 8 Nov [1931] (10/38), inspired by Zinnser, *Science, 74* (23 Oct 1931), 402: "Institutional rivalry, bidding for support, has had a tendency to foster the submission of results to public and inexpert applause before they have been passed upon in the forum of technical criticism." Tuve and Urey also thought themselves victims of an irresponsible press; Tuve to Hafstad, 4 Jan 1931 (MAT, 8) and Brickwedde to Urey, 12 Dec 1931 (Urey P, 1).

development of these powerful new weapons in the war on cancer
[he meant cyclotrons] is a splendid example of the fruitfulness of
the active interest and support of the government in medical
research [the government had had nothing directly to do with
financing cyclotrons], for it may be truly said that the National
Advisory Cancer Council has greatly accelerated the day in our
generation when countless cancer sufferers may be benefitted by
these new radiations."[105] The fallen Doctor Arrowsmith himself
became a journalistic object, attaining the frontispiece of *Time* in
1937 and the insides of *Scientific American* in 1940.[106]

A Solution

An enduring feature of "science" reporting during the 1930s
was atomic energy. Scientists had raised the possibility from the
time of the discovery of radioactive decay among the natural elements in 1902. The measurement of the masses of hydrogen and
helium to unusual precision in 1922 by Francis Aston of the
Cavendish Laboratory in Cambridge, England, gave perhaps the
first substantial ground for hope or fear that civilization might run
or destroy itself by exploiting the atom. Aston found that four
atoms of hydrogen have a greater mass than one of helium; should
it be possible to synthesize helium from hydrogen, Einstein's law
of equivalency between mass and energy promised that something
noteworthy would ensue. According to his calculations, Aston
said, the hydrogen in a pint of water could yield enough energy to
drive a steamship across the ocean and back; or "the transmutation might be beyond control and result in the detonation of all
the water on the earth," a possibility he considered "interesting"
but remote. If all went well, "there would be literally no limit to
the material achievements of the human race."[107]

Most reputable physicists, Max Planck for example, wrote of
the need, desirability, and eventual practicality of "the liberation

105. Sproul to Lawrence, 29 Sep, and reply, 4 Oct 1932 (40/49); L. Hektoen to
Lawrence, 5 and 11 Oct 1938, requesting press release, and reply (quote), 20 Oct
1938 (13/29).

106. *Time, 30:2* (1 Nov 1937), *Sci. Am., 163:2* (Aug 1940), 68–71.

107. Aston, *Nature, 110* (25 Nov 1922), 705. Cf. Fermi, *CP, 1,* 33–4, a text of
1923, and O.M. Corbino's famous speech of 1929 (Segrè, *Fermi,* 66–7), for similar
considerations.

of atomic energy."[108] Scientists who thought practical atomic energy farfetched or impossible found it awkward to protest, since they would open themselves to the difficult duty of proving an impossibility. Millikan was one of the few physicists who then talked regularly with the Creator and could know the impossibility of the release of useful or destructive atomic energy. One of the favorite arguments of those who favored a moratorium in physical research turned on the likelihood that unsupervised physicists might discover a way and unprepared society might blow itself to bits. No chance, said Millikan. His observations of cosmic rays, which showed that the sort of synthesis Aston contemplated could not occur on earth, and his calculations about radioactive decay, which showed that atomic disintegration could not power a popcorn machine, demonstrated sufficiently that God had made the universe proof against destruction by inquisitive physicists. "There is not even a remote likelihood that man can ever tap this source of energy at all."[109]

Millikan was gainsaid by the Compton brothers: by Arthur, who discredited Caltech's conception of cosmic rays, and by Karl, who, in January 1933, predicted the arrival of useful atomic energy—or, at least, of "the most exciting and far-reaching developments in the whole history of science"—within a generation.[110] The venue for this utterance could not have been more public or more appropriate: the Century of Progress Exposition, whose centerpiece was a huge robot, signifying science, pushing a man and a woman into the future. The high-tech razzle-dazzle of the exposition incorporated a "philosophy of showmanship for the contributions of science and their applications" developed by a committee headed by Frank Jewett in cooperation with the NRC. No mundane signal like the radio call that activated the Diesel at the San Francisco fair would do for Chicago: the Century of

108. Planck, *Die Woche* (1931), 1419–20, and *Ernte, 13:7* (1932), 33; for other examples, Badash et al., APS, *Proc.,* *130* (1986), 198–202, and Weart, *JP, 43* (1982), suppl., 306–8.

109. Millikan, *Science, 68* (28 Sep 1928), 282–4, quote on 248, and *Scribner's, 87* (Feb 1930), 121.

110. Karl Compton, *New York Times,* 30 Jan 1933, 15, and *Techn. rev., 35* (Feb 1933), 190; for the battle of the cosmic rays, De Maria and Russo, *HSPS, 19* (1989), 226–61.

Progress Exposition came to life at the command of forty-year-old light from the star Arcturus collected at several observatories and forwarded by Western Union. As F.K. Richtmeyer, dean of the Graduate School of Cornell, said in celebration, "Scientific research pays large dividends." Would the study of the atom do so? When? "There is no telling," according to the dean, "when a scientific Columbus is going to discover another America."[111]

Compton's prediction and others even less responsible excited what Lawrence called a "newspaper ballyhoo." That was too much for the great proprietor of the nucleus, Lord Rutherford, who thundered before the British Association in the fall of 1933 that "anyone who says that with the means at present at our disposal and with our present knowledge we can utilize atomic energy is talking moonshine." The thunder made the front page of the *New York Herald Tribune*, with some echoes by American physicists. The editors of *Scientific American* applauded Rutherford's statement, although, they said, it was hardly scientific to rule out the possibility. Against them stood Lawrence, who by then was shooting at nuclei with projectiles from his cyclotron. To him, according to the *Tribune*, release of useful energy is "purely a matter of marksmanship."[112] This view of the matter appears to have attracted the attention of Einstein. At a conference arranged by the new National Association of Science Writers, he disparaged the efficacy of Lawrence's artillery. "'You see,' he said, with his characteristic sense of humor and picturesque expression, 'it is like shooting birds in the dark in a country where there are only a few birds.'"[113]

Against the public declarations of Rutherford and Einstein, and within an environment not favorable to an economic transformation of the sort that atomic power was expected to bring ("It appears doubtful...whether coal mining or oil production could survive after a couple of years"),[114] Lawrence thought it prudent

111. Rydell, *Isis*, 76 (1985), 527–31, 534; the words about showmanship are Jewett's.

112. Lawrence to Swann, 28 Sep 1933 (17/3); A.F., *Nature, 132* (16 Sep 1933), 432–3, quoting Rutherford; *New York Herald Tribune*, 12 Sep 1933, 1, quoted in Badash et al., APS, *Proc., 130* (1986), 203; *Sci. Am., 149* (Nov 1933), 201.

113. Interview at AAAS meeting, Dec 1934, quoted by Dietz, *RSI, 7* (1936), 4–5.

114. Szilard, memo of 28 Jul 1934, in Weart and Szilard, *Szilard*, 39; same sen-

to blunt his many hints that nuclear physicists might open the atom for business. In the early 1930s these hints came primarily in direct connection with victualling his laboratory. In a memorandum of accomplishments written in 1933, for example, he pointed to data indicating the release of energy in disintegration, and hinted, "This is a matter of great scientific interest and eventually may have a practical application." In asking for the renewal of a student's fellowship, he wrote that the experiments it supported "have the possibility of contributing knowledge which may ultimately lead to utilization of [the] vast store of atomic energy." Nor did he disdain to make the same point when requesting the loan of a pound of beryllium from an industrial supplier: "If a means can be found for stimulating on a large scale the explosion of beryllium nuclei, a practically unlimited store of energy is thereby made available."[115]

By the mid 1930s Lawrence was in great demand as a speaker. He often took the opportunity to hint at practical atomic energy. The "University Explorer," a radio show originating at the University of California and carried by the networks, asked him in 1936 what the future might hold. "I'm almost afraid to guess," he replied. "Speaking officially, as one scientist, I can only say that we are going to continue our studies." And unofficially? "Certainly we are much closer to [the release of subatomic energy] than we were a few years ago....If we can discover a method of starting chain reactions...the problem would be solved." Again, speaking officially, at commencement exercises at the Stevens Institute of Technology in 1937: "It is only of interest [here, at this engineering school, where speculation has no place] to indicate the present state of knowledge with proper humility." Current knowledge made the release of atomic energy appear "fantastic;" no one had any idea how to do it and the second law of thermodynamics might forbid it. And yet, unofficially: "It is conceivable that in our lifetime this great principle [the transformation of mass into energy] will play a vital role in technical developments which at

timent in Dietz, *RSI, 7* (1936), 4–5.

115. Lawrence, "A brief statement concerning the Radiation Laboratory" [12 Sep 1933] (20/13); letters to E.B. Reed, secretary, Commonwealth Fund, 7 Dec 1934 (10/18), on behalf of Bernard Kinsey, and to L.L. Stott, Beryllium Products Corporation, 14 Feb 1935 (2/28).

the moment are beyond our dreams—for such has been the history of science."[116]

By this time, 1937, the nucleus had become the enduring fad of physics. A marginal topic before the crash, it was absorbing 10 percent of the efforts of physical scientists, as reported in *Science Abstracts*, during the depths of the Depression. The increase of interest and labor, as expressed by the content of the the *Physical Review*, was so sharp that even the numbers that represent it are dramatic: in 1932, 8 percent of its articles, letters, and abstracts concerned nuclear physics; in 1933, 18 percent; in 1937, 32 percent.[117] Men beginning their careers in physics in the early 1930s saw the nucleus as (to quote one of them) an unstudied frontier, a research land of opportunity, a gold field. The young Otto Robert Frisch, a Viennese who had studied all over Europe, wrote of the nucleus—the *unerforschtes Neuland*, the *Goldfelder*—like a Forty-niner.[118] The metaphor was most apt. Although several of the key discoveries that excited interest in the nucleus during the Depression were not made in the United States, the Americans, with their big machines and high pressure, very quickly dominated the field. And the field dominated them. A physicist inexpert in nuclear physics in 1937 was, according to one of them, Arnold Sommerfeld, who had taught atomic theory to all of Europe, "completely uneducated by American standards."[119]

At the head of the American expeditionary force against the nuclear citadel stood Ernest Lawrence. He was the quintessential American leader, active, youthful, optimistic. He had built while others idled, sustained while others retrenched, inspired while others doubted. "The trade of a 'cyclotroneer,'" wrote one of his students in an unintended double entendre, "is one which has experienced no depression."[120]

116. "University Explorer," no. 109, 9 Feb 1936 (40/15); Lawrence, *RSI, 8* (1937), 313; cf. Lawrence, "The work" [1937], 5.
117. Weiner, *PT, 25:4* (1972), 47, and in Thackray and Mendelssohn, *Science and values*, 210.
118. Frisch to "Bavitschkerl," 17 Jul 1933 (Frisch P). Cf. Frank H. Spedding to G.N. Lewis, 1 Dec 1934 (Lewis P): "This field is moving so rapidly that one becomes dizzy contemplating it."
119. Sommerfeld to Einstein, 30 Dec 1937, in Pauli, *Briefwechsel, 2*, 118.
120. Kurie, *Jl. appl. phys., 9* (1938), 691.

Was the cyclotron "a product of the land of the Golden Gate"? The question was put to Donald Cooksey, then Lawrence's alter ego, by a reporter at the Golden Gate International Exposition of 1939. Cooksey was standing in the exposition's Hall of Science, before a full-scale model of the latest cyclotron. The model used steel balls accelerated by gravity to knock apart a representation of a lithium atom, "which is unable to defend itself against such vigorous attack, and is blown to bits." But is it Californian? "It literally and truly is," Cooksey replied. "Practically coincident with the opening of this marvelous Exposition [on San Francisco Bay] is the culmination of years of development [at Berkeley] of the world's largest cyclotron." Attempting to reduce its capacity to his audience's, Cooksey estimated that it would shoot more "atomic bullets" in one second than all the real bullets that all the machine guns in the world could fire off during the entire life of the exposition. "Or more than one gun could fire in three million years."[121] How this machine and its predecessors came to be created in California, how their sizes and uses grew, how they and their makers spread throughout the land, and how the machines and the men went off to war, are the main subjects of this book.

121. Cooksey, interview by Mutual Broadcasting Company, 17 Feb 1939 (4/22).

II
A Million Volts or Bust

1. SOME PRELIMINARIES

The naturally occurring radioactive isotopes emit nuclei of helium (α particles), fast electrons (β particles), and penetrating x rays (γ rays). The energies of these particles and rays are usually given as multiples of the "electron-volt," the energy of motion an electron acquires in falling through a potential difference of one volt (two-thirds the voltage of a standard flashlight battery). An electron-volt is an absurdly small measure, a little over a million-millionth of the "erg," the basic energy unit of the physicist; and the erg itself amounts to less than the footfall of a fly. To fix ideas and notation, an alpha particle from the naturally occurring isotope polonium with atomic weight 210 (Po^{210}) has an energy of 5.3 million electron-volts (5.3 MeV) and a velocity of 1.6 billion centimeters a second ($1.6 \cdot 10^9$ cm/sec), about one-twentieth the velocity of light. The individual polonium α particle, although projected with tremendous speed, has a negligible mechanical effect on the surface that stops it. A great many of them, however, are noticeable.

A noticeable quantity is 37 billion, roughly the number of disintegrations occurring each second in a gram of radium; this rate of collapse, called a curie, is the basic measure of radioactivity. If each of these billions of disintegrations resulted in an alpha particle as energetic as polonium's, all together they would carry away some $3.5 \cdot 10^5$ erg, or a little less than a hundredth of a calorie, each second. An ampere of alpha particles is the equivalent of almost a hundred million curies (10^8 Ci). A milliamp (mA) or even a microamp (μA) of alpha particles greatly exceeds the flux from available natural isotopes: 1μA = 80 Ci, 1 mA = $8 \cdot 10^4$ Ci; a microamp playing on a surface would heat it at the rate of almost one calorie a second, a milliamp at almost a thousand.

Since the nucleus occupies about as much of an atom as the earth does of the sphere whose radius is its distance from the sun, the chance that an alpha particle will strike a nucleus head on before it comes to rest is very small. But only in such collisions can the particle force its way into the nucleus; hence, if forced entry is the experimenter's goal, he would do well to furnish himself with a microamp rather than with a curie of projectiles. The μA has an even greater advantage than the preceding numbers indicate; the radiation from a natural source goes in all directions, whereas the outpouring from an artificial source may be tightened into a directed beam.

In 1919 Sir Ernest Rutherford unintentionally introduced alpha particles into nitrogen nuclei and, what was much more difficult, recognized what he had done. His initial objective was to study collisions between alpha particles and other light nuclei. As source he used a descendent of radium, RaC (Bi^{214}), with a maximum strength of 0.08 Ci. That was strong enough to make the effect under investigation—knocking on, or driving forward, one nucleus by another—easy to detect. The knock-on particles were caught on a screen coated with a material that glowed where hit; and the source could be brought so close to the detector that as many as 40 or 50 flashes a minute could be spotted in the field of a microscope pointed at the screen. In this way, with hydrogen gas as target, Rutherford found swift knock-on protons, and, with oxygen and nitrogen, somewhat slower bumped ions, as he had expected.[1] A slight anomaly occurred in nitrogen, however. Despite every precaution against traces of hydrogen in the experimental space, protons even swifter than knock-on hydrogen ions persistently obtruded. It appeared to Rutherford that a few alpha

1. Rutherford, *CP,* 2, 549–50, 564–5, 583 (1919).

particles from RaC, making particularly intimate contact with nuclei of nitrogen, had driven out protons from the heart of the atom. He believed rightly that he had disintegrated the nucleus and wrongly that the process resembled a successful shot at marbles, in which both the impinging and the struck balls end up outside the target area.[2] He continued this work in collaboration with James Chadwick, an excellent experimenter trained at the Cavendish Laboratory, who had perfected his physics in Germany as a prisoner of war and who had returned to Cambridge, where he became in time the associate director of the laboratory. By enlarging the field of their microscope and viewing the disintegration protons at right angles to the direction from source to target, Rutherford and Chadwick showed that alpha particles could knock protons from most light elements and that the yield increased sharply with the energy of the bombarding particles.[3]

In these experiments perhaps ten alpha particles in a million made a collision that resulted in a detectable disintegration proton. Since the average source was 0.05 Ci and the solid angle at the microscope about 0.0002 steradians at most, the maximum number of countable protons was one a second, or, taking the efficiency of the screen into account, around one a minute per mCi. That sufficed to detect disintegration protons but not to give satisfactory quantitative measurements of their speeds, penetration, angular distribution, or yield as a function of the energy of bombardment. Until late in 1924, and probably for some time afterwards, Rutherford and Chadwick continued to suppose that they were engaged in a game of marbles, and tried to fit their meager quantitative results to irrelevant billiard-ball mechanics. But that year, in one of those triumphant syntheses that distinguished the Cavendish, one of its senior members, P.M.S. Blackett, succeeded in photographing in a Wilson cloud chamber (an invention of the laboratory) the trajectories of the particles participating in productive collisions of the type discovered by Rutherford. In Blackett's beautiful pictures—he obtained records

2. Rutherford, *CP*, *2*, 583–90 (1919); cf. Stuewer in Shea, *Otto Hahn*, 23–8.

3. Rutherford and Chadwick, in Rutherford, *CP*, *3*, 48–9, 53–4 (1921), 68–9 (1922), 110–1, 113–4 (1924); reviewed in Rutherford, Chadwick, and Ellis, *Radiations*, 281–301.

of eight disintegrations in 23,000 photographs presenting 400,000 tracks—no trace appeared of an alpha particle leaving the site of a productive collision (plate 2.1). The mental picture required that it continue on; the optical evidence indicated that it disappeared, swallowed by the target nitrogen nucleus. If so, the resultant nucleus, after expulsion of the swift proton, should be an isotope of oxygen.[4] In keeping with the convention used in the 1930s, we shall write the reaction as $N^{14}(\alpha,p)O^{17}$ and refer to it as an (α,p) transformation.

The problem of accounting for the energy in nuclear transformations became pressing and frustrating. Here another bit of Cavendish pioneering enriched and complicated the proceedings. Following up work he had done before the war as assistant to J.J. Thomson, Francis Aston perfected his system of separating the isotopes of the light elements (in very small quantities, to be sure!) by electric and magnetic fields. From the trajectories of the ions of the various isotopes he could calculate their masses; which, by 1930, he was reporting to a ten-thousandth of the mass of a proton (m_p). Now 0.001 m_p is about 1 MeV: hence Aston's measurements of isotopic masses appeared to be just accurate enough to be used in working out the energy balance in nuclear transformations provoked by the input of a few million electron volts. The upshot: the numbers obtained by Aston, Blackett, and Rutherford and Chadwick did not agree, and prompted the dispiriting hypothesis that the normal internal states of nuclei of the same isotope differ energetically.[5]

If only the statistics were better! If only the sources were not so weak! If only the geometry of the experiments could be improved! Laments of this character enliven the pages of Rutherford and Chadwick.[6] To improve upon the work of nature appeared to

4. Blackett, *PRS, A107* (1925), 350–1, 356; Rutherford, *CP, 3*, 136–8 (1925).

5. Rutherford and Chadwick, in Rutherford, *CP, 3*, 218–24 (1930); Rutherford, Chadwick, and Ellis, *Radiations*, 302–9.

6. E.g., Rutherford and Chadwick, in Rutherford, *CP, 3*, 48 (1921), "so few in number and so feeble in intensity;" 75 (1922); 136 (1925), "unfortunately, on account of the small number of particles;" 166 (1927), "unfortunately, the amount of disintegration...is very small..., an elaborate technique is required to obtain quantitative results;" 220 (1930); and, in Rutherford, Chadwick, and Ellis, *Radiations*, 306, "information is not sufficient to test accurately the balance between the energy and mass changes."

require a machine capable of producing at least a microamp of light positive ions and of accelerating them to a few million volts; if a nucleus propelled an alpha particle at 5 MeV, would it not be necessary to hurl it at 5 MeV to return it? Already before the war Rutherford had judged the creation of machines operating at the highest possible voltage to be "a matter of pressing importance" for the study of beta rays;[7] but for many years it proved impractical to make insulators that could hold much more than seven or eight hundred thousand volts (700 or 800 kV) or accelerating tubes that could withstand even half that much. Subsequent improvements in electrical equipment, x-ray tubes, and insulating materials, the closer ties between industry and academy forged during the war, and the newly important field of nuclear transformation gave new urgency to the matter.

Rutherford properly took the lead in promoting development of million-volt accelerators. In 1927, in addressing the Royal Society of London as its president, he challenged his audience to fulfill his long-time wish for "a copious supply" of projectiles more energetic than natural alpha and beta particles. His appeal received wide attention and many proposals for realizing it came forward. Progress was slow at the laboratory level. In 1930, while his associates struggled to make a source of a few hundred thousand volts, Rutherford asked big electrical industry for help in raising the "puny experiments in the laboratory" to nature's scale. "What we require [he said, at the opening of a new High Tension Laboratory at Metropolitan-Vickers Electrical Company] is an apparatus to give us a potential of the order of 10 million volts which can be safely accommodated in a reasonably sized room and operated by a few kilowatts of power. We require too an exhausted tube capable of withstanding this voltage....I see no reason why such a requirement can not be made practical."[8]

7. Rutherford, quoted in H.A. Lorentz to Comité scientifique Solvay [1914] (Langevin P).

8. Rutherford, *Nature, 120* (3 Dec 1927), 809–10, excerpting Rutherford's address later printed in *PRS, A117* (1928), 300–16; speech at Metro-Vick, quoted in Eve, *Rutherford,* 338.

2. HIGH TENSION

Were the million volts the only criterion—were size of current and steadiness of operation of little importance—the quest would have been over soon after it started. Several techniques for obtaining very high, fleeting voltages across discharge tubes succeeded by or before 1930. None gave rise to a useful beam of positive projectiles. Nonetheless they are worth attention, since some of the problems they raised, and workers they employed, recur in our story. In addition to these impulsive methods, two others for obtaining steady high potentials directly were under development in 1930. For some purposes they held an advantage over the cyclotron.

The Impulsive Way

Nature provides big voltages gratis to anyone bold enough to play with lightning. In the summers of 1927 and 1928, three members of the physics institute at the University of Berlin hung an antenna between two mountains in the Italian Alps 660 meters apart. In their definitive arrangement (fig. 2.1), a string of heavy-duty insulators, provided free by their German manufacturer, kept the antenna and the probe line attached to it from ground; the potential reached by the antenna was controlled and measured by the air space between a metal sphere dangling from the probe line and an earthed sphere hanging beneath it. During thunderstorms, the antenna rose to a very high potential over ground, sending sparks between the spheres across as much as 18 meters of pure Alpine air. The intrepid experimenters calculated that, in that case, they were dealing with 15 million volts. The two who survived the experiment, Arno Brasch and Fritz Lange, returned to Berlin to construct a tube that might stand up to a good fraction of this voltage.[9]

In the comparative safety of the Allgemeine Elektrizitäts-gesellschaft's research laboratory in Berlin, Brasch and Lange

9. Brasch and Lange, *Zs. f. Phys.*, *70* (1931), 10–1, 17–8. Their unfortunate companion, Curt Urban, was killed during the experiments of 1928; Livingston, *Particle acc.*, 2, misdates the experiment and Livingston and Blewett, *Particle acc.*, 22, kills the wrong man.

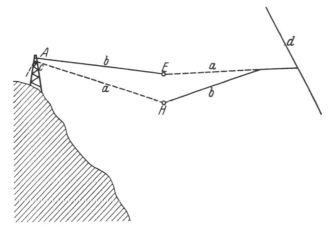

FIG. 2.1   Brasch and Lange's lightning catcher. E and H are the spheres between which the discharge occurs; AE, the antenna; a,a, insulators; b,b, conductors; d, a grounded wire. Brasch and Lange, *Zs. f. Phys., 70* (1931), 17.

tested designs for a sturdy discharge tube with the help of an impulse generator able to reach 2.4 MV. Its principle, the "Marx circuit," may be clear from figure 2.2. A transformer at the center of the hexagon to the right of the diagram delivers rectified direct current via the indicated diodes to the string of n capacitors C, which charge in parallel each to the potential V. When V suffices to drive a spark across the gaps F, the capacitors connect briefly in series, the voltage on the last plate rises to nV, and a spark jumps to the top electrode of the constantly pumped special discharge tube figured on the left.[10] Such generators served the electrical industry to test the characteristics of insulators and other equipment during flashovers. The grandest ever made, which could manage 6 MV, was built by General Electric in 1932.[11]

Another industrial high-voltage instrument adapted to nuclear physics was Southern California Edison's million-volt cascade transformer at Caltech. This object did not satisfy Rutherford's requirement of convenient size: it filled a room 300 square feet in area and 50 feet high.[12] As we know, it was adapted to the

10. Brasch and Lange, *Zs. f. Phys., 70* (1931), 29–30.
11. Livingston and Blewett, ibid., 21.
12. Lauritsen and Bennett, *PR, 32* (1928), 850–1; Kargon in Shea, *Hahn,* 77; Ising, *Kosmos, 11* (1933), 144–7, for industrial forerunners of Lauritsen's design.

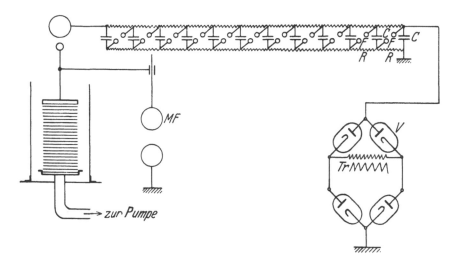

FIG. 2.2 Brasch and Lange's discharge tube and impulse generator. Voltage from the transformer Tr multiplied by the string of capacitors discharges across the constantly pumped laminated tube. Brasch and Lange, *Zs. f. Phys.*, *70* (1931), 30.

production of x rays and ion beams by C.C. Lauritsen, who had been so inspired by a lecture by Millikan that he gave up making radios and enrolled at Caltech in 1926, at the age of thirty-four, as a doctoral student. He developed a very fine experimental technique, precise in measurement and simple in style. European visitors judged him to be one of them, free from "the technological extravaganzas that Americans like so much." Brasch, who had reason to know, rated Caltech's engineer turned physicist "an uncommonly good experimenter" with outsized electrical apparatus. Lauritsen's first big challenge as a superannuated graduate student at Caltech was similar to what Brasch and Lange faced at the same time: to make a tube that could stand up to a million-volt generator.[13]

13. Lauritsen, "Stipulation" to Patent Office, Sep 1933 (Lauritsen P, 1/8); Holbrouw in Weart and Phillips, *History*, 87; quotes from, resp., W. Elsasser to Joliot,

A third group—Lawrence's chum Tuve and his co-workers at the Department of Terrestrial Magnetism at the Carnegie Institution of Washington—had come to a similar situation with still another approach to high potential. Tuve's Ph.D. thesis, completed at Johns Hopkins in 1926, was a measurement of the height of the ionosphere via radio pulses, using a method devised by Carnegie's theorist Gregory Breit. Tuve and Breit continued this work; but they also wanted something novel to mark Tuve's arrival, and decided to try to force the ordinary Tesla induction coil up to several million volts in the service of atomic and nuclear physics. The rationale offered the officers of the institution for this undertaking in the department's report for 1926/27 was that the Tesla coil might be driven to 30 MV to make cosmic rays in the laboratory. Further formal justification for the development of a program so obviously alien to the department's mission was found in the consideration that, as Tuve put it, "the problems of terrestrial magnetism will never reach a really satisfying solution until we possess an adequate understanding of the basic phenomena of magnetism itself." How else secure this understanding than by knocking the nucleus apart? On this flimsy pretext, the department allied itself with the U.S. Navy, which loaned it transformers, condensers, and a glass blower, and steamed forward to create one of the most important laboratories in the world for nuclear physics during the 1930s.[14]

The Carnegie's coil in its definitive form consisted of a primary of a few turns of copper tubing wrapped into a flat spiral around a secondary of many thousand turns of fine insulated wire wound in a single layer on a pyrex tube (plate 2.2). The primary was driven by the discharge across a spark gap of a very big condenser the navy had used in an old wireless transmitter (fig. 2.3). If tuned to the frequency of the oscillating primary discharge, the secondary could acquire a greater peak potential than the air around it could sustain. By immersing it in oil under pressure, however, Breit and Tuve arranged that the secondary held all the primary delivered,

---

13 Sep 1936 (JP, F28), and Brasch to Szilard, 27 Jan 1939 (Sz P, 29/306).

14. Abelson, *PT, 35:9* (1982), 90–1; CIW, *Yb, 26* (1926/7), 169, and *31* (1931/2), 229; Tuve to Lawrence, 13 Nov 1926 (MAT, 4), remarks to the AAAS, 1931 (MAT, 8), and, *JFI, 216* (1933), 1–2; J.A. Fleming to F.W. Loomis, 16 Dec 1931 (MAT, 8).

some 5.2 MV (they thought) with a secondary half a wavelength long. Although this estimate appears to have been too high, whatever they had would have driven an effective beam of alpha particles had they had any way to apply their coil. And so, like Lauritsen and Brasch and Lange, having solved the problem of high tension to their satisfaction, they set out, in 1928, to make a tube to withstand it.[15]

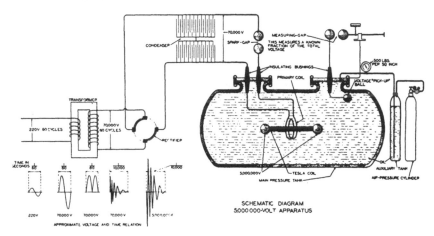

FIG. 2.3 Schematic of the installation of the Carnegie Institution's 5=mV Tesla coil. Breit, Tuve, and Dahl, *PR, 35* (1930), 56.

Lauritsen fielded the first effective tube, which required a scaffolding fourteen feet high of good California redwood for its support (fig. 2.4). Its fundamental features were adapted from the high-potential x-ray tubes developed at General Electric by W.D. Coolidge, who reached some 350 kV in 1926 by shielding the glass walls with a copper tube (fig. 2.5). In this way he defeated the buildup of charge in the walls from electrons driven into them; and hence also the discharges to the walls that punctured the tubes and made the main obstacle to increasing the voltage of x rays. To go beyond 350 kV, Coolidge recommended cascading two or more tubes (fig. 2.6), passing the electron beam through thin windows that would act as anodes for one tube and cathodes for the

15. CIW, *Yb, 26* (1926/7), 169, and *27* (1927/8), 208; Breit and Tuve, *Nature, 121* (7 Apr 1928), 535–6; Breit, Tuve, and Dahl, *PR, 35* (1 Jan 1930), 53–7 (the definitive coil). Livingston and Blewett, *Particle acc.*, 17, doubt the 5.2 MV.

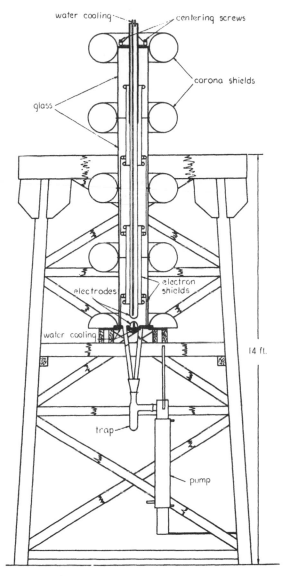

FIG. 2.4 Cal Tech's high-voltage x-ray installation. The corona shields are attached to the metal rings at the joints between the gas-pump cylinders constituting the tube; the very long cathode reaches almost to the tube's bottom. Lauritsen and Bennett, *PR, 32* (1928), 852.

next and stop positive ions that might otherwise gain high energy and make difficulty. Lauritsen's tube had four segments, to correspond with four cascaded transformers; each tube was twenty-eight inches long and twelve inches in diameter, and made, appropriately to its location, from the glass cylinders then used in pumps in gas stations. The segments joined at steel rings, which also supported the internal wall shields and, externally, circular slips of tin foil to protect against corona losses. The high-potential end of the cascade transformers fed the long central electrode of the tube (a three-inch steel pipe extending to within an inch of the earthed target) through a water resistor. The tube itself was continually pumped to retain a good vacuum. By August 1928 Lauritsen and Bennet could put 750 kV across the tube with no difficulty, and obtain x rays capable of penetrating over 2 cm of lead.[16]

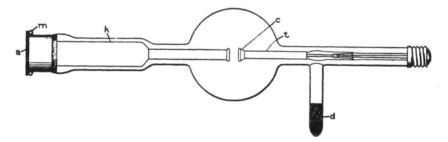

FIG. 2.5   Coolidge's first design for a high-voltage x-ray tube. Its chief feature is the metal shields around the electrodes, k and t, which prevent buildup of charge on the glass. Coolidge, *JFI, 202* (1926), 696.

Lauritsen patented the design. General Electric had supported Coolidge's work in the hope that very penetrating x rays might be especially effective against cancer. A cheap and efficient high-tension plant had commercial as well as humanitarian possibilities, whence the considerable interest of both industry and medical philanthropy around 1930 in large numbers of volts. As Lauritsen wrote in his patent application that year, his tube, then able to operate at a million volts, could serve "as the full equivalent of

16. Lauritsen and Bennett, *PR, 32* (1928), 851–3, 857; Coolidge, *JFI, 202* (1926), 695–6, 719–21.

radium in the treatment of disease, or for therapeutic purposes."
A gram of radium then cost $60,000 to $70,000; an x-ray plant of
moderate potential, about $30,000. Rewards could be high. And
hopes. Many victims of cancer basked briefly in the million volt
x rays of the Kellogg Radiation Laboratory.[17]

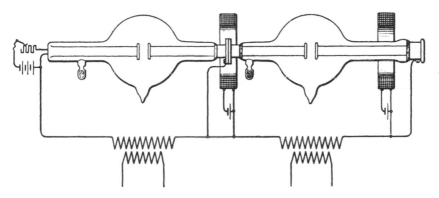

FIG. 2.6   Coolidge's cascade. Two tubes of the type shown in fig. 2.7
are joined together so that the anticathode of one becomes the anode of
the other. Coolidge, *JFI, 202* (1926), 720.

Meanwhile Brasch and Lange were adapting Coolidge's design
for use with the AEG impulse generator and the Alpine lightning
factory. Late in 1929 they had a porcelain tube over seven feet
long, with walls an inch thick studded with 300 nickel rings. They
solved the problem of stray electrons not with complete shielding
of the walls, as Lauritsen provided, but by enough rings of
sufficient capacity to prevent large buildup of unwanted voltages.
This great studded stick—made so long in order to space out the
longitudinal potential drop—could stand a surge of 1.2 MV. Nine
months later, in August 1930, they miniaturized to three feet by
immersing the tube in oil to cut out corona discharges from the
now high-potential external walls into the air. Careful study
showed that the hurdle to going higher was a creep of electrons
along the walls between the metal rings. They nipped the creep in
a new tube composed of alternating rings of paper, aluminum, and

17. Benedict Cassen to Lauritsen, 8 Oct 1932 (Lauritsen P), on cost of x-ray
plants; Tuve, memo, 23 Jan 1932 (MAT, 8), re radium; supra, §1.2.

rubber of different widths. The spacing of the metal disks remained close, while the creeping distance between them increased enough to discourage the most persistent electron.[18]

Although the vapor pressure of the rubber prevented them from achieving a very high vacuum, Brasch and Lange got spectacular results: the laminated heap withstood the maximum impulse of the generator, 2.4 MV, with transient currents of 1,000 amps; the zippy cathode rays thus produced made x rays that could penetrate 10 cm of lead, and offered a new medical possibility in themselves. Rather than apply very hard x rays against deep cancers, why not try short bursts of million-volt cathode rays, which deliver most of their energy as they come to rest? "Then the irradiation can work very effectively deep within the body without doing so much damage to the parts near the surface." Brasch and Lange's main goal, however, was to adapt their tube to the acceleration of positive ions. They managed to obtain a stream of hydrogen ions at 900 kV, which they thought too low to provoke the transformation of elements. They returned to the Alps to continue their alchemy.[19]

The Department of Terrestrial Magnetism was most impressed by "the spectacular performance" of Brasch and Lange's laminated tube. "They are to be congratulated without reserve," wrote Tuve and his associates, whose own handiwork did not perform quite so well. They had stayed close to Coolidge's design, multiplying segments and decreasing size by immersing the whole in oil. By 1929 their Tesla coil was driving a tube with six segments at 850 kV and one with fifteen segments (all seven feet of it) at 1.4 MV; later, by heatworking the glass to remove bubbles, they operated a twelve-segment tube about a yard long at what they thought was 1.9 MV.[20] Just before Tuve's group learned about the spectacular performance in Berlin, they succeeded in detecting cathode rays and x rays from the tube and fixing their energies at

18. Brasch and Lange, *Nwn, 18* (1930), 16, 765–6, and *Zs. f. Phys., 70* (1931), 21–9. The laminations are suggested in fig. 2.3.

19. Brasch and Lange, *Zs. f. Phys., 70* (1931), 30, 33, 35, and *Nwn, 21* (1933), 82–3; Badash et al., APS, *Proc., 130* (1986), 202.

20. Tuve, Hafstad, and Dahl, *PR, 36* (1 Oct 1930), 1262 (quote); CIW, *Yb, 27* (1927/8), 209, and *28* (1928/9), 214; Tuve, Breit, and Hafstad, *PR, 35* (1 Jan 1930), 66–7; Tuve, Hafstad, and Dahl, *PR, 35* (1 June 1930), 1406–7.

1,250 to 1,500 kV. It remained to accelerate protons, and also to develop a current in the tube sufficiently strong to permit measurement of the intensity, as well as detection of the existence, of the penetrating radiations. The Tesla coil as used in Washington, with its very brief duty cycle, was not a competitor for the medical purse.[21]

In the report of their work for the year 1929/30, Tuve's group pointed out that the tube perfected with the aid of the Tesla coil already outperformed it; and they called for a new generator "in order to adapt this new tool effectively to the studies in atomic and nuclear physics for which it was developed."[22] The tube won them a prize from the AAAS and some attention: page 1 in the *New York Times* and an advertisement in *Time* that with their x rays—reported as equivalent in gamma radiation to $187 million worth of radium—they might split atoms and cure cancer. Lawrence congratulated them on "such a fine recognition," meaning the prize, not the puff.[23] Their disappointment with the performance of their Tesla coil evaporated in this sunshine. They tried to accelerate protons, succeeded on December 8, 1931, and directed the beam overambitiously to shattering the elements. Nothing detectable happened; so small was the beam that "further attempts seeking evidence of nuclear disintegration with this set up [were] discontinued." Lawrence again offered encouragement. The tube, he reminded Tuve, was "a terribly important thing!!!—as I have emphasized to you many times (and you try modestly to give Coolidge the credit for)." It remained to find a way to make the tube produce something other than a prize.[24]

21. Tuve, Hafstad, and Dahl, *PR, 36* (1 Oct 1930), 1262; CIW, *Yb, 29* (1929/30), 256–7, and *30* (1930/1), 291.

22. CIW, *Yb, 29* (1929/30), 257.

23. *New York Times,* 4 Jan 1931, 1; *Time, 17:1* (12 Jan 1931), 44; Tuve, Hafstad, and Dahl, *PR, 37* (1931), 469, and statement of 19 Feb 1935, re the prize of 1931 (MAT, 14/"lab. letters"); Lawrence to Tuve, 4 Jan [1931] (MAT, 8), adding that Lauritsen had expressed "keen admiration for your work."

24. CIW, *Yb., 30* (1930/31), 292–3; ibid., *31* (1931/32), 231, 229 (last two quotes); Lawrence to Tuve, 15 June [1932] (MAT, 4); Tuve, Hafstad, and Dahl, *PR, 39* (1932), 384–5. Cf. Tuve to Lauritsen, 30 Mar 1933 (Lauritsen P, 1/8), advising against trying a Tesla setup for disintegration experiments.

The Old-Fashioned Way

And one was provided. The inventor, Robert Jemison Van de Graaff, conceived a grand idea at the onset of his career, as a Rhodes scholar at Oxford, where he arrived with an engineering degree from the University of Alabama and professional experience with the Alabama Power Company. From Oxford, where he earned his Ph.D. in physics in 1928, he went to Princeton as a National Research Fellow, and soon had a prototype accelerator working at 80 kV. Its principle would have been plain to Benjamin Franklin. An endless belt of a good insulating material running vertically between two pulleys picks up electricity from a point discharge at the bottom and delivers it, again by point discharge, to a large insulated spherical conductor at the top (fig. 2.7). The lower electrical spray comes from any rectified source. The upper spray continues, irrespective of the potential attained by the sphere, which exerts no electrostatic force at its internal surface, until the field at the external surface suffices to break down the air. In a later refinement (fig. 2.8), a second set of points adroitly placed removed electricity of the unwanted sign from the sphere on the belt's downward journey and eliminated the rectified source by connections that allowed the amplification of any slight charge present on the belt. Since the practical limit to the potential of the sphere is the dielectric strength of the air, the way to millions of volts was to increase the sphere's radius (and so lessen its external field at a given potential) and the dielectric constant of the surrounding medium (and so raise the field at which breakdown occurs). To make good on the second possibility, the entire machine must be encased in a vessel that can be evacuated or filled with a gas or fluid under pressure. The first small pressurized model operated in 1932.[25]

By mid August 1931 Van de Graaff could charge a brass sphere mounted on a glass stick to about 750 kV. Between two such spheres, one positive and one negative, an inspiring potential difference of 1.5 MV could be maintained. It, and the trivial cost,

25. E.A. Burrill, *DSB, 13,* 569–70, and *PT, 20:2* (1967), 49–50; H.A. Barton, D.W. Mueller, and L.C. van Atta, *PR, 42* (1932), 901; Van de Graaff, Compton, and van Atta, *PR, 43* (1933), 152–5.

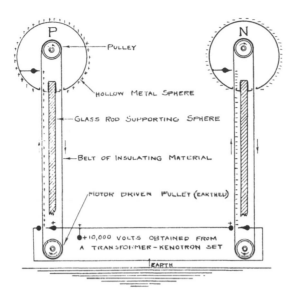

FIG. 2.7   Principle of the Van de Graaff generator. Charge sprayed on the endless silk belt at the bottom leaves by corona discharge at the top; it is derived in the first instance from a transformer. Van de Graaff, Compton, and Van Atta, *PR, 43* (1933), 152.

about $100 for the entire outfit, inspired Karl Compton, who brought Van de Graaff to MIT as a research associate (he became associate professor in 1934) and arranged some heady publicity. The newly formed American Institute of Physics held a dinner for scientists and journalists at the New York Athletic Club; the machine, in an alcove in the dining room, looked like "two identical rather large floor lamps of modernistic design." Van de Graaff demonstrated for his supper, and also for Paramount and Pathé news; he allowed that he saw no difficulty going to 10 MV with two balls each 20 feet in diameter on towers 20 feet tall. Compton misguessed that this big machine would provide alpha particles in a current "so enormously larger than that from radium, that the experiment opens up the possibility of transmutation of the elements on a commercial scale;" and he miscalculated that it could be done for a few hundred dollars.[26]

26. Quotes from *New York Times,* 6 Nov 1931, 1, 6, and 11 Nov 1931, 1, 17; J. Boyce to Lawrence, 17 Nov 1931 (3/8); Van de Graaff, *PR, 38* (1931), 1919–20;

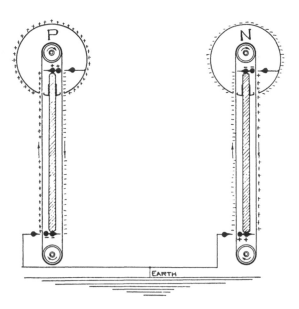

FIG. 2.8 An improved Van de Graaff generator. The points are arranged so that the belt charges the sphere when going down as well as when going up; the system works with any stray charge, no transformer being required. Van de Graaff, Compton, and van Atta, *PR, 43* (1933), 153.

There remained what Lawrence, who recognized Van de Graaff as a competitor, called the "old problem of a high vacuum tube." Those who thought they had solved the problem regarded the matter differently. As one of Lauritsen's students wrote him after witnessing one of Van de Graaff's demonstrations, "His scheme is really very good and actually works...[and] would make a very fine combination with one of your tubes." Lauritsen eventually did build a Van de Graaff machine.[27] Tuve's group rushed to do so. In September 1931 Tuve drove Van de Graaff and his easily portable equipment to Washington and hooked it up to the segmented tube. With a charging current of 40 μA and 600 kV on the spheres hooked in parallel (plate 2.3), the tube carried a

Van de Graaff, Compton, and van Atta, *PR, 43* (1933), 154.

27. Lawrence to Cottrell, 21 Nov 1931 (5/3); Cassen to Lauritsen, 12 Nov 1931 (Lauritsen P, 1/8); Nahmias to Joliot, 21 Sep 1937 (JP, F25), on progress of installation of Lauritsen's Van de Graaff.

proton beam of not quite a millimicroamp (mμA), not enough to burn a hole in cardboard, but enough to make tracks in a cloud chamber. Tuve studied the working apparatus closely and got a spark to his nose for his curiosity. It did not discourage him from reaching higher, for 1.4 MV, half the theoretical value of the maximum potential restricted by the dielectric strength of normal air surrounding a sphere two meters in diameter. (A useful rule of thumb: the theoretical maximum in MV equals the radius in feet.) Simultaneously, Coolidge, supported by GE, planned a high-current version at 1 MV, and Van de Graaff, seconded by the Research Corporation, went forward with one requiring two 15-foot spheres.[28]

The two-meter sphere, which cost $700, worked well with the million-volt tube insofar as it could be tested outdoors, where it sparked and fluttered under bombardment by bugs and dust and threw lightning bolts that reduced its redwood base to splinters. The Carnegie Institution's Department of Terrestrial Magnetism had no place for the wonder it had built. We read in its annual report for 1931/32: "A highly satisfactory equipment for the production of high energy particles, particularly of high-speed protons, is thus ready for use as soon as operating space becomes available."[29] During the late fall of 1932, while awaiting the construction of suitable housing off-site (the Carnegie Institution's executive committee approved a building fund in January 1933), Tuve and his associates made a version about one meter in diameter for the space that had belonged to the Tesla coil. With this machine, their desire to do nuclear physics was at last requited. It consisted of two hollow hemispheres of aluminum joined by a short cylindrical section containing the belt, one pulley, an ion source, and the high potential end of a segmental discharge tube. The belt brought about 180 to 200 μA; the sphere held 400 to 600 kV; the tube transmitted as much as 10 μA of proton beam, constant in energy to perhaps 3 percent, to the target. The x rays

28. Tuve to Lauritsen, 11 Dec 1931 (Lauritsen P, 1/8); Cottrell to Lawrence, 11 Nov 1931 (5/3); Boyce to Lawrence, 17 Nov 1931 (3/8); Fleming to van de Graaff, 19 Dec 1931 (MAT, 8); Tuve, *JFI, 216* (1933), 26; Wells, *Jl. appl. phys., 9* (1938), 677–80.

29. CIW, *Yb, 31* (1931/2), 230; Tuve, Hafstad, and Dahl, *PR, 48* (1935), 317; Fleming to J.M. Cork, 5 Dec 1932, on costs.

incidentally produced drove the experimenters into a hut outside the twelve-inch concrete walls of their laboratory, where they worked by remote control. (Their conservative value for the safe tolerance of radiation was ten times the dose from cosmic rays.) In the operation of the tube, Tuve's group had some advice from Lawrence, whose own investigations had by then acquainted him with the ability of coaxial cylindrical electrodes to focus a beam and with the excellences of certain "Apiezon" oils made by Metropolitan-Vickers as the working fluid of vacuum pumps. With this setup the Department of Terrestrial Magnetism was able to set right some sloppy results reported by Lawrence's Radiation Laboratory.[30]

In 1933 the two-meter machine found a home and Van de Graaff's 15-foot giant threw its first sparks in a disused blimp hanger in Round Hill, Massachusetts. Tuve's photogenic apparatus (plate 2.4), with four belts and two concentric shells (the inner, one meter in diameter), reached 1.2 MV under favorable conditions. Lawrence visited it and was impressed. "I must say that Tuve's apparatus is performing better than I expected," he wrote the Research Corporation after his inspection. "Seeing Tuve's apparatus perform makes me much more enthusiastic about van de Gr[a]aff's outfit than I was before."[31] By then, November 1933, the cyclotron could give more volts, but at far less current, than Tuve's "outfit." As for Van de Graaff, his 1.5 MV model gave a charging current almost a million times Lawrence's beam, as his patron Compton liked to observe, and his giant one held promise of another factor of ten (plate 2.5 and fig. 2.9).

"Experience to date indicates that there is in sight no unsurmountable obstacle to the construction of [10 MV Van de Graaff] generators." When Compton spoke these words—which were realized long after the war, by a technology not available in the

30. Tuve to Cottrell, 24 Jan 1933 (MAT, 4); Cottrell to Lawrence, 16 Sep 1932 (5/3); Tuve to Lauritsen, 11 Nov 1932 (Lauritsen P, 1/8), claiming 800 kV; Tuve to E.W. Sampson, MIT, 22 May 1934 (MAT, 13/"lab letters"), re Lawrence's advice; Tuve, Hafstad and Dahl, *PR*, *48* (1935), 318–20, 329–32, 337; infra, §4.1.

31. Tuve, Hafstad, and Dahl, *PR*, *48* (1935), 321–5; Lawrence to Poillon, 20 Nov 1933 (15/16A). Lawrence based his opinion on Cooksey's negative report on the performance of Tuve's machine in January; Cooksey to Lawrence, 25 Jan 1933 (4/19).

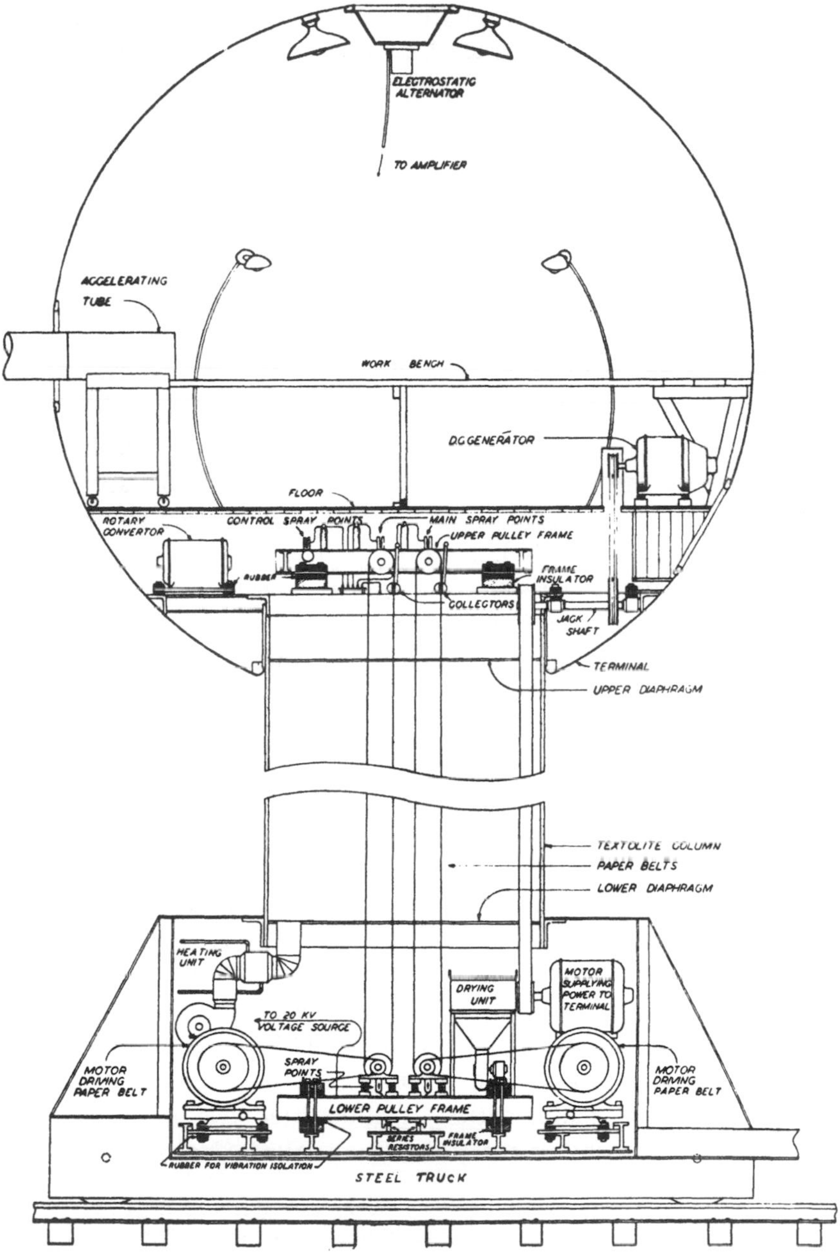

FIG. 2.9   An insider's view of the 15-foot generator. It delivered 1.1 mA to the accelerating tube under a tension of 5.1 MV. Van Atta et al., *PR, 49* (1936), 762.

1930s—Van de Graaff's original model was showing off at Chicago's Century of Progress Exposition, "producing millions of volts for the enlightenment of the visitors."[32] But neither this nor any other million-volt plant was the first to accomplish the purpose of all, and bring that enlightenment to physicists vouchsafed by the disintegration of the atom.

The English Way

Like Van de Graaff and Lauritsen, John Douglas Cockcroft began professional life as an engineer, in his case in 1920, with a degree from the University of Manchester. He spent the next two years as a college apprentice at Metropolitan-Vickers, working with large transformers and strong insulators. And then, like his counterparts, he changed his field and place of study, to mathematics and Cambridge. In his second year there he began to frequent the Cavendish Laboratory, where he did postgraduate work after gaining a second bachelor's degree in 1924. Metro-Vick continued to give him partial support, on the understanding that he would do some research work for them; and, reciprocally, he served the Cavendish as a "spare-time, honorary electrical engineer."[33]

In 1926 Cockcroft's experimental space was invaded by another man from Metro-Vick, T.E. Allibone, who had been inspired by Blackett's demonstration of disintegration and Coolidge's design for high tension to try his hand at artificial sources. With the advice of colleagues at the High Voltage Laboratory at Metro-Vick, he chose the Tesla coil; and it appears to have been the progress he had made with it during 1926/27 that prompted Rutherford to speak optimistically of the prospects of artificial sources before the Royal Society. Another would-be atom splitter then arrived, Ernest Walton, who came from Trinity College, Dublin, with a degree in mathematics. Walton tried two methods of a type that will occupy us presently. Both failed. Nor did Allibone

32. Compton, *Science, 78* (21 Jul 1933), 50–2; *Science service*, in *Science, 77* (2 June 1933), suppl., 9, quote, on the exposition; Livingston and Blewett, *Particle acc.*, 65–6, on MIT's postwar 12-MV generator.

33. R. Spence, *DSB, 3*, 328–9; Hartcup and Allibone, *Cockcroft*, 17–21, 24–5, 31 (quote from Cockcroft), 36.

fulfill Rutherford's hopes, although he did succeed, with the help of a transformer loaned by Metro-Vick, in developing tubes that could stand 450 kV in air and 600 kV under oil. By then, the end of 1928, Cockcroft had finished the research on molecular beams that constituted his doctoral work. Notwithstanding the failures with which his room was strewn, he decided to try to shatter nuclei himself.[34]

Cockcroft was a businesslike man. He took up atom splitting because he knew it to be practicable. In contrast to all the other would-be splitters, he kept his attention fixed upon the goal: to make particles with energies sufficient to penetrate nuclei, not with energies above a million volts. In estimating the minimum requirement, he followed calculations in a manuscript that circulated in the Cavendish in December 1928. Its author, George Gamow, a young Russian working in Niels Bohr's Institute of Theoretical Physics in Copenhagen, explained that, according to the then new wave mechanics, a charged particle making a head-on collision with a nucleus has a chance of entering even if it does not have as much energy as would be required to do so by the older physics, on which the estimate of millions of volts had been based. Cockcroft deduced from Gamow's equations that protons would be better agents than alpha particles and that a proton of 300 kV would be about one-thirtieth as efficient against boron (in fact, beryllium) as an alpha particle from polonium would be against aluminum. In January 1929 Gamow came to Cambridge to talk, Rutherford approved Cockcroft's project, and Cockcroft and Walton teamed up to produce protons at 300 kV in sufficient quantity to overcome the low probability that any of them would effect a disintegration.[35]

Cockcroft chose 300 kV as his first goal because he judged it to be within easy reach of the art of vacuum tubes. The arrangement he and Walton devised is shown in plate 2.6. Metro-Vick provided much of the technology: the 350 kV transformer, shown

34. Hartcup and Allibone, *Cockcroft*, 38–9; Allibone, *PRS, A282* (1964), 447–8.

35. Hartcup and Allibone, *Cockcroft*, 40–3; the odds from boron in Cockcroft's calculation of 1928/29 are those for beryllium in Cockcroft and Walton, *PRS, A129* (1930), 478. Cf. Hendry, 15–7; Allibone in Hendry, *Cambridge physics*, 156–61, and in *PRS, A282* (1964), 448; Cockcroft, "Development" [1937], 1–2; Gamow, *Zs. f. Phys., 52* (1928), 510–5.

schematically at the left, a model custom-made (but later marketed for x-ray plants) to fit the cramped space of the Cavendish; the rectifiers A, designed by Allibone; the pumps, constantly at work on the rectifiers and discharge tube, invented by Metro-Vick's C.R. Brush and run on his Apiezon oil of miraculously low vapor pressure. F is a 60 kV transformer that energizes the little canal ray tube (atop the discharge tube) that Cockcroft and Walton used as a source of protons; the entire transformer F stands at 300 kV above ground. The discharge tube itself, a bulb with two steel pipes as electrodes, terminated in a small experimental space and a hookup to the vacuum system. About 1 μA of 280 kV protons survived the trip down the tube to slam into targets of lead or beryllium. "Very definite indications of a radiation of a non-homogeneous type were found," Cockcroft and Walton wrote in August of 1930, without saying what they indicated.[36] Very probably they had disintegrated beryllium. Not knowing what evidence to look for—they expected to find gamma rays—they concluded that Gamow was mistaken and sought higher potentials.[37]

A move to a larger room with a higher ceiling allowed them to build new apparatus to a plan of Cockcroft's. The design multiplies voltages by an intricate set of condensers and switches. In the setup of figure 2.10, where all condensers have equal capacities, the action begins with the closing of the dotted switches S_1, S_2, S_3. That links K_3 and X_2 in parallel, at the potential E of the constant source. Next the dotted switches open and the solid ones close, causing X_2 to divide its charge with K_2 and bringing each to $E/2$. The switches are again reversed, leaving K_3 and X_2 at E, K_2 and X_1 at $E/4$. Reverse connections again: we have K_3 at E, K_2 and X_2 at $5E/8$, K_1 and X_1 at $E/8$. And so on until the upper plate of K_1 is $3E$ above ground. In practice Cockcroft and Walton used a low-frequency alternating current for E and rectifying diodes as switches; since they could now make rectifiers for 400 kV, and since each rectifier had to stand twice the voltage E, they used a 200 kV transformer as source, four rectifiers, and four con-

36. Cockcroft and Walton, *PRS, A129* (1930), 479–89, rec'd 19 Aug; Hartcup and Allibone, *Cockcroft*, 43–6; Allibone in Hendry, *Cambridge physics*, 161–2.
37. Cockcroft, "Development," 2–3.

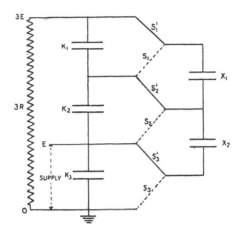

FIG. 2.10   Principle of the voltage multiplier constructed at the Cavendish. Cockcroft and Walton, *PRS, A136* (1932), 620.

densers, in the hope of attaining a final drop of 800 kV. The experimental tube (fig. 2.11) came in two segments; the middle electrode, maintained at half the total potential, carried a little diaphragm, G, to stop stray electrons. The successful proton beam, 10μA at 710 kV, passed from the evacuated tube through a window of mica into the experimental space.[38] What happened there started an era in nuclear physics.

Metro-Vick continued to play a part. The firm patented Cockcroft's voltage multiplier, although, as it turned out, a German, Heinrich Greinacher, had anticipated him, and only the British observed his rights. Cockcroft's circuit could power a high-potential x-ray plant, and in hard x rays, as we know, there was hard cash. What Metro-Vick had in mind in patenting the voltage multiplier appears from a request from George McKerrud, lieutenant to A.P.M. Fleming, the company's director of research, to Cockcroft, to entertain F.L. Hopwood, a member of the Grand Council of the British Empire Cancer Committee. "Will you please see Allibone and fix up to have a really good show going,

38. Cockcroft and Walton, *PRS, A136* (1932), 619–30, rec'd 23 Feb 1932; Hartcup and Allibone, *Cockcroft*, 46–7.

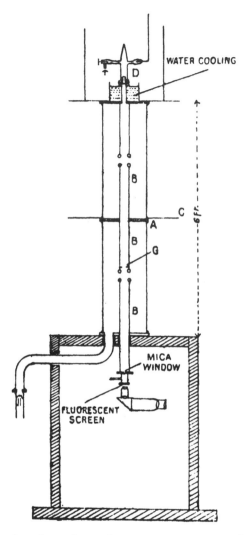

FIG. 2.11 Accelerating tube and target arrangement of the Cockcroft-Walton machine. The source is at D; C is a metallic ring joint between the two sections of the constantly pumped tube. The mica window closes the evacuated space. Cockcroft and Walton, *PRS, A136* (1932), 626.

because...the Cancer Research people...are, as you know, a very rich organisation so that we hope that they will contribute towards the support of the work of the future and probably order some tubes to be made." Hopwood left Cambridge convinced that

Metro-Vick could produce "the super x ray tube." McKerrud congratulated Cockcroft on his successful soft sell. "The lunch at St John's was an invaluable detail in the scheme." Hopwood's institution, St Bartholomew's Hospital in London, commissioned Metro-Vick to make it a million-volt x-ray plant based on Cockcroft's patented voltage multiplier.[39]

3. MAGNUM PER PARVA

The high-tension accelerators stretched the power of insulators and the nerves of physicists to the breaking point. They also taxed the finances and furnishings of the few institutes in which they were developed: they demanded unusual electrical service, special apparatus and safety precautions, and, above all, space, a commodity more precious even than money in the laboratories of the time. An obvious way to relieve the tension on men, material, and money was to accelerate particles in several steps, each requiring only a moderate electrical force. The high-voltage energy would be accumulated on the particles, not on the apparatus.

The idea, we say, was obvious. It occurred to several, probably to many, physicists and electrical engineers during the 1920s. But between theory and its realization stood the usual malevolence of the inanimate and the conservatism of persons who are not inventors. Three examples will illustrate the variety of problems encountered in realizing the principle of acceleration by steps: the proto-betatron, the proto-linac, and the cyclotron.

The Electrical Vortex

The beam of charged particles undergoing acceleration constitutes a current that can be considered the secondary ciruit of a transformer. Recall the watchword of the faith of the electrodynamicist: $F = evH/c$, where F signifies the force exerted on a particle carrying a charge e and moving with velocity v by a

39. Allibone in Hendry, *Cambridge physics*, 154, 165, 172; McKerrud to Cockcroft, 5 June and 14 Jul 1930, and Hopwood to N.R. Davis, 11 Jul 1930 (CKFT, 20/59). For European forerunners of Cockcroft's multiplier, see Ising, *Kosmos, 11* (1933), 155–7.

magnetic field of strength H, and c stands for the speed of light.[40]
For the formula to hold, H must be perpendicular to the plane
defined by v and r. Since F stands at right angles to v, it can neither speed up nor retard the motion of the particle; instead it
pushes it constantly toward a fixed point, forcing the particle to
describe the arc of a circle. The situation appears in figure 2.12.
The greater the velocity v, the larger the radius r of the circle must
be so that the tendency of the particle to fly off at a tangent can be
countered by an inward magnetic push. The balance of tendency
($mv^{2/r}$) and push (evH/c) gives the most important relation we
shall have to consider, the "cyclotron equation,"

$$v/r = eH/mc, \qquad (2.1)$$

where m is the mass of the particle.

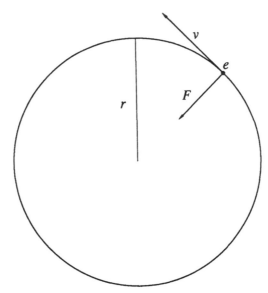

FIG. 2.12 The force on an electron circulating in a magnetic field. The
field H is perpendicular to the plane of the paper; the force $(ev/c)H$ is
perpendicular to the velocity v and directed toward the orbit's center.

40. The factor c is required to write the watchword in electrostatic units, which
we use exclusively throughout; H is then in gauss.

In October 1927, an electrical engineer named Rolf Wideröe presented his doctoral dissertation at the Technische Hochschule, Aachen. In it he described experiments based on the cyclotron equation and the transformer principle. By increasing H in time, Wideröe developed an electric force in the direction of v (the transformer principle); and by constructing specially shaped poles for the electromagnet producing H, he made it possible for particles to remain in circular orbit with constant radius (the cyclotron equation). As he observed, to obtain the necessary centripetal and tangential forces from the same electromagnet, its poles must provide a field that is half as strong at the electrons' orbit as within it—a condition easier to state than to realize.[41] Wideröe called this device a beam transformer; it can be applied practically only to *electrons*, which, because of their small mass, can attain useful energies under the electric force produced by the time change in H. Evidently the entire acceleration must be accomplished during that part of the cycle of the alternating current energizing the electromagnet in which the electromagnetic force acts in the desired sense.

That is the theory. In practice even the partial cycle was more than Wideröe could use, since he did not succeed in holding his electrons to circular orbits within the evacuated glass doughnut in which he tried to accelerate them. The best he could do, even with several extra coils arranged to compensate for unevenness in the H field, was to guide the electrons around a circuit and a half before they ran into the walls of their doughnut.[42] He thus confirmed the prediction of his former professor, Wolfgang Gaede, an expert in vacuum technology, who had refused to allow the project in his institute at the Technische Hochschule in Karlsruhe on the ground that it was sure to fail. That had driven Wideröe to Aachen, to the more optimistic Walter Rogowski, an expert on cathode-ray tubes and oscilloscopes. As Wideröe later explained

41. Wideröe, 1928, in Livingston, *Development*, 107. A magnetic field H perpendicular to a circle of radius r and changing at the rate of dH/dt within the circle gives rise to a tangential electric force $(er/2c)$ (dH/dt) and hence to a tangential acceleration $dv/dt = (er/2mc)$ dH/dt. To balance the resultant increase in "centrifugal force," a changing field H' *at the orbit of magnitude (er/mc) dH'/dt* is required by the cyclotron equation. Hence $H' = H/2$.

42. Livingston, *Development*, 113; Wideröe, *Europhysics news*, *15*:2 (1984), 9.

his failure to implement his idea, which had come to him as early as 1922, "The theory of the stabilizing forces acting on the orbit had not yet been developed sufficiently."[43]

No more did the electrical vortex succeed in Cambridge, where Walton tried to implement Rutherford's suggestion of acceleration in the "electrodeless discharge." Here electrons in an evacuated tube without electrodes serve as carriers of the secondary current, the primary being generated in a coil wrapped around it. A supplementary magnet supplied the additional H field, which, according to Walton's detailed and correct calculations, would hold the electrons to a tight circular path within the tube. Walton charged a condenser to 40 kV and discharged it through the coil, setting up oscillations that produced the required varying magnetic field. If all went as calculated, the electrons would attain an energy of 536 keV in a quarter of a cycle. They declined to be regulated, ran into gas molecules, and scattered into the tube wall. Walton gave them up and joined Cockroft in multiplying voltages.[44]

The idea was so good and so obvious that several other physicists flirted with it before Cockcroft and Walton's success with straight tubes and high voltages in 1932. The friendly rivals from South Dakota each tried his hand. Working with Breit, Tuve set up an apparatus similar to Wideröe's, but independently of Wideröe. Tuve and Breit made an important improvement by injecting the electrons into their accelerator at high speeds via an electron gun and claimed to have obtained an acceleration to 1.5 MeV, but they nevertheless ended no more successfully than Wideröe; "No provision has been made [they wrote] to repeat the process very often." Lawrence thought that he could correct the fault of Walton's design, which did not make sufficient provision for axial focusing, with extra coils to create a field that would drive errant electrons back to their orbital plane. He had his assistants realize his design. On June 10, 1931, he tried it, with no better luck than his predecessors. The same success would have attended the efforts of Leo Szilard, had he attempted to make flesh the transformer-accelerator for which he applied for a

43. Wideröe, *Zs. angew. Phys.*, 5 (1953), 187.
44. Walton, Cambr. Phil. Soc., *Proc.*, 25 (1929), 469-81. Cf. Kerst, *Nature, 157* (1946), 90–3, and Walton in Hendry, *Cambridge physics*, 51.

German patent in January 1929.[45] There were other patents as well. One of them, secured by Max Steenbeck, a physicist at the laboratories of the Siemens electrical company in Berlin, perhaps referred to a machine that worked.[46]

The method can be made to work. After a decade's experience in focusing beams of fast particles in other sorts of accelerators and in electron microscopes, physicists managed to construct a magnetic system capable of steering and speeding electrons on circular orbits within evacuated tubes. Szilard proposed to have a try in 1938, in collaboration with a physicist at the Clarendon Laboratory at Oxford, James Tuck; their unimplemented design was perhaps the most promising put forward before the first conspicuous success. That was achieved in 1939 and, on a larger scale, in 1940 by Donald Kerst at the University of Illinois.[47] He had the assistance of a former student of Oppenheimer's, Robert Serber, in computing the motions of the electrons. The business continued at General Electric's research laboratory, where a machine for 22 MeV and then for 100 MeV came into existence during the war. Just after the war, Kerst's "betatron" had an important influence on machines built in Berkeley.[48]

Resonance Acceleration

Wideröe had a second failure to announce in completion of his thesis. This one involved the acceleration of heavy ions by modification of a technique proposed, but not implemented, by a Swedish physicist, Gustav Ising. In Ising's plan, positively charged particles fly down a straight evacuated glass tube through a series of hollow cylindrical electrodes, each separately connected to one pole of a spark gap in an oscillatory circuit (fig. 2.13). The particles are drawn from the source into the first electrode when

45. Breit, Tuve, and Dahl, CIW, *Yb, 27* (1927/8), 209; Lawrence to Cottrell, 10 June 1931 (RF/1944/RC Misc); Szilard, *CW, 1,* 555–61.

46. Slepian, U.S. patent 1,645,305 (1922); Smith (Raytheon), 2,143,459 (1931); Steenbeck, 2,103,303 (1937), German patent 698,867 (1935), and *Nwn, 31* (1943), 234–5; Wideröe, *Zs. angew. Phys., 5* (1953), 199; Lawrence to A.F. Maston, 16 Feb 1938 (15/17A).

47. Szilard, *CW, 1,* 568–81; Szilard to Tuck, 21 Oct 1938 (SzP, 17/198); Kerst, *PR, 60* (1941), 47–53, and *Nature, 157* (1946), 93–5.

48. Serber, *PR, 60* (1941), 53–8; E.E. Charlton and W.F. Westendorf, *Jl. appl. phys., 16* (1945), 581–2.

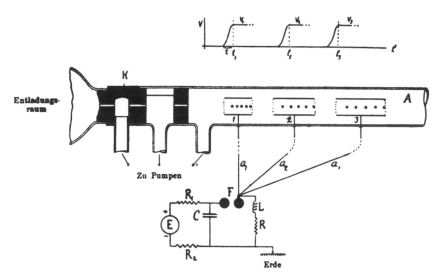

FIG. 2.13 Ising's proposal for a linear particle accelerator. The high-frequency field is supplied by a discharge across the spark gap F; K is the cathode; a_1, a_2, a_3, connections to the drift tubes. Ising, *Kosmos, 11* (1933), 171.

its potential has the correct sign and size; while in it they feel no electrical force; on escaping from it they are again accelerated, the direction of the field in the gap between the first and second electrodes having meanwhile altered to the correct sense. The business in principle can be continued through any number of gaps and electrodes.[49]

Wideröe turned to Ising's method after failure of his proto-betatron in the hope of having something that worked to describe in his thesis. He replaced the old-fashioned spark vibrator with an up-to-date vacuum-tube oscillator, and theory with practice. He contented himself with only two accelerating steps, which did indeed give the ions twice the energy they would have gained from falling through the maximum potential across the oscillator.

49. Ising, *Arkiv för matematik, astronomi och fysik, 18* (1924), 1–4, and *Kosmos, 11* (1933), 170–3.

Nevertheless Wideröe regarded his process as unpromising. He was not interested in particle accelerators, but in what he called "kinetic-voltage transformers," devices for generating currents of high-energy particles useful to engineers. He got currents so small, however, that he despaired of realizing anything much above a milliamp, which ruled out his invention as "a technical generator for high direct voltages."[50] But so puny a flow, of no consequence to engineers, was enough and more than enough for physicists, for, as we know, a milliamp of the right sort of particles exceeds the output of alpha rays from eighty kilograms of radium.

In Wideröe's linear accelerator (fig. 2.14), or "linac" to use the later term of art, sodium or potassium ions from the heated filament at A fall through 20 kV across the gap I when the oscillating potential U_b reaches its maximum negative value on the metal tube BR. In the time the ions float through the tube, which shields them from electrical forces, the potential on it switches to maximum positive value, and the emerging ions drop through 20 kV across gap II. On emerging from the grounded tube S they move under an electrostatic force between the plates at K to strike the fluorescing screen P a distance a below the axis. From measurement of a, Wideröe could confirm that the ions at P had energies of 40 keV. Much higher potentials could be reached, he thought, by multiplying the number of steps, increasing the voltage U_b, and using heavier ions.

This last consideration, which is perhaps not obvious, played an important role in the early history of accelerators. Wideröe's primary improvement over Ising, using a vacuum-tube oscillator to generate the high-frequency potential U_b (the spark gap in fig. 2.14 is for calibration, not generation), avoided the wide band and inconstant energy of the spark circuit, but brought a limitation of its own. It could not operate effectively above a frequency of 10^7 Hz, or about ten million oscillations a second. It was this limit that directed Wideröe's attention to heavy ions rather than to protons. A singly charged ion of mass Am_H (m_H is the mass of the hydrogen atom) has a velocity of about $10^9/\sqrt{A}$ cm/sec at 1 MeV.

50. Wideröe in Livingston, *Development*, 93, 106, and in *Europhysics news, 15:2* (1984), 9.

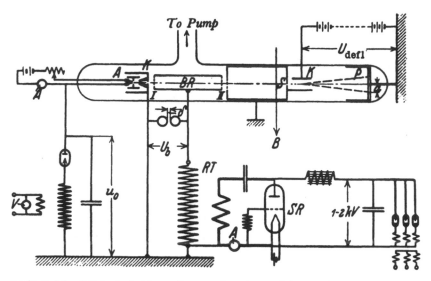

FIG. 2.14 Wideröe's linac for acceleration of heavy ions. The high-frequency field is supplied by the circuit containing the tube SR. Livingston, *Development*, 102.

In the last steps of its acceleration it would travel $100/\sqrt{A}$ cm during one period at 10^7 Hz, near the frequency limit of the oscillators then available. To accommodate protons ($A = 1$), Wideröe would have needed an evacuated acceleration tube many meters in length filled with electrical equipment and maintained at a pressure of less than a millionth of an atmosphere (10^{-7} mm Hg). These specifications were technically impractical. For cesium ions, which Wideröe proposed as particularly favorable for attaining 1 MeV, the apparatus would be about a meter in length.

Wideröe's thesis was published in the issue of the *Arkiv für Elektrotechnik* of December 17, 1928. On the very same day—the resonance is scarcely credible—Szilard applied for a patent on a similar device at the Reichspatentamt in Berlin. In contrast to the engineer, the physicist explicitly intended his machine to accelerate particles that might disintegrate atoms; as Szilard later explained himself, he rightly saw that the nucleus was the frontier in physics and wrongly guessed that its disintegration would lead quickly to "practical application of very great importance." But he was also busy preparing an application for a British patent on a scheme of refrigeration that he and Einstein had invented, and he devoted little care to the practical details of his linac. Its principle appears from figure 2.15: ions accelerated between the anode 13 and the perforated cathode 11 enter the accelerator tube 9 sectioned by the grids 1, 2,...6. At S is an oscillator, one of whose leads supplies the odd-numbered grids and the other the even-numbered, which are therefore 180° out of phase with one another. In the ideal case, an ion passes through a grid only when all the grids instantaneously reach zero potential and when the potential is rising on the grid just passed and falling on the grid next in line. The ions are not shielded from the ac field by metal cylinders, as in Wideroë's version, and no doubt the wires in the grids would thin the beam to uselessness long before it had passed the 100 grids that Szilard specified for acceleration to 2 MeV. He planned to work at a hundred million cycles per second, but did not say where he would procure an oscillator. The patent for the linac was not granted.[51]

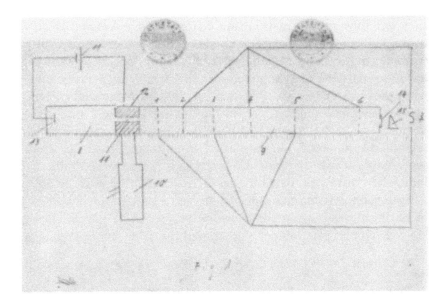

FIG. 2.15   Szilard's linac. The rf oscillator S is at the far right; the source is roughly indicated by the circuit 11, 12, 13. Szilard, *CW, 1,* 553.

51. Szilard, "Beschleunigung von Korpuskeln," in *CW, 1,* 545–53; Szilard to Fermi, 13 Mar 1936, ibid., 729; cf. Weart and Szilard, *Szilard,* 11.

Another inventor, perhaps independent, of the Wideröe linac was Jean Thibaud, a young physicist who worked with classic radioactive substances in Maurice de Broglie's private laboratory in Paris. Impatient with what nature and radiochemistry provided, Thibaud hoped to reach 10 MeV in small steps; he managed to get to 145 keV by pushing positive ions through no fewer than 11 successive gaps between cylindrical electrodes with an oscillator going at $3 \cdot 10^6$ Hz. To proceed to alpha-particle energies, however, to compass "the liberation of intra-nuclear energy, an outcome of undoubted general utility," Thibaud recognized that he would need a tube ten meters long and a very powerful and dangerous oscillator, capable of handling 100 kW. "To avoid these difficulties," he said, "I have tried another method."[52] It was the method of the cyclotron. Thibaud did not disclaim credit for its invention or for making it operate independently of Lawrence.[53]

The cyclotron solved the problem of tube length and made possible the resonant acceleration of protons and other light particles. In it the ions spiral out from the center of an evacuated shallow drum, suffering a kick each time they cross a gap between electrodes arranged along a diameter of the drum (fig. 2.16). Between kicks the particles are maintained in circular orbit by a magnetic field. Several people say that they thought of some such scheme before Lawrence did, but for various reasons did not publish it. Wideröe recalled that one of Rogowski's assistants asked whether the ion beam could not be curled around and made to spiral repeatedly through the same electrode gap. "I answered him, as I remember quite well, that I thought that it might be possible but it seemed relatively impracticable to me and I thought the possibility of hitting the gap again could not be very great."[54] Denis Gabor, later a Nobel prizewinner for something else, thought of the

52. Thibaud, CR, 194 (1932), 360-2, quotes; Pestre, Physique et physiciens, 70, 78.

53. Thibaud in Solvay, 1933, 71-5, and in Congrès int. électr., Comptes rendus, 2 (1932), 965, says that he succeeded in November 1930, before unambiguous resonance was achieved at Berkeley.

54. Wideröe, Jena U., Wiss. Zs., 13 (1964), 432-5; Europhysics news, 15:2 (1984), 9; and interview with David Judd, 21 Sep 1961 (18/24), quote; cf. Paul, Aesth. and sci., 51. Cockcroft, "Development," 2, hints that Walton may have had the idea of the cyclotron too.

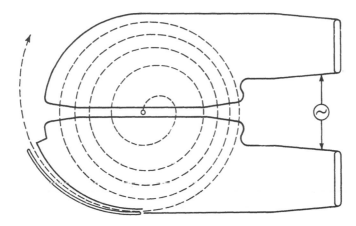

FIG. 2.16   Simplified schematic of a cyclotron, showing the spiral path of an accelerated ion. Livingston and Blewett, fig. 6.4.

cyclotron as early as 1924, or so he said. But he had other things to do. "My Dr.-Ing. study [on oscilloscopes like Rogowski's] was already in progress, and I let it go." He had another opportunity a few years later. On January 5, 1929, the irrepressible Szilard filed for a patent on, yes, a cyclotron, which he constructed, on paper as usual, by bending the accelerator tube of his linac into a drum sectored by grids. The accelerated particles were cycled back through the grids by a magnetic field. Szilard recalled late in life that he had offered the exploitation of his inventions to Gabor, who then served the Siemens electrical empire, but neither Gabor nor Siemens appears to have seized the opportunity.[55]

The most circumstantial account by a near inventor we owe to Max Steenbeck, who finished up an adventurous career in physics as head of the board for scientific research (Forschungsrat) of the German Democratic Republic. When a doctoral student at the University of Kiel in 1927, Steenbeck conceived the idea of a cyclotron while trying to illustrate certain properties of electrical forces to a particularly stupid student. He worked out a numerical example—acceleration of protons to MeV energies in an orbit of

55. Gabor to Lindemann, 20 Mar 1956 (FAL, D/80); Allibone, Roy. Soc. Lond., *Biog. mem., 26* (1980), 111; Szilard, "Korpuskularstrahlröhre," in *CW, 1,* 554–63; Szilard to W.B. Mann, 27 Feb 1952 (SzP, 95/1092).

20 cm under a magnetic field of 14,000 gauss—exactly the experiment done four years later in Berkeley. It was not tried in Kiel. An assistant to Walther Kossel, one of Steenbeck's professors, said it would be too costly; the other professor, Hans Geiger, said that it would be too risky. Before finishing his degree, Steenbeck went to work for Siemens in Berlin. He told his co-workers what he had in mind. They insisted that he publish the idea, no doubt as a precaution for possible future patent applications. Steenbeck agreed, but reluctantly, since the calculation was so elementary. His chief returned his manuscript with a note that he wished to speak further about it. Steenbeck interpreted the note as dissuasion (the chief had probably intended only to request a patent search) and, following his inclination, dropped the matter. Siemens lost the cyclotron again.[56]

These several anticipations of practical resonance accelerators do not detract from, but rather enhance, the achievement of Ernest Lawrence. As Szilard said of Lawrence's success, "The merit lies in the carrying out and not in the thinking out of the experiment."[57] Lawrence came across Wideröe's paper around or before April 1, 1929, while glancing through current journals. (The relevant issue of the *Archiv für Elektrotechnik* reached the University library on January 23, 1929.) This inspirational encounter may seem, and is said to have been, a matter of chance. The library had just opened a subscription to the *Archiv*, Wideröe's issue being only the fourth received; the *Archiv* catered primarily to "engineers working scientifically at electrotechnology" and to physicists only in so far as they took an interest in technical problems; and the *Archiv* was written in German, a language that Lawrence did not understand. The historian, who seeks the causes of things, prefers to think that Lawrence sought the journal for a definite purpose. Lawrence and Beams had cited an article in the *Archiv* by Rogowski and others in their paper on Kerr cells; similar references later appeared in theses by Lawrence's students and in a paper he wrote with one of them.[58] It may well be that Lawrence

56. Steenbeck, Jena U., *Wiss. Zs.*, *13* (1964), 437, and *Impulse und Wirkungen*, 556.

57. Szilard to O.S. (Otto Stern?), n.d., in Szilard, *CW*, *1*, 728.

58. Beams and Lawrence, *JFI*, *206* (1928), 176; theses by Washburn and by Dunnington (1932); Lawrence and Dunnington, *PR*, *35* (1930), 396–407; Dunnington, *PR*, *38* (1931), 1511, 1545. Intended audience and date of receipt of the

asked the library to subscribe to the *Archiv*. He later told Wideröe that he had picked up the issue with the fateful article to pass the time at a boring meeting. Having this succor to hand was not a matter of chance.[59]

Lawrence later gave another explanation. He said that in 1928, when he moved up, and far, from assistant professor at Yale to associate professor at Berkeley, he decided to shift his attention from the effete and played-out photoeffect to the robust new field of nuclear physics. His interest in apparatus drew him to the problem of high-voltage machines. He was already familiar with Tuve's struggle with the Tesla coil and had decided that a million volts could be reached directly only with great difficulty and large apparatus. He would have looked through the *Archiv* for exactly what he found there, a low-potential way to high energy. Lawrence would soon be building machines rated by his contemporaries as gigantic; but he recommended his first magnetic resonance accelerator as needing only "relatively modest laboratory equipment."[60]

It is sometimes said that Lawrence's cyclotron is Wideröe's two-step bent into a circle. The proper analogy to Wideröe's apparatus, however, is a single ring, within which the ions circulate under a magnetic field H (which Wideröe did not require) as they accelerate from periodic knocks, from the high-frequency field F. To shield the ions between knocks, each of the electrodes has to fill almost half the ring. As the ions speed up, they come to the accelerating gaps between electrodes at shorter and shorter intervals; to remain in step with them, the frequency of the field must increase. Similarly, H must grow to keep the ions on their circle. Such a system was then not practicable. It became the dominant method after the war when implemented in, among other machines, the Bevatron, in the sixth generation of Berkeley cyclotrons.

Wideröe issue appear from the cover of the copy in the Berkeley library.

59. Wideröe, Jena U., *Wiss. Zs., 13* (1964), 432–5. The date April 1 appears in a formal statement by Thomas Johnson, 15 Sep 1931 (10/1), written for possible use in patent fights; Lawrence says "early in 1929."

60. Lawrence, "Nobel lecture," in Livingston, *Development*, 136–7; Lawrence and Livingston, *PR, 40* (1932), 20, in ibid., 119, quote.

Lawrence saw that a looser analogy to Wideröe's ion accelerator might be possible with constant control field H and oscillator frequency f. The possibility appears from the cyclotron equation (2.1): the frequency of rotation of an ion of charge e_i, mass m_i, in the field H is *independent of the radius of the orbit*:

$$f = v/2\pi r = (e/m)_i(H/2\pi c). \qquad (2.2)$$

It appears that nature conspires in favor of cyclotroneers. Taking, then, not an annular tube, but a shallow drum or cylinder, as the playground of the ions, Lawrence worked out, on paper only, the scheme indicated in figure 2.17. The iconographer might consider it a combination of Wideröe's circle for electron acceleration with his mechanism for kinetic voltage transformation, a creative misunderstanding abetted and even made possible by Lawrence's inability to read Wideröe's text. "I merely looked at the diagrams and photographs," Lawrence said, in accepting the Nobel prize for his invention.[61]

In the figure, A and B represent hollow half-cylinders of metal, which are alternately charged to a maximum of around 10 kV by a radio oscillator. The close connection with Wideröe appears in, among more obvious features, Lawrence's retention of the word "tube," which Wideröe used appropriately for his straight glass accelerating vessel, for the cylindrical, metallic tank of the cyclotron.[62] "Cyclotron," perhaps Lawrence's first term for his invention, also echoed the notion of tube, in analogy to "radiotron," "thyratron," and "kenotron." (In formal parlance, Lawrence insisted upon "magnetic resonance accelerator;" "cyclotron" rose from slang to the official name in 1936.)[63] To return from the name to the substance of things: Ions liberated in the gas above a filament near the center of the cyclotron enter the gap between the

61. Livingston, *Development*, 137.

62. E.g., Livingston, *Production*, 11; Lawrence to Livingood, 18 Feb 1932 (12/11).

63. McMillan to L.D. Weld, 8 May 1952 (12/31); Lawrence to Tuve, 23 Sep 1933 (3/32), introducing "cyclotron;" to Cockcroft, 26 Jan 1935 (5/4), to Bainbridge, 6 Feb 1935 (2/20), to Johnson, 10 Jul 1935 (10/1), to Chadwick, 31 Jan 1936 (3/34), still a nickname; Henderson to Lawrence, 6 May 1935 (9/16), and Lawrence, McMillan, and Thornton, *PR, 48* (1935), 493, "a sort of laboratory slang."

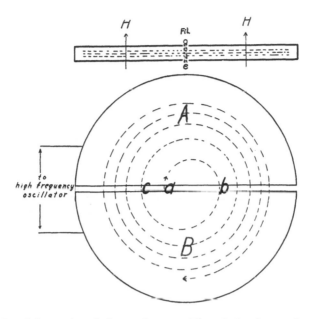

FIG. 2.17 Schematic of the cyclotron. The circle shows the meridian plane; ions enter at a, accelerate as they cross the gap at b, c, etc., between the two dees A, B. The inset at the top shows the vacuum chamber or "tube" in elevation; *H* the magnetic field, "fil" the ion source; the dotted lines indicate particle orbits. Lawrence and Livingston, *PR, 40* (1932), 23.

cylinders when the electric force there can pull them in the direction shown by the arrow at *a*. On entering *A*, which shields them from the radio-frequency (rf) field *F*, the protons revolve in a semicircle *ab* under *H*, which is perpendicular to the plane of their orbit, and reemerge at *b* just when *F* has reversed itself to pull them into *B*. The acceleration across *b* brings the ions into a wider semicircular orbit *bc* within the shielding electrode *B*, whence they emerge at *C* again in step with the rf field. They gradually spiral out under the guidance of *H*, acquiring 10 or 20 kV from *F* at each crossing of the gap, and so reach the circumference after a hundred or so turns with an energy of a million volts.[64]

64. That Lawrence thought out the general principles of the cyclotron in roughly this form during the spring of 1929 appears from the testimony of Johnson (10/1).

Thus the theory. There were reasons to doubt the possibility of its implementation. Only an inventor could think that the beautiful synchronization could last, that the ions could be kept going for one hundred turns without colliding with other molecules or flying from the median plane into the walls of the electrodes or going astray in crossing the gap. Lawrence planned to put his hollow electrodes, or dees as they were later christened, inside a very good vacuum, which would reduce the average distance between collisions to about the length of the spiral path; but that brought the problem of maintaining vacuum seals under the stresses created by the radio-frequency field and the guiding electromagnet. As for keeping the ions in the median plane through the dees, that appeared to be, and for a time was, the greatest problem of all. But here again nature unadorned favored the cyclotroneer.

Lawrence did not move quickly to implement his plans. No doubt the difficulties just described, and the refusal of his friend Thomas Johnson of the Bartol Foundation, then teaching in Berkeley, to join him in the work, gave him pause. But he had a strong positive reason for shelving the idea. He had spent the time since his arrival at Berkeley in the late summer of 1928 building up apparatus to continue his studies of the photoeffect. He had three doctoral students, "a relative importance...that I could never have attained at Yale in years," and a need to get results, to justify the reputation for clever and productive experiments that he had brought, and that had brought him, West.[65] One of his students, Niels Edlefsen, confirmed a curiosity earlier studied by Lawrence, that the probability of ionization of potassium vapor increases with the frequency of the light after reaching a local maximum at the series limit. Lawrence tossed the problem of explaining the rise beyond the limit over to Oppenheimer, Berkeley's new part-time theorist; it fell to his graduate student Melba Phillips, who took the problem of photoionization in potassium vapor for her thesis. The business was too much for Oppenheimer's people too, a "pandora's box," according to its historian, "of ultraviolet photoabsorption, photoionization, and ARPES [angle-resolved photoelectron spectroscopy]."[66] Lawrence

65. Johnson, statement, 15 Sep 1931 (10/1); Lawrence to his parents, 16 Sep 1928 and 7 Jan 1929 (10/38).

66. Lawrence to Hull, 30 Aug 1928 (7/28); Lawrence and Edlefsen, *PR, 33*

continued the experiments and found excitement, if not hope, in the box. "The longer I am in scientific research work," he wrote his parents, "the more fascinating it becomes."[67]

In January 1930 Lawrence began to look more favorably on his "proton merry-go-round." He had had two useful stimuli, one reassurance that it might work, the other notice that he might be anticipated. The reassurance came from Otto Stern, the world's expert in the handling of beams of hydrogen atoms, then providentially visiting Berkeley from his base in Hamburg. Stern thought the spiral beam had a chance and urged Lawrence to take it. Lawrence acknowledged the debt: "Probably if you hadn't urged me so strongly I would not have started the development of the method until some time later." And some time later: "It was your enthusiasm one evening during dinner...that stimulated me to try the development of the method." The stimulus from competition came from Tuve and company, who in the *Physical Review* for January 1, 1930, announced that they could operate their Tesla coil under oil under pressure at 5 MV with 120 sparks/sec. Should they succeed in driving alpha particles along at 5 or 10 MeV and at such a rate, they said, they would have radiation equivalent to that from 2,600 grams of radium. They had found the way to high energy, and found it "without much difficulty."[68]

Lawrence did not work alone. It was one of his strengths. From the time of his arrival at Berkeley, he published only two independent research reports, in 1933 and 1935. He accordingly did not attempt to build the merry-go-round himself, but enlisted Edlefsen, who had just finished his thesis on photoionization of alkali vapors; Edlefsen had a teaching assistantship in the Physics Department and reluctantly stayed on to help make the protons spin. By the end of February 1930, work had begun. "I have started an experimental research based on a very interesting and

(1929), 265 (abstract of paper for APS meeting, 8 Dec 1928); *PR, 33* (1929), 1086–7 (abstract, APS meeting, 18–20 Apr 1929); *PR, 34* (1929), 1056–60 (rec'd 28 Aug 1929); M. Phillips, *PR, 39* (1932), 552, 905–12; Jenkin, *Jl. spect., 23* (1981), 256–60, quote.

67. Letter of 7 Jan 1929 (10/38).

68. Lawrence to Stern, 12 Sep 1931, requesting statement for possible use in patent litigation, and 6 Feb 1934 (16/49), quotes; Breit, Tuve, and Dahl, *PR, 35* (1930), 51; Breit and Tuve, *Nature, 121* (1928), 525.

important idea," Lawrence wrote his parents. "If the work should pan out the way I hope it will it will be by all odds the most important thing I will have done. The project has fascinating possibilities."[69] In March he was working "night and day on...producing million volt protons without using high potentials. If this turns out it will of course constitute an important piece of work."[70] But it was not turning out.

During the summer of 1930 Lawrence lazed playing tennis while his graduate students, now half a dozen or more, were "making up for my laxity in activity along research lines."[71] Then a breakthrough came. Edlefsen left to become assistant professor of irrigation investigations in the University's Agricultural Experiment Station.[72] In a carefully worded statement, he and Lawrence described their method, implied that it had succeeded, and proposed parameters for an instrument capable of accelerating protons to a million volts. There is considerable doubt that Edlefsen achieved any resonance acceleration at all. But Lawrence wanted to believe and his natural optimism resonated with his enthusiasm. At the end of their statement, which Lawrence read to a meeting of the National Academy of Sciences on September 19, 1930, he let all qualifications go: "Preliminary experiments indicate that there are probably no serious difficulties in the way of obtaining protons having high enough speeds to be useful for studies of atomic nuclei."[73]

69. Childs, *Genius*, 146–7 (re Edlefsen); Lawrence to his parents, 23 Feb [1930] (10/38). Stern remembered seeing a preliminary "Magnetapparatur" around March 1930 (Stern to Lawrence, 2 Nov 1931 [16/49]); McMillan, "Brief history," 1, misdates the building of the Edlefsen-Lawrence cyclotron to January 1930.

70. Lawrence to Akeley, 21 Mar 1930 (1/12).

71. Lawrence to Boyce, 4 Aug 1930 (3/2); cf. Lawrence to Akeley, 4 Apr 1930 (1/12).

72. F. Adams to Lawrence, 15 Mar 1930 (7/1).

73. Lawrence and Edlefsen, *Science, 72* (10 Oct 1930), 376–7, in Livingston, *Development*, 116. The proposed parameters: $f = 1.5 \cdot 10^7$ Hz, $H = 15$ kG, R (dee radius) = 10 cm.

4. DOUBLE PLAY

In the fall of 1930 Lawrence found what he needed to make accelerators: two graduate students, very capable, hardworking, and dependent on his good opinion. One, David Sloan, we have already met; the other, M. Stanley Livingston, who came to the Physics Department with an M.A. from Dartmouth in the fall of 1929, sought a subject for a doctoral thesis. Both had financial support: Sloan his Coffin Foundation fellowship, Livingston a teaching assistantship. Livingston took up Edlefsen's problematic cyclotron; Sloan, Wideröe's proven, but useless, ion accelerator. Their working during 1930/31 demonstrated the practicability not only of both devices, but also of Lawrence's as yet tentative method of organizing work: simultaneous development of complementary instruments or parts by graduate students or postdocs, each of whom had responsibility and considerable independence on his own project. They were inspired to outstanding work by a spirit of cooperative competitiveness and by the bittersweet satisfaction of laboring at the edge of technology. "It is rather remarkable," Lawrence wrote one of his old professors, "how physics is attracting the best much as engineering used to do."[74]

The straight way proved the quicker. Sloan and Lawrence used mercury ions in a tube with eight electrodes driven by a 75-watt oscillator, with whose robust personality Lawrence had become acquainted during his summer at GE. They immediately got the expected eightfold amplification, 90-keV ions with the oscillator operated at 11 kV, some five times its rating. They extended the tube with thirteen more electrodes and got ions of 100 keV, then of 200 keV, exactly what they expected, with less than 10 kV on the oscillator (fig. 2.18). They reported this performance to the National Academy of Sciences in December 1930 and recommended their installation for its simplicity and economy. "An entirely practicable laboratory arrangement," viz., an evacuated tube 2.3 meters long containing electrodes run at 25 keV would give mercury ions of 125 kV.[75] To what purpose? Lawrence and Sloan

74. Lawrence to Akeley, 4 Apr 1930 (1/12); Lawrence to Coffin Foundation, 25 Feb 1930 (16/27).
75. Lawrence to C.R. Haupt, 8 Nov 1930 (9/2); Lawrence and Sloan, NAS, *Proc.*, *17* (Jan 1931), 68–9. The 21-electrode apparatus was run at wavelengths of 118 m and 85 m.

suggested studies of the properties of the high-speed rays, but did not stop to undertake them themselves. Instead, Sloan built a bigger tube, with thirty electrodes (plate 2.7). By the end of May 1931 he had passed the great artificial barrier and could boast of 0.01 μA of mercury ions with energies exceeding a million volts.[76]

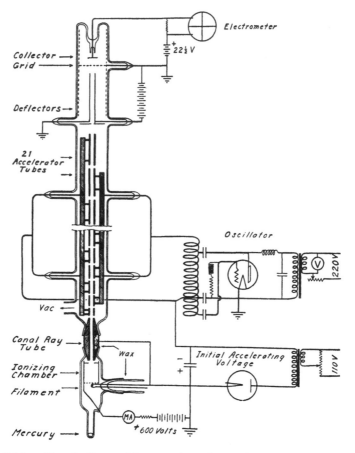

FIG. 2.18 Sloan's linac. Mercury ions from the source at the bottom accelerate through the canal ray tube and then through the 21 accelerating tubes to the collector at the top. Lawrence and Sloan, NAS, *Proc., 17* (1931), 68–9.

76. Lawrence to Cooksey, 26 May [1931] (4/19); to Darrow, 3 Jun 1931 (6/9).

Although the long linear accelerator had only a short life at Berkeley during the 1930s, it contributed much more to the Laboratory than the inspiring achievement of a million volts. It gave experience in working with a powerful commercial oscillator that could put as much as 90 kV on an electrode; in the definitive version of the 30-electrode tube, Sloan got ions of 1.26 MeV, or an average gain of 42 kV per electrode. It also forced attention on the problem of the synchronization and focusing of the beam. Synchronization—keeping the beam in step with the rf field at the first short electrodes—was resolved by trial-and-error adjustments, which gave important information about the fates of ions that do not enter a gap when the peak voltage spans it. Focusing turned out to be easy: it was another case where nature favors the particle accelerator. Figure 2.19 shows the lines of force between neighboring cylindrical electrodes. Besides suffering acceleration along the tube, an ion is driven toward the axis of the accelerating system during transit of the first half of a gap and away from it during transit of the second. If it enters the gap at the optimum time, near the peak voltage, it will speed up and spend less time in the second than in the first half of the gap. The net result of the minuet is motion toward the axis: the openings of the electrodes act as lenses, which inhibit ions from leaving the accelerating system.[77]

The discovery and exploitation of automatic focusing opened the possibility for ever longer linear resonant accelerators. Sloan started on a new tube, with thirty-six electrodes, each able to hold 80 kV, to be driven by two oscillators working at a wavelength of 27 m. He expected to have mercury ions at 4.5 MeV. If successful, he would go to the next mystical threshold, 10 MeV. That would require an accelerating system forty feet long fed by eight power oscillators.[78] Such a system would no longer be the convenient laboratory accelerator advertised by Sloan and Lawrence, but a big project in radio engineering. Sloan did not proceed immediately to quicker quicksilver. By November 1931 he had begun to construct an x-ray tube of novel design, which was to

77. Sloan and Lawrence, *PR, 38* (1931), 2026 (rec'd 19 Oct 1931).
78. Ibid., 2030–2.

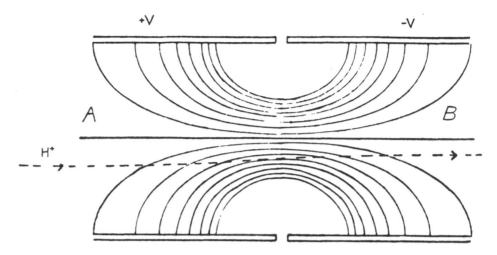

FIG. 2.19   Electrostatic focusing across the dee gap. AB is the median plane; the curves are the lines of electric force at potential difference $2V$ between the dees; the errant hydrogen ion traverses the path indicated back toward AB. Lawrence and Livingston, *PR, 40* (1932), 29.

play an important part in the technique and financing of the Laboratory.[79] Sloan finished the 36-electrode tube with the help of a graduate student, Wesley Coates, in 1932. It gave mercury ions with energies up to 2.85 MeV. Coates loosed these ions on various targets and discovered that soft x rays left the collision sites.[80] A similar investigation by another graduate student, Leo Linford, disclosed that each impacting mercury ion drove out around ten electrons, a result not uninteresting to designers of high-voltage vacuum tubes.[81]

79. Lawrence to Cottrell, 21 Nov 1931 (5/3); infra, §3.2.
80. Sloan and Coates, *PR, 46* (1934), 539–42; Coates, ibid., 542–8. Coates first glimpsed the soft x rays using 2.85 MeV mercury ions in the fall of 1932; Lawrence to Boyce, 1 Oct 1932 (3/8).
81. Linford, *PR, 47* (15 Feb 1935), 279–82.

While Sloan pushed Wideröe's technique to a million volts, Livingston struggled to reproduce the resonance experienced by Edlefsen. The earliest experiments recorded in Livingston's notebook took place on September 20, 1930. He had built a cylinder of metal 4 inches (10 cm) in diameter, about the size of Edlefsen's instrument, to serve as vacuum chamber; and he had access to a small magnet, capable of a maximum of 5,500 gauss, to control the spiralling ions. During October and November he found only feeble indications of resonance, and they came at values of the field H that depended upon the potential V of the oscillator. That violated the sound doctrine of equation 2.1. Livingston decided that his predecessor's success had been an illusion of faith.[82]

The obstacles Livingston had to overcome may best be indicated by following the path of ions that succeed in completing the spiral course. Obstacle 1. The source must give off an ion current i_S strong enough that the survivors will constitute a detectable current i_D at the collector (fig. 2.20). Livingston assumed that fewer than one in a thousand ions would run the course; to get an i_D of a few microamps he would need an i_S of a few milliamps. He did not know how to produce milliamps of protons. Lawrence wrote to the Forschungsinstitut of the Allgemeine Elektrizitätsgesellschaft, which he was told had perfected a proton source, but insufficient information returned and Livingston had to work with hydrogen molecule ions (H_2^+). These he generated in the center of the chamber by firing electrons from a radio-tube filament into the hydrogen gas that filled the chamber.[83]

Obstacle 2. The pressure P within the chamber must be small enough to make collisions between ions infrequent and large enough to accumulate an adequate i_S. In practice a vacuum pump worked continuously, maintaining a pressure around 10^{-5} mm Hg,

82. Livingston, "Workbook," 16–7 (39/9); Livingston, *RSI, 7* (1936), 55, Livingston, *Development*, 117, and Livingston in Weart and Phillips, *History*, 255; Childs, *Genius*, 158. Livingston, *Production*, 17, traced the dependence of H on V to the primitive detecting system with which he began.

83. Lawrence to Ramsauer, 10 Nov and 17 Jan 1931 (15/1); Livingston, *Production*, 12–14. Wideröe (in Livingston, *Development*, 103) had found generation of ion beams the most difficult part of kinetic voltage transformation.

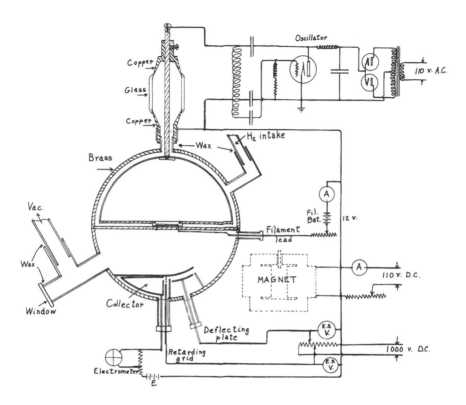

FIG. 2.20 Livingston's cyclotron. Livingston, *Production* (his doctoral thesis), fig. 3.

at which the average distance between collisions equalled the length of the spiral course.[84]

Obstacle 3. The grid across the entrance to the dee, which Lawrence and Livingston wrongly thought necessary for shielding, must not soak up a sensible fraction of the beam; the final arrangement used parallel slits, which opposed less metal to the ions than a rectangular screen.[85]

Obstacle 4. An electrostatic force D must be established between a special plate and the dee at the periphery of the spiral to deflect the beam (if any) into the collector. Since the value of the deflecting force measures the energy of the ions, it is

84. Livingston, *Production*, 15–6.
85. Ibid., 12–3.

important that the passage into the cup be so restricted that only ions in a narrow range of energies can be focused on it by D; on the other hand, the passage must be wide enough to admit an i_D capable of registering itself, and, ultimately, of inducing nuclear transformations.

On December 1, 1930 Livingston recorded his first success: with a newly designed detector, he had killed the dependence of H on V. "At last we seem to be getting the correct effect."[86] He then tried what values of oscillator frequency f, dee potential V, and pressure P gave the sharpest and strongest rise in i_D as he brought the electromagnet through the value $H = 2\pi fmc/c$ calculated for resonance. With $f = 2.5 \cdot 10^6$ and $V = 300$, he got a rise around $H = 3300$, where theory placed it. If the rise did record the presence of resonantly accelerated H_2^+ ions, they each had an energy W of 6 keV:

$$W = mv^2/2 = 2\pi^2 mf^2 R^2 = (e^2/2mc^2)R^2 H^2, \qquad (2.3)$$

which, for H_2^+ and $R = 4.8$ cm, the radial distance to the entrance to the collector, is $9.6 \cdot 10^{-9}$ erg or 6,000 eV. These ions had accelerated in twenty steps over a spiral path of ten complete turns.[87] That was most encouraging if true. Over the Christmas holidays Livingston borrowed a magnet capable of almost 13 kilogauss (kG), over twice the limit of 5.5 kG with which he had been working. On January 2, 1931, he got good resonance at 12.4 kG, $V = 1,800$ volts, for a calculated energy W of 70 keV attained in forty-six steps. With the magnet at its maximum, 12.7 kG, $W = 80$ kV, Livingston detected resonance at this setting with a little less than 1,000 volts on the dee, indicating eighty-two crossings of the dee gap, or forty-one entire turns. That was enough for a thesis, if not for disintegration.[88]

86. Livingston, "Workbook," 19–21 (1, 3, 13 Dec 1930, in 39/9). Cf. Lawrence to F.A. Osborn, 8 Nov 1930 (12/12).

87. Livingston, "Workbook," 25 (16 Dec 1930).

88. Livingston, "Workbook," 27, 40 (23 Dec 1930, 2 Jan 1931, in 39/9); Livingston, Production, 14–5. McMillan's statement, "Brief history," 1–2, that Livingston got to 80 keV on January 2, is not confirmed by the "Workbook."

Although Livingston marshalled his evidence very well, it did not amount to a demonstration. The resonances came sharply at the predicted places, to be sure; but i_D showed other bumps too, which often exceeded the resonance peak. Figure 2.21 gives a typical best case; but as appears from figures 2.22 and 2.23, the curve of i_D against H depended sensitively on P and D, and on V as well. In the drawings D and B are resonance peaks, the former for ions that travel through a complete circle in one period of the rf field, the latter for ions that go through a half turn in a period and a half. Livingston explained A and C as consequences of background ionization, photoeffects, partially accelerated ions, scattering, and so on, and supported his special pleading by arguments that need not be repeated. What was needed was a clean measurement of W; but to obtain the value of the deflecting potential D ($=W$) that made i_D a maximum required a collimation so tight

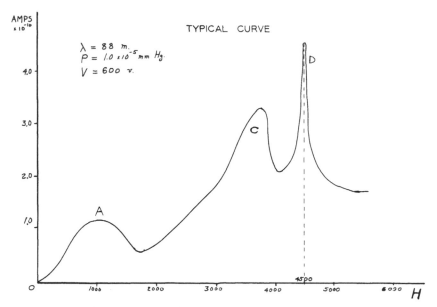

FIG. 2.21   Livingston's thesis results, 1: the ordinate is i_D, the abscissa the magnetic field H. Livingston, *Production*, fig. 5.

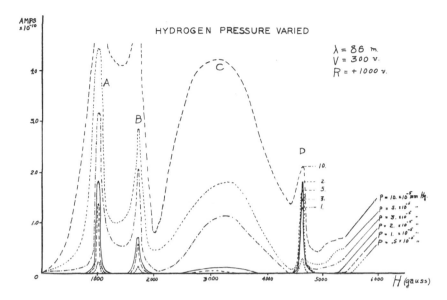

FIG. 2.22   Livingston's thesis results, 2: i_D against H at several pressures in the "vacuum" chamber. Livingston, *Production*, fig. 8.

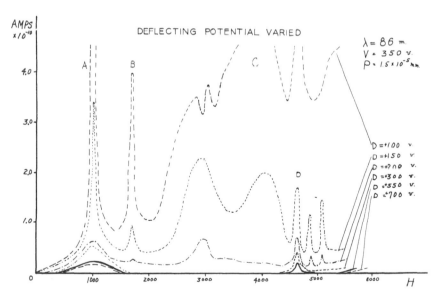

FIG. 2.23   Livingston's thesis results, 3: i_D against H for various values of the deflecting potential D. Livingston, *Production*, fig. 11.

that the beam entering the collector became too weak to measure.[89]

The point about inferring rather than measuring the energy of the particles bothered Lawrence. "We can make them spin around alright," he wrote Swann in January, "but we have not been able to determine how many times and therefore what speeds we have been able to produce."[90] Although by April 1931, when it was time to declare results, the difficulty had not been overcome, Lawrence and Livingston had no doubt that they had demonstrated the way to high energy for light ions. Their declarations differ subtly, however, in accordance with their places, personalities, and ambitions. Livingston, concluding his thesis: "There appear to be no fundamental difficulties in the way of obtaining particles with energies of the order of magnitude of one million volt-electrons." Lawrence, concluding a talk to the American Physical Society: "There are no difficulties in producing one million volt ions in this manner."[91]

A test was already being implemented. In a move typical of his later practice, Lawrence had started on the next bigger machine as soon as experiments with the one in hand gave the first indications of success. On January 9, a week after Livingston's ions had appeared to reach 80 keV, Lawrence wrote around for advice about building a magnet with the specifications he and Edlefsen had given: $R = 10$ cm, $H_{max} = 15$ kG. He received useful information from Kenneth Bainbridge, then at the Bartol Research Foundation, who had just built a similar magnet to practice mass-spectroscopy in Aston's manner. According to Bainbridge, materials would cost around \$700.[92] Lawrence turned to the University for money, an advance, he called it, on (indeed almost over) his next year's research allowance. The University's Board of Research, still headed by Armin Leuschner, recommended the advance, which the new president of the University, Robert

89. Livingston, *Production,* 19–25, and Livingston in Weart and Phillips, *History,* 256.

90. Lawrence to Swann, 24 Jan 1931 (17/3).

91. Livingston, *Production,* 28; Lawrence and Livingston, *PR, 37* (1931), 1707.

92. Lawrence to Swann, 9 Jan 1931, and Bainbridge to Lawrence, 20 Jan 1931 (7/19). Livingston, "Workbook," 69 (n.d.), mentions 9-inch pole faces, 8.5-inch path diameter ($R = 10.8$ cm).

Gordon Sproul, approved, setting a pattern for the early development of the Laboratory. Both had recently had the opportunity to
take Lawrence's measure. The preceding semester, with the
advice and prompting of G.N. Lewis, they had defeated an
attempt by Northwestern University to lure him away. Lawrence
had emerged from the negotiations with a full professorship, at a
high salary and a low age, and with easy access to the chief financial and administrative officer of the University.[93]

Another pattern began then too: the estimated budget fell far
below the true cost. Lawrence acquired the additional $500 he
needed from another familiar source, the National Research Council, from which he had had a grant-in-aid for his photoelectric
studies.[94] His second cyclotron still served the express purpose of
bringing physicists to high energy at low cost: no unusual funding
was required. Manufacture of the magnet was entrusted to
Federal Telegraph, which brought it to the Physics Department on
July 3, 1931, just after Lawrence had returned from the symposium on the production of high-energy particles at the American
Physical Society meeting in Pasadena, where he heard Tuve
describe the acceleration of protons and H_2^+ ions to a million volts
by a Tesla coil. Tuve had depreciated that great barrier as a
"moderate" voltage. Lawrence went back to Berkeley hoping to
set protons spinning in less than two weeks.[95]

He did. He wrote on July 17: "The proton experiment has succeeded much beyond our fondest hopes!!! We are producing
900,000 volt protons in tremendous quantities—as much as 10^{-8}
ampere. The method works beautifully." The published
announcement followed familiar lines. "With quite ordinary
laboratory facilities proton beams having great enough energies [to
effect disintegration] can readily be produced." All it takes is a
magnet with pole pieces nine inches in diameter, a maximum field

93. Seidel, *Physics research*, 358–60; Lewis to Sproul, 27 Aug 1930 (Lewis P).
94. Lawrence to Vernon Kellogg, NRC, 27 Mar 1931, and Kellogg to Lawrence,
12 May 1931, announcing the grant (46/24R); Childs, *Genius*, 161, errs twice by
having the NRC finance the magnet for $1,000. Cf. Lawrence to D.C. Miller, 27
Feb 1931 (12/23): "Unless we can get some outside funds to build a magnet, we
must hold up the work until the University is able to supply additional funds."
95. *PR, 38* (1931), 579–80; Lawrence to F.G. Cottrell, 3 Jul 1931 (5/3); Tuve,
Hafstad, and Dahl, *PR, 39* (1932), 384–5.

of 15 kG, and a few accessories, like the 500-watt short-wave power oscillator Lawrence borrowed from Federal. One gets ions of over 500 keV, an amplification of at least 100, and full conviction that a million volts are around the corner and ten million less than a dream away.[96] In August 1931 Lawrence was awooing in New Haven. He received a telegram from the secretary of the Department: "Dr Livingston has asked me to advise you that he has obtained 1,100,000 volt protons. He also suggested that I add 'Whoopee'!"[97]

There remained the problem of direct measurement. By November Lawrence and Livingston had obtained a current i_D strong enough to allow the necessary collimation by applying to their instrument the very important discovery made by Sloan and Lawrence that grids worsen the focusing of the circulating beams. Lawrence alerted two main doubters of cyclotronics, Lauritsen and Tuve, to his latest accomplishment. "Recently we have put our experiments on a very sound basis by proving quantitatively by electrostatic deflection that the particles we are measuring are the expected high speed protons or H_2^+ ions."[98] To get high final voltages, Livingston had to put a relatively high potential on the dee. He could not go beyond an amplification of seventy-five steps. According to Lawrence, Livingston thought that he had struck the relativistic limit, where increase of particle mass with velocity destroys the synchronization expressed in equation 2.1. In fact, imperfect regularity in the controlling magnetic field was the culprit. Lawrence suggested placing small soft-iron shims shaped as in figure 2.24 where they would do the most good. "A little work with this led to the most gratifying results," he wrote Tuve soon after the new year. On January 9, 1932, Lawrence and Livingston attained protons at 1.22 MeV with only 4,000 volts on the dee, an amplification of over 300.[99]

96. Lawrence to Cooksey, 17 Jul [1931] (4/19), quote, and to Cottrell, same date (5/3 and 5/16); Lawrence and Livingston, *PR, 38* (1931), 834 (letter of 20 Jul), specifying pole pieces 9 inches in diameter.

97. Rebekah Young to Lawrence, c/o Dr. George Blumer, 3 Aug 1931 (12/12); cf. Lawrence to Cooksey, 10 Aug [1931] (4/19).

98. Lawrence to Lauritsen, 10 Nov 1931 (10/36), quote, and to Tuve, same date (7/34). Cf. Livingston, *RSI, 7* (1936), 57.

99. Lawrence, "Notebook," 6 and 9 Jan 1932 (39/2); Lawrence to Tuve, 13 Jan 1932 (17/34).

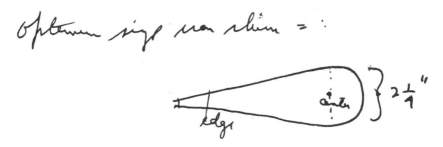

FIG. 2.24 Lawrence's earliest designs for cyclotron shims. Lawrence, "Notebook," 6 Jan 1932 (39/2).

It was time to write up. One knows the format. The principle of magnetic resonance acceleration had proved itself with "quite modest laboratory equipment," which functioned best with nature's cheap gridless focusing and a few shims that looked like tears. With a magnet costing $1,200 and with a 20-kW power oscillator, protons could be spun around 300 times to an energy of 1.22 MeV. To be sure the final current was small, only 0.001 μa. But by improving the source, multipying the shims, putting in two dees with 50 kV between them, and using a magnet giving 14 kG between pole pieces only 114 cm in diameter, it would be "entirely feasible" to go to twenty-five million volts.[100] The technical achievement was mainly Livingston's; the inspiration, push, and, above all, the vision of future greatness, were Lawrence's. His professorship gave him the place to stand, and Livingston gave him the lever, to move the world of physics. Compare the case of Thibaud, who, also in 1932, had made a little cyclotron, which he ran in the gap of a 10-kG magnet with 20-cm (8-inch) poles; he took as his ion source a discharge tube that fed the vacuum chamber through a narrow canal. Thibaud therefore could maintain his ion source at a pressure considerably higher than the pressure in the chamber and he obtained a proton current a thousand times larger than Livingston's. It is not clear that he caused his ions to resonate; it is certain that he had no one like Lawrence to push him along; and no further progress was made in cyclotroneering in France for several years.[101]

100. Lawrence and Livingston, *PR*, *40* (1932), 20, 28–9, 32–4 (rec'd Feb 1932).
101. Thibaud in Solvay, 1933, 71–5, and Congrès int. électr., *Comptes rendus, 2*

When this account was completed, in February 1932, Lawrence had long since sought a magnet to raise his energy by a factor of ten. It is almost superfluous to add that he had begun raising money for it the preceding July, as soon as protons resonantly accelerated to 500 keV appeared in the collector of the second cyclotron. It was "practically certain," he said then, that he could get to 10 MeV, or even to 20 MeV, if he could put his hands on a suitable magnet and oscillator. He reckoned he needed $10,000, or possibly $15,000, for the purpose. That took him out of the range of University research funds and NRC grants-in-aid. Another angel had to be found, he wrote the man he hoped would act the part, "if we are to proceed immediately towards the goal of 20,000,000 volts."[102]

(1932), 966–7; Lawrence to Thibaud, 18 Mar 1933 (17/4), requesting details about the prolific proton source.

102. Lawrence to Cottrell, 17 Jul 1931 (5/3 and 15/16), quote; same to same, 20 Jul 1931 (5/3).

III
Foundations of the Rad Lab

1. THE PLACE

The View from Above

The angel above the usual academic heaven to whom Lawrence first appealed was Frederick G. Cottrell, who had grown rich combating the powers of darkness. In 1906, while an assistant professor of chemistry at Berkeley, Cottrell had invented an electrostatic precipitator that removed noxious particles from the smoke of a local Dupont plant. The trick worked for other industrial polluters too; and in 1914 Cottrell endowed a not-for-profit business, the Research Corporation, to "serve the growing number of men in academic positions who evolve useful and patentable inventions from time to time in connection with their work and [who] without looking personally for any financial reward, would gladly see these further developed for the public good." Its charter required its board of directors, which included T. Coleman Dupont, Elihu Thomson, and other powerful men of American industry, to seek out and support research projects that might lead to profitable patents. Cottrell expected that proceeds from working or leasing the patents would repay the Corporation's investment in the research behind them and replenish its capital fund.[1]

The conservatism of the Corporation's board and the indifferent return on its patents during the war kept it from fulfilling its

---

1. Cameron, *Cottrell,* 151–60; Coles, in *Cottrell symposium,* xi–xvi, quote.

charter until the year of the crash, when it gave the Smithsonian Institution a grant to study the biological effects of radiation. Then, under an aggressive president, Howard A. Poillon, who wished to cut a figure in the world of scientific giving, and with the advice of Cottrell, who was retained as a consultant, the Corporation collected what it could from its precipitators and shopped for patentable material at research universities. By the end of 1932 it had given $120,000 in aid of research to universities and other public institutions.[2] At the time the Corporation began to stir, the universities were awakening to the predicament into which the new relations between science and industry, or research and commercialization, had placed them. Who if anyone should protect the patentable results flowing from academic laboratories? Should not the universities whose facilities produced the results? And ought not educational institutions to prosecute their rights with what vigor they could manage during the downturn in the economy, to offset losses in endowment income and state allocations?[3] But then, would it not be compromising, and even immoral, to obtain royalties from inventions made at tax-exempt institutions supported by public monies or private gifts? The farmer whose taxes underwrote agricultural research at a state university would not be pleased to pay for it again through royalties on the patents it furnished. Would not the farmer—and everyone else—decide to leave a university so conducted to make its own way? "Why should gifts intended for the general welfare play the rôle of capitalizing a business?"[4]

These questions were aired frequently during the early 1930s, in the pages of professional journals, in a report by the Committee on Patents of the AAAS, in a Symposium on Patents at the Patent Office, and at the NRC.[5] A minority preached that universities should have nothing to do with patents—the proximity to commer-

2. Cameron, *Cottrell,* 273, 280–2, 285, 292, and RC, Board of Directors, minutes, 534, 556, 589, 699–700, 710–11, 734–5, 742–3, 782, 791–3, 796, 848–9, 864–5, 969 (RC).

3. Gray, *Harper's, 172* (1936), 540, 542; Cottrell, Am. Inst. Chem. Eng., *Trans., 28* (1932), 224.

4. Gregg, *Science, 77* (1933), 259.

5. E.g., Am. Inst. Chem. Eng., *Trans., 28* (1932), 183–234, the Patent Office's Symposium; and AAAS, Comm. Patents, *Protection,* which quotes copiously from the journals.

cialization would demoralize the professoriate, threaten impartiality and openness, promote jealousy, ruin academic standards, debase learning, and exterminate the race of Faradays and Maxwells who did science for the love of it and made discoveries that transformed the world.[6] Another minority held that "there is nothing laudatory in the fact that [Faraday and Maxwell] and their colleagues failed to visualize the vast network of power utilities and communication systems which have evolved from their investigations" and urged scientists to pursue whatever financial gain their discoveries might bring them; "'Truth for truth's sake' is a delusion of so-called savants."[7] The majority favored some sort of protection. The AAAS's Committee on Patents, on which Cottrell served, urged patenting by or on behalf of universities, and not primarily for profit. That would be counterproductive: "A scientist who is impelled only by a motive of profit is far less likely [than others] to make any important contribution to knowledge." The main interests of both the inventive professor and his university should be to control the products resulting from his research: to oversee their quality, price, and advertising; to protect the public; and to return enough to pay for the oversight and, perhaps, to support further research.[8]

The AAAS's Patent Committee drew attention to two practices that enabled universities to obtain some measure of control and return without entering into business directly. One set up a special board within the institution, as Caltech, MIT, and the universities of Illinois and Minnesota had done. The other left the administration in the hands of a special corporation, of which the cynosure was the Wisconsin Alumni Research Foundation (WARF), established in 1925 with a Wisconsin professor's patent on vitamin D as its capital. WARF put its profits back into research at the university; and by 1937 it had given about $700,000 and an endowment that yielded over $125,000 a year. This success inspired other hard-pressed universities. By 1936 Cincinnati, Columbia, Cornell, Iowa State, Lehigh, Penn State,

6. Gregg, *Science, 77* (1933), 259; Gray, *Harper's, 172* (1936), 544, 548; Flexner, *Science, 77* (1933), 325; Withrow, Am. Inst. Chem. Eng., *Trans., 28* (1932), 207.

7. Foote, *RSI, 5* (1934), 59–60.

8. AAAS, Comm. Patents, *Protection*, 10–1 (quote), 12–4, 18–22; Ryan, Am. Inst. Chem. Eng., *Trans., 28* (1932), 209; Gray, *Harper's, 172* (1936), 543.

Purdue, Rutgers, and Utah had similar organizations.[9] California was exceptional among the great land grant colleges in not having an equivalent of WARF.

The year that WARF began, the president of the University of California charged the Board of Research to formulate a patent policy. An overstrict system resulted: any member of the University who perfected a patentable invention was obliged to bring it to the attention of the president, who would appoint a special board to advise him what to do with it. Neither the institution nor its faculty gained much from this procedure. Six years later, in 1931, when WARF was assisting research at Wisconsin at $1,000 a day, Leonard Loeb, then chairman of the Patent Committee of the Board of Research, decided that more might be done in California. He approached Cottrell. Now Cottrell had originally wished to give his patents to the University and had set up the Research Corporation only when it became clear that no public body acceptable to him had the mandate or enterprise to accept them.[10] It now appeared that the relationship earlier proposed could be reversed and the Research Corporation set up as the holding company for the University. Discussions between Cottrell and the Board of Research in the late spring of 1931 seemed promising to both parties. Armin Leuschner, the board's head, tried the alliance with a successful application to the Research Corporation for $5,000 for general support of work in the natural sciences at Berkeley.[11]

These pourparlers had the endorsement of president Sproul, who met with Cottrell and inclined to put promising material "at the disposal of the Research Corporation or [its partner] the Chemical Foundation for consideration as to whether or not it desires to secure patents." Sproul believed in a symbiosis of capitalistic and academic industry. "This is the age of science and

9. AAAS, Comm. Patents, *Protection*, 22; Gray, *Harper's*, *172* (1936), 542–3; Nicholls, *Jl. farm econ.*, *21* (1939), 496.

10. Board of Research, minutes, 23 Apr and 15 Dec 1925, 26 Jan 1926, 25 Mar, 5 and 22 Apr, 2 May, and 7 Oct 1927 (UCA, CU/9.1); Gray, *Harper's*, *172* (1936), 541; Cottrell, Am. Inst. Chem. Eng., *Trans.*, 28 (1932), 222–3.

11. Cottrell to Poillon, 6 Mar and 5 Apr, 1931, and RC Board of Directors, minutes, 16 Jan, 11 Mar, and 23 Apr (RC); Cottrell to Poillon, 7 Jul 1931, in Cameron, *Cottrell*, 289; Palmer, Pat. Off. Soc., *Jl.*, *16:2* (1934), 122.

democracy," he said in his inaugural address in 1930, "an age of strain and steel, electricity, chemistry, and science." "The work of the laboratory capitalized in the factory and by industry has built up a great civilization." Sproul had graduated from Berkeley in engineering in 1913 and returned with two years' experience to begin the climb from the Comptroller's Office to the presidency. He declared that no one belonged on the faculty who did not do research; for research, especially scientific and engineering research, research that might eventuate in patents, was the pump of progress. "Endlessly going over old lessons is a narcotic to progress."[12] He would have liked, though he could not have afforded, a faculty of Lawrences. To him, the aggressive policies of the Research Corporation fit the circumstances of the University perfectly: the strain of the age, the obligation to research, the cut in the budget. He rescinded the patent policy of 1926, in order, he said, to give faculty members full freedom of action. But he recommended that the action go to the Research Corporation.[13]

And the Corporation was aggressive. To work its major asset, it set up a worldwide cartel so restrictive and unfair in its practices that the U.S. government felt obliged to force the American branch to change its ways. The Corporation fought in the courts as vigorously as General Electric, pushed development of the inventions entrusted to it, and hunted out promising professors. It was more combative than WARF, which did not always behave gently either (the Corporation restricted the number of licenses of its primary patents and eventually also had to alter its ways). But WARF's ties to the university moderated its commercialism.[14] The Research Corporation had fewer inhibitions. Lawrence was to adopt its methods as well to profit from its support.

Cottrell became Lawrence's agent as well as his angel. The first job was to secure a big old Poulsen magnet, one of a pair of derelicts belonging to Federal Telegraph. At the prompting of its

12. Sproul to Poillon, 4 Sep 1931 (25/2); Pettitt, *Twenty-eight years*, 3–9, 23, 57, and 195, 199 (first two quotes), 42 (last quote); Cottrell, "Diary and letter file," 3 Sep 1934 (RC).

13. Sproul to Archie Palmer, 14 June 1932 (RC); Palmer, Pat. Off. Soc., *Jl.*, *16:2* (1934), 123.

14. Vaughan, *U.S. patent system*, 161, 307–8; Nicholls, *Jl. farm econ.*, 21 (1939), 496–8.

former employee and Lawrence's colleague in the School of Engineering, Leonard Fuller, Federal seemed willing to give one to the University. But the magnet required much reworking to answer Lawrence's purpose. Not then believing that the cyclotron had sufficient promise to justify a very large investment by the Research Corporation, Cottrell recommended that Lawrence stop in Saint Louis in the spring of 1931 on his way to a meeting of the NAS; the Radiological Research Institute there might give something toward refurbishing the magnet. The president of the institute, E.C. Ernst, had an interest, but no principal, to invest in Lawrence's projects. Cottrell next suggested the Chemical Foundation, set up by the government in 1920 to administer 5,000 German chemical patents seized during World War I. The Foundation was intended to protect American chemical industry while it developed domestic strength on German innovations, and in the shade of a high tariff wall, it earned royalties that amounted to nearly $9 million by 1932. Its charter obliged it to spend its income "for the advancement of chemical and allied science and industry," a mission very liberally interpreted, or stretched, by its chief officers Francis Garvan and William Buffum. Its adventure of greatest interest to us was the licensing of two patents it held on high-voltage x-ray tubes. When GE brought a suit for infringement against the license, the Foundation and the Radiological Research Institute countersued, with the consequence that both suits were dismissed and the field opened to further development.[15]

This connection may have been important when, on Cottrell's urging, Poillon went to see the executives of the Chemical Foundation some time in July 1931 to work out cooperative support for Lawrence's project. The connection: the quashing of GE's patents on x-ray tubes had saved Federal Telegraph over a million dollars. "[Buffum] is going to approach Federal and see how appreciative they are and suggest that they not only donate the magnet, but condition it." Then, Poillon continued to Cottrell, Lawrence would need but $7,000, which the Research Corporation

15. Lawrence to Ernst, 30 May 1931 (15/16) and 9 June 1931 (RC); to Cottrell, 3 Jul 1931, and Ernst to Lawrence, 30 June 1931 (RC); Palmer and Garvan, *Aims and purposes*, 49, quote; Anon., *Chem. ind.*, *36*: suppl. (1939), 139–43.

and the much richer Chemical Foundation could manage.[16] Federal did not care to square its obligation in this way. Poillon decided to contribute $5,000 and Buffum $2,500 of the $12,000 that Lawrence deemed necessary to rebuild, transport, and set up the magnet. Sproul put up the difference.[17]

From the standpoint of the Research Corporation and the Chemical Foundation, the most promising of the applications of magnets that Lawrence had in mind in the early summer of 1931 was the acceleration not of positive ions but of electrons, in the manner tried with no conspicuous success by Walton and by Tuve and Breit. Lawrence explained in a long memorandum, written in June, how he would improve on Walton's arrangement by multiplying magnets to hold the accelerating electrons on their paths. Lawrence's electrons would have energy high enough to penetrate the marketplace. "[They] carr[y] us straight into the field of high voltage x ray tubes and as such may prove medically and economically highly important," Cottrell wrote Poillon after studying Lawrence's memorandum. He also reminded the head of the Research Corporation that the head of the Chemical Foundation had "very definite ideas on the need for the development of x ray facilities." Following his policy and his penchant, Cottrell urged Lawrence to patent his scheme and to keep quiet about it until he had.[18]

After the pledge by the Research Corporation and the Chemical Foundation of the money ′ toward the cyclotron—and after Lawrence had been told to patent it—Cottrell returned to the business of the early spring, "closer and more effective teamwork in general on patent matters" between the University and the Corporation. The goal might be secured in stages. "The Research

16. Cottrell to Poillon, 7 Jul 1931, and Poillon to Cottrell, 31 Jul 1931 (RC).

17. Poillon to Lawrence, 6 Aug 1931 (15/16): $4,000 for rebuilding, $1,000 for moving, $1,000 for a motor generator set, $1,500 toward operations for 1931/2; Leuschner to Sproul, 24 Sep 1931, and Sproul to Poillon, 8 Oct 1931 (25/2); Lawrence to Richtmeyer, 24 Jul and 18 Aug 1931 (25/2). The University paid for installation ($3,300) and for power; Leuschner to Sproul, 23 Sept 1931 (25/1). In a creative misinterpretation, Lawrence thought that the Research Corporation and the Chemical Foundation would pay for everything; Lawrence to Cooksey, 10 Aug [1931] (4/19), and to Poillon, 6 Aug 1931 (RC).

18. Lawrence to Cottrell, 10 June 1931, enclosure, and Cottrell to Poillon, 7 Jul 1931 (RC).

Corporation [should] study the patents and University of California negotiations with the purpose of seeing if it could not more effectively handle exploitation in eastern territory at least as an entering wedge, and if all worked well the Research Corporation could eventually take over the whole works."[19] The regents' secretary, R.M. Underhill, the University's comptroller, R.C. Nichols, and Sproul agreed to this procedure; and Lawrence immediately assigned his patent rights to the Corporation.[20]

Lawrence himself was an important property to the leadership of the small philanthropies that had pledged to support him. Cottrell outlined a grand future: "I believe Lawrence is a mighty promising young man for us to keep in touch with and develop in connection with our major plans....Lawrence's work seems to me a very good peg on which to hang definitely a concrete proposal of cooperation between not only Chem. Found. & R.C. but also with Max Mason & *Rockefeller Foundation*, and thus bring to a *head* the tentative contacts already started. This particular work of Lawrence's is right in Max Mason's own field and will keenly interest him technically I am sure, which is an added advantage for the present purpose. Even if you [Poillon] and Buffum feel that you want to cover the complete needs of this particular line of research so as to feel freer with regard to the patent matters developing out of it, I still think it would be well for Lawrence to have a visit with Max Mason and for you all to talk over the larger plan." If the Rockefeller Foundation would play, the Carnegie Corporation and the California State Legislature might join the game. An empire might be built on precipitators. Lawrence was to make the enterprising Research Corporation his headquarters during his many trips to New York to raise money for his Laboratory.[21]

There was a snag, however. Professors did not have the spirit of cooperative enterprise of the industrial research laboratory. Cottrell to Poillon: "I find there is greater unwillingness on the part of people to tell each other their innermost ideas and note

19. Cottrell to Poillon, 9 Sep 1931 (RC).
20. Sproul to Poillon, 4 Sep 1931, F.C. Stevens to A.O. Leuschner, 21 Sep 1931, and Lawrence to Poillon, 22 Sep 1931 (RC).
21. Cottrell to Poillon, 21 Jul 1933 (RC); Lawrence to John Lawrence, 14 Apr 1936 (11/16).

that it is extremely prevalent among scientific men, especially when it comes to do with their colleagues." Nor did they immediately see the point of patents. Poillon to Cottrell: "If he [Lawrence] is one of the men that we are going to make awards to from time to time, it seems that we should develop his protective instincts."[22]

The View from the Ground

When the Research Corporation came into his life, Lawrence shared the inhibitions of the academic scientist against securing a personal financial interest in his discoveries or inventions. This inhibition arose at the end of the nineteenth century, when applications of electricity made physicists newly useful and threatened the identification of professors with high culture and disinterested speculation. Patents and professors should not mix. "Working as he does with public funds, directing as he does the minds and hands of students, it is, to say the least, scarcely honest [for a professor] to go with the results of such work to the Patent Office." Thus a British periodical for applied science editorialized in 1884. As for the Germans, "It is well known throughout the world," they said, "that the physical laboratories of Germany have no windows looking towards the patent office."[23] In the United States, *Popular Science Monthly* castigated people who would degrade science to a "low, money-making level," and memorialists praised defunct physicists who had resisted the blandishments of industry. That was their morality and their opportunity: "Nature turns a forbidding face to those who pay her court with the hope of gain, and is responsive only to those suitors whose love for her is pure and undefiled."[24]

These inhibitions—"a gesture of repugnance toward money-making as a practice inconsistent with intellectual integrity"—remained strong in the 1930s despite the increasing integration of academic science and industrial development. Hale built it into

22. Letters of, resp., 14 Sep 1931 and 6 Aug 1931 (RC).
23. *The electrician, 13* (1884), 454; Münsterberg, *Science, 3* (1896), 162, anent the discovery of x rays.
24. Texts from circa 1900 quoted in Heilbron in Bernard et al., *Science, technology, and society,* 67.

his quest for endowment for academic science; donors need have no worry, he said, that their gifts would enrich their recipients; "the men of science...may be counted upon to devote their efforts to the advancement of knowledge without thought of personal gain." The unconventional Szilard's constant attendance on the Patent Office caused his friends to warn him of "an opposition to you [from the British physicists] on account of taking patents." "It is not customary [he allowed] to take out patents on scientific discoveries." In defending Steenbeck's claim to the invention of the betatron, the German industrial physicist Carl Ramsauer apologized for the "unusual" character of the documentation, "a prior patent application rather than a [scientific] publication."[25] But there was the difficulty that anyone might seek to patent a process openly described in the scientific literature after adding some small improvement to it; scientific ethos and self-protection did not run in parallel. A way out of the difficulty was to raise patenting to a social responsibility, according to the following formula. "In spite of the fact that I think that university people should not be interested in patents," Urey wrote his collaborator in the elecrolytic separation of deuterium, "for the protection of pure science...it would be wise to patent this process."[26] That was the way Lawrence came to think under the tutelage of the Research Corporation.

The University gave him no guidance. The cognizant body, the Board of Research, demonstrated its level of leadership and vigor of oversight in its opinion about what it understood to be a patent on a process for making radium from sodium. Leuschner to Lawrence: "We do not feel ourselves quite competent to judge the ethical side of the question. Your and Mr Poillon's own knowledge concerning the practice of pure physicists in this regard I think should be sufficient to guide you in whatever action you wish to take." And Leuschner to Poillon: "We have faith in the Research Corporation because it is a non-profit corporation and

25. Quotes from, resp., Gray, *Harper's, 172* (1936), 539; Hale, ibid., *156* (1928), 243; Polanyi to Szilard, 11 Nov 1934 (Sz P, 17/197, and Weart and Szilard, *Szilard*, 40); Szilard to Segrè, 1 Apr 1936, in Szilard, *CW, 1,* 732; Ramsauer, note in *Nwn, 31* (1943), 235.

26. Urey to E.W. Washburn, 3 May 1933 (Urey P, 5). Cf. Connolly, *Science, 86* (1937), 38–45.

supports research where its funds, according to its own judgment, are best applied."[27] By the time the ersatz radium came up, Lawrence was ready to put himself in Poillon's hands. He had put off patenting the cyclotron as unbefitting despite the Corporation's urging. Then he learned from John Slater, the new head of MIT's physics department, that an engineer at Raytheon had hit on the idea of a "proton merry-go-round" independently of all its other inventors. Raytheon called in a former student of Slater's, Eugene Feenberg, to calculate the machine's practicality; and, on receiving assurance of its promise, applied for a patent. "It would never occur to me," so the physicist's ethos spoke through Slater, "to patent such a thing." "It never occurred to me to patent the work we are doing either," Lawrence replied, "and I am doing so only at the urgent request of the Research Corporation and the Chemical Foundation, who apparently have a better perspective of practical things than we have."[28]

The work and immediate practical problems then fully occupied his time. His magnet, valued at $25,000, weighed over eighty tons. Where to put it? While Federal deliberated over making the gift, Lawrence toured the campus in search of a firm floor. Since he proposed to work on an engineering scale, he faced the challenge of winning a foothold in the preserves of the engineers. Civil Engineers politely declined to house him; Mechanical Engineers robotically refused him the ideal space they underutilized in the Mining Building; and he turned to Cottrell not only for the money to move and refurbish the magnet, but also for somewhere to put it.[29] Sproul supplied the firm floor. On August 26, 1931, also the day that Molly announced their engagement at the New Haven Lawn Club, Lawrence was assigned the Civil Engineering Test Laboratory, "a large frame structure with several substantial concrete piers in the rooms" (plate 3.1), for his experiments. We take this far-seeing decision of Sproul's as the foundation act of the Radiation Laboratory.[30]

27. Leuschner, for Board of Research, to Lawrence, 10 Oct 1934 (35/77), and to Poillon, 23 Nov 1933 (RC); Owens in Marcy, *Patent policy*, 658.

28. Sproul to Poillon, 4 Sep 1931 (25/2); Bremer in Marcy, *Patent policy* (1978), 558; Slater to Lawrence, 4 Sep 1931, and reply, 8 Sep 1931 (35/3).

29. Lawrence to Cottrell, 10 June 1931 (15/16), and to Akeley, 21 Mar 1930 (1/12).

30. Lawrence to Poillon, 27 Aug (quote) and 8 Oct 1931 (15/16); Lawrence to

Sproul's act freed Lawrence's operations from control and even supervision by the Department of Physics, although the Laboratory remained an integral part of the Department until 1936 and a satellite until 1939. From the beginning the Laboratory had a research budget exceeding the Department's, which remained just under $12,000 from 1930/31 to 1932/33, and fell to $8,000 in the worst Depression year, 1933/34, while the Laboratory's expenditures continually increased.[31] Lawrence spent his money without overscrupulous accounting and, when he required more, raised it from outside the University or by dealing directly with Sproul. Although Lawrence's rapid rise and independent base inspired jealousy in some of his fellow seekers after truth, in general his relations with his colleagues in the Physics Department were cordial, if not close. No senior member of the Department besides Lawrence steadily worked in the Laboratory during the 1930s.

Lawrence did not find it easy to consolidate his domain. The usual administrative burden of establishing a new institution in old surroundings was increased in his case by the weight of the magnet. "It is one hell of a job getting things moving," he wrote Cooksey in December, in the technical language of administrators. "I guess the new magnet is too damn heavy."[32] He got it moved first to the Pelton Waterwheel Company in San Francisco to rebuild the poles (plate 3.2).[33] He saw to the renovation of the new laboratory and to the eviction of most of its tenants. The mapping division of the Forest Service and French phonetics remained to soak up radiation (fig. 3.1).[34] And he laid industry under contribution. Federal Telegraph gave 650 gallons of transformer oil (value $227.50); American Smelting and Refining Company lent lead for shielding; and Federal supplied old, gassy, reject 20-kW oscillator tubes.[35]

Tuve, 27 Aug 1931 (MAT, 8). The Laboratory received its name from the regents on 12 Jan 1932, which Birge, *History, 4*, xi, 22, takes as its birthday.

31. Birge, *History, 3*, app. xvi: A; infra, §5.1.

32. Lawrence to Cooksey, 23 Sep 1931 (4/19).

33. Lawrence to Poillon, 29 Sep 1932, 9 Jan 1933 (15/16); to Joliot, 20 Aug 1932 (10/4).

34. Lawrence to Sproul, 10 Aug 1936 (25/3); Seaborg, *Jl., 1*, 12.

35. Lawrence to Leuschner, 4 Apr 1932 (46/23R); to Federal Telegraph, 8 Sep 1932 (7/10); to Poillon, 29 Sep 1932 (15/16); to B.C. Tuthill (Federal), 2 Feb 1933 (25/1); to Livingston, 22 Oct 1935 (12/12).

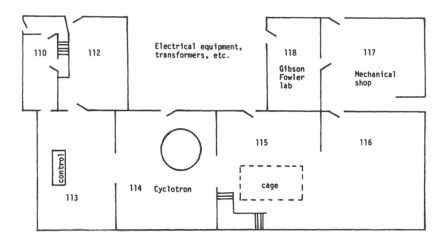

FIG. 3.1 Layout of the ground floor of the old Radiation Laboratory. Room 118 belonged to Chemistry. Seaborg, *J1, 1,* 12.

Lawrence became very adept at scrounging. He always needed power tubes, power transformers, and just plain power. In 1932 he wanted tubes for an x-ray machine. Federal sold them at $330 a piece, discounted. Lawrence offered $225 and the thought that if the Laboratory made a successful high-voltage x-ray plant with them, Federal's fortune would be made too. He tried the same ploy with the Deforest Radio Company: "The engineering development of our method will lead to a considerable oscillator tube business. In view of this you may feel disposed to furnish us tubes for our experimental development at considerable discount." Deforest allowed 10 percent; Federal agreed to accept $225, and charge the rest to charity; Lawrence stayed with Federal.[36]

Transformers provided an unlikely subject of comedy. GE loaned three 25-kW, and PG&E three 75-kW transformers, total value $2,000, in the fall of 1931, for three months or so. That was not to know Sproul or Poillon, who badgered the companies into extending the loans and then selling at a giveaway discount.[37]

36. Lawrence to L.E. Replogle, Deforest Radio Company, 8 Sep 1932, and reply (6/15); to Federal, 7 Jan and 2 Feb 1933, and reply (25/1).

37. Lawrence to Leuschner, 18 Sep 1931 (25/1); to GE, draft request by Fuller (25/2); to Cottrell, 21 Nov 1931 (15/3); GE to Lawrence, 22 Sep 1931 and 5 June 1933 (7/28); Leuschner to Poillon, 16 June 1933, and reply, 7 July 1933 (25/1, 15/16A); Lawrence to Poillon, 3 June and 16 Sep 1933 (15/16A); Sproul to GE, 27 Sep 1933, and reply, 22 Dec 1933 (7/8).

On power, however, PG&E would not budge. In reply to Leuschner's pleas, the Company's president observed that PG&E was the second largest taxpayer in the state and saw no reason to abate its charges to an institution it already supported beyond its desires. "It seems to me the cost of experiments coming in the category of 'pure science' ought to come out of the funds of the University." For years Sproul had to top off Lawrence's power bill from his emergency fund.[38]

With gifts in kind and discounts Lawrence stayed within his first-year budget, that for 1931/32, even though he decided to rebuild the magnet more extensively than he had expected. Federal's gift was asymmetric; to provide a sufficiently uniform field for the cyclotron, however, it needed symmetrical poles on either side of the gap in which the vacuum tank would sit (plate 3.3). The Laboratory accordingly procured the answering pole from the remaining derelict. By November 1931, with almost everything bought or ordered, given and discounted, Lawrence had $600 left from the $7,500 from the Research Corporation and the Chemical Foundation. That, he thought, would get him through the rest of the academic year 1931/32, on the big cyclotron project at least.[39] But he had other things in mind as well.

2. X RAYS THE BERKELEY WAY

Sloan's Tube

While Lawrence was moving into his new domain, an old friend, Joseph Boyce from Chicago, came to see what sort of physics went on in California. He reported his findings to Cockcroft. The work of C.C. Lauritsen and C.D. Anderson at Caltech, he said, was something in its way; "but the place on the Coast where things are really going on is Berkeley." Lawrence had in mind or hand no fewer than six different machines for throwing atomic projectiles. "On paper this sounds like a wild damn fool program,

38. A.F. Hockenblamer to Leuschner, 18 Sep 1931 (25/1).
39. Lawrence to Poillon, 27 Nov 1931 (46/25R); to Cottrell, 7 Dec 1931 (5/3). Cf. Lawrence to J.E. Henderson, 14 Apr 1933 (9/5), giving $7,000 for the cost of refurbishing and setting up the magnet.

but Lawrence is a very able director, has many graduate students, adequate financial backing, and in his work so far with protons and mercury ions has achieved sufficient success to justify great confidence in his future." Boyce itemized Lawrence's armamentarium: the second and third cyclotrons, the mercury-ion linac, a larger linac for protons, a Van de Graaff, and what Boyce misidentified as a Tesla coil. It is with this last apparatus, which Ralph Fowler of Cambridge thought the most interesting new apparatus at Berkeley, that we are now concerned.[40]

After the 30-stage mercury linac opened fire in the fall of 1931, Lawrence put Sloan to work on a resonant transformer as an alternative to the cyclotron for producing fast protons. The general idea of this apparatus appears from figure 3.2, which represents Sloan's final design. The secondary coil supports the water-cooled secondary tube. When the tube goes strongly negative, protons rush into it from the grounded ion source; they emerge half a cycle later, when the now positive tube drives them into the bombarding chamber. The protons in effect fall twice through the high potential of the secondary (800 kV in this design). When the device was designed, Livingston had not yet reached 1 MeV; Sloan's scheme promised to do so and, what was extremely important, to give a beam far more intense than could be expected from the cyclotron.[41]

Two features of the design in figure 3.2 distinguish it sharply from the Tesla coil used by Tuve's group. First, the secondary is part of a single oscillating circuit rather than a separately tuned one. The advantage of the design is efficiency, bought at the expense of high-power radio engineering, in which Sloan had the advice of Fuller. The second distinctive feature is that the heavy copper secondary coil is in effect an antenna placed inside the evacuated acceleration chamber and supported at a voltage node from the copper roof of the internal tank wall: the all-metallic connection obviated the need for insulation, the failure of which

40. Boyce to Cockcroft, 8 Jan 1932 (CKFT, 20/4); Fowler to Rutherford, 15 Jan [1933] (ER).
41. Sloan and Lawrence, PR, 38 (1931), 2022, in Livingston, Development, 152; Sloan, PR, 47 (1935), 71; Lawrence to Cottrell, 7 Dec 1931 (5/3).

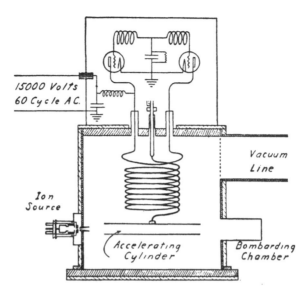

FIG. 3.2 Sloan's resonance transformer for doubling the energy of positive ions. Sloan, *PR, 47* (1935), 67.

haunted the usual methods of producing and holding high voltages.[42]

In November, having planned a tube that incorporated the entire oscillating system in the vacuum to eliminate corona discharge, Sloan turned to something that would make a better Christmas present for Lawrence and Cottrell. This was "a helluva x-ray outfit," with an intense beam at 100 kV, which, in Lawrence's opinion, could easily be hardened tenfold. "I feel quite sure [he wrote Cottrell in December] that ere long we will be producing million volt x rays."[43] Lawrence was right in fact but wrong in timing. It took two years of hard work to develop the generator of figure 3.3 into a plant that could be operated continuously at 800 kV.

Sloan's invention opened a new set of technical challenges and financial opportunities for the Laboratory. It promised to give the

42. Sloan, *PR, 47* (1935), 62–6. The oscillator worked at 6 MHz.
43. Lawrence to Lauritsen, 10 Nov 1931 (10/36), to Cooksey, 1 Dec 1931 (4/19), quote, and to Cottrell, 7 Dec 1931 (5/3).

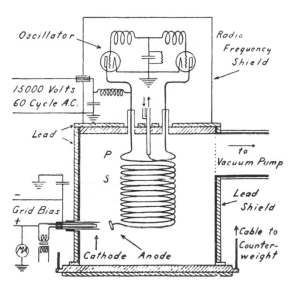

FIG. 3.3   Sloan's x-ray generator. The high-energy x rays arise at the anode attached to the secondary when it stops electrons from the cathode. Sloan, *PR, 47* (1935), 65.

same sort of beam that Lauritsen obtained from his huge installation, but at a fraction of the cost. It gave a beam with over three times the energy of the output of the largest x-ray plants in use in hospitals with about the same demands on space and power (about 200 kW). And it marked the beginnings of interdisciplinary work at the Laboratory. To proceed, however, it was necessary to raise the amounts necessary to harden Sloan's hundred-thousand volt rays to Lawrence's canonical million.

Lawrence turned to the Research Corporation with the tactics it had taught him. "I know that the General Electric Company would be only too glad to get behind the project because it has immediate commercial possibility, but, of course, I hope this can be avoided." He thought $500 or $1,000 would bring the matter well forward.[44] Sloan's personal needs were covered by his fellowship, which GE renewed. Lawrence longed to proceed; Sloan had no desire to study for his exams; the work stalled for want of a

44. Lawrence to Cottrell, 7 Dec 1931 (5/3), quote; to W.W. Buffum, 21 Jan 1932 (46/15R).

$400 pump and a little something for contingencies. Vacation was coming, freeing Sloan from his already minimal studies; perhaps, at the end of the summer, he would have a commercially viable machine. And perhaps not. If not, Lawrence held out, Sloan might be able to generate million-volt x rays another way, by bombarding light elements with the intense ion beam from the x-ray machine when adapted to accelerate protons. "This is a possibility that may turn out to be much more important than the production of x rays by electrons....The medical applications of these latter considerations are certainly of considerable importance." Poillon returned a check for $500.[45]

During July 1932 Sloan succeeded in producing plenty of hard x rays without the intervention of protons, x rays powerful enough to penetrate a centimeter of lead or half an inch of steel. Birge, now department chairman, esteemed Sloan's tube "the most important of the discoveries of the radiation laboratory," not excluding the cyclotron; he recognized it as "mainly a commercial proposition;" and he worried that it would result in a "patent war or something with G.E."[46] The University announced Sloan's success to the national press and a mixed pilgrimage of humanitarians and promoters trekked to Berkeley. GE's San Francisco representative came right over and declared that the tube had "very great commercial value, not only for medical work but for the examination of steel welds." The home office quickly confirmed his interest.[47] Meanwhile the chief engineer of the Kelley-Koett Company, a major constructor of x-ray apparatus in the Midwest, put in an appearance. He was followed by a representative of Westinghouse, then entering the x-ray business in

45. Lawrence to Coffin Foundation, 25 Feb 1930 and 24 Feb 1931, and to dean, Graduate Division, 16 Nov 1933 (16/27); Lawrence to Poillon, 12 Apr 1932, and to Buffum, 15 Apr 1932 (quote), and Poillon to Leuschner, 20 Apr 1932 (15/16). Cf. Lawrence and Livingston, *PR, 40* (1932), 20, in Livingston, *Development*, 119 (Feb 1932): "The development of an intense artificial source of gamma radiation...would be of considerable value for nuclear studies."

46. Birge to H. Sponer, 8 Oct 1932, and to Jenkins, 18 Sep 1932 (Birge P, 33).

47. Lawrence to Poillon, 4 Aug 1932 (15/16); "Safer x rays," *New York Herald Tribune*, 2 Aug 1932, 10 (7/11); Coolidge to Lawrence, 24 Aug 1932 (7/28). Coolidge did not understand at first how much Sloan's design differed from his spark coil and from Tesla's coil; Lawrence to Coolidge, 6 Oct and 4 Nov 1932 (7/28).

search of a market for its power tubes, by the officer in charge of x-ray work in army hospitals, and by the roentgenologist at the University hospital in San Francisco, Robert Stone, who crossed the Bay, saw Sloan's rays burn through steel, and desired to use them on his patients.[48]

Sloan, J.J. Livingood (a postdoc from Princeton "very anxious to get into [the] nuclear racket," for whose services Lawrence paid nothing), and others improved the apparatus of figure 3.3 until it gave out rays of perhaps 700 kV. Then Lawrence, like the ingénue in *The Importance of Being Earnest*, outdid his wooers; he advised Poillon to rush commercialization by engineering the tubes in the Research Corporation's laboratories even while Sloan and his unpaid helpers continued development at Berkeley. "Not only do I feel that the method is superior for the production of radiations above a half million volts, but also I am inclined to think that very inexpensive outfits can be manufactured for the production of radiation in the region of 200 or 300 kilovolts. I am told that there is a very big market for such deep therapy outfits."[49]

Hospitalization

Two major research hospitals decided to try what Sloan's tube could do. The first commission came from Francis Carter Wood, the director of the Institute of Cancer Research of Columbia University, who had encouraged Lauritsen's work and knew of Berkeley's alternative through Buffum. The Chemical Foundation dispatched its physicist, Frank M. Exner, to help Sloan build an improved machine, for which Wood's Institute would pay $5,725 from a fund given by Charles Crocker, the son of one of California's railroad barons. How improved? Lawrence told Exner he wanted to push to a million volts; Exner told Wood, and Wood objected. His institute had long experience treating cancer with the gamma rays of radium, at 2 MeV, and they had not "done any miracles." Lawrence thought about it from the "physical point of view" (and also as an entrepreneur with a balky

48. Lawrence to Poillon, 4 Aug 1932 (15/16), and to Buffum, 25 Aug 1932 (3/38); "Harold" to Tuve, 5 Jan 1931 (MAT, 3), re Westinghouse.

49. Lawrence to Poillon, 4 Aug 1932 (15/16) and 18 Aug 1932 (RC); Boyce to Lawrence, 27 Jan, and reply, 9 Feb 1932 (12/11), re Livingood.

client) and agreed. "There is not much point in x rays above a half a million volts for therapy purposes."[50] But the Sloan tube was to be developed for the art, as well as for medicine and patents, and also to support life in the Laboratory. Livingston, having graduated, needed a job to sustain him for his essential work on the cyclotron. Lawrence looked everywhere and found nothing. The obvious solution to all difficulties would be to employ Livingston part-time on x-ray money. The goal would stay a million volts.

Everything fell into place with the commissioning of a million-volt plant by the University's Medical School. The initiative came from Stone and the money—some $12,000 for plant and installation—from William H. Crocker, brother to Columbia's patron and a regent of the University. Livingston became to Stone's machine what Exner was to Wood's. Therefore he was expected—so flimsy were the professional qualifications of radiologists then—to run it as well as to make it. His relevant training consisted of two weeks with Lauritsen to learn technique. He then set dosages for the human guinea pigs at the University hospital.[51]

The estimated costs of the twin Crocker machines included no profit, overhead, or salaries. Their 20-kW Federal power oscillators cost $300 each, after a 20 percent discount, and each machine needed four, or, in the new design, six. Economy was necessary. Sloan salvaged a ton of lead plates from old storage batteries, melted them down, and remolded them into shielding blocks. Safety demanded an additional 6,000 pounds of lead, which Lawrence begged as an additional loan from the American Smelting and Refining Company. Efficiency required an instrument maker, whose salary had to be found. To begin development of the hospital machines, Lawrence needed $4,000 immediately, or so he told the Chemical Foundation. His reckoning omitted the salaries of Sloan and Livingood.[52]

50. Wood to Lawrence, 29 Aug 1932, and Lawrence to Wood, 3 Sep 1932 (9/21); Lawrence to Exner, 19 Aug 1932 (9/21); infra, §8.3.

51. Lawrence to Buffum, 25 Aug 1932 (3/38), to Stone, 20 Sep 1932 (16/28), to Henderson, 7 Jan 1933 (9/6), and to Cooksey, 1 Jan 1934 (4/19); W.H. Crocker to Sproul, 11 Jan 1933 (UCPF, 337/366).

52. Lawrence to Federal Telegraph Company, 8 Sep 1932, and reply, 15 Sep 1932 (7/10); Lawrence to Buffum, 25 Aug 1932 (3/38); to Wood, 3 Sep 1932 and 10 Jan 1933 (9/21); to Boyce, re Livingood, 1 Oct 1932 (3/8).

By December 1932 the prototype hospital machine was working beautifully. It could perform briefly at over a million volts, and at 750 kV gave out 14 milliamps so steadily as to melt its water-cooled copper target. It far exceeded their fondest hopes, Lawrence told Wood, being much superior to Coolidge's cascaded tubes or Lauritsen's set of gas-pump cylinders, and, moreover, compact and inexpensive, like the first cyclotron. It was a light that could not be hidden, the equivalent of half the world's supply of radium. "Though we have most of the tube plastered with lead (7,000 pounds in all), we see x rays almost everywhere in the lab with a fluoroscope." "There is no question now [December 1932] but that we have the most powerful x ray tube in the world and that our outfit is incomparably superior to any other, and promises to revolutionize x ray technology. I say this unqualifiedly."[53]

This last disclosure was directed to Poillon, who was trying to patent Sloan's invention when he received it. Poillon had not looked upon the work at Berkeley with unmixed satisfaction. It was the old problem with academics: they innovated, jerry-rigged, experimented, always improving and refining, indifferent to the business side of things. "I have told Sloan and Lawrence [Poillon complained to his patent lawyer] that we wanted this device brought to a certain state of perfection at Berkeley, where it could be engineered for production. They seem to be very anxious to get somebody else to do that secondary step.... This of course does not suit our book and I have warned them about directing the attention of others to it lest the others start the development and the patent situation becomes very cloudy." Poillon tried to silence the publicity that the commissioning of the hospital plants had generated.[54]

Poillon's impatience did not deflect Lawrence. Having done the physics, if not the engineering, of million-volt x rays, the Laboratory proposed to replace the heart of the machine, the expensive commercial oscillator, with cheaper devices of its own manufacture. That required patience from Stone and Exner and

53. Lawrence, "A brief summary of the work of the Radiation Laboratory," attached to Lawrence to Poillon, 5 Jan 1933 (15/16A); quotes from, resp., Lawrence to Cooksey, 11 Dec 1932 (4/19), and to Poillon, 13 Dec 1932 (15/16).

54. Poillon to A.P. Knight, 11 Oct 1932 (RC); Exner to Lawrence, 7 and 8 Feb 1932 (9/21).

money from Poillon and Buffum. Lawrence asked for $500 to $1,000. Poillon, hard-pressed and unenthusiastic, returned $700. "Money as you know is extremely scarce."[55] Fuller took over the direction of the project and assigned an engineering student to help design a practical 200-kW oscillator. A satisfactory design emerged, in which the oscillator shared the vacuum of the discharge tube; Lawrence stopped development of instrumentation for the cyclotron and threw his disposable resources into making and improving the new tube.[56] By June 1933, having spent $550 of the $700, Fuller, Sloan, and company had an oscillator more powerful than Federal's; by September they had licked the remaining vacuum problems; by December Stone's machine was ready for installation, and Sloan, exhausted, was ready for the hospital.[57]

Livingston and Sloan set up the San Francisco machine (plate 3.4). It performed beautifully, at a lower cost for power than expected. Crocker's $12,000 bought a building ($5,000), the machine ($4,000), Livingston's salary ($1,000), and accessories ($2,000). Running at 800 kV and 10 milliamps, it outdid all other x radiators, including Lauritsen's.[58] The business was, in fact, a great success. It showed how a subsidized university, with clever men, out-of-work or underpaid postdocs, and, if it wished, no overhead charge, could outdo industry. As Lawrence pointed out, GE had recently built an x-ray unit for a Chicago hospital that cost $65,000, faltered at 600 kV, and took much longer to install than Sloan's machine. GE's standard tubes cost $25,000, Kelley-Koett's 600 kV tubes $29,000. The Crocker brothers got a great bargain.[59]

55. Lawrence to Stone, 17 Jan 1933 (16/4); to Poillon, 10 June 1933, and Poillon to Lawrence, 11 Feb 1933 (15/16A).

56. Lawrence to Poillon, 18 Feb and 14 Apr 1933 (15/16A); to Sproul, 24 Jan 1933 (16/42).

57. Lawrence to Poillon, 14 Apr, 3 June, 16 Sep, 4 Dec 1933 (15/16A); to Cooksey, 3 May 1933 (4/19); to Exner, 12 Sep 1933 (9/21); to dean, Graduate Division, 16 Nov 1933 (16/27).

58. Lawrence to Fowler, 28 Dec 1933 (7/2); to Exner, 17 Nov 1933 and 18 Jan 1934 (9/21). Sloan resigned his University fellowship in January 1933 to work on the installation at the Medical School; Birge to M. Deutsch, 7 Jan 1933 (Birge P, 33).

59. Lawrence, "The Sloan x ray machine" (16/28).

Exner began to set up the Columbia machine after his return to New York in late spring of 1933. He eventually had as his helper Wesley Coates, who had completed his doctorate under Lawrence in 1933 with work on the soft x rays he found in targets struck by fast heavy ions from the old mercury linac. Coates went to Columbia to build a machine on Sloan's principles for the acceleration of protons. He then became the physicist at the Crocker Research Laboratory of Columbia Presbyterian Medical Center, where the Sloan x-ray machine had been placed. One day in 1937 the machine declined to hold its rated voltage. Coates peered in to diagnose the trouble, brushed a high-potential line, and fell a martyr to high energy.[60]

The Laboratory continued its efforts to improve Sloan's tubes well into 1935. Wood decided that experimentation with rays comparable in penetrating power with radium's gammas might be valuable after all and made available ample funds for building oscillators large enough to drive the tank at the highest voltage it could take. He paid salaries to Sloan and the instrument maker, E.W. Lehmann, and various shop costs, amounting to upward of $1,700.[61] During this work, Dr. Walter Alvarez of the Mayo Clinic came to town. Lawrence took him to the University's Medical School. "He was exceedingly enthusiastic about it," Lawrence wrote Poillon, "and wants to start negotiations." Lawrence urged the Research Corporation to greater activity: "There is absolutely no question but that you should push the commercial development as rapidly as possible."[62] As symbol and agent of this alliance, Lawrence's student Harry White went to work for the Research Corporation after learning to run the Sloan machine in San Francisco in order to promote it "and possibly...other by-products of the Radiation Laboratory work in the future."[63] Several pests then attacked Sloan's x-ray plants. For one,

60. Lawrence to B. Davis, 5 May 1933, and Exner to Lawrence, 23 Mar 1937 (4/4); Coates, PR, 46 (1934), 542–8; Lawrence to Poillon, 14 Apr 1933 (15/16A).

61. Exner to Lawrence, 14 Mar and 3 May 1934, and Lawrence to Exner, 21 Apr 1934 (9/21); Exner to Lawrence, 8 Apr 1934 (25/1); Lawrence to Poillon, 30 Oct 1934 (15/16A).

62. Quotes from, resp., Lawrence to Poillon, 6 Jul and 7 Nov 1934 (15/16A). Another interested party was Allibone; A.P.M. Fleming to Lawrence, 19 Feb 1935 (13/3).

63. Lawrence to Poillon, 5 and 22 Jul 1935 (15/17).

Berkeley's homemade oscillator appeared to infringe RCA patents; the existing plants switched to commercial tubes, which would have driven the cost of future installations to around $25,000.[64] For another, Sloan suffered a back injury that made him a semi-invalid from 1935 to 1937 and unable to carry development further.[65] And another of the Research Corporation's investments, the Van de Graaff generator, showed itself capable of producing million-volt x rays more economically than the Sloan tube.[66]

As Birge had said, the Laboratory's development of high-voltage x radiators was largely a commercial venture. It was not the sort of thing that won prizes in physics. After crediting the Research Corporation and the Chemical Foundation with the initiative for developing Sloan's ideas into an effective generator of x rays, Lawrence wrote in his report to them for 1932: "From the point of view of physicists the most interesting aspect of the Sloan apparatus is its tremendous effectiveness in producing intense beams of two [!] million volt protons."[67] In his correspondence with physicists, Frédéric Joliot, for example, Lawrence emphasized not the x rays but the protons; and even when, in justifiable pride, he mentioned the fearsome power of Sloan's first machine, he specified the primary purpose of the second as the creation of milliamps of million-volt protons.[68] It was the first task Sloan took up after recovering from overwork. By February 1934 he had observed his protons, but not in the numbers desired. It took well into the year before this, the "primary purpose" of his work, was realized.[69]

64. Lawrence to Poillon, 15 Mar and 30 Oct 1930, and Poillon to Lawrence, 23 Oct and 15 Nov 1934 (9/21); Exner to Lawrence, 26 Jan 1935 (9/21); Sloan, interview, 6 Apr 1976 (TBL); Lawrence to F.C. Blake, 19 Nov 1936 (16/28), estimating the cost of a Sloan apparatus.

65. Lawrence to Exner, 31 Jan 1935 and 27 Jan 1937; to Wood, 28 Aug 1935 and 27 Oct 1936 (all in 9/21). Sloan did not recover entirely until 1939; Lawrence to Poillon, 11 Jul 1939 (15/18).

66. Stone and Aebersold, *Radiology, 29* (1937), 297–304; Lawrence to F.C. Blake, 19 Nov 1936 (16/28); J.G. Trump to Poillon, 7 Oct 1936 (15/17), re a 1.2-MV Van de Graaff x-ray machine at MIT.

67. Lawrence, "A brief summary," attached to Lawrence to Poillon, 5 Jan 1933 (15/16A).

68. Lawrence to Joliot, 20 Aug 1932 (10/4); to H.A. Barton, 8 Sep 1932 (2/25); to J.A. Fleming, 10 Dec 1932 (3/32).

69. Lawrence to Exner, 1 Feb 1934 (9/21); Sproul, report for 1932, quoted in Birge, *History, 4*, 22.

3. THE RADIO AND THE CYCLOTRON

Going on the Air

The distinctive feature of the Radiation Laboratory's approach to particle accelerators was reliance on radio technology. Many of the Laboratory's earliest workers, including Lawrence, had been radio hams in their youth; they continued the sport on a grand scale by a ham link between the Laboratory and one of its earliest satellites at the University of Michigan.[70] The heart of Livingston's first cyclotron was a pair of off-the-shelf Radiotrons, air-cooled, rated at 75 watts each and arranged in a standard radio circuit, with the single dee in place of an antenna. The analogy is not idle. The cyclotron could interfere with commercial broadcasting or police communications; the Laboratory listened to itself on the radio so as to correct spillage into others' air waves and "prevent investigations by the Radio Commission which may result in our having to put in elaborate and expensive protection devices." It is said that Lawrence used to tune his home radio to the cyclotron's operating frequency to monitor its, and his students', performance.[71] The second cyclotron required a 20-kW water-cooled oscillator in another standard radio circuit; the third cyclotron used two 20-kW tubes, which also drove Sloan's x-ray machine.[72] As we know, their cost provoked the Laboratory to make their own and to enter still more deeply into the art of radio engineering. This move so far from ordinary physics was a surprise even to people familiar with Lawrence's methods.[73] The Laboratory was so filled with radio waves that its members could light a standard electric bulb merely by touching it to any metallic surface in the building.[74] Many cyclotron laboratories were to eke out their resources by cannibalizing old radio parts.[75]

70. Cork to Lawrence, 21 Aug 1936, and Lawrence to Cork, 26 Aug 1936 and 9 Feb 1937 (5/1).

71. Lawrence, memo to Purchasing Dept., 31 Jul 1935; Childs, *Genius*, 251.

72. Terman, *Radio-engineering*, 228, 253, 258–9; Livingston, *Production*, 13–4, and fig. 3; Lawrence and Livingston, *PR*, *40* (1932), 28, in Livingston, *Development*, 127; Lawrence, "Workbook," 139 (39/4).

73. Cooksey to Lawrence, 6 May 1933 (4/19).

74. Alvarez, *Adventures*, 42.

75. E.g., disused tubes from a radio station, at Rochester, Dubridge to Lawrence, 20 Sep 1935 (15/26); an old radio transmitter from the navy, at Har-

The maximum energy given particles by a cyclotron depends not on the power but on the frequency of its oscillator. From the fundamental equation 2.1,

$$\text{Maximum energy} = (mv^2/2)_{max} = 2\pi^2 mf^2 r_{max}^2 = 1.23\cdot 10^{13} mf^2 R^2 \text{ eV,}$$

where R is the radial distance to the collecting cup. Livingston achieved million-volt protons with the second cyclotron with the cup at 11.5 cm and the oscillator at 20 MHz. This last number represented close to the practical limit on frequency: to have gone higher would have required, first, a more intense magnet (since the frequency is proportional to the field strength) and, second, a tube capable of changing polarity over twenty million times a second and delivering around twenty thousand watts of power. The first requirement would have pushed the art of magnet design, the second that of power oscillators, to or beyond the edge of available technology. The only immediate way to increase the energy of the protons by an order of magnitude was to increase R *threefold.* Hence the great value to the Berkeley cyclotroneers of Federal's derelict magnet, which could be rebuilt to give a field of appropriate intensity between pole pieces 27.5 inches (70 cm) in diameter.

The Federal magnet, like the oscillator tubes, was a product of radio technology, the Poulsen generator, which had at its heart a periodic arc between electrodes in hydrogen. When the arc struck, it carried the oscillatory discharge from a large condenser, which fed an antenna; when the discharge current diminished sensibly, the arc went out and a battery recharged the condenser, which, when full, relit the arc (fig. 3.4). The magnet assisted the extinction of the arc: it made the ions carrying the current run in a curved path from one electrode to another; they therefore could not create by collisions any fresh ions along the straight line between the electrodes; and consequently not enough carriers were available there to continue or restart the arc when the potential across the gap at A (fig. 3.4) no longer sufficed to ionize the gas

Lawrence, 20 Sep 1935 (15/26); an old radio transmitter from the navy, at Harvard, Hickman to J.R. Wier (GE), 1 May 1937 (UAV, 691/60/3); out-of-date radio tubes for ionization gauges at Berkeley and elsewhere, Lawrence to H.W. Edwards, 18 Jan 1934 (7/2).

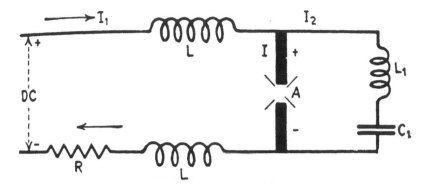

FIG. 3.4   Schematic diagram of the electrical arrangement in the Poulsen arc. I_1, the charging current; I, the rf current through the resonant circuit when the arc strikes. The choke coils L keep the rf current from the battery circuit. Heilbron, *Museoscienza, 22* (1983), 16.

near the electrodes. In a word, the field insured that the arc went out without flickering (plate 3.5). The objective was to produce an undamped signal of almost constant frequency rather than the broadband output characteristic of the spark transmitters of early wireless.[76]

We are familiar with Federal's growth under navy contracts, and with the culmination of their relations in the commissioning of four 1,000-kW generators, two for each end of a radio link between the United States and its expeditionary forces in France. Their magnets could deliver 18,000 gauss.[77] The war ended before the huge antenna towers—second only in height to the Eiffel tower—could be completed in France. The navy withdrew, leaving France with half a radio station and Federal with four 80-ton magnets. In 1919 the French government decided to proceed with the Lafayette station, as they called it in memory of the American alliance, in order to communicate with its empire in Southeast Asia. Its signal received in San Francisco was four to eight times stronger than the signals from other major European transmitters. That did not recommend completion of the American end of the

76. Heilbron, *Museoscienza, 22* (1983), 13–24.
77. Howeth, *Communications*, 241, 243, 246, 253; Anon., *Radio review, 2* (1921), 85–90.

link to the navy, however, which now favored development of vacuum-tube oscillators. That left Federal with two war-surplus magnets. Everyone switched to tubes except the Dutch, who constructed a 3,600-kW Poulsen generator to drive an antenna stretched between two hills in Java, which gave the Dutch East Indies a voice audible in Amsterdam.[78]

Lawrence's big magnet, being a piece of high technology, required professional engineering help in its metamorphosis into a tool of science. Fuller advised about renovating the pole pieces and about windings and power supplies; and he procured the assistance of an employee at Federal, Gilbert W. Cattell, who inquired into all sorts of details: the best sort of paper insulator for the windings, of oil for cooling, of cables for connecting, and so on.[79] Cattell wound and insulated the coils at Federal before the magnet went to Pelton for machining (plate 3.6). Toward the beginning of February 1932, the foreman at Pelton's machine shop, Henry Nelson, carted its handiwork across the Bay and erected it in the Radiation Laboratory. Pelton did an excellent job, the pole faces parallel to four-thousandths of an inch, the field homogeneous up to 18 kG.[80] Lawrence hoped to have the cyclotron itself in operation at the end of February and shortly thereafter to pass "the next milestone," protons with energies above 3 MeV.[81] In March he worked "night and day" with Livingston and a graduate student, James Brady, on the machine. "I have neglected everything else—even my fiancée has suffered."[82] Molly's suffering did not make the machine go. In April they thought that they had cured the leaks that the huge magnetic forces kept springing in the cyclotron tank. Still no results. Lawrence went East in May, to Molly and marriage. Livingston

78. Ibid., 93, 579; Howeth, *Communications*, 136, 208–9; Hooper, *Electr.*, *18* (1921), 1112–3.

79. Correspondence between Lawrence and Cattell, Nov and Dec 1931 (25/2); Fuller, interview, 72, 142–5 (TBL).

80. Pelton to Lawrence, 14 Sep 1931, 29 and 30 Oct 1931, and Lawrence to Pelton, 2 Feb 1932 (25/2); Lawrence and Livingston, *PR, 45* (1934), 608.

81. Lawrence to Poillon, 9 Jan 1932 (15/16), to Tuve, 13 Jan 1932 (17/34), to Livingood, 18 Feb 1932 (12/11), and to Buffum, 21 Jan 1932 (46/15R); cf. Lawrence to Tuve, 10 Nov 1931 (17/34), expecting to have the machine going before Christmas.

82. Lawrence to Haupt, 11 Mar 1932, and reply, 15 Mar 1932 (9/2).

labored on; the chairman of the Physics Department, Elmer Hall, alarmed at his appearance, counselled a long rest.[83]

Vacation cured what slavery had not. By mid September Livingston had overcome his various difficulties and produced hydrogen-molecule ions at 1.6 MeV. "When he gets up in the 3,000,000 volt range [Lawrence wrote Cottrell] he intends to stop and bombard various elements with them before going to higher energies." A week later he had reached 3.6 MeV at about one μA, always with H_2^+, since the lighter protons would have required an oscillator with impractically high frequency. Lawrence and Livingston thought they saw their way clear to 5 or 6 MeV, but they decided to "hesitate now awhile on the road to higher voltages [to] do some experiments."[84]

Stopping for physics was perhaps a pleasure. It was certainly a financial necessity. Lawrence had started planning for a larger apparatus long before Livingston had got a beam. He did not plan to carry off the Poulsen arc from Java, but to enlarge the pole pieces and vacuum chamber of the new cyclotron from 27.5 to 37.5 inches. He floated this bubble in April 1932. Poillon indicated that he would incline toward granting the $1,100 needed when asked for it. But Lawrence had overreached. Poillon had the same month pledged $2,500 to the Laboratory to enable it to make much needed detectors, a cloud chamber with cinema camera ($1,500) and a magnet to analyze particle beams ($1,000). Somehow Lawrence thought that the Research Corporation and the Chemical Foundation would provide another $800 for instruments and supplies; but when he asked for it in August, he was told that the University should pay for such things. Lawrence already had spent almost the entire allocation for expenses and power for AY 1932/33. It was only August! He would have to close down the Laboratory, he said, unless his hard-fisted backers allowed him to pay for his electricity from the $1,000 granted for

83. Lawrence to Poillon, 12 Apr 1932 (3/28, 5/16); Hall to Lawrence, 16 June 1932 (8/8).

84. Lawrence to Cottrell, 22 Sep 1932 (5/3), and to Cooksey, 29 Sep 1932 (4/9); Livingston, *PR, 42* (1933), 441–2 (3 Oct 1932). Parameters: R = 10 inch (25.4 cm), H = 15,250 gauss, f = 11.6 MHz. Cf. Lawrence to Barton, 29 Sep 1932 (2/25), and to Boyce, 1 Oct 1932 (3/8).

the analyzing magnet. As an inducement to Poillon, Lawrence dropped—only temporarily—the unfunded proposal to enlarge the pole pieces and vacuum chamber. Poillon allowed the redirection of funds.[85]

Broadcasting upon the Waters

The creation of the 27-inch cyclotron called for an unusual blend of faith, energy, and entrepreneurism. How unusual the combination was may be gathered from the reluctance of other physicists to follow Lawrence's lead. The nuclear physicists in and around the Cavendish at first considered the cyclotron to give too small a current to be useful.[86] They next depreciated it as "ticklish to adjust," a view common among English physicists as late as 1935, although Cockcroft had seen for himself, in 1933, that "all the trick lies in correcting for inhomogeneities of field around the gap by inserting 'shims' or pieces of sheet iron." He had witnessed the operation; the trickster was Livingston, "who does most of the work."[87] A closer observer saw the same thing. "Lawrence does no actual experimental work anymore," Birge wrote another member of the Department. "It keeps him busy just bossing all the men working with him!"[88]

The first Englishman to see the 27-inch run, Ralph Fowler, came to Berkeley in January 1933, when the machine was producing 2.5 MeV hydrogen molecule ions. Neither the machine nor its products interested Fowler. "Probably only trivial stuff," he wrote Rutherford. Six months later Cockcroft did not see that the cyclotron had opened up any important areas of investigation not accessible to the Cockcroft-Walton accelerator. "We can get in long before California in this field [he wrote his co-inventor] and there are a lot of points to be cleared up [about nuclear

85. Poillon to Leuschner, 6 Apr 1932, and to Lawrence, 30 Aug and 8 Sep 1932; Lawrence to Poillon, 26 Aug and 5 Sep 1932 (46/25R, 15/16). The cost of enlargement was projected at over $2,000 when Lawrence next mentioned it to Poillon, in a letter of 23 Jan 1934.

86. Cockcroft to Gamow, 29 Sep 1932 (CKFT, 20/10), and to Lawrence, 17 Sep 1932 (5/4); cf. McKerrud (Metropolitan-Vickers) to Cockcroft, 27 Feb 1932 (CKFT, 20/60).

87. Cockcroft to Walton and Dee, [24 June 1933], "ticklish," and to Rutherford, 22 Jul 1933, "trick" (ER); Lawrence? to Chadwick, 31 Jan 1936 (3/34).

88. Birge to Jenkins, 18 Sep 1932 (Birge P, 33).

processes]."[89] There is no doubt that the 27-inch was a temperamental machine and that the maintenance of its vacuum occupied and irritated many members of the laboratory. But where visitors saw unreliability, and nonvisitors doubted that cyclotrons worked at all, the natives appreciated difficulties overcome and augured a rich harvest when they turned their machine to physics.[90]

Ever zealous and generous in his cause, Lawrence told Cockcroft that he could easily reproduce the Berkeley machine with a magnet that could be had for the asking. He referred to the Lafayette arcs, then about to be junked in favor of vacuum-tube oscillators and to decommissioned 500-kW arcs procurable from navy surplus. "[They] could be obtained for a song and transport," Cockcroft wrote Rutherford, "if Oliphant [Marcus Oliphant, a prominent nuclear physicist at the Cavendish] shows any enthusiasm in this direction." Oliphant did not show enthusiasm, preferring to go to a million volts in the old, direct, dependable, one-step way. Not until 1936 did the Cavendish decide to create a cyclotron, which it did from scratch, following the plans of Berkeley's then newly designed 37-inch machine. The delay put it two generations of accelerators behind Berkeley at the end of World War II.[91]

While the Cavendish was losing its opportunity, Frédéric Joliot, son-in-law and heir apparent to Mme Curie, inquired of Lawrence what it might cost to build a cyclotron. The expense of the large magnet worried him. Lawrence replied with news about the Lafayette monsters and the necessary modifications of the pole pieces.[92] Joliot applied to the engineer at the station and received all the consideration he could have wished. One of the magnets was being dismounted. "It is only a matter of hauling it to the Ecole Normale Supérieure." The refurbishing could be done in the station's shop, which could also supply 20-kW water-cooled

89. Fowler to Rutherford, 15 Jan 1933, and Cockcroft to Walton [24 June 1933] (ER).

90. Cf. Varney, PT, 35:10 (1982), 27; Rasetti, Viaggi, 3 (1936), 78, complains about the irregular functioning of the 27-inch; Kurie to Cooksey, 4 Mar 1934 (10/21), says that "no one really believes [Lawrence's] cyclotron works."

91. Cockcroft to Walton, [24 June 1933], and to Rutherford, 22 Jul 1933 (ER); Oliphant to ER, 7 Jan 1934 (ER); Crowther, 186, 230; infra, §7.2.

92. Joliot to Lawrence, 14 June 32, and Lawrence to Joliot, 20 Aug 1932 (10/4).

oscillator tubes. The director of the French wireless service authorized the visit.[93] Then Joliot dropped the initiative, because, he later said, he could not get permission to rebuild the magnet. But it is probable that the size of the task, the demand for expertise in radio engineering, and the prospect of dragging an 80-ton magnet through the narrow streets of the Latin Quarter combined to defeat his interest.[94] Like his Cambridge colleagues, Joliot waited until 1936 to begin constructing a cyclotron, and he then required help from Berkeley to correct the many mistakes in its design.

No European center made use of Poulsen arc magnets except the Centre anticancéreux in Marseilles, which made a cyclotron from a small unit decommissioned from the telegraph service in Lyon.[95] The Dutch were the last to consider the option, in 1940, when they asked Lawrence for advice on cannibalizing their Batavian transmitter, which could attain 20 kG.[96] The war put an end to the plan. Cyclotroneering qualities were more easily found in the New World than in the Old. In 1932 New York University inquired into the fate of the 500-kW arcs at Annapolis and reserved one at its decommissioning in June 1934. In 1933 Stanford obtained the mate to Lawrence's magnet from Federal on the understanding that it would be released to the first institution that succeeded in raising the money needed for conversion. In 1934 Cornell tried to obtain one of the decommissioned Annapolis magnets. Columbia, too, was interested; the navy had but one magnet to give, the one that NYU had reserved but had since relinquished; Columbia, which had been frustrated by the navy's sale as scrap of another Poulsen magnet it had coveted, won the prize.[97]

93. Ingénieur en chef, Station radiotélégraphique, Croix d'Hirns, to Joliot, 29 Oct 1932, 4 and 13 Jan 1933; Directeur du service de la TSF to Joliot, 17 Jan 33 (JP, F28).

94. Joliot, preface to Nahmias, *Technique*. Possibly also Joliot could not acquire the pole piece from the second magnet, which he would have needed to make the first symmetrical, and could not raise the money to construct another.

95. Nahmias, *Machines*, 27. Segrè proposed to utilize a magnet from a Poulsen installation in Italy, but it had been dismantled by the time he asked about it (personal communication).

96. J. Clay to Lawrence, 23 Jan 1940 (3/20).

97. Breit to Lawrence, 6 May 1932 (2/15), re NYU; Lawrence to J.E. Henderson, 14 Apr 1933 (9/5), re Stanford; Livingston to Lawrence, 20 Sep 1934 (12/12),

The last cyclotrons built around a Poulsen arc came into existence in 1951. The circumstances were unusual. The Japanese had begun to make cyclotrons as early as the British and the French. At the end of World War II, they had three. The U.S. Army, newly afraid of nuclear physics, threw them all into the ocean. Lacking resources to buy or build a substitute, physicists at the Nishina laboratory in Tokyo scrounged for parts to reconstruct what they had lost. "Fortunately we have a magnet which was originally used for a Poulsen arc generator."[98]

4. IMPORTED PHYSICS

The year 1932 began at the Radiation Laboratory with the 11-inch cyclotron at 1.2 MeV, the 27-inch cyclotron little more than a magnet, and the Sloan x-ray tube a gleam in its inventor's eye. The staff of cyclotroneers had tripled. Besides Livingston there were now James Brady, who had recently completed his doctoral thesis on Lawrence's discontinued topic, the photoelectric effect in alkali vapor, and remained on, on the payroll of his new employer, Washington University of Saint Louis, to learn something about his professor's new lines of work; Milton White, a Berkeley graduate of 1931, who earned his Ph.D. in 1935; and Malcolm Henderson, from Yale, who already had a Ph.D. and, like Livingood, a private income, which allowed him to contribute his services in an unhurried way to the Laboratory. White worked on the 11-inch cyclotron; Brady, at first, on the 27-inch. When the big machine began to operate, Henderson replaced Brady, who then, with White, had the effrontery to check whether Lawrence and Livingston had, in fact, obtained high-speed resonant protons from the smaller machine. They did not find confirmation easy. By the summer of 1932, however, they had measured the penetrating power of the very meager beam and found it to answer expectations for protons of a million volts or so. This work did not enjoy the glory of publication because, according to Brady, he and

re Cornell; *Science service,* 25 Jan 1939, and other documents (Pegram P, 27/"cyclotrons"), and infra, §6.2, re Columbia.

98. T. Yasaki to Lawrence, 24 Oct 1951 (9/43), and F. Yamasaki et al. to Lawrence, 24 Oct 1951 (9/38), quote.

White thought it advisable not to give the impression that they had doubted their professor's claim.[99]

In their hurry to build accelerators, Lawrence and his associates did not find the time to provide themselves with the detectors needed to register the effects of any disintegrations they might provoke. That was only reasonable: the accelerators did give a beam that could be improved; there lay a proven road to progress. Lawrence's group therefore lagged behind Tuve's, which had developed excellent detectors—Geiger counters, linear amplifiers, a novel cloud chamber—during 1931. Not that Tuve's people preferred detection to acceleration. The dilatoriness of their overlords at the Carnegie Institution and their indifference to the possible market value of their x rays (the Carnegie did not pursue patents) had reduced them to small jobs while they awaited permission to rip out their Tesla coil and install their 1-meter Van de Graaff.[100] When the age of disintegration began, therefore, their combination of detectors and high-current accelerator made theirs the preeminent laboratory for nuclear physics in the country. Or so Johnson, a good friend of both Lawrence's and Tuve's, judged the situation early in 1932. "It looks as if he [Lawrence] has some nice work under way but as far as the general technique for working in the field is concerned I cannot help but feel that you are way ahead."[101]

Lawrence's old friends from Yale, Donald Cooksey and Franz Kurie, came to Berkeley in the summer of 1932 to help shorten the lead. They had been asked before—Kurie for the summer of 1931, Cooksey for the spring semester 1932—but had not been able to accept.[102] Both of them were accomplished performers on the sorts of instruments the Laboratory needed most. Kurie's instrument was the cloud chamber. Cooksey was a one-man band, curator of precision instruments in Yale's Sloan Laboratory, an

99. Birge, *History, 4*, xi, 14, 25; xii, 7, 26; Brady, interview with James Culp, Oct 1981, and *PT, 36:3* (1983), 11, 13.

100. Tuve to Lauritsen, 11 Nov 1932 (MAT, 8); Tuve, *JFI, 216* (1933), 13–8; Johnson to Tuve, 6 June 1932 (MAT, 8); J.A. Fleming to Fleischmann Laboratories and to John Colt Bloodgood, 16 Feb 1931 (MAT, 9). The installation of the Van de Graaff was approved on 8 Oct 1932; it took about a month.

101. Johnson to Tuve, 6 Jan 1932 (MAT, 8).

102. Lawrence to Cooksey, 17 Jul [1931] (4/19); Birge to Cooksey, 2 Oct 1931 (Birge P, 33).

excellent machinist, and a connoisseur of fine design. They set up with the help of an improbable character, a retired officer of the Italian Navy, Telesio Lucci, who volunteered his services to the Laboratory. The reinforcements from Yale left the Laboratory with the rudiments—but only the rudiments—of the instrumentation of nuclear physics. Both later returned, Cooksey for a lifetime.[103]

Disintegration at Last

Lawrence's boys—he called all his staff, including his senior Cooksey, "boy," there being no women among them—pursued instrumentation with such urgency during the summer of 1932 because of the many great novelties in nuclear physics discovered that spring. In February, Cockcroft and Walton sent to the Royal Society of London a paper describing the acceleration of hydrogen ions down an eight-foot tube connected with their voltage multiplier. They managed to get around 700 kV across the tube and a current of perhaps 10 μA down it; and they mentioned preliminary experiments on fast protons entering a chamber separated from the evacuated tube by a thin mica window. The experiments were not very interesting. Cockcroft and Walton measured the range, or distance of penetration, of the protons through various gases. Their technique for detecting protons was identical to that perfected by Rutherford in the counting of alpha particles: they looked for scintillations where protons struck a fluorescent screen placed in the experimental chamber (plate 3.7). They promised more measurements of the same character and certain improvements in the apparatus by which they hoped to coax it to its design potential of 800 kV.[104]

Here we see again the compulsion of the instrument builder: to perfect the apparatus and to make it the subject of study. For two months after completing their apparatus, Cockcroft and Walton did not search for disintegration products. It is said that they did

103. Birge, *History*, 4, xi, 14, 23–4; xii, 8, 30, 32; Kurie to Cooksey, 4 Mar 1934 (10/21); Lawrence, recommendation for Kurie, 24 Feb 1932 (Simon Flexner P, B/F365/"NRC Fellowships"); Kurie, *Naval res. rev.*, 9:2 (1954), 13; Lawrence to D.L. Webster, 13 Apr 1934 (18/2).

104. Cockcroft and Walton, *PRS, A136* (1932), 626–9; *Nature, 129* (5 Feb 1932), 242, letter of 2 Feb.

not look immediately for material products of disintegration because they expected that much or all of the energy would be carried off by gamma rays, in keeping with the widespread preconception that (as will appear) had protected Joliot and Curie from discovering the neutron. Rutherford became impatient; Cockcroft and Walton should stop playing with their protons and look for disintegration products in the old way, with a fluorescent screen.

On April 14, 1932, they placed a thin film of lithium in the experimental chamber to receive protons and saw flashes of light on a screen that caught particles liberated from the target. A few elementary precautions ruled out the possibility that scattered protons caused the flashes. Two days later they knew the cause: a lithium nucleus that captures a proton can disintegrate into two alpha particles.[105] They had reached their goal: they had split the atom, by a reaction almost the inverse of the process discovered by Rutherford a dozen years before. He had then used alpha particles from natural sources to knock protons from nitrogen nuclei; they had used protons from an artificial source to make alpha particles from lithium. "When one learns that protons and lithium nuclei simply combine into alpha particles," Bohr wrote, on hearing the news from Rutherford, "one feels that it could not have been different although nobody has ventured to think so."[106]

Cockcroft and Walton subjected various targets to beams of various energies and tracked the disintegration particles with a fluorescent screen, a cloud chamber, and an advanced electronic detector developed at the Cavendish. They found that about half their beam was protons and the rest hydrogen-molecule ions; that the threshold for disintegration of lithium was astonishingly low, 125 kV; that the number of disintegrations increased rapidly with the energy of the incident protons, there being one for a billion at 250 kV, and ten per billion at 500 kV; and that the range of the emergent alpha particles, 8.4 cm of air, was consistent with the conservation of energy and momentum, the equivalence of mass and energy, and the reaction $Li^7(p,\alpha)\alpha$.

105. Cockcroft and Walton, *Nature, 129* (30 Apr 1932), 649, letter of 16 Apr, and *PRS, A137* (1932), 229; Hartcup and Allibone, *Cockcroft*, 50–3.

106. Letter of 2 May 1932, in Eve, *Rutherford*, 356, answering Rutherford to Bohr, 21 Apr 1932 (ER).

Boron and fluorine also gave off alpha particles under proton bombardment and other elements, even silver and uranium, seemed to yield a few. As for agreement with Gamow's theory, "which was largely responsible for stimulating the present investigation," it was not at all good. The theory gave as probability for the entry of a proton into a lithium nucleus one in six at 600 kV and three in a hundred at 300 kV; measurement suggested one in a million in the first case and five in a hundred million in the second.[107] But with elements heavier than iron, theory was much more stingy than fact: there should have been nothing from silver or uranium. In Tuve's opinion, it was this disintegration of heavier elements—"by far the most striking feature..., a feature which is distinctly disconcerting from the standpoint of all present-day theoretical ideas of the nucleus"—that had the first claim on the attention of physicists. Cockcroft also felt the difficulty. He appealed to Gamow, who could only suppose that the silver and uranium had contained some light impurity that gave the evidence of disintegration. Cockcroft did not like to lose a discovery or yield to theory so easily. "I always believed it possible for a really good theoretical physicist to explain any experimental result and now you fail me in the first test."[108]

The Berkeley group knew about Gamow's theory by the time Livingston got the 11-inch cyclotron going. He mentioned it in his thesis as the stimulant to efforts to reach high potentials and Lawrence understood that Livingston had given him what he needed to crack nuclei. "These protons have enough energy for nuclear disintegration experiments and it will not be long probably before we are disrupting atoms. But even more exciting are the possibilities of the large magnet, which at this very moment is being installed in our new radiation laboratory."[109] The moment was January 1932, four months before Cockcroft and Walton got their first counts.

The main reason that the Berkeley group did not succeed in splitting atoms before Cambridge is that they did not try. (As

107. Cockcroft and Walton, *PRS, A137* (1932), 229–42.

108. Tuve, *JFI, 216* (1933), 28–9; Gamow to Cockcroft, 7 Sep 1932, and reply, 29 Sep 1932, in Hartcup and Allibone, *Cockcroft,* 55. Cf. Bloch to Bohr, 12 Aug 1932 (BSC).

109. Livingston, *Production,* 4; Lawrence to Akeley, 13 Jan 1932 (1/12).

much might be said about Lauritsen, then busy irradiating cancer victims, and Tuve, who had not yet set up his Van de Graaff.) Another reason, as Lawrence freely admitted, was that neither he nor Livingston "kn[e]w very much about nuclear theory."[110] They had not even a wrong idea about what to look for. The news that lithium yielded to particles with the piddling energies attainable at Cambridge was disappointing. "When Cockcroft and Walton...disintegrated lithium with only a few hundred thousand volts, the thought naturally suggested itself that perhaps our efforts to obtain very high voltages were hardly worth while." But soon enough, Lawrence wrote Poillon, in a curious mixture of apology and triumph, Berkeley had results more interesting than Cambridge's.[111]

Lawrence learned while preparing for his honeymoon that he had been anticipated by Cockcroft and Walton. He wired Brady, who turned the beam of the 11-inch cyclotron onto a crystal of LiF procured from the Chemistry Department.[112] He found only a few hints of disintegration: he had too weak a current and too insensitive a detector. As late as mid August, the Berkeley group, although enriched by Cooksey and Kurie, had nothing decisive. "Unfortunately [Lawrence wrote Cockcroft] our beam of protons is not nearly as intense as yours although of higher voltage."[113] Just before the group disintegrated—Brady to Saint Louis and Cooksey and Kurie to New Haven—they got counts with a beam of 700 kV protons originating in a filament source constructed by Cooksey and detected with his Geiger counter. In September a new group—Lawrence, Livingston, White, and Henderson—took over the experiments, increased the intensity of the proton beam, and got alpha particles in numbers that agreed, more or less, with the results of Cockcroft and Walton.[114] Thus Berkeley's labor-intensive effort, extending over six months, confirmed the discov-

110. Lawrence to Slater, 19 Feb 1932 (12/12).
111. Lawrence to Poillon, 3 June 1933 (15/16A).
112. Childs, *Genius*, 170–1, 181, 191–3; Birge, *History, 4*, xi, 14, and App. xvii (c); Birge to Lawrence, 7 May 1932 (Birge P, 33).
113. Lawrence to Cockcroft, 20 Aug 1932 (5/4).
114. Lawrence to Barton, 8 Sep 1932 (2/25), to Cooksey, 13 Sep [1932] (4/19), and to Cottrell, 22 Sep 1932 (5/3); Cooksey and Henderson, *PR, 41* (1932), 392; Lawrence, Livingston, and White, *PR, 42* (1932), 150–1 (letter of 15 Sep); Kurie, *Naval res. rev., 9:2* (1956), 13.

ery made by a pair of Cambridge physicists in two days. The delay was not a consequence of too many cooks at the broth, but rather of the inexperience of the Berkeley group with detectors and of the difficulty of operating the cyclotron with a steady and useful current. Cockcroft and Walton forced the Laboratory to make its accelerators into instruments of research.

The Deuteron and the Neutron

The combined performance of several generations of chemists on the balance had resulted by World War I in a detailed knowledge of the relative weights of atoms. In particular they had fixed the weight of a hydrogen atom as 1.0077 on a scale defined by taking the weight of oxygen to be 16. The recognition of the existence of isotopes around 1914 opened the possibility that hydrogen or oxygen or both might be mixtures of atoms of different weights. The point was examined by Francis Aston, whose improved mass spectrograph of 1927 disclosed that the weight of a hydrogen atom was 1.00778 (O = 16).

That appeared to be quite satisfactory until Berkeley chemists challenged Aston's assumption that oxygen has but one isotope. Soon evidence from band spectra confirmed the challenge and allowed estimates of the relative abundance of O^{16} and O^{18} in ordinary oxygen. When the heavier isotope is taken into account, the weight of ordinary oxygen exceeds that of O^{16}. Aston's ratio between the weights of ordinary hydrogen and ordinary oxygen, 1.00778:16.00000, could be saved by supposing hydrogen too to be complex. Berkeley's specialist on physical constants, Birge, computed the consequences of the supposition based on the relative abundances of oxygen isotopes (O^{16}:O^{18}::630:1); and he found that everything came into order if about one hydrogen atom in 4,500 has a weight twice that of the more plentiful type.[115]

A Ph.D. from Lewis's school of thermodynamics and chemistry at Berkeley, Harold C. Urey, had a system of classifying isotopes that required the existence of deuterium—to give heavy hydrogen

115. Aston, *Nature*, *123* (1929), 488; Giauque and Johnston, ibid., 318, 831; Birge and Menzel, *PR*, *37* (1931), 1669–71; Bleakney and Gould, *PR*, *44* (1933), 265-8.

the name he gave it after a struggle with his colleagues and professors of Greek that lasted longer than his search for the new element.[116] Encouraged by Birge's numbers, he undertook to separate the graver from the lighter hydrogen atoms. Thermodynamic calculations suggested that evaporating liquid hydrogen around its critical point would enrich the liquid in heavy atoms at the cost of the vapor; a still set up at the low-temperature facilities of the National Bureau of Standards by F.G. Brickwedde effected a sufficient separation to allow detection of very faint spectral lines characteristic of deuterium from a fraction of the slightly enriched liquid. The detection made a stir and brought Urey a Nobel prize. "A great scoop for America," said Aston, who, like Birge, had graciously missed the discovery. Urey and Brickwedde gave a sample of the liquid to Aston's rival, K.T. Bainbridge, who worked to more than Cavendish accuracy. Bainbridge reported a mass for deuterium of 2.0126, where $O^{16} = 16$.[117]

While Urey was evaporating heavy hydrogen, his former professor Lewis was trying to separate the isotopes of oxygen. It was tedious and unrewarding work. The separation of hydrogen isotopes, with their pronounced difference in mass, appeared easier and, what was better, "much more interesting and important." Following up an observation by E.W. Washburn, chief of the Chemical Division of the National Bureau of Standards, that the water in old electrolytic cells was richer in deuterium than tap water, and guided by his feeling for the thermodynamics at work, Lewis effected a far better separation than did Urey or Washburn and estimated relative abundance at 1:6,500 as against his former student's modest 1:30,000.[118]

116. Among the many rejected possibilities were pycnogen, barydrogen, barhydrogen (Urey's original preference), and diplogen; correspondence of Urey and Brickwedde, May and June 1933 (Urey P, 1), and Stuewer, *AJP, 54* (1986), 206-18.

117. Urey, *JACS, 53* (1931), 2872; Urey, Brickwedde, and Murphy, *PR, 40* (1932), 2-3, 6-7; Aston, quoted in *Science service,* press release, 20 June 1933 (Lewis P); Bainbridge, *PR, 41* (1932), 115; Birge to Urey, 15 Mar 1932 (Urey P), offering congratulations, "even if I should have discovered it myself."

118. Washburn and Urey, NAS, *Proc., 18* (1932), 496; Lewis and Macdonald, *Jl. chem. phys., 1* (June 1933), 341-4 (rec'd 15 Apr), quote on 341. Diffusion through metals, not electrolysis, had been Lewis's first plan for separation; Lewis to Polanyi, 13 June 1933, and to E.O. Kraemer, 5 Oct 1933 (Lewis P).

Some numbers speak louder than words. In January 1933 Urey's best sample of heavy water contained only 1.2 percent deuterium oxide, about four times the concentration Washburn had managed. They had nothing better in April, when Lewis had 30 percent. "I am very curious to know how G.N. Lewis succeeded...with so little effort," Urey wrote his collaborator. "I suppose that he used some electrolytic method and succeeded in hitting the correct conditions exactly."[119] Urey had had no luck with electrolysis. It was uncanny the way everything that Lewis tried succeeded, although he had no surer explicit knowledge than Urey of the electrochemistry at work. On February 28, 1933, the University of California News Service announced Lewis's discovery, which the national press advertised under the whimsical headline, "Record weight of water achieved by California chemist."[120]

The news brought requests from all over for samples to use in place of ordinary hydrogen in the favorite experiments of the petitioners. One could scarcely fail to find something publishable about the behavior of a brand-new substance with a familiar chemistry but unknown physical properties. Lewis gave deuterium freely, almost as fast as he could make it.[121] Several important experiments, as well as many trivial ones, sprang from his generosity. The earliest request came from Bainbridge. Then Otto Stern telegraphed from Hamburg for ammunition for his newly perfected beam machine for measuring magnetic moments. The sample arrived at the end of July. "It was really fabulously nice of you to have answered my cry of need for isotopic water so promptly," Stern wrote, especially since their own attempts to make heavy water had failed.[122] Then there was water for electrolytic studies, for the Stark effect, for spectroscopy, for overvoltage, and for Lewis's own investigations of the physical and chemical

119. Urey to Washburn, 17 Jan and 3 Apr 1933, quote, and Washburn to Urey, 5 Dec 1932, 15 Apr and 10 May 1933 (Urey P, 5).

120. Fowler to Rutherford, 5 Apr 1933 (ER); Urey to Lewis, 17 Apr 1933, and Fowler to Lewis, 23 Oct 1933 (Lewis P); releases of 28 Feb and 2 Mar 1933 (Lewis P, "Science service").

121. Lewis to Fowler, 17 Jan 1934, and to Foster, 26 Jan 1934 (Lewis P).

122. Bainbridge to Lewis, 13 Mar 1933 (Lewis P); Frisch to Meitner, 27 Mar, 14 May, 26 Jul 1933, and to Houtermans, 24 June 1933 (Frisch P); Stern to Lewis, 7 Aug 1933 (Lewis P). Cf. J. Rigden, *HSPS, 13:2* (1983), 341–4.

properties of deuterium.[123]

The experiments to conjure with came from the mating of the deuteron and the accelerator. As Fowler watched a 50 percent solution of heavy water brew in Lewis's still, the Radiation Laboratory was planning "to repeat Cockcroft and Walton on Li and B with H particles [deuterons]—also to make He^4 by shooting H^2 at $OH^2 H^2$ ice. All nice and wild of course but exciting." With characteristic openness and generosity, Lewis sent Rutherford "some heavy water to play with," although the Cavendish physicists, with Lauritsen and Tuve, were the world's only competitors of the Berkeley bombardeers.[124] Their competition proved most fruitful.

The deuteron immediately attracted sustained theoretical interest. It was to nuclear physics what the hydrogen atom was to the old quantum theory: the simplest system available for calculation and experiment. According to the systematics of isotopes from which Urey had drawn evidence for the existence of deuterium, the deuteron should have consisted of two protons and an electron. Just before he announced his discovery, however, another intervention from Cambridge simplified the picture. The long-sought neutron, which Rutherford had believed in and which his students had looked for for a decade and more, had at last put in an appearance.[125]

The course of its discovery contains several points of general interest. To go back no further, in 1930 two physicists in Berlin, Walther Bothe and Hans Becker, discovered that beryllium gives off very penetrating rays, more penetrating even than the hardest gamma rays from radium, when hit by alpha particles from

123. Requests from, resp., T. Heyrovsky (Prague), 24 May and 6 Jul (1933); J.S. Foster (Montreal), 7 Jul 1933, and Lewis's reply, 26 Jan 1934; L. Wertenstein (Warsaw), 8 Nov 1933; and T.M. Lowry (Cambridge), 12 Jan 1934 (Lewis P); Cabrera and Fahlenbrach (Madrid), Soc. esp. fis. quim., *Anales, 32* (1934), 538–42 (magnetic susceptibility); Lewis et al.'s notes in *JACS, 55* (1933), e.g., 3503–4 (biochemistry of heavy water), 3057–9 (density), 4730–1 (viscosity and dielectric constant). Cf. "Uses of deuterium" [1934] (Urey P, 5).

124. Fowler to Rutherford, 22 Mar and 5 Apr 1933 (ER); J.A. Fleming to H.S. Taylor, 4 Jan 1934 (MAT, 13/"lab. letters").

125. For prehistory, see Kröger, *Physis, 22* (1980), 184–90, and Langer and Rosen, *PR, 37* (1931), 1579–82, who assimilated a hypothetical neutron to a collapsed hydrogen atom, and the energy of collapse to Millikan's cosmic radiation.

polonium.[126] They supposed that this "beryllium radiation" consisted of gamma rays, for which, indeed, they had been looking. The discovery interested Irène Curie, who worked in her mother's Institut du radium in Paris, where there was more polonium, and less physics, than in Berlin or Cambridge. At the end of 1931 Curie had found beryllium rays so penetrating that, were they gamma rays, as she assumed, their energy would have been between 15 and 20 MeV, three times that of the alpha particles that brought them forth.

Meanwhile a student of Chadwick's at the Cavendish had found that the beryllium rays emitted in the direction of the incident alpha radiation were far more plentiful than those emitted in the opposite direction, an asymmetry suggestive of the collision of particles. As this asymmetry came to light in Cambridge, Curie and Joliot found that when they placed a screen of wax in the path of the beryllium radiation, a detector behind the screen registered protons with energies of around 4.5 MeV. That caused them to increase their estimate of the power of the radiation: were it electromagnetic and obedient to the equations of the Compton effect, it would require about 50 MeV to liberate protons of 4.5 MeV. The discrepancy of a factor of three in their estimates of the energy of the radiation did not encourage belief in their interpretation of its nature.[127]

In February 1932, a month after this second set of results from Curie's Institute had come to hand, Chadwick declared the discovery of the neutron. A colleague, Norman Feather, had scavenged enough polonium from old radon tubes (mainly from the Kelly Hospital in Baltimore) to make a source nearly as powerful as the Parisian one. Excellent electronic detectors stood ready. With all this it was not difficult to show that the beryllium radiation knocked protons from other materials than wax and that everything made sense if it consisted of neutral particles of mass close to the proton's. As Fowler summed up for the jury of

126. For the following story and references, see Feather in Hendry, *Cambridge physics*, 31–41; Badash, *AJP, 51* (1983), 886; and the recent careful study by Six, *Rev. d'hist. sci., 55 (1988)*, 8–12, 18–22.

127. Frisch to Meitner, 6 Mar 1932 (Frisch P), points to the factor of three as bothersome but "not so tragic" as the assumption by Curie and Joliot of an implausibly improbable mechanism, a Compton effect involving an entire atom.

theorists: "It seems to be absolutely correct and the case is good enough to hang anyone on, but perhaps not yet scientifically certain."[128]

By the time the news reached Berlin, Bothe and Becker had found that beryllium does emit gamma rays under alpha bombardment, but of a much more modest energy, some 5.1 MeV, than Curie and Joliot required. (In fact the energy is more modest still, since the gamma ray comes from the decay of an excited state of C^{12}, some 4.4 MeV above the ground state.) Thus in a year and a half the unknown beryllium radiation, which physicists first tried to absorb in a way that did the least damage to received ideas—hence problematic gamma rays rather than an entirely new particle—was unravelled to yield an important discovery and a warning about the complexity of nuclear reactions.

Because of its complexity, the Bothe-Becker radiation was not adapted to a clean determination of the neutron's mass. Chadwick therefore had recourse to the transformations of boron and lithium, which he supposed to occur as follows: $B^{11}(\alpha,n)N^{14}$ and $Li^7(\alpha,n)B^{10}$. Chadwick set the velocity with which the neutrons leave the interaction equal to the maximum velocity of the protons they knocked out of the paraffin screen. With this assumption and the known energy of alpha particles from polonium, conservation of momentum gives the energy of the recoil atoms and conservation of relativistic energy the mass of the neutron. For the latter calculation the isotopic masses of the participating atoms must be known accurately. Here the then recent—and subsequently invalidated—work of Aston was essential. From the first transformation, of boron, $m_n = 1.0066$ (O = 16) with a probable error of 0.001. From the second, of lithium, for which the neutron's velocity had not been measured precisely, $m_n = 1.0072$. Hence the sum of the masses of neutron and proton (Aston's value) appeared to be 1.0072 + 1.0078 (m_p) - 0.0005 (m_e) = 2.0139, or some 0.0013 mass units greater than Bainbridge's value for the mass of the deuteron. On the natural supposition

---

128. Fowler to Bohr, 1 Mar 1932 (BSC); Curtis Bowman, Kelly Hospital, to Tuve, 10 Sep 1931 (MAT, 4). Cf. Segrè, *Fermi*, 69–70, on the independent unpublished theoretical discovery of the neutron by Ettore Majorana.

that d = p+n, the deuteron would have the meager binding energy of just over 1 MeV; should it consist of two protons and an electron, it would be more securely bound, by 0.0025 mass units or 2.5 MeV.[129]

Although, as soon appeared, Chadwick's estimate was based on a mistaken reaction and inaccurate isotopic masses, its consequence, that the deuteron possesses a positive binding energy, however meager, seemed an essential postulate. How else could the deuteron and nuclear physics hold together? Whatever one's views about the nature of the neutron—whether it is simple and singular, or a tight union of a proton and an electron—its weight could not be much less than the proton's if they were to join together in a stable deuterium nucleus.[130] Against the Cavendish's heavy neutron and its secure deuteron, Lawrence and his Laboratory placed a light neutron and an unstable nuclear physics.

Scurrying for Position

The discoveries of the neutron and deuteron and the demonstration of nuclear disintegration gave the few laboratories in the world able to do "high-energy physics" great scope for maneuver. They chose according to the opportunities afforded by their machines and circumstances, and by the temperaments of their directors. The Cavendish decided to work accurately and slowly, with very light elements and very modest energies Carnegie's Department of Terrestrial Magnetism (DTM) worked from Cavendish energies up to a million volts or so and also concentrated on lighter elements; they sought to make their beams as homogeneous and their targets as clean as possible to provide a firm basis for further work. The Kellogg Laboratory at Caltech worked with a similar energy range, but not so energetically as the

129. Chadwick, *PRS, A136* (1932), 692, and Solvay, 1933, 100–2. Reviewing other evidence in February 1933, Bainbridge put the neutron's mass between 1.0057 and 1.0063; Bainbridge, *PR, 43* (1933), 367–8, and *PR, 41* (1932), 115.

130. Theorists only gradually came to agree on the elementary character of the neutron, which, however, experimentalists appear to have assumed almost from the start. When theorists agreed to regulate the relations between neutron and proton by Fermi's theory of beta decay, they could not admit a neutron lighter (and hence stabler) than the proton. But at first this difficulty did not bother many people. Cf. Chadwick, Solvay, 1933, 102–3, and Bromberg, *HSPS, 3* (1971), 309–23.

DTM. The Radiation Laboratory at Berkeley played its long suit, went to higher energies and heavier targets as quickly as it could, experimented sloppily, and opened and discovered vast new territory. Outclassed for exact measurement by high-tension machines with straight tubes and relatively strong currents, the cyclotron could compete only by shooting faster projectiles than the opposing artillery could field.

Following the detection of disintegration at Berkeley at the end of the summer of 1932, Lawrence's group, which then consisted of himself, Livingston, and White, thought that their minuscule beam of protons, a mμA (a billionth of an ampere), liberated more alpha particles from lithium than Cockcroft and Walton had reported.[131] This cornucopia, sanctioned by Oppenheimer's calculations from Gamow's theory, turned out to be a mistake, an error of reckoning, as Cockcroft observed, by a factor of sixty. With proper arithmetic, the Berkeley results came into rough agreement with Cambridge's. Meanwhile Henderson had taken over the 11-inch cyclotron and Cooksey's Geiger counter. He drove the measurements up to the energy limit, with protons of 1.23 MeV, and made a discovery: the probability of interaction of a proton with a lithium nucleus does not increase with energy of bombardment above 400 keV.[132]

Early in the new year, White came back on the machine, put boron in place of lithium, got a surprisingly high yield, and was anticipated by Cockcroft and Walton.[133] Then Livingston joined the game, with Lawrence. They took on aluminum, which, they reported, gave off alpha particles and, like lithium and boron, had an energy threshold above which the probability of proton absorption became constant. But they had run too fast and taken analogy, rather than Cockcroft and Walton, as their guide. Aluminum's "alpha particles" turned out to be soft x rays. "We are finding it a dickens of a job to make sure whether radiations are protons or alpha particles or gamma rays or the Lord knows

131. Lawrence, Livingston, and White, *PR, 42* (1932), 150–1.

132. Cockcroft to Lawrence, 3 Oct 1932, and Lawrence to Cockcroft, 4 Nov 1932 (5/4); Henderson, *PR, 43* (1933), 98–102 (letter of 10 Dec 1932).

133. Lawrence to Henderson, 7 Jan 1933 (9/6); Cockcroft and Walton, *Nature, 131* (7 Jan 1933), 23 (letter of 22 Dec 1932); White and Lawrence, *PR, 43* (1933), 304–5 (letter of 27 Jan).

what," Lawrence had written Boyce a week before sending his misidentification of the aluminum x rays into print.[134] The error probably came to light when Livingston started "disintegrating 'to beat hell'" with the 27-inch cyclotron. In January 1933, having figured out how to shim the magnet, he knocked apart lithium and carried Henderson's curve up to 1.5 MeV.[135] But from aluminum the most conspicuous products were soft x rays.

For a month or so in February and March 1933, Lawrence believed that the initial era of machine design at the Laboratory had ended. "Therefore for some time in the future [we] shall devote most attention to the experimental study of the nucleus rather than the development of experimental technique." He was not looking forward to it. "We have decided not to be in any hurry about it and have settled down to patient, painstaking experimental study."[136] They were released from this uncongenial line of work by Lewis's method of collecting heavy water.

Tuve followed up the earliest of the discoveries, that of the neutron, as soon as he heard about it. He tried to assemble a polonium source equivalent to Chadwick's, but found it hard going. Feather had left few old radon tubes on the East Coast and was asking for more. "I have the greatest respect and warm regard for Cambridge physicists, and feel like a small boy speaking up in church when I deflect anything toward myself that might have helped them," Tuve wrote the radiologist at Kelly Hospital, insisting, however, that charity begin at home. His eagerness to confirm and extend Chadwick's discovery contrasts with Lawrence's initial reaction, that the neutron "has no particular effect on our experiment excepting to emphasize that there is a fascinating world of phenomena that will be accessible when we will have developed our method for producing swiftly moving protons."[137]

134. Livingston and Lawrence, *PR*, *43* (1933), 369 (letter of 11 Feb); Lawrence to Cooksey, 2 Feb and 18 Mar 1933 (4/19), to Barton, 13 Mar 1933 (2/25), and to Boyce, 2 Feb 1933 (3/8); cf. McMillan and Lawrence, *PR*, *47* (1935), 348.

135. Lawrence to Cooksey, 11 Dec 1932 (4/19), quote; to Henderson, 7 Jan 1933 (9/6). Livingston was using hydrogen-molecule ions, which he could get up to 4 MeV by February 1933, as projectiles; Lawrence to Cooksey, 2 Feb 1933 (4/19).

136. Resp., Lawrence to G.G. Kratschmar, 9 Feb 1933 (10/8), and, almost verbatim, to Kurie, same date (10/21); Lawrence to Barton, 13 Mar 1938 (2/25).

137. Tuve to F.W. West, 4 Mar 1932 (quote), and to G. Failla, Memorial Hospital, New York, 29 Feb 1932 (MAT, 4); Lawrence to G.W. Cattell, 16 Mar 1932 (25/2).

Tuve's group assembled enough polonium to develop a technique for detecting neutrons with their linear amplifier and improved cloud chamber. When the Van de Graaff came on line, Tuve searched for "artificial neutrons" using protons and alpha particles as projectiles, but found very little even at the maximum potential, some 750 kV, that he could reach in the spring of 1933. Nevertheless, he expected that even at moderate energies a Van de Graaff generator would be a very spectacular neutron source. On Gamow's theory, each μA of alpha particles on beryllium would give only 6 percent of Chadwick's yield (from 5.2-MeV polonium rays) at 800 kV; but twice that yield at 1,200 kV, 15 times it at 1,600 kV, over 50 times it at 2,000 kV.[138] Lauritsen's group checked the prediction as far as they could go; at 950 kV and 30 μA, their tube put out neutrons as copiously as the largest polonium source in use anywhere. Then they tried the same experiment with heavy water from Lewis. This time the artificial output exceeded the natural a hundredfold.[139] Simultaneously Lawrence was obtaining large yields of neutrons from deuterons shot at beryllium by the cyclotron.

When the 1-meter Van de Graaff started working, Tuve repeated Cockcroft and Walton's experiments using the target and detection scheme indicated in figure 3.5. He and his group confirmed the presence and the ranges of the alpha particles reported in the (p,α) reactions investigated at Cambridge.[140] They then began a series of painstaking bombardments of light elements by protons that led them to a fine discovery. These measurements determined the "excitation function," the yield of a nuclear reaction as a function of the energy of the incident particles. As far as they could go, they found all Berkeley's thresholds wrong.[141] That was not their fine discovery. Their strong, homogeneous beam made possible detection of narrow reaches of energy at which

138. Tuve to Lauritsen, 29 Mar 1933 (Lauritsen P, 1/8); Tuve, *JFI, 216* (1933), 36–8.

139. Crane, Lauritsen, and Soltan, *PR, 44* (1933), 514, letter of 3 Sep, and ibid., 692–3, letter of 30 Sep.

140. Tuve to Lawrence, 24 and 30 Jan 1933 (MAT, 4).

141. Fleming to Robert D. Potter, *New York Herald Tribune*, 20 Jan 1933 (MAT, 8); Tuve and Hafstad, *PR, 45* (1934), 651–3.

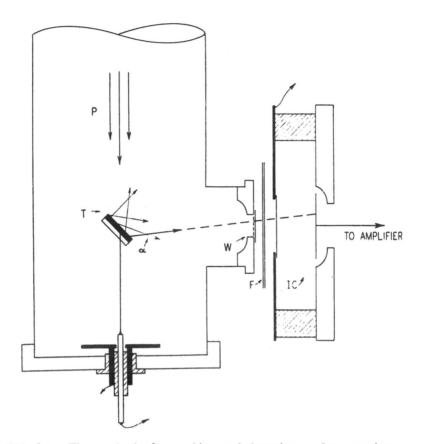

FIG. 3.5 The method of provoking and detecting nuclear reactions at the Carnegie Institution. P is the incident proton beam; W, a window closing the vacuum space; F, an absorber; IC, an ionization chamber. Tuve, *JFI, 216* (1933), 32.

impacting protons are particularly effective in exciting or disintegrating their targets. They found this "resonant excitation" in a characteristic way, by trying to clear up discrepancies between their excitation functions for carbon and Lauritsen's.[142] En route, they provided strong evidence that the apparent disintegrations of elements of medium atomic weight reported by Cockcroft and

142. Tuve to W.H. Wells, 7 Feb 1935 (MAT, 14/"lab. letters"); Hafstad and Tuve, *PR, 48* (1935), 306–9; Breit, *RSI, 9* (1938), 69–70; Abelson, *PT, 35:9* (1982), 92. Cf. Ofstrosky and Breit, *PR, 49* (1936), 22.

Walton—the disintegrations that Tuve had regarded as eminently subversive of theory—originated, as Gamow had suggested, in contamination of the targets.[143] Their view proved correct.

Berkeley experimenters contributed little to this line of work or to the precise determination of the energies involved in nuclear reactions. These energies, measured primarily at the Cavendish and at Caltech, concerned hydrogen, helium, lithium, beryllium, and boron. The measurements established the basis of a system of isotopic masses more accurate than Aston's. One outcome of this system, which constantly underwent revision during the 1930s, was Hans Bethe's theory of the source of the sun's heat. None of this painstaking work will engage us further.[144] It was not the sort of thing that Lawrence's boys did.[145] As Bethe said of another sort of demanding measurement of importance in nuclear physics, the dependence of the range of a fast particle on its energy, "it is not likely to be solved in the ordinary course of research because it is too complex for a graduate student and would involve too much time for a more advanced scientist."[146]

Accuracy is not everything, even in science. In the first two years of its existence, the Rad Lab built a machine that outdid all others in the acceleration of charged particles. It made the most powerful commercial x-ray tube in the world. It developed a linear accelerator. It received national attention. The dilution of effort and the demands of institution building resulted in work that was scarcely fastidious. But they did not cost Lawrence what he valued most: victory in the race for high energy.

143. Hafstad to Ladenburg, 8 June 1933 (MAT, 8).
144. Bainbridge, *PR, 44* (1933), 123; Bethe, correspondence with Bainbridge, Bonner, and Brubacker, and Cockcroft, 1935–37 (HAB, 3); Oliphant, Kempton, and Rutherford, *PRS, A149* (1935), 406–16, and *PRS, A150* (1935), 241–58, in Rutherford, *CP, 3*, 397–406, 407–23; Bonner and Brubaker, *PR, 49* (1936), 19–21; Oliphant, *PT, 19:10* (1966), in Weart and Phillips, *History* 185–6.
145. An exception is Lawrence, Henderson, and Livingston, *PR, 46* (1934), 324–5.
146. Bethe to Miss Skinner, American Philosophical Society, 23 Jan 1937 (HAB, 3).

PLATE 1.1 "The 13th Labor of Hercules." Official poster of the Panama-Pacific International Exposition, 1915. Benedict, *World's fairs*, 114.

PLATE 1.2 Pelton's 20,000-horsepower turbine exhibited at the Panama-Pacific International Exposition was created for hydroelectric power production in California. Todd, *Story, 4*, facing 178.

PLATE 1.3 LeConte Hall, new home of the Physics Department of the University of California, around 1930. University Archives, TBL.

PLATE 1.4 Lawrence as a young associate professor. University Archives, TBL.

PLATE 2.1   Cloud-chamber tracks of the reaction N(α,p)O. The sheaf of lines are tracks of alpha particles; the short jagged spur on the far left, the track of a recoiling oxygen nucleus; the fine straight line running from the collision fork, that of the ejected proton. Blackett, *PRS, A107* (1925), plate 7, no. 1, facing 361.

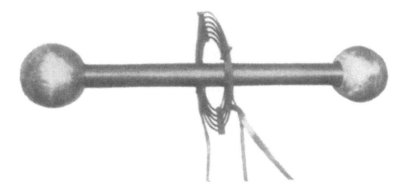

PLATE 2.2   Photograph of the Carnegie Institution's Tesla coil. The primary consists of two or three turns of copper tubing, the secondary of 5,000 to 7,000 turns of silk-covered wire. Breit, Tuve, and Dahl, *PR, 35* (1930), 56.

PLATE 2.3   Test of Van de Graaff's first setup at the Carnegie Institution. Tuve, *JFI, 216* (1933), 26.

PLATE 2.4 The Carnegie Institution's two-meter Van de Graaff; Dahl
is on the ladder, Tuve in the suit. The business end of the discharge tube,
deflecting magnets, and pumps are under the floor. Tuve, Hafstad, and
Dahl, *PR, 48* (1935), 322.

PLATE 2.5   Van de Graaff's 15-foot generator at MIT's Round Hill Experiment Station. The spheres stood 43 feet above the ground; their steel trucks ran on a railroad track to make possible changes in the striking distance. Tuve, *JFI, 216* (1933), 34.

PLATE 2.6   Cockcroft-Walton machine.   A, capacitor stack; B, rectifying system of kenotrons with corona shields; C, pumping system; D, transformer system; E, accelerating tube.   Cockcroft and Walton, *PRS, A129* (1930), facing 490.

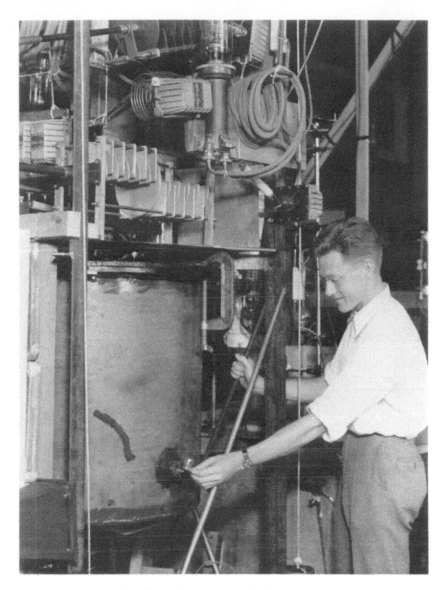

PLATE 2.7   David Sloan in the Radiation Laboratory about 1931.
LBL.

PLATE 3.1   The old Radiation Laboratory looking west toward the Campanile. LBL.

PLATE 3.2   1,000-kw   Poulsen   arcs   being   machined   at   Federal
Telegraph's plant in Menlo Park.   LBL.

PLATE 3.3   The 27-inch cyclotron as installed in the Radiation Laboratory. The glass vessels supply the source. LBL.

PLATE 3.4 David Sloan and J.J. Livingood at work at the Sloan x-ray tube at the University of California Hospital in San Francisco in 1932–33. LBL.

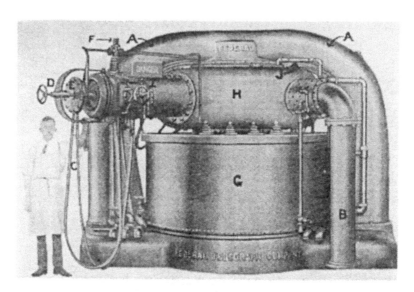

PLATE 3.5   Federal's fully assembled 1,000-kV arc generator. A is the magnet yoke; B, an exhaust pipe; C, the hydrogen supply; G, the tank containing the coil; H, the tank containing the arc. Heilbron, *Museoscienza, 22* (1983), 17.

PLATE 3.6   Lawrence and Livingston standing beside the Federal Telegraph magnet at Pelton Waterwheel Company, where it was machined for use in the Radiation Laboratory. Courtesy of Lois Livingston.

PLATE 3.7   Cockcroft and Walton's corner of the Cavendish. The tall transparent cylinder in the center is the discharge tube; the other cylinders are stacks of condensers and rectifiers. The curtained box is the observation center. Cockcroft and Walton, *PRS, A136* (1932), 625, plate 11.

PLATE 4.1   The members of the Solvay Congress of 1933. LBL.

Research and Development, 1932–36

1. A FRUITFUL ERROR

A Claim

The Laboratory's first sustained appearance on the international scene was the result of mixing heavy water with resonance acceleration. The mix gave rise to the false idea, which Lawrence defended resourcefully, that the deuteron is unstable. This hypothesis of deuteron disintegration had many unpleasant implications for nuclear theory. It therefore immediately excited incredulity and opposition in the wider world of physics. The persistence of the Laboratory in its error spread its reputation for sloppiness and forced others to make an important discovery; and it illustrated the strengths and the weakness of the Laboratory's style of teamwork centered entirely on Lawrence's research program.

Lewis and Lawrence first loosed deuterons into Livingston's 27-inch cyclotron in April 1933. Their expectation that "heavy protons" would be better projectiles than light ones was quickly fulfilled.[1] On May 3 Lawrence wrote his close colleagues—Boyce, Cooksey, and Oppenheimer—that "H-twotrons" blasted lithium into alpha particles with ranges in air of 8.2 cm and 14.8 cm, and knocked slow alpha particles from boron, magnesium, nitrogen,

---

1. Lawrence to Poillon, 14 Apr 1933 (15/16A). The deuterons, at 1.3 MeV, came from H^1H^{2+} ions accelerated to 2 MeV; Lewis, Livingston, and Lawrence, *PR, 44* (1 Jul 1933), 55 (letter of 10 June).

and aluminum. The yield from lithium was ten times larger with deuterons than with protons. Lawrence hurried to capitalize on the new projectile and the new machine between the end of the term and early June, when he was to leave for vacation and teaching at Cornell. "I have been working night and day in the laboratory since class ended," he wrote Oppenheimer. "You will pardon me for dictating this. I am doing it while proctoring an examination. I am really trying to conserve time."[2]

Explaining the action of deuterons on lithium to Cockcroft, Lawrence supposed that the swifter alpha particles came from Li^7, according to $Li(d,n)2\alpha$, and the slower from Li^6 via $Li(d)2\alpha$. A week later he changed his mind and identified the fast alpha particles with the disintegration of Li^6. Both processes, whichever was which, went well even with deuterons of modest energy. To coax alpha particles from aluminum, magnesium, and nitrogen, however, required energies above a million volts, well out of Cockcroft's range.[3] Every element bombarded with deuterons of such energies "disintegrated." It was time for a news release. The release: lithium, beryllium, boron, nitrogen, fluorine, aluminum, and sodium all suffered "transitions" at Berkeley; "at this rate of progress, one dares not guess what will be achieved in nuclear physics within a few years."[4] Within a few weeks the rate was retrograde; Lawrence still could not say for certain from which lithium the fast alpha particles came, nor whether anything else but nitrogen, and possibly beryllium and boron, disintegrated under bombardment by deuterons of 1.3 MeV.[5]

Then came an extraordinary find. Not stopping to resolve the problems they had raised, the Berkeley group found that every element gave off protons under bombardment by deuterons with more than 800 keV of energy. And more than that: the protons all

2. Lawrence to Boyce (3/8), Cooksey (4/19), and Oppenheimer (14/9), all 3 May 1933; Lawrence to Boyce, 2 Feb 1933 (3/8), for Lawrence's summer plans.

3. Lawrence to Cockcroft, 4 May 1933 (5/4); to Darrow, 10 May 1933 (6/9).

4. Lawrence to Houtermanns, 12 May 1933 (8/11); Science service, 20 May 1933 (Lewis P), quote.

5. Lewis, Livingston, and Lawrence, PR, 44 (1 Jul 1933), 55–6 (letter of 10 June), 317 (abstract for June meeting of the APS); Henderson also participated in the work. Cf. Lawrence to Cooksey, 2 June 1933 (5/4).

had the same range, 18 cm, irrespective of the material from which they came. "I am almost bewildered by the results," Lawrence wrote. But only almost. He reasoned that such homogeneous protons could scarcely come from so heterogeneous a set of elements, especially from the highly repulsive (to deuterons) nuclear charges of atoms as heavy as gold; and he declared that they must originate not in the targets but in the projectiles. "We have strong evidence that we have disintegrated the deuteron itself." So Lawrence wrote Bainbridge, asking for his latest values of the masses of H^2 by return airmail.[6] The numbers were needed to calculate the mass of the neutron from the hypothesis of deuteron disintegration.

In the easiest hypothetical case, a deuteron of energy 1.33 MeV hits a gold nucleus, which is too heavy to be moved much by the blow and splits into an 18-cm proton and a neutron of unknown velocity. Lawrence's group had some shaky evidence that in the explosion of the deuteron, the constituents received equal momentum and energy. Now an 18-cm proton has an energy of 3.6 MeV; the kinetic energy produced in the explosion is accordingly (7.20 − 1.33)MeV = 0.0063 mass units. This, together with the masses of the proton and the neutron, must equal the mass of the deuteron, which Bainbridge had fixed at 2.0126 from his analysis of Lewis's water. The indicated arithmetic produced the startling answer, $m_n \approx 1.000$. Thus was born the short-lived light Berkeley neutron.[7]

Lawrence's idea may be reconstructed from an improved calculation made after his return from Cornell and from later correspondence. The calculation rested on experiments by Henderson and Livingston, which seemed to show that the proton came off with all the kinetic energy of the deuteron and that a bombarding energy of 1.2 MeV, not 1.33 MeV, was enough to give protons of 18 cm. Therefore, in the explosion the proton gets 3.6 − 1.2 = 2.4 MeV and, to conserve momentum, the neutron does too. (The new numbers raised m_n slightly, to 1.0006). The

6. Quotes from, resp., Lawrence to Cockcroft, 2 June 1933 (5/4), and to Bainbridge, 3 June 1933 (2/20).

7. Lawrence, Livingston, and Lewis, *PR, 44* (1 Jul 1933), 56 (letter of 10 June); Bainbridge, ibid., 57 (letter of 14 June).

model: the deuteron tends to explode in the field of a nucleus at the place where it experiences the strongest force for the longest time, at rest at its closest approach. The neutron and proton receive equal and opposite momenta; in addition, the charged proton is pushed out by the electrostatic force of the nucleus, which returns to it the energy lost by the parent deuteron in penetrating to the place of its demise.[8]

Lawrence advertised the light neutron at a symposium held at Caltech in May 1933 in honor of Bohr. Always hoping for revolution, Bohr welcomed Lawrence's news as a "marvellous advancement." Millikan, the master of self-advertisment, praised Lawrence's work as "altogether extraordinary, and most intelligently announced."[9] On the way to Cornell Lawrence stopped at the Century of Progress Exposition in Chicago, where the American Physical Society held, on its own assessment, "perhaps the most important scientific session in its history." Aston, Bohr, Cockcroft, and Fermi, among others, were there as guests of the AAAS to help celebrate. In a special session on nuclear disintegration, Cockcroft discussed the blasting of light elements by his voltage multiplier, Tuve described the blasting of middling elements by his Van de Graaff generator, and Lawrence claimed the blasting of everything by the cyclotron. Bohr presented a public commentary, during which he wrapped himself up in the wave-particle duality and the microphone cord. It was not a hard act to follow. "It was much easier, and much more pleasant [wrote the reporter from *Time*] to understand round-faced young professor...Lawrence...tell how he transmuted elements with deuton bullets."[10]

Lawrence stole the show. The *New York Times*'s correspondent, William L. Laurence, introduced his namesake as the proprietor of a "new miracle worker of science," which, when whirling deuterons, liberated about ten times the energy it was fed. "The newest developments give only an inkling of what lies in store for man when and if he finally succeeds in unlocking what

8. Livingston, Henderson, and Lawrence, *PR, 44* (1 Nov 1933), 781–2 (letter of 7 Oct); Darrow to Lawrence, 28 Nov 1933, and answer, 9 Dec 1933 (6/9); Boyce to Lawrence, 19 Feb 1934, and answer, 27 Feb 1934 (3/8).

9. Childs, *Genius*, 199.

10. *PR, 44* (1933), 313–4; "Complementarity in Chicago," *Time, 22* (3 Jul 1933), 40.

Sir Arthur Eddington calls the 'cosmic cupboard of energy'."[11] Lawrence explained that the cracking of the deuteron released a little of that vast store of atomic energy of which Aston had given an inkling in his famous overstatement that half a glass of water could drive the *Mauretania* across the Atlantic. According to the Laboratory's measurements and theories, the neutron weighed less than the sum of the weights of the proton and the electron, which Lawrence understood to be its constituents. Hence, according to the relativistic equivalence of mass and energy, the combination of a proton and an electron should yield energy. Here was another source—along with the disintegrating deuteron—of perpetual motion.[12] Lawrence did not bother readers of the *Times* with the worry that hydrogen atoms might, or rather should, collapse spontaneously to neutrons and blow the world apart.

A Doubt

Meanwhile the Cavendish Laboratory had started to project deuterons. Rutherford had been most eager to procure heavy water for his experiments even before the startling discovery of the efficacy of deuteron bombardment. When Fowler returned to Cambridge from Berkeley early in May 1933 without any, he was "nearly lynched," he told Lewis, who forthwith forwarded a protective ampule containing as much deuterium as Lawrence's group had used in their disintegration experiments. Rutherford promised through Fowler not to hurry any results he might obtain into print, "so as to give Lawrence plenty of chance to get in first."[13] He received the news that deuterons had smashed atoms of lithium, nitrogen, magnesium, and beryllium. It reminded him of better times and of the Maori warrior on his baronial shield. "I should like to congratulate Lawrence and his colleagues for the

11. Seidel, *Physics research*, 381, from William L. Laurence, "New 'gun' speeds breakup of atom: 'Deuton's bullet frees ten times its own energy,' scientists are told," *New York Times* (20 June 1933), 1. By our count the factor was five (7.2/1.33) for the deuterons that disintegrated; it is, of course, scarcely worth mentioning in comparison with Laurence's exaggerations.

12. Laurence, "Neutron 'weighed' by Prof. E.O. Lawrence, proves lighter than its component parts," *New York Times* (24 June 1933), 1.

13. Fowler to Lewis, 9 May 1933, and Lewis to Rutherford, 15 May 1933 (Lewis P).

prompt use they have made of the new club to attack the nuclear enemy....These developments make me feel quite young again."[14]

Rutherford hurled his clubs from a proton accelerator that had been made for him by Oliphant to follow up the experiments of Cockcroft and Walton. This accelerator may stand as a symbol of Rutherford's methods in contrast with Lawrence's. Rather than go to higher energies than Cockcroft and Walton had reached, Rutherford opted for lower; theirs could attain 800 kV, his only 200 kV. He wanted to examine the thresholds of proton-induced nuclear reactions, identify the products, and estimate yields as functions of the energy of bombardment. To achieve his purposes, he directed that his new machine have a large proton current; a directive so well executed that Oliphant and Rutherford had for their experiments about 1,000 times the single microamp of Cockcroft and Walton's early experiments.[15]

The Oliphant-Rutherford accelerator, though short on energy, was by no means the cheap, jerry-rigged contraption of string and sealing wax dear to Cavendish mythology.[16] Metropolitan-Vickers designed, built, and contributed the oil-diffusion pump that created the vacuum, which was produced and maintained with the help of Apiezon oils and greases; the accelerating system incorporated the voltage-multiplier circuit perfected by Cockcroft and Walton with the help of Metro-Vick's engineers and a 100-kV transformer bought from Metro-Vick at what Rutherford thought the extravagant price of 85 pounds; and the detectors, which recorded the ionization created in a special chamber by the disintegration products, used a linear amplifier, thyratron tubes, and a purely electronic counting system then just invented at the Cavendish by C.E. Wynn-Williams and his co-workers.[17] The Cavendish was then far ahead of Berkeley in electronics and vacuum technology and in integrating the work of academic

14. Rutherford to Lewis, 30 May 1933 (Lewis P).
15. Rutherford to Hevesy, 3 Apr 1933, in Eve, *Rutherford*, 370; Oliphant and Rutherford, *PRS, A141* (1933), 259–81, in Rutherford, *CP, 3,* 329–50.
16. Crowther, *Cavendish*, 242–4, and Cockburn and Ellyard, *Oliphant*, 49–50, both relying on Oliphant.
17. Oliphant and Rutherford, *PRS, A141* (1933), in Rutherford, *CP, 3,* 330–4; Allibone in Hendry, *Cambridge physics*, 158–9, 161–2, 169; Wynn-Williams in ibid., 142–7; Hendry, ibid., 114–9.

physicists and industrial engineers.

By February 1933 Oliphant and Rutherford were extending Cockcroft and Walton's results, provoking the disintegration of lithium with protons of only 100 kV. "It is a great show! But who would have thought that anything would happen at 100,000 volts, except perhaps Rutherford?"[18] In June, about the time the heavy water came to hand, they presented an account of their work on proton disintegration. They found that lithium's threshold stood at 20 kV and boron's at 60 kV; that beryllium's could not be determined because of its very small yield; and that in elements heavier than boron, except for a trace at fluorine, even the most energetic protons available, at 200 kV, did not stimulate disintegration. Oliphant and Rutherford traced apparent reactions in heavy metals such as gold to disintegration of boron impurities from the glass walls of their pyrex discharge tube.[19] This was an important warning. Lawrence did not heed it: he was too busy running through the periodic table, too eager to accept the astounding, to take the time to track down subtle effects.

When he ran Lewis's water against lithium, Oliphant detected particles of 13.2-cm range, which he identified with Berkeley's particles of 14.8 cm. "The Professor seems very happy."[20] Soon Oliphant and Rutherford confirmed the existence of rays that penetrated to 8.2 cm, which they showed to be alpha particles with the maximum energy possible in the reaction $Li^7(d,n)2\alpha$.[21] Walton informed Cockcroft, then visting the Laboratory, of the good general agreement of Cambridge's results with Berkeley's. Cockcroft found himself curiously placed: he sat in the camp of one of his competitors while his partner, Walton, sat in the other. He decided that the most attractive subject for them was the prolific 18-cm proton, which Oliphant and Rutherford could not excite and Lawrence's company could not stop to study. "We ought to be able to get many of these [Cockcroft advised Walton]

18. Fowler to Bohr, 15 Feb [1933], in Hendry, *Cambridge physics*, 107.

19. Oliphant and Rutherford, *PRS, A141* (1933), in Rutherford, *CP, 3,* 335–7, 343, 349–50; Rutherford to Hevesy, 3 Apr 1933, in Eve, *Rutherford,* 370.

20. Dee to Cockcroft, 10 Jun 1933 (CKFT, 20/4), quote; Rutherford to Boyle, 28 Jul 1933, in Eve, *Rutherford,* 374.

21. Dee to Cockcroft, 7 Jul 1933 (CKFT, 20/7); Oliphant, Kinsey, and Rutherford, *PRS, A141* (1933), 722–33, in Rutherford, *CP, 3,* 354–6, 358–60.

and I hope you will be able to [borrow] some of the Professor's H_2^2O. I think that after Dee gets the Boron tracks you might go straight on to that with the Wilson chamber as we can get in long before California in this field and there are a lot of points to be cleared up."[22] For example, the origin of the 18-cm proton.

The Cavendish work, and Rutherford's congratulations to Lawrence on the "fine reward for his labour in developing his accelerated [!] system," pleased the Berkeley group and probably helped to harden their belief in what Lewis called "the most important discovery so far[:] the essential instability of the H^2 nucleus and the low mass of the neutron." Cockcroft was also pleased at the confirmation obtained using the Oliphant-Rutherford accelerator. That had given him the hope that with the Cockcroft-Walton machine it would be possible to detect the 18-cm protons, for which Lawrence gave a threshold of 700 kV. "If so [Cockcroft wrote Rutherford] it [sic] will find a whole lot more work to be done with the present apparatus."[23]

To make all this work possible, the Cavendish required a steady supply of heavy water. Rutherford detailed a visitor, Paul Harteck, from the Kaiser-Wilhelm-Institut für Chemie, to the task. The Cavendish apprentice system in this respect paralleled Berkeley's: Harteck had been told to help with electronic counters, but was reassigned because of his knowledge of chemistry when Rutherford decided to domesticate the manufacture of deuterium. The change did not please Harteck. "I must take on the production of heavy water at the wish of the high Lord," he wrote his patron, K.F. Bonhoeffer. "If you know anything [about it], write me soon, for with the Lord everything must go very quickly....You must hurry, since heavy water evidently seems to be no rarity in America."[24] The gift from Lewis took the heat off Harteck, who had not succeeded. An application to Lewis for information elicted a full answer that did not help; and by the end of June,

22. Walton to Cockcroft, 20 June 193[3] (CKFT, 20/35); Cockcroft to Walton [24 June 1933] (ER).

23. Lewis to Rutherford, 12 Jul 1933, and Cockcroft to Rutherford, 22 Jul 1933 (ER).

24. Harteck to Bonhoeffer, 28 Apr and 12 May 1933 (Harteck P). Cf. Harteck and Streibel, *Zs. anorg. und allgem. Chemie, 194* (1930), 299.

Harteck could only imagine that there was some trick to the business that had been withheld from him. Rutherford then lost half the original sample. Had Lewis then turned off his water, the Cavendish deuteronomers would have been out of business until the end of October, when Harteck managed to make enough heavy water at sufficient concentrations for their needs.[25]

There was already good reason to worry that Berkeley water did not give the same results in England as at home. Walton and Dee saw only a few 18-cm protons, nothing like the profusion the Berkeley group had reported. "I noticed Lawrence's views about the nature of these tracks," Rutherford wrote Lewis at the end of July, "but we are at the moment not inclined to view with favour the conversion of a deuton into a neutron of mass about 1. However, it is too early to take definite views."[26] The principal parties then relaxed for the summer, to prepare for more definite views in the fall.

The Error

At Berkeley preparation included improvements in the cyclotron, which made possible production of 0.02 μA of 3 MeV deuterons, and curing "vacuum troubles and other misfortunes" that kept the machine down for most of September.[27] When the streams began to flow, they called forth showers of neutrons from everything they hit, a confirmation most agreeable to the minds, but also threatening to the bodies, of the cyclotroneers.[28] (The Laboratory was so full of stray neutrons that an investigator quirky enough to have tested the fillings in his teeth might have discovered artificial radioactivity.) Another set of doubters then

25. Harteck to Lewis, 4 June, and Lewis to Harteck, 23 June 1933 (Lewis P); Harteck to Bonhoeffer, 9 and 28 June 1933 (Harteck P); Rutherford to Lewis, 27 Jul, 10 Aug, 21 and 30 Oct 1933 (Lewis P); Cockcroft to Rutherford, 23 Aug and 29 Sep 1933 (ER). Imperial Chemical Industries had trouble with the plant it built not long after Harteck got his first heavy droplets; Fowler to Lewis, 3 Feb 1934 (Lewis P).

26. Darrow to Lawrence, 11 Oct 1933, Livingston to Darrow, 20 Oct 1933, and Lawrence to Darrow, 20 Nov 1933 (6/9); Rutherford to Lewis, 27 Jul 1933 (Lewis P).

27. Lawrence to Cooksey (5/4) and to Tuve (3/32), 23 Sep 1933.

28. Lawrence to Barton, 28 Sep 1933 (2/25), to Beams, 4 Oct 1933 (2/26), and to Tuve, 9 Oct 1933 (3/32); Lewis to Rutherford, 5 Oct 1933 (ER).

entered the game. Richard Crane, a junior collaborator of Lauritsen's at Caltech, came to Berkeley, collected some heavy water, dribbled it on a beryllium target in Lauritsen's machine, and got "enormous quantities of neutrons." That was of course most gratifying, "in entire agreement with our expectations," Lawrence wrote Cockcroft, "though the precise interpretation is as yet ambiguous." The point of imprecision was whether Crane's neutrons came from the disintegration of the beryllium target or of the deuteron projectile. In either case, however, Lawrence thought that Crane's evidence favored a value of the neutron mass close to unity.[29]

Once the cyclotron returned to work, Livingston and Henderson found quantitative evidence of the deuteron's instability. They counted the number of "disintegration" protons (some 40,000 per minute registered in their ionization chamber) and then the number of "recoil" protons reported by the same chamber when covered with a wax-coated lead screen (12/min.). Their previous estimate of the probability of the conversion of neutrons to protons in wax suggested that 40,000 neutrons would make around 12 protons. Hence neutrons and protons appeared in equal numbers, which would be necessary if they came from the breakup of deuterons. They realized that this agreeable agreement had "profound theoretical implications" through its relevance to the value of the neutron mass, which they set at 1.0006. Their report, signed also by Lawrence, appeared on November 1.[30]

At Cambridge preparation included completing reports for the Solvay conference to be held in Brussels at the end of October. Cockcroft had responsibility for reviewing particle accelerators and Chadwick for the state of knowledge about neutrons. Lawrence contributed by sending Cockcroft information about the cyclotron and by bringing his latest evidence for disintegration to the Solvay meeting in person. The invitation to attend, at his own expense for travel, was a great honor; Lawrence was but the eighth American so distinguished since the conferences began in 1911

29. Quotes from, resp., Lawrence to Tuve (3/32) and to Cockcroft (5/4), 23 Sep 1933.

30. Livingston, Henderson, and Lawrence, *PR*, *44* (1933), 782; Lawrence to Tuve, 9 Oct 1933 (3/32); Lawrence to R.C. Gibbs, 9 Feb 1933 (9/16), on Henderson.

and the only one in 1933. He declared himself "surprized and tremendously pleased,"[31] though the invitation came very late, some six weeks before the meeting (he owed it to Peter Debye, a member of the Solvay scientific committee, who had visited Berkeley the previous summer and taken a fancy to the Sloan x-ray tube).[32] Lawrence went to this most august of physicists' gatherings (fig. 4.1), the first international meeting he had ever addressed, to correct the opinions of Chadwick, Cockcroft, and Joliot, who also had a candidate for the neutron mass. "I particularly want to make some rather extensive remarks on Cockcroft's report."[33]

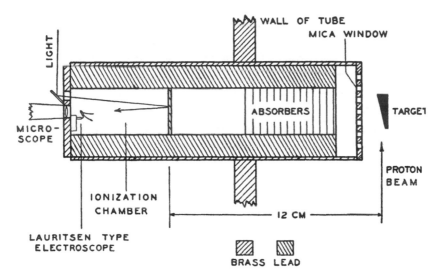

FIG. 4.1 McMillan's apparatus for studying the absorption of gamma rays by various materials. The rays scatter at right angles to the proton beam from the target and enter through the mica window. McMillan, *PR, 46* (1934), 868.

31. Lawrence to Cockcroft, 2 Jun and 16 Sep 1933 (4/5); to Beams, 16 Sep 1933 (2/26); and to Swann, 28 Sep 1933 (17/3), quote.

32. Lawrence to Cooksey, 5 Oct [1932] (4/19); Debye to Langevin, 8 Oct 1932 and 16 Sep 1933 (Langevin P, "Solvay 1933").

33. Lawrence to Langevin, 4 Oct 1933 (Langevin P, 75). Each participant was asked to state the points in others' reports on which he wished to speak; Bethe to Langevin, 10 Oct 1933 (Langevin P, "Solvay 1933").

Cockcroft ended his report with an unenthusiastic review of Berkeley work. He accepted the alpha particles from the bombardment of lithium and presented the data about the 18-cm protons, but declined to entertain the hypothesis of disintegration. "It is rather superfluous to discuss further the nature of the transformations with proton emission until we have more experimental information." And how to get the information? From improved Cockcroft-Walton machines. The weak current the cyclotron brought to the target, one-thousandth the flux from the "direct" Cambridge method, might well wash out the advantage of the greater efficacy of its faster particles. "Our present information does not suffice for prediction."[34]

Lawrence's extensive comments centered on the cyclotron method and its latest achievements—hydrogen-molecule ions of over 5 MeV, deuterons of 3.6 MeV, a promise of protons at 3.5 MeV, evidence of the disintegration of heavy hydrogen in the fields of target nuclei, and numbers that made the neutron's mass unity. These last remarks made Lawrence himself the object of a bombardment. Heisenberg observed that if disintegration occurred in the electric field of a nucleus, the yield should decline for heavy targets since the deuteron's penetration, and hence the rate of change of force on it, must decrease with increasing atomic number (Z); for (very) high Z the field would appear to the deuteron to change adiabatically and produce no disintegration at all. That being the case, added Bohr, we might suppose that the deuteron splits after entering a nucleus; but then the speed of the ejected proton should increase with atomic number, like the nuclear Coulomb field, contrary to Lawrence's results.[35]

Then came the experimentalists. Rutherford said that he had found no neutrons from lithium under deuteron bombardment. Chadwick reaffirmed the value of the neutron mass at between 1.0067, which he deduced from the hypothetical reaction $B^{11}(\alpha,n)N^{14}$, and 1.0072, which he had deduced from $Li^7(\alpha,n)B^{10}$. Joliot and Curie came forward with a neutron still heavier than Chadwick's. Their careful examination of decay products of

34. Cockcroft, Solvay, 1933, 50–5; cf. Oliphant, *PT, 19:9* (1966), in Weart and Phillips, *History*, 181.

35. Solvay, 1933, 71, 72.

alpha-bombardment of boron and other light elements had disclosed quantities of positive electrons along with neutrons. They supposed that these particles came away simultaneously, according to the reaction $B^{10}(\alpha,ne^+)C^{13}$, and constituted when together the familiar proton. In place, therefore, of Chadwick's initial conception, that $n = p + e^-$, they now proposed $p = n + e^+$. So much, and much more, was tied up in the question of the neutron mass. Calculations based on the transmutation of B^{10} made $m_n = 1.012$. This big mass had the advantage of accounting for the stability of the hydrogen atom, since it prevented the spontaneous union of its proton and electron into a neutron; but it made the decay of the neutron into a proton, electron, and neutrino energetically possible.[36] The only difficulty that Joliot and Curie saw with their fat neutron was the conflicting experience in Berkeley. They were prepared to compromise: "It is not impossible that it will be necessary to suppose the existence of neutrons with different masses," theirs being the elementary one and Lawrence's a condensed combination of the elementary with an electron-positron pair.[37]

Lawrence responded to these challenges by invoking suppositious gamma rays, whose inclusion in the energy balance would lower the neutron mass. Chadwick denied the gammas and insisted, against Joliot and Curie, that the neutrons came from the more plentiful isotope B^{11}, with the mass he had assigned them. After this exchange the theorists could only feign hypotheses and await the outcome of the squabble.[38] As perhaps no one outside France expected, victory eventually fell to Joliot and Curie.

36. Solvay, 1933, 77, 101–2, 155–6; Curie and Joliot, *CR, 197* (1933), 237 (17 Jul 1933), in *Oeuvres,* 417–8; Langevin, "Rapport sur les titres et travaux de M. Frédéric Joliot" (Langevin P, 73/2); Pauli to Joliot, 1 Feb 1934, in Pauli, *Briefwechsel, 2,* 271.

37. Joliot, Solvay, 1933, 156. This was to recur to the hypothesis of A. von Grosse, *PR, 43* (1933), 143, who supposed neutrons of various masses to explain the energy spctrum in beta decay. Cf. Pauli to Heisenberg, 14 Jul and 30 Sep 1933, in Pauli, *Briefwechsel, 2,* 185, 216.

38. Solvay, 1933, 165–9; Pauli to Heisenberg, 17 Apr 1934, in Pauli, *Briefwechsel, 2,* 316. Chadwick had wanted to talk about the mass and nature of the neutron (letter to Langevin, 13 Oct 1933, Langevin P), but, apparently, not to listen.

On his way back to Berkeley, Lawrence stuck his head into the Cambridge lions' den. Chadwick bared his claws, to such effect that his behavior needed explanation. It was found in the consideration that he had been the effective director of the Cavendish for some time and was too overworked to observe the niceties of philosophical combat. With the other lions, especially Cockcroft and Rutherford, who licked his chops over Berkeley's "broth of a boy," Lawrence got along well. They merely roared in a friendly way against the hypothesis of deuteron disintegration and pointed their paws at the possibility that Lawrence had contaminated his targets and tank.[39]

Back at home Lawrence mobilized Lewis and switched Livingston and Henderson from trying to withdraw a beam from the cyclotron to clearing up the enigma of the 18-cm protons. Livingston arranged a target holder that would make possible bombardment of many samples in succession to test possible contamination.[40] Working night and day through the Thanksgiving holiday, Lawrence and his group found the yield of protons from deuterons to be unaffected by their efforts to clean up their targets. "Perhaps before long the evidence will be such as to convince the most skeptical, including those at Caltech and even Chadwick."[41]

The Caltech team, Crane and Lauritsen, suggested several possible complications in the analysis of Berkeley's experiments (for example, that the neutron found in deuteron bombardment of lithium might come from $Li^7(p)2\alpha$ followed by $Li^7(\alpha,n)B^{10}$), and could find no trace of the 18-cm protons.[42] Then Tuve's group, which had the only machine then capable of checking Berkeley's results above a million volts, entered the picture. Lawrence had visited the Carnegie Institution on the way back from Brussels. "I persuaded Tuve to investigate the origin of the 18 cm protons and

39. Edward Pollard to Lawrence, 6 Dec 1933, and reply, 20 Dec 1933 (14/30); Lawrence to Cockcroft, 20 Nov 1933 (5/4); Cockburn and Ellyard, *Oliphant*, 74. Rutherford to Lewis, 30 Oct 1933 (Lewis P): "He is a broth of a boy, and has the enthusiasm which I remember from my own youth."
40. Lawrence to Cockcroft (5/4), to Cooksey (4/19), and to Darrow (6/9), all 20 Nov 1933.
41. Lawrence to Gamow (7/25), to Poillon (15/16A), and to Haupt (9/2), quote, all 4 Dec 1933.
42. Crane and Lauritsen, *PR, 44* (Nov 1933), 783–4 (letter of 14 Oct).

the hypothesis of the disintegration of the deuteron right away," Lawrence wrote Cockcroft. "I want to get the matter cleared up as soon as possible and it will be a great help if Tuve, with his independent set up, will investigate the problem."[43]

The experiments went forward everywhere at an American pace. Lewis prepared two samples of calcium hydroxide, one with ordinary and the other with heavy hydrogen. Under bombardment by 3 MeV deuterons the ordinary target showed nothing extraordinary, whereas the heavy target yielded a cornucopia of 18-cm protons. What could be clearer? The bombarding protons broke up the bound deuterons. Lawrence dispatched this "unambiguous proof of d[e]uton disintegration" to the Cavendish; "It would seem now that even Chadwick will agree."[44] The same message went to the East Coast, to Pollard at Yale ("these recent observations definitely rule out the possibility of impurities") and Beams at Virginia ("the deuton is energetically unstable and disintegrates into a proton and a neutron"); to all other physicists through the *Physical Review;* and, of course, to the Research Corporation. "We have proved beyond any reasonable doubt that the deuton explodes when struck hard enough....This first definite case of an atom that itself explodes when properly struck is of great interest, not only as a possible source of atomic energy, but especially because it is not understandable on contemporaneous theories....[It] promises to be a keystone for a new theoretical structure."[45] The flaw in the hydroxide experiment, which we shall reveal in a moment, was no more subtle than the hint about cheap energy. So far did hope, ambition, impatience, and a need for benefactors drive Lawrence from the objectivity he would have claimed as the first virtue of the scientist.

When the Cambridge atom splitters returned from Brussels, they had a fountain of dilute heavy water and samples of almost

43. Tuve to Lawrence, 30 Jan and 12 Sep 1933 (3/32); J.A. Fleming to Lewis, 9 May 1933 (Lewis P); Lawrence to Cockcroft, 20 Nov 1933 (5/4).

44. Quotes from, resp., Lawrence to Walton, 20 Dec 1933 (18/1), and to Fowler, 28 Dec 1933 (7/12); Lawrence to Rutherford, 20 Dec 1933 (ER); Lewis, Livingston, Henderson, and Lawrence, *PR, 45* (1933), 497.

45. Lawrence to Pollard, 20 Dec 1933 (14/30), to Beams, 28 Dec 1933 (2/26), and to Poillon, draft enclosing report for the year 1933, 15 Dec 1933 (15/16A); Lewis, Livingston, Henderson, and Lawrence, *PR, 45* (1934), 242-4 (rec'd 3 Jan).

pure deuterium created in their absence by the finally successful, and consequently now esteemed, Harteck. His achievement came in good time, since Lewis's still had run dry and he could supply nothing until just before the Solvay Congress.[46] Now, with sufficient stock to hand, Rutherford and a research student, A.E. Kempton, returned to the master's old game, let alpha particles from polonium plunge through deuterium gas (the inverse of deuterons on helium), and found no evidence of fast disintegration protons.[47] Cockcroft and Walton sent deuterons against copper and gold and likewise detected no protons.

In the middle of December, Rutherford gave a speech at the Royal Society summarizing the latest evidence. He and Oliphant had at last found neutrons from deuterons on lithium, and Lauritsen neutrons from deuterons on beryllium; but whereas everyone associated these neutrons with nuclear transformations, Lawrence plumped for what Rutherford dismissed in a letter to Bohr as an "exothermal nucleus," and, together with Livingston, offered very shaky evidence that ruled out the reaction $Li^7(d,n)2\alpha$ as the source of the Cambridge neutrons.[48] Another Cavendish man, D.E. Lea, countered the hypothesis of the deuteron's instability by ascribing the hard gamma rays he observed from wax irradiated by neutrons to the spontaneous, endothermic formation of deuterons.[49]

46. Harteck to Bonhoeffer, 21 and 24 Oct, 5 and 24 Nov, 2 Dec 1933 (Harteck P); cf. "Discussion on heavy hydrogen," PRS, A144 (1934), 27 (Rutherford), 10–1 (Harteck); Lewis to Rutherford, 5 Oct 1933 (Lewis P).

47. Rutherford and Kempton, PRS, A143 (1934), 724–30, in Rutherford, CP, 3, 377–83.

48. Rutherford, Nature, 132 (23 Dec 1933), 955–6; Oliphant, Kinsey, and Rutherford, PRS, A141 (1933), 722–33, in Rutherford, CP, 3, 358; Lawrence and Livingston, PR, 45 (1934), 220 (letter of 3 Jan). "Exothermal nucleus" comes from Rutherford to Bohr, 3 Jan 1934 (ER). Lawrence and Livingston's evidence: there appeared to be too many neutrons in comparison with the number of alpha particles to agree with the supposed reaction.

49. Lea, Nature, 133 (6 Jan 1934), 24. Lea's observations did concern deuteron synthesis, but not, as he thought, by direct action of the fast neutrons with the wax protons. Only after they have been slowed by collisions in the wax do the neutrons become very effective for making deuterons. Thus Lea, and Chadwick and Goldhaber, who questioned his results, missed the grand discovery of thermal neutrons later made by Fermi, who mentioned Lea's work. Goldhaber in Stuewer, Nuclear physics, 93–4, and in Hendry, Cambridge physics, 191–3; Fermi, CP, 1, 757.

Lawrence sought protection behind his big gun. Your conclusions are "hardly justified," he told Cockcroft, since the Cavendish experiments had run at under 600 kV.[50] Beginning at 700 kV things became more interesting. Lauritsen had then recently found many neutrons from fast deuterons on beryllium, carbon, and even copper. That, Lawrence crowed to Livingston, amounted to unquestionable corroboration of their experiments. "Chadwick will have to come down off his high horse now." The word at the Laboratory was that Lawrence had "clinched his mass of the neutron—though the evidence [as Kurie rightly objected] is not as clear as I'd like to see." Why did Rutherford not find fast protons from alpha particles on deuterium gas?[51] Lawrence tried to convince the reigning theorist in the business, Gamow, whom he had met at the Solvay conference; Gamow found the fog of conflicting experimental findings too dense to penetrate and offered to "come to California and try to split nuclei by pure theory."[52]

The Fruit

The Cavendish experimenters instead resolved the matter by pure experiment. At the end of February 1934, Cockcroft reported that they had at last detected fast protons from deuteron bombardment of copper, iron, gold, and yttrium, at energies down to 200 kV. Also, and the eventual key to the problem: "Oliphant is getting queer results with H^2 and H^2." Oliphant and Harteck had made up targets like NH_4Cl containing heavy hydrogen. When bombarded with protons, the targets behaved as if constituted entirely of ordinary hydrogen. When shelled with deuterons, they gave off protons with a range of 14.3 cm, which the Cavendish researchers tacitly identified with Berkeley's 18-cm rays. And this irrespective of the deuterons' energy, as long as it exceeded 20 keV. At 100 keV the effect was too large for recording on the

50. Cockcroft to Lawrence, 21 Dec 1933, and response to Cockcroft, 12 Jan 1934 (4/5).
51. Livingston to Lawrence, 23 Jan 1934, and Lawrence to Livingston, 26 Jan 1934 (12/12); Kurie to Cooksey, 6 and 21 Mar 1934 (10/21).
52. Lawrence to Gamow, 28 Dec 1933, and Gamow to Lawrence, 12 Jan 3[4] (7/25).

sensitive detector of the Oliphant-Rutherford apparatus, so great in fact that Oliphant and Rutherford at first ascribed it to a burst of x rays. Analysis of range and energy, confirmed by cloud chamber pictures, showed that the protons arose from the reaction $d+d \rightarrow H^3+p$. They also found the neutrons that Lawrence had advertised. Rutherford discerned their origin in the competitive reaction, $d+d \rightarrow He^3$ n.[53] Oliphant broke the news gently. "We suggest very tentatively that your results may be explained as due to the bombardment of films of D and of D compounds....I hope these results are of interest to you."[54]

The puzzle Lawrence started had therefore the simple solution of contaminated apparatus: deuterons stuck in the walls of the cyclotron chamber and in the targets gave the plentiful d-d reaction when bombarded by deuterons; the fast protons, apparently from everything, which had prompted the disintegration hypothesis, did in fact come from deuterons, not from the target, but in consequence of fusion, not disintegration. The accelerator men at Caltech and Carnegie also offered plentiful evidence that Berkeley's results came from synthesis of deuterons with pervasive contaminants. Lauritsen and Crane showed that the fast protons from deuterons on carbon came from $C^{12}(d,p)C^{13}$, an observation extended by Cockcroft and Walton to explain, as a consequence of oil films on metals, why deuterons appeared to knock protons from copper and tungsten. Tuve and Hafstad repeated all the Berkeley experiments with carefully controlled beams on immaculate targets and found that the Berkeley experimenters had got nothing right. With a chamber contaminated with deuterons, however, they had no trouble recovering Berkeley errors.[55] The demonstration with the hydroxide targets, which Lawrence had thought so convincing, also failed through ubiquitous deuterons.

53. Cockcroft to Lawrence, 28 Feb 1934 (4/5); Oliphant to G.P. Thomson, 9 Mar 1934 (GPT, D/1), on Dee's confirmation with the cloud chamber; Dee, *Nature, 133* (14 Apr 1934), 564; Oliphant, Harteck, and Rutherford, *Nature, 133* (17 Mar 1934), 413, in Rutherford, *CP, 3,* 364–5, expanded in *PRS, A144* (1934), 692–703, in *CP, 3,* 386–96; Cockburn and Ellyard, *Oliphant,* 53–5.

54. Oliphant to Lawrence, 12 Mar 1934 (10/16).

55. Lauritsen and Crane, *PR, 45* (1934), 345–6 (letter of 15 Feb); Cockcroft and Walton, *PRS, A144* (1934), 704–9, 717–9; Tuve and Hafstad, *PR, 45* (1934), 651–3 (letter of 14 Apr).

As Lauritsen and Oliphant independently explained it to Lawrence: the proton beam playing on $Ca(DH)_2$ liberated deuterons by collision or exchange; these deuterons then joined the bombardment and produced the 18-cm protons by d-d synthesis.[56]

The hunters horsed to catch the hare let loose from Berkeley were formidable: the Cavendish from the top down and the experienced teams and big machines at two of the best-endowed physics research institutes in the United States. Resolution of the conflicting results required very careful experiment in previously unexplored regions of multiple, competing nuclear reactions; and it required, and abetted, discovery of the occult d-d fusion process and the isobars of mass three. That detection and analysis of deuterium fusion was not so easy as its plentiful yield might suggest appears from the struggles of Otto Frisch, who proposed to Bohr that they build a big source of neutrons utilizing the d-d reaction. He had a machine available in December 1934. As late as the following March he had not detected a single neutron, although he commanded 60 kV, far above the threshold reported by Oliphant, Harteck, and Rutherford. He did succeed eventually, as did G.P. Thomson, who also had trouble reproducing the Cambridge results.[57]

In accordance with the social expectations of science, Lawrence confessed himself chagrined at the "stupidity" of his error, and Cockcroft softened the blow with the falsehood that "for a long time Rutherford and Chadwick were nearly convinced that you were right" and the truth that the experiments of Lawrence, Lewis, and Livingston had brought "an important enlargement of the field of nuclear research."[58] Lawrence diagnosed his stupidity as a consequence of the great productivity of the cyclotron, which

56. Lawrence to Cockcroft, 14 Mar 1934 (5/4), and Lewis, Livingston, Henderson, and Lawrence, *PR, 45* (1934), 497 (letter of 15 Mar), re Lauritsen's suggestion; Oliphant to Lawrence, 12 Mar 1934, and Lawrence to Oliphant, 5 June 1934 (10/16). Kurie to Cooksey, 16 Mar 1934 (10/21): "Lauritsen has thrown the final monkey wrench in the disintegration hypothesis."

57. Frisch to Meitner, 25 Mar, to Bohr, 29 Jun, and to Rausch von Traubenberg, 28 Dec 1934; to Meitner and to his father, 10 Mar, and to Jacobsen, 5 Apr 1935; and to Meitner, 27 Oct 1937 (Frisch P); Thomson to Oliphant, 5 Mar 1935 (GPT).

58. Quotes from, resp., Lawrence to Cockcroft, 14 Mar 1934, and Cockcroft to Lawrence [1934] (4/5), and Cockcroft and Walton, *PRS, A144* (1934), 704.

discouraged "methodical, quantitative, measurements;" a healthy point of view that enabled him soon to dismiss the experiments that had prompted his error as "of the character of a preliminary survey."[59] He had persisted in his errors, however, in the face of warnings from many sides of the likelihood of contamination.[60]

This flouting of good advice and good procedure irritated some of Lawrence's closest associates. Tuve wrote Cockcroft: "From our own experiments we feel that the important issue is rather one of judgment and point of view rather than of the errors in technique which can give rise to such a situation." Lawrence had botched everything so badly that the corrective, as Tuve wrote in an internal report, "will be a difficult thing to present in public." Kurie wrote Cooksey: "The Englishmen are doing what seem at this distance to be clean experiments, but Ernest and Malcolm [Henderson] are too excited to go slowly." These expressions of disappointment appear to reflect a worry that Lawrence's way of doing science and his rising celebrity in the United States might compromise American physics just as it was assuming world leadership. Cockcroft agreed that Lawrence's style did not suit nuclear physics. "There is a real danger of the subject getting into a mess, and I feel that the only thing to do is to delay publication until we are reasonably sure."[61]

Tuve did not hide his irritation from Lawrence. Their different institutional settings and personal ambitions had become more important in determining their scientific ethos than their similar backgrounds and almost identical scientific interests. In the spring of 1934 Tuve advised Lawrence to withdraw his faulty claims formally, a step not only ethically sound but also useful to Tuve, who would be spared the obligation of a public exposé. Lawrence replied petulantly. In the summer of 1934 Tuve carried the attack

59. Quotes from, resp., Lawrence to Rutherford, 10 May 1934 (ER), and to Beams, 13 Apr 1934 (2/26); Lawrence, *PR, 47* (1935), 17; Oliphant, *PT, 19:9* (1966), in Weart and Phillips, *History*, 182–4.

60. E.g., in letters from Boyce, 23 Jan 1933 (3/8), Cockcroft, 30 Mar 1933 (5/4), and Fleming, 10 May 1933 (3/32).

61. Tuve to Cockcroft, 18 Apr 1934, and Tuve, "Memorandum," 3 Apr 1934 (MAT, 5); Cockcroft to Tuve, 30 Apr 1934 (CKFT, 20/80), the latter also in Hartcup and Allibone, *Cockcroft*, 65; Kurie to Cooksey, 4 Mar 1934 (10/21). Cf. Beams to Lawrence, 7 Apr 1934 (2/26), and Tuve, Hafstad, and Dahl, *PR, 48* (1935), 316, in Livingston, *Development*, 29.

to Berkeley, to a meeting of the American Physical Society. Either the hospitality or the prevailing genius of the place caused the secretaries who reported the discussions to nod: they attributed to Tuve the irenic solution that his and Lawrence's results could be brought into harmony by considering the different apparatus and energy with which they worked. Not a mention of contamination or haste. Tuve could not allow that to pass, since it implied that not Lawrence, but he, had been wrong. He therefore wrote to *Science*, which had published the irenic report, to point out that the Berkeley group had abandoned their major claims and withdrawn many of their results, "which have been sought for but not verified," and which, by confusing the subject, had "delayed somewhat the examination of these questions."[62]

To recover his standing, Lawrence thought to make a few "precise and trustworthy measurements." The deuteron business brought home forcefully, he told Cockcroft, that most of the Laboratory's fast data had no value. Accordingly, Lawrence made an elaborate study of the excitation and decay products of the useful isotope radiosodium, a novelty whose discovery will occupy us presently. Up to the maximum energies he used (1.9 MeV), deuterons seemed to induce transformations in sodium in the amounts predicted by Gamow's theory. But continuation of the experiments to aluminum and to higher energies in collaboration with a new recruit to the Laboratory, a postdoc from Princeton named Edwin McMillan, who knew how to work exactly, showed excitation larger than Gamow allowed. The departure of their measurements from a theory that worked well for protons and alpha particles interested Oppenheimer and his student Melba Phillips, who had just finished failing to explain the curious rise of absorption beyond the series limit in Lawrence's old reliable measurements on the photoeffect.[63]

When a fast deuteron enters a heavy atom, Coulomb repulsion slows its proton and so increases the relative velocity of its constituents. Under these circumstances, the nucleus may capture the

62. Tuve to Lawrence, 17 Apr 1934, and response, 20 Apr 1934 (17/34); Tuve to Darrow, 28 Jul 1934, and to Cattell, 4 Aug 1934 (MAT, 13/"lab. letters"); Ward, *Science, 80* (20 Jul 1934), 49, and Tuve, ibid., 17 Aug 1934, 161–2, quote.

63. Lawrence to Cockcroft, 14 Mar 1934 (5/4), and to R.W. Ditchburn, 13 May 1933 (6/16).

neutron. The released proton then makes an appearance in the world. Encouraged by this attractive possibility, which occurred to several physicists, Lawrence, McMillan, and another new post-doc, Robert Thornton from McGill, pushed on to silicon and copper. Their results agreed exactly with the calculations of Oppenheimer and Phillips for a binding energy of the deuteron—about 2 MeV—that agreed very well with values obtained from mass-energy balances in nuclear transformations as measured by others.[64] Here was a point for the underdog. "I am amazed at the agreement," Lawrence wrote Rutherford. And also incurable: with the new mechanism, deuterons with available energies might pro-voke nuclear reactions "much further up the periodic table than one could ever have hoped for."[65]

There was another point for Lawrence's side of the story. Maurice Goldhaber, an émigré research student at the Cavendish, suggested to Chadwick an idea for an experiment that he had brought from Berlin: the photodisintegration of the deuteron. Goldhaber recalled that Chadwick showed little interest in the matter until told that if successful it would give data from which the neutron's mass could be deduced.[66] Telltale protons announced the disintegration of deuterons in heavy water irradiated by the hardest photons then available, the gamma rays of 2.6 MeV from ThC", of which the Cavendish had a good supply. With a preliminary determination of the kinetic energy of the proton, Chadwick and Goldhaber had a value of the neutron's mass without making assumptions about the mass of complex nuclei:

$$m_n = m_d - m_p + E_\gamma - 2E_p \approx 1.0081.$$

Their value, almost midway between Chadwick's and the Joliots', is close to the modern one (1.0085). Since the mass of the proton was known to be 1.0078, it followed that Chadwick's neutron was radioactive. Its decay was first observed in 1948.[67] The discovery

64. Oppenheimer and Phillips, *PR, 48* (1935), 500–2. Cf. Pollard to Lawrence, 6 Dec 1933 (14/30); Oppenheimer to Bethe, "27 Nov," and Fleming to Bethe, 7 June 1935 (HAB, 3).

65. Lawrence, McMillan, and Thornton, *PR, 48* (1935), 494; Lawrence to Rutherford, 17 Apr 1935 (15/34), quote; Lawrence, *Ohio jl. sci., 35* (1935), 404.

66. Goldhaber in Stuewer, *Nuclear physics,* 84–8, summarized in Goldhaber in Hendry, *Cambridge physics,* 190–1; Bethe to F.W. Loomis, 5 May 1938 (HAB, 3).

67. Chadwick and Goldhaber, *Nature, 134* (1934), 237, disconfirming Lauritsen

of the photodisintegration of the deuteron gave rise to a new branch of nuclear investigation, to which the Berkeley Laboratory, which preferred to shoot particles it could accelerate, contributed very little. For a time the (γ,n) reaction appeared to be a promising source of neutrons.[68]

And finally, a point to be dilated later, the discovery of the d-d reaction, although instigated by cyclotron experiments, did not much recommend cyclotrons to nuclear physicists. Certainly it helped confirm the opinion of Rutherford, who had outplayed Lawrence with beams ten or even a hundred times less energetic than Berkeley's, that the Cavendish had no need for a cyclotron. More generally, it made possible a cheap substitute for the energetic neutron beams that, for a brief time, had been the peculiar preserve of large accelerators. It was now only necessary to have a supply of heavy water and a Cockcroft-Walton or Van de Graaff that could develop 100 or 200 kV, and one had a copious current of fast neutrons almost equal in energy courtesy of the peculiar sociability of deuterons.[69]

2. A FRUITFUL BUSINESS

The episode with the deuterons exposed weaknesses in the work of the Radiation Laboratory that were only partly corrected before its mobilization in World War II. The pressure for quick results to encourage financial backers continued, with consequent hype and hurry. Errors plagued output: as Lawrence anticipated in

---

and Crane, *PR*, *45* (1934), 550–2 (letter of 24 Mar), whose careful study of $Li^7(d,n)2\alpha$ had recovered Chadwick's earlier value (m_n = 1.0068), and Curie and Joliot, *Nature*, *133* (12 May 1934), 721, whose review of (α,ne^+) reactions had confirmed their own value (m_n = 1.010). Cf. Ladenburg, *PR*, *45* (1 Feb 1934), 224–5, and *PR*, *45* (1 Apr 1934), 495; Chadwick, Feather, and Bretscher, *PRS*, *A163* (1937), 366–75, giving m_n = 1.0090. Chadwick and Goldhaber, *PRS*, *A151* (1935), 479–93, predicted the decay of the neutron; the first observations took place in reactor laboratories after World War II. Feld in Segrè, *Exp. nucl. phys., 2* (1953), 217–8.

68. Among the first to follow up the demonstration by Chadwick and Goldhaber were Szilard and the chemist T. Chalmers, who recommended radium gamma rays on beryllium. Szilard and Chalmers, *Nature*, *134* (29 Sep 1934), 494–5, in Szilard, *CW*, *1*, 145–6.

69. Cf. Feld in Segrè, *Exp. nucl. phys., 2* (1953), 380–4.

answering Cooksey's condolences over the death of the disintegration hypothesis, the Laboratory's mission and method guaranteed mistakes. "I have gotten over feeling badly," he wrote. "We would be eternally miserable if our errors worried us too much because as we push forward we will make plenty more." He worried a bit about the consequences. As a palliative he proposed that meetings of theorists should always include leading experimentalists, who could certify the value of the data, "Theoretical physicists," said he, forgetting his own persistence in error, "so often are liable not to appreciate which experimental observations are trustworthy."[70]

The proportion of solid results did increase, however, owing partly to Lawrence's resolution, which he sometimes kept or imposed on others, to finish and write up one research project before rushing to another; and owing largely to the appearance in the Laboratory of experienced researchers, who could design and carry through their own projects. Their presence brought something of the Cavendish pattern to the Laboratory. Just as Cockcroft and Walton, Oliphant and Rutherford, Chadwick and Lea, Walton and Dee, formed the nuclei of small groups that tackled similar problems from different points of view, so now Franz Kurie and Edwin McMillan, National Research Fellows for 1933/34, helped break the Laboratory from exclusive preoccupation with Lawrence's programs for physical research and machine design.

Kurie spent the first six months of his fellowship year fashioning a fine cloud chamber on the principle of Tuve's instrument, which worked "well, damned well as a matter of fact."[71] McMillan spent the same time disenchanting himself with the research project that had brought him to the Physics Department and considering what he might do in the Laboratory. Both found rich research lines under the inspiration of a discovery that also established the purpose, and secured the financing, of the prewar Laboratory. This prepotent discovery was made in France by

70. Lawrence to Cooksey, 12 Mar 1934 (4/19); to J.A. Fleming, 2 Jul 1935 (3/32), resp.

71. Kurie to Cooksey, 4 Mar 1934 (10/21), quote; Kurie, *RSI, 3* (1932), 655–7; Dahl, Hafstad, and Tuve, *RSI, 4* (1933), 373.

Curie and Joliot as they ploddingly reexamined the reactions on which they had based their generous estimate of the mass of the neutron.

Induced Radioactivity

Lawrence was not the only one to suffer for his hypothesis in Brussels. Curie and Joliot ran into formidable opposition to their conception of (α,ne$^+$) reactions—the supposition that boron and aluminum transform under alpha bombardment, with the simultaneous emission of a neutron and a positive electron. In their view, (α,ne$^+$) paralleled (α,p) and demonstrated that the proton consists of a neutron and a positron.[72] No one doubted the presence of the positive electrons: but, to avoid the heavy neutron and the complexity of the proton, most Solvay participants preferred to place the origin of the positron outside the bombarded nucleus. The subterfuge appeared to work for beryllium, which emits gamma rays as well as neutrons under alpha bombardment, for the gammas might later convert into pairs of positive and negative electrons. But as Curie and Joliot pointed out, this explanation could hardly hold for aluminum, which, according to their experiments, did not emit gamma rays under alpha bombardment and gave out very few (if any) negative electrons in comparison with its positives. Consequently they held to (α,ne$^+$), only to be shot down by Lise Meitner, who had found no neutrons from alpha irradiation of aluminum.[73] Her faulty observation, which she later retracted, was to the Joliots what Chadwick's assurance was to Lawrence. They went back to their laboratory to prove their opponents wrong.

They thought that they could strengthen their argument by showing that neutrons and positrons appeared together and in equal numbers regardless of the energy of the incident alpha particles. Altering the incident energy required inserting absorbers between the polonium source and aluminum target; showing the associated production meant registering the positron on a Geiger counter and a conversion proton (from the neutron) in a cloud

72. Curie and Joliot, *JP, 4* (1933), in *Oeuvres,* 444–54, esp. 452–3.
73. Solvay, 1933, 173–7.

chamber. All went as expected down to a certain energy, at which the conversion protons stopped, but not the positrons. Here Joliot confirmed the suspicion that Thibaud had expressed at the time of the Solvay Congress, that some radioactive bodies can emit positive electrons.[74] Joliot next tried the experiment with alpha particles of full energy. After the irradiation he removed the polonium source altogether. Still the positrons appeared for their allotted three minutes.

Joliot had made the aluminum radioactive by hitting it with alpha particles. He had discovered a two-step process, an (α,n) reaction resulting in the creation of a new, unstable isotope of phosphorus followed by a positron decay to a stable isotope of silicon. The two-step achieved the same end as the single, straightforward, old-fashioned reaction (α,p) would have procured. The intermediate product in the two-step brought not only confirmation of the Parisian heavy neutron, but something much more important, and altogether new: artificially created radioactive substances, which could be identified chemically by the carrier technique developed to analyze the products of natural radioactive decay. With the help of his wife, Joliot demonstrated that the three-minute activity followed the chemistry of phosphorus and that the fourteen-minute activity produced by (α,n) on boron followed that of nitrogen. They brought a vial of one of their new creations to old Madame Curie, then dying of leukemia. Joliot described the scene. "I can still see her taking [it] between her fingers, burnt and scarred by radium....This was without doubt the last great moment of satisfaction in her life."[75]

News of the discovery did not provoke much satisfaction when it reached Berkeley via *Time* and *Nature*. Lawrence, Livingston, and Henderson spent the weekend of February 24/25, 1934, repeating the experiments of Joliot and Curie in their own way, with deuterons from the cyclotron in place of alpha rays from polonium. "To our surprise we found that everything we bombar-

74. Breit to Tuve, 9 Oct 1933 (MAT, 12/"spec. letters").

75. Joliot, quoted in Goldsmith, *Joliot-Curie*, 57; Curie and Joliot, *CR, 198* (15 and 29 Jan 1934), 254, 559, in *Oeuvres*, 515-9, and *Nature, 133* (1934), 201, 721 (reaffirming the neutron mass), in *Oeuvres*, 520-1. Cf. Amaldi, *Phys. rep., 111* (1984), 109.

ded...is radioactive." And also to their chagrin. "We have had these radioactive substances in our midst now for more than half a year. We have been kicking ourselves that we haven't had the sense to notice that the radiations given off do not stop immediately after turning off the bombarding beam."[76] It was not that the effect hid near the limit of detection: for aluminum it overpowered the Geiger counter. That made missing it—and the Nobel prize awarded to Joliot and Curie the following year—particularly galling. Later Lawrence's junior collaborators recalled what they remembered of their feelings. Thornton: "We looked pretty silly. We could have made the discovery at any time." Livingood: "We felt like kicking our butts."[77]

According to the standard apologies, the Laboratory missed the discovery because the same switch operated the cyclotron and the Geiger counter, and so turned off the means of detection with the initiating beam. It may be doubted that the equipment was so peculiarly wired. And even if it were, the fact that no accelerator laboratory thought to make substances radioactive remains to be explained. The Cavendish had looked for delayed activity in aluminum, among other elements, during the 1920s, with natural sources of alpha particles; they had found nothing, because, since neither the neutron nor the positron had yet been noticed, they had no idea what to look for, and sought to detect short-lived proton or alpha emitters with scintillation screens. As one frustrated investigator wrote, more truly than he knew, "It is very unfortunate that time did not permit of further experiments with a wide variety of elements and with devices for the detection of radiation of other kinds." Despite their larger sources and greater knowledge, accelerator builders did not reopen the matter.[78] It was not a question of labor-saving switches, but of labor-saving thinking. One expected either transmutation to known, stable species, or reduction to fundamental pieces of nuclei, but not the creation of brand-new radioelements.

76. Lawrence to Beams (2/26), to J. Boyce (3/8), quote, both 27 Feb 1934; Kurie to Cooksey, 4 Mar 1934 (10/21).

77. Davis, *Lawrence and Oppenheimer*, 60.

78. Shenston, *Phil. mag.*, *43* (1922), 938–43, quote on 943; Blackett, *PRS, A107* (1925), 357; Rutherford, Chadwick, and Ellis, *Radiations*, 312–3.

Joliot and Curie had raised the possibility that deuterons might create artificial activities in their announcement of their discovery in *Nature*. They gave $C^{12}(d,n)N^{13}$ as an example. Four groups stood ready to follow up the suggestion: Cockcroft's, Tuve's, Lauritsen's, and Lawrence's. Cockcroft at first preferred his original projectile and made N^{13} by stuffing a proton into C^{12}. Later he and his associates confirmed (d,n) reactions on boron, carbon, and nitrogen at energies under 600 kV. Tuve did not interrupt his investigations of Berkeley's mistakes to follow up Joliot and Curie's suggestions. Lauritsen did. He sent preliminary results on (d,n) reactions for publication on the same day that Lawrence did.[79]

The difference in research objectives between Caltech and Berkeley deserves notice. Henderson, Livingston, and Lawrence examined fourteen elements, from lithium to calcium, under bombardment by 1.5 MeV protons and 3 MeV deuterons; they noticed signs of proton activation only in carbon and supposed the ubiquitous deuteron activation to arise via $(d,e^+\gamma)$ reactions. They gave few and only rough quantitative data, for example, a half-life of the boron activity of about two minutes. In an unpublished lecture, Lawrence conceded that none of the measurements could stand up to the "very significant experimental findings" of Crane and Lauritsen.[80] The Caltech group limited its initial studies to 0.9 MeV deuterons on beryllium, boron, and carbon, understood that the activities they created arose from (d,n) reactions, showed that the half-life of the activity made from carbon agreed with that of N^{13} as given by Joliot and Curie, had their colleagues Carl Anderson and Seth Neddermeyer confirm the existence of positrons in the decay of N^{13} by observations with the Caltech cloud chamber, showed that the gamma ray found at Berkeley probably came from electron-positron annihilation, and determined the half-life of the boron activity to be ten times as large as Berkeley made it. Where Lauritsen's group gave careful and reliable

79. Joliot and Curie, *Nature, 133* (1934), 201–2, and in *Oeuvres,* 521; Cockcroft, Gilbert, and Walton, *Nature, 133* (1934), 328 (letter of 24 Feb), and *PRS, A148* (1934), 225–40 (rec'd 26 Sep); Crane, Lauritsen, and Harper, *Science, 79* (1934), 234–5 (letter of 27 Feb).
80. Lawrence, "Outline of lecture on artificial radioactivity," n.d. (40/16).

information about a few features of the new terrain, Lawrence's characteristically bolted through an impressionistic survey.

The usual tendency in the Radiation Laboratory may have been strengthened in this case by the increasing difficulty in maintaining the hypothesis of deuteron disintegration and by Lawrence's desire to assimilate their earlier results to the great Parisian discovery and insinuate an anticipation of it. "Indeed, in the light of our recent experiments in which neutrons and protons were found to be emitted from many elements when bombarded with deutons, the possibility presented itself that in these nuclear reactions [!] new radioactive isotopes of many of the elements might be formed." So Henderson, Livingston, and Lawrence hinted in the *Physical Review* in 1934. Later and in private Lawrence may have claimed more. A representative of the Rockefeller Foundation recorded this remark: "[Lawrence] said that they had discovered artificial radioactivity before Joliot and Curie did, but wishing to be overly sure [!] of their results, did not publish and were taking time to repeat the work."[81]

Some Physics Fallout

Lewis had very probably been the instigator in the deuteron experiments.[82] An extremely clever man with a secure reputation as a chemist, he had little to lose by backing poor physics; a hasty man, guided by smell and inspiration, he was the worst sort of collaborator for Lawrence. Also, he had ideas about atomic and nuclear structure that differed in principle from those physicists entertained. He had sponsored a static atom, in which electrons stand at the vertices of a polyhedron centered on their nucleus, in competition to Bohr's dynamic-electron model. His colleague Wendell Latimer had extended the scheme to the nucleus, which he supposed to consist of as many alpha particles as possible joined together in equilateral pyramids. He had no place for neutrons; they come to life outside nuclei, by couplings of protons

81. Henderson, Livingston, and Lawrence, *PR*, *45* (1934), 428-9 (letter of 27 Feb); Crane and Lauritsen, *PR*, *45* (1934), 430-2 (letter of 1 Mar); Amaldi, *Phys. rep.*, *111* (1934), 115-18; Frank Blair Hansen, "Trip report," 3-13 Apr 1938 (RF, 1.1/205).

82. Lawrence to Potter, Pierce, and Scheffler, 3 June 1935 (35/7).

and neutrons, couplings easily broken and reformed, in his opinion, so as to make hydrogen atoms, deuterons, mass-three helium, and so on.[83] It was a qualitative, tinker-toy world, in which one part—such as the strength of the stick holding the deuteron together—could be changed without doing violence to other parts. Even before the detection of heavy water, a student of Berkeley's nuclear family foresaw the exploitation of the cyclotron to check the chemists' physics. "It may be," he wrote, "the research by Professor Lawrence and Dr Livingston will offer means of proving or disproving it."[84]

Fowler was astonished at the eagerness and confidence with which Berkeley chemists did physics. "If they do any chemistry it's kept well out of sight."[85] The fiasco of his work with Lawrence did not discourage Lewis. In 1936 he offered an explanation of neutron scattering that was as disruptive to theory as exploding deuterons. Bethe reviewed the manuscript for the *Physical Review*: "I think it is an extremely instructive example of the dangers of purely qualitative arguments." Lewis's former confederates at the Laboratory would not follow him: "The effect [for which Lewis argued] is so feeble, and the instruments so barbaric (he doesn't want to hear about counters) that no one believes him here."[86] With this rejection, Lewis ceased to play a direct part in the work of the Laboratory.

Kurie did not allow himself to be drawn into the search for activities induced by deuterons. He took on instead the elucidation of the mechanism of neutron activation. He studied closely the forked tracks created in his cloud chamber during neutron irradiation of nitrogen. In contrast to Rutherford's prompt reaction $N^{14}(\alpha,p)O^{17}$, Kurie thought he saw the delayed reaction $N^{14}(n)N^{15} \rightarrow B^{11} + \alpha + Q$, where Q designates the energy carried away in gamma rays and N^{15} is an "intermediate nucleus." Kurie reported this first piece of careful physics done with cyclotron

83. Lewis, *Valence*; Latimer, *JACS, 53* (1931), 981–90, and *JACS, 54* (1932), 2125–6; Kohler, *HSPS, 3* (1971), 343–76.

84. G.A Pettitt, *California monthly, 27* (1931), 18–21.

85. Fowler to Rutherford, 22 Mar [1933] (ER).

86. Bethe to Buchta, 6 May 1936 (HAB, 3); Nahmias to Joliot, 28 Apr 1937 (JP, F25); Seaborg, *Jl., 1,* 242; Lewis and Schutz, *PR, 51* (15 June 1937), 1105.

beams and a good detector at a meeting of the American Physical Society in Berkeley in June 1934.[87] In the fall he gave a seminar on his work. "For the first time in my life [it] was not a recital of numbers but of ideas." Everyone seemed convinced, except Oppenheimer, who worried about the powerful gamma ray that, if the conservation laws held, must be emitted in the formation of the intermediate nucleus. "Robert says that the evidence is well explained by it but he 'wishes it were not so' "[88] Kurie published his hypothesis and measurements—the sort of paper Lawrence was "proud to have from the lab"—and it was not so. As Bethe laid down the law, in 1937: "This [intermediate excited nitrogen nucleus] has no justification either theoretically or experimentally, and has subsequently been discarded." Apparently Kurie had overinterpreted his tracks.[89] His work was not to end the Laboratory's stream of flawed physics.

At the same meeting in Berkeley of the American Physical Society at which Kurie spoke about delayed disintegrations, McMillan discussed preliminary results of his study of gamma rays excited by 1.15 MeV protons driven against fluorine. He had taken up the subject on the advice of Oppenheimer, who had two objects in mind. For one, the energy balance in nuclear reactions could not be struck without knowledge of the amount carried away by high-frequency radiation. For another, and of greater interest to Oppenheimer, gamma rays from some artificially induced reactions might well be more energetic than any from natural souces; if so, they would permit a check of the theory of pair production—the materialization of a gamma ray into a positron and an electron in the field of a nucleus—at higher energies than previously available. Oppenheimer and one of his students, Wendel Furry, had a calculation of pair production in such regions in hand.[90]

McMillan's experimental arrangement occupies figure 4.2. The Lauritsen electroscope consisted of a quartz fiber suspended from a wire and carrying on its free end a crosshair viewed against a

87. Kurie, *PR, 46* (1934), 324; Lawrence to Gamow, 19 May 1934 (7/25).

88. Kurie to Cooksey, 18 and 27 Sep 1934 (10/21).

89. Kurie, *PR, 47* (1935), 98–105; Lawrence to Cooksey, 29 Dec 1934 (4/19); Livingston and Bethe, *RMP, 9* (1937), 338, 339 (quote), 341.

90. McMillan, *PR, 46* (1934), 325, 868, 870.

scale in the microscope eyepiece. Foils of various metals allowed determination of the absorption of the gamma rays from the target as a function of atomic number Z, and thereby distinction of the portion owing to pair production (which increases as Z^3) from contributions from the photoeffect and the Compton effect. McMillan took elaborate precautions not to be duped by contaminants. The best results came from fluorine, which produced fine energetic gamma rays, some 5.4 MeV, in the reaction $F^{19}(p,\alpha)O^{17}$. Pair production by these rays agreed perfectly with the curvature of the tracks of the most energetic photoelectrons they produced in a cloud chamber at Caltech.[91] McMillan's were the first experimental results in nuclear physics obtained at the Laboratory and controlled by a quantitative theory that have stood up under bombardment from other investigators.

Tracer Business

While McMillan and Kurie went their independent ways and Lawrence's group continued firing deuterons, another capital discovery arrived from Europe. Fermi had reasoned that because they carry no electrical charge, neutrons should be able to gain admission to nuclei more readily than protons, deuterons, or alpha particles; and that they were the only way to activate nuclei heavier than phosphorus, where even Berkeley's deuterons could not penetrate.[92] A methodical man, Fermi began with hydrogen as a target, then lithium, and so on, at first with no luck. He was not discouraged before reaching fluorine, which released electrons when struck by the neutrons from his modest radon-beryllium source. That was on March 25, 1934. Fermi then mobilized his collaborators, Edoardo Amaldi, Franco Rasetti, and Emilio Segrè, who raced at California speeds to procure and bombard specimens of all the elements in the periodic table. By July they had reached uranium and detected no fewer than forty artificial isotopes of the

91. McMillan, *PR, 46* (1934), 871–2; Henderson, Livingston, and Lawrence, *PR, 46* (1934), 38; Crane, Delsasso, Fowler, and Lauritsen, *PR, 46* (1934), 531.

92. Cooksey had suggested the effect Fermi sought the year before: "I suppose that the neutron in the H^2 is the boy that when given an introduction in the company of a proton raises all this merry hell." Cooksey to Lawrence, 6 May 1933 (4/19).

Joliot-Curie type, but which decayed by emitting negative rather than positive electrons. Of particular interest, for reasons to appear, was Na^{24}, with a half-life of fifteen hours, which Fermi's group could create either from Al^{27} via (n,α) or from Mg^{24} via (n,p).[93]

The discoveries of artificial radioactivity and of the capacity of neutrons to effect transformations beyond the reach of deuterons coalesced several disconnected ingredients of Lawrence's research, machine building, and fund-raising into an enduring whole. Even before Fermi's discovery, at the Solvay Congress of 1933, Lawrence had emphasized the importance of deuteron bombardment as a source of neutrons: with a current of only 0.01 μA the Laboratory had a source that appeared to be more powerful than any likely to be obtained from natural radioelements. One standard estimate, used by Fermi, gave 1,000 neutrons/sec as the output of one mCi of Rn-Be; another, used by Joliot and Curie, made the efficiency of nuclear reactions induced by alpha particles from natural radioactive sources about 10^{-6} or 10^{-7}; hence 0.01 μA of deuterons, or $6.3 \cdot 10^{10}$ particles/sec, would give rise to around 10,000 neutrons/sec, which would have required the radon from 10 grams of radium. This was an exaggeration. Later elaborate measurements by Amaldi and Fermi and by Amaldi, Hafstad, and Tuve raised the yield from a mCi of Rn-Be to 25,000 neutrons/sec and the conversion efficiency of 1 MeV deuterons on beryllium to $3 \cdot 10^{-5}$, whence 0.01 μA of million-volt deuterons would give as many neutrons as 70 mCi of Rn-Be. Fermi did his first experiments with 50 mCi of radon and at times used 700 mCi, which his supplier, G.C. Trabacchi, drew from a gram or so of radium at the Istituto di sanità pubblica in Rome.[94] Fermi's source of

93. Segrè in Fermi, *CP, 1,* 639–41; Fermi, *Ric. sci., 5* (1934), 283, in *CP, 1,* 645–6 (dated 25 Mar 1934), announcing (n,α) reactions on F and Al; *Nature, 133* (1934), 757, letter of 10 Apr, describing activities of two dozen elements; Fermi, Amaldi, D'Agostino, Rasetti, and Segrè, *PRS, A146* (1934), 483–500, in *CP, 1,* 732–47, esp. 746–7 (rec'd 25 Jul 1934). Amaldi, *Phys. rep., 111* (1934), 130, considers the Rome group to have been "probably the first large physicists' team working successfully for about two years in a well organized way."

94. Lawrence, intervention in Solvay, 1933, 68. Cf. Livingston, Henderson, and Lawrence, *PR, 44* (1 Nov 1933), 782–3 (letter of 7 Oct); Fermi, *Ric. sci., 5* (1934), 283, in *CP, 1,* 645–6, and *Nuovo cim., 11* (1934), in *CP, 1,* 715; Amaldi, Fermi, Rasetti, and Segrè, *Nuovo cim., 11* (1934), in *CP, 1,* 725; Fermi, Amaldi,

neutrons compared well with the cyclotron vintage 1933, but it could not compete in total output with later versions or with any other artificial source of a μA of deuterons above a million volts. Still, a natural source retained its usefulness in situations where constancy, reproducibility, and good geometry were especially important.

By April 1934 the Laboratory was busy following up Fermi's results, which greatly complicated inventorying nuclear reactions. "Because of the magnitude of the radioactivity induced by neutrons it was immediately apparent [Lawrence wrote Gamow] that we should study thoroughly the phenomena before trying to untangle other nuclear reactions produced by proton and deuteron bombardment." And there was another reason, as Lawrence wrote another interested party, with interests much different from Gamow's. "We are not unmindful [he told Poillon] of the possibility that we may find a substance in which the radioactivity may last for days instead of minutes or hours, in other words, a substance from which we could manufacture synthetic radium. The probability is not at all remote at the present time. I am very glad that we have a patent on the cyclotron."[95]

With the help of Henderson and Livingston, Lawrence got tremendous amounts of radioactive aluminum, copper, silver, and fluorine; owing to a new set of dees made wider at the center to accommodate a larger ion source, the cyclotron was putting out 0.7 μA of 3 MeV deuterons, which shook over 500,000,000 neutrons/sec from its beryllium target. The poor estimate then still accepted, 1 mCi of Rn-Be gives 1,000 neutrons, implied that the cyclotron had the value as a neutron source of half a kilogram of radium. (Rutherford then estimated the equivalent of Oliphant's machine as a tenth of a gram of radium).[96] When

D'Agostino, Rasetti, and Segrè, PRS, A146 (1934), in CP, 1, 733, 745; Amaldi and Fermi, Ric. sci., 7 (1936), in CP, 1, 887, and PR, 50 (1936), in CP, 1, 892, 937–8; Jaeckel, Zs. f. Phys., 91 (1934), 493; Paneth and Loleit, Nature, 136 (1935), 950; Amaldi, Hafstad, and Tuve, PR, 51 (1937), 896.

95. Lawrence to Poillon, 3 Mar 1934 (15/16A); cf. Lawrence to Kast, 13 Apr 1934 (12/32).

96. Kurie to Cooksey, 4 Mar 1934 (10/21); Lawrence to Beams, 13 Apr 1934 (2/26), to Cooksey, 21 May 1934 (4/19), to Oliphant, 5 June 1934 (14/6); Livingston, Henderson, and Lawrence, NAS, Proc., 20 (1934), 470–5; Rutherford to Fermi, 20 June 1934, in Amaldi, Phys. rep., 111 (1984), 133.

Franco Rasetti arrived in Berkeley in September 1935 to investigate artificial neutron sources, the cyclotron could generate 9 μA of 3.5 MeV deuterons, and therewith ten billion neutrons per second. Rasetti was flabbergasted by the "enormous superiority" of an artificial method that yielded what, by the natural way and the standard conversion, would have required the radon from kilograms of radium. In Rome he had made a certain activity by placing a silver target on top of a 500 mCi source; in Berkeley he got the same amount of activity with the target fifteen feet from the cyclotron wall.[97]

Lawrence did not find it necessary to trouble his backers with the names of Joliot, Curie, and Fermi. In writing Poillon and Ludwig Kast, president of the Macy Foundation, he claimed, what was true, that he had been the first to induce radioactivity in a range of elements by deuteron bombardment and allowed them to infer that he had discovered artificial radioactivity. A similar invitation was offered in the news that "we have found that an analogous effect is produced by neutron rays." This skillful reporting brought $2,300 from the Macy Foundation and $5,000 from the Research Corporation for working up materials to make radioactive substances on a large scale.[98] In the summer of 1934, therefore, when the Depression had forced substantial cuts on the Physics Department and Sproul went begging for money for the University, Lawrence found himself in excellent financial shape to perfect his machine for a purpose that had not been dreamed of when he and his associates built it: the creation of new radioelements and their manufacture in quantities sufficient for biomedical research. Before biology, however, comes chemistry. The physicists and machine builders had to transmute themselves into part-time chemists to separate out the various activities created by their neutrons and deuterons and to judge the possible utility of their handiwork for biologists.[99]

97. Rasetti, *Viaggi, 3* (1936), 77–8 (Aug–Oct); Lawrence to Cockcroft, 12 Sep 1935 (4/5).

98. Lawrence to Kast, 3 May 1934 (12/32), and to Poillon, 15 and 26 Mar 1934 (15/16A); Kast to Lawrence, 23 Apr and 5 June 1934 (13/32); Lawrence to Sproul, 20 Feb 1936 (20/19).

99. Lawrence to Beams, 17 Sep 1934 (2/26); Pettitt, *Twenty-eight years,* 40–1, 63.

It remained to find something useful. The plum fell to Lawrence, in September 1934, just after he had lamented to Beams that "the nuclear reactions we have been studying are not particularly novel."[100] Perhaps ignorant that Fermi's group had made the active isotope of sodium, Na^{24}, by (n,α) on aluminum and (n,p) on magnesium, Lawrence did the same, by deuteron bombardment of table salt. But whereas Fermi's group had merely reported the existence and half-life of the product, Lawrence, from his special perspective, immediately emphasized the properties that might make it useful for "the biological field." These included its convenient half-life (fifteen hours), its nontoxic chemical character, and the energetic gamma ray that accompanied its disintegration.[101] And, what Lawrence did not make explicit, the fact that the valuable Na^{24} was isotopic with ordinary sodium made it unnecessary to remove the radioactive atoms from their parent for application. The target and the converted atoms could be administered together.

The essential property of radiosodium for biological research and medical application was the gamma ray emitted by the excited magnesium atoms produced by the decay of Na^{24}. Lawrence established that the reactions at play are $Na^{23}(d,p)Na^{24}$, $Na^{24} \rightarrow Mg^{24*} + e^-$, $Mg^{24*} \rightarrow Mg^{24} + \gamma$; and he estimated that the gamma ray had an energy of over 5 MeV. He was pleased with this result, which made the gamma ray from radiosodium more than three times as hard as the hardest ray from radium, and which showed that he was still capable of solid scientific work. But not precise work. As one of his students, Jackson Laslett, soon showed, his determination of the gamma ray energy erred by excess by about 30 percent.[102] (In fact, the energy of the most energetic gamma ray from Na^{24} is under 3 MeV.) A gamma ray of 3 MeV was none the less a very useful item in physical research; members of the Laboratory used it to study pair production and the photodisintegration of the deuteron. It also retained its

100. Birge, *History, 4*, xi, 12; Lawrence to Beams, 17 Sep 1934 (2/26).

101. Kurie to Cooksey, 27 Sep 1934 (10/21); Lawrence to Livingston, 1 Oct 1934, and answer of 12 Oct (12/12), mentioning Fermi's work; Lawrence, "Notebook," 27 Sep 1934 (40/14), and *PR, 46* (1934), 746.

102. Lawrence, *PR, 47* (1935), 25; Lawrence to Cockcroft, 12 Feb 1935 (4/5).

promise for the health sciences. Lawrence's hopes could easily sustain a reduction of 30 percent. He wrote in December 1934: "We have succeeded in producing radioactive substances that have properties superior to those of radium for the treatment of cancer, and probably before long we shall make available to our medical colleagues useful quantities of radiosodium."[103]

The rate of production rose quickly. Within two months of first production, more than a mCi of radiosodium had been made and improvements in manufacture were under way that would increase the rate a hundredfold. Two years later Lawrence could make 200 mCi a day, with a current of only 1 μA, and he looked forward to multiplying the yield a thousandfold. With 20 μA, a day's product of radiosodium emitted the equivalent in gamma radiation of 100 mg of radium.[104] These production levels made an impact. Fermi supposed that Lawrence had slipped by a factor of a thousand and had meant to announce a μCi; Lawrence silenced his doubts by sending him a letter containing a mCi of Na^{24}. Wilhelm Palmaer, president of the Nobel Committee on Chemistry, highlighted the promise of a cornucopia of radiosodium during the ceremonies in which Urey and Joliot and Curie received their prizes. In another happy omen, the Rockefeller Foundation, the greatest patron of biophysics in the 1930s and eventually Lawrence's most generous private benefactor, advertised radiosodium as the exemplar of the cost-effective service to mankind it liked to support.[105]

The salt to make the equivalent of a gram of radium cost less than a penny (in fact much less, since the Myles Salt Company of Louisiana donated crystals of rock salt), and the power for the eight-hour exposure in the cyclotron less than $2. Putting the same point a different way, *Science Service* headlined its report of the meeting of the American Physical Society of December 1936, at which the Laboratory's Paul Aebersold announced Berkeley

103. Lawrence to Cooksey, 4 Nov 1934 (4/19), on pair production; Nahmias to Joliot, 27 June 1937 (JP, F25), on photodisintegration; Lawrence to E.B. Reeves, Commonwealth Fund, 7 Dec 1934 (10/18).

104. Lawrence to Akeley, 14 Nov 1934 (1/12), and to Cooksey, 12 Sep 1936 (4/5); Lawrence and Cooksey, *PR, 50* (1936), 1140.

105. Segrè, *Ann. rev. nucl. sci., 31* (1981), 7; Palmaer in *Prix Nobel en 1935*, 38, and *Nobel lectures, chemistry, 2*, 337; Rockefeller Foundation, "Trustees confidential bulletin," Dec 1937, "Atom smashing and the life sciences" (RF).

production levels, "Machines of science produce radiation equal to $5,000,000 worth of radium." The calculation: the biological effect of the neutrons from deuterons shot at beryllium targets in the cyclotron equalled that of the gamma rays from 125 mg of radium. The reporter estimated the price of radium at $40 a mg and the cost of the cyclotron at under $100,000. "Thus as a radiation source the machine turned in a 50-to-one investment." And more: to get an equivalent neutron yield from a Rn-Be source would have required 10 kg of radium. That should answer "people who urge more practical scientific research and bemoan the apparently wasted ingenuity of those scientists who probe the hearts of atoms."[106]

Radium was passé. Lawrence advised his correspondents against investing in any of the stuff. Radioisotopes from the cyclotron, he said, would soon drive down the price of radium and supplant it in clinical use.[107] This advertisement attracted the attention of Bernard Lichtenberg, director of the Institute of Public Relations in New York. It did not seem good public relations to him, and he complained to Sproul. A mild reprimand went forward. "In explaining the work of the Radiation Laboratory," Sproul's assistant wrote Lawrence, "make sure the listeners realize that radium and radio-active salts are not the same."[108]

Lawrence reserved his more extravagant claims for presentation to his backers in private. He usually spoke modestly about the Radiation Laboratory in public, and allowed his audience to imagine what the future might hold. A lecture given at several colleges and universities under the sponsorship of the Sigma Xi in 1935 is representative. "I hesitate to express views [about the future]," Lawrence said. "I leave it to you to estimate the advantages for radiation therapy and biological research of radioactive substances having practically any desired chemical and physical properties." Two years later, in a second round of Sigma Xi lectures given at ten institutions from Virginia Polytechnic to Oregon State College during May 1937, he could point to the prospects for

106. Myles Salt to Lawrence, 26 Nov 1934, and Lawrence's response, 4 Dec 1934 (11/24); *Science service,* 23 Dec 1936, re Aebersold's talk.

107. E.g., Lawrence to G.M. Schrum, University of British Columbia, 7 Nov 1936 (2/29), and to Earl R. Crowder, M.D., Evanston, Ill., 4 May 1937 (5/5).

108. G.A. Pettitt to Lawrence, 11 Mar 1937 (20/19).

biological research of machine-made radiophosphorus (P^{32}) and radioiron (Fe^{59}), the former found by Joliot and Curie in 1934, the latter a discovery of Berkeley cyclotroneers.[109] And he gave his audience a new basis for estimating the possibilities of his products. He had fresh samples of radiosodium airmailed to him for each performance. He called up volunteers, fed them radiosodium, and followed the course of the activity in their blood with a Geiger counter he carried with him. This "vaudeville," as he called it, held attention; no ear-witness could doubt "that we can make really strongly active substances."[110] Berkeley colleagues, including Oppenheimer, served as guinea pigs in local demonstrations, and would-be cyclotroneers elsewhere copied the show for their purposes. Lawrence was pleased to provide the main ingredient for these performances.[111]

Some indication of the impressions desired and the advertisments volunteered may be gathered from a radio interview in the spring of 1939 to which Lawrence brought his hot sodium. He passed a Geiger counter over it. Click, click. He asked his interviewer, Hale Sparks, to put the counter behind his back. Click, click. "You mean to say [Sparks cried] that the radiation is actually passing through my body now?" "Yes." Sparks: "Then the cyclotron has an unlimited future despite its great achievements of the past?" "Yes, indeed."[112]

109. Lawrence, "Artificial radioactivity" (July 1935), 16 (40/17), quote, and *Ohio jl. sci., 35* (1935), 405; letter to S.B. Arenson, 24 Sep 1937 (40/17); Newson, *PR, 51* (1937), 624–7; Livingood and Seaborg, *PR, 52* (1937), 135.

110. Lawrence to McMillan, 5 and 11 May 1935, quotes (12/30); to M. Henderson, 3 May 1935 (9/6); and to Boyce, 18 Dec 1935 (3/8); McMillan, *PT, 12:10* (1959), in Weart and Philipps, *History*, 263.

111. E.g., Robley Evans to Cooksey, 25 June 1937, and to Lawrence, 17 Feb 1937, and Lawrence to Evans, 28 Jan and 23 Feb 1937 (7/8); S.J. Simmons to Lawrence, 8 and 20 Oct 1939 (16/24).

112. Lawrence to H.A. Scullen, 21 Apr 1937 (40/17), and to Cooksey, 12 May 1937 (4/21); "University Explorer," "Adventures in science," 15 Apr 1939, 16 (40/15); George Volkoff to R. Cornog, 10 Mar 1941 (5/2).

3. BUSINESS

The discovery of Na^{24} brought Lawrence into a new relationship with patents. In the case of the cyclotron, he had been persuaded to seek protection in order to block a possible monopoly by Raytheon or other firms. In the case of artificial radioelements, he participated in an effort to secure a monopoly position in a future radio-pharmaceutical industry. Here he and the Research Corporation were stymied by the difficulties in their case and by the practices of the U.S. Patent Office. A few European physicists did manage to secure patents on nuclear processes. The competitive environment influenced the direction of the work undertaken at the Laboratory, but did not restrict its openness to visitors or lessen its generosity to colleagues.

A Corner on the Market

Early in January 1932 the Research Corporation's lawyer, A.P. Knight, completed a draft application for a patent on the cyclotron, an instrument to produce high-speed ions by successive impulses "in a compact or relatively small apparatus." The ions, according to Knight, might be "utilized in any suitable manner, for example, for application to the disintegration or synthesis of atoms, or for general investigations of atomic structure, or for therapeutic investigations or applications." (Here the lawyer foresaw applications that Lawrence, still mired in machine design, probably had not; lawyers might be useful adjuncts to research teams.) Knight claimed injecting, accelerating, focusing, deflecting, and extracting the ions as patentable. The patent examiner rejected them all, as was the custom, enforced greater precision in language, and allowed the claims in September 1933.[113] The patent was granted in February 1934.

Shortly after the Research Corporation received Lawrence's assignment of rights in the cyclotron, its leaders visited Berkeley

113. Draft application, 5 Jan 1932, filed with changes, 26 Jan, as "Method and apparatus for the acceleration of ions" (no. 589,033); correspondence between Knight and the Patent Office, 1932/33; Knight to Lawrence, 7 Jul 1932 and 15 May 1933; patent no. 1,948,384, 20 Feb 1934; all in 35/2. Cf. Vaughn, *Patent system*, 23.

to see the cyclotron in action and to encourage its inventor to stay alert to patentable material developed in the Laboratory. Lawrence became almost an agent for the Research Corporation. In September 1933 he proposed patenting the water-cooled anode at the tip of the coil in Sloan's x-ray machine; in 1934 he wanted to patent a cheap cloud chamber for lecture use; in 1938 he brought Beams' ultracentrifuge to the Research Corporation; in 1939/40 it was the turn of a new oscillator developed by Sloan and Lauriston Marshall, and in 1941 of an application to industrial radiography by a Berkeley engineer. Lawrence wrote the engineer, for whom he interceded with the Research Corporation, without a hint of the old physicist's ethos: "If you are going after any patents along this line, you have my blessing."[114]

The apparently fallow years in the preceding recital were, in fact, Lawrence's busiest and most vexatious time with patent affairs. A week after he had identified Na^{24}, he recommended his discovery to Knight and Poillon as "almost ideal for biological work," a novelty that "might ultimately supersede radium in usefulness." He added that by running fast enough, Knight could get an application to the Patent Office before news of Lawrence's type of radiosodium appeared in print. Poillon was ecstatic: the discovery fulfilled half the prophecy he had made six months earlier to the president of the Chemical Foundation, that Lawrence's results qualified him for a quick Nobel prize and that his further work would help bring about "the production of synthetic radium." Poillon ordered Knight to run.[115] Knight had the application in hand a week later; it emphasized that in Lawrence's process the radioactive material was chemically identical with the target and, moreover, that the chemical sodium does not harm the human body. The emphasis on the chemical identity of target and

114. Poillon to Lawrence, 23 Feb 1932; Lawrence to Knight, 16 Sep 1933, both in 35/2; correspondence with Central Scientific Company and others, Jul–Aug 1934 (10/7), and Lawrence to Cooksey, 7 Jul 1934 (4/19); Lawrence to Poillon, 22 Jan 1938 (15/17A), and 27 Apr 1939 (15/18); Lawrence to John E. Dorn, Mechanical Engineering, Berkeley, 1 Mar 1941 (19/24). Infra, §10.2 for the Sloan-Marshall tube.

115. Lawrence to Poillon, 29 Sep 1934 (35/7); Poillon to Knight, 1 Oct 1934, and to Buffum, 3 Apr 1934 (RC): "Both these statements are of such great importance that I forbid their being published lest it affect my standing as a Doctor."

product aimed to elude Fermi's patent on Na^{24} made by (n,α) on aluminum. This was perfectly proper, since patent law made a clear distinction between process and product: controlling a product does not mean controlling all ways to make it, and vice versa.[116]

The patent examiner observed that the novelty claimed was production via (d,p) and that Crane and Lauritsen had priority of publication of results obtained in that manner. "There is no invention in utilizing the well known effects of deuteron bombardment upon any particular light metal," he held, and rejected all Lawrence's claims.[117] Although Knight insisted that the work of Crane and Lauritsen had nothing to do with Na^{24}, and although Lawrence declared that excitation by (d,p) was first observed at Berkeley, the examiner held firm. Meanwhile the Laboratory had found that (d,p) could make another important radiosubstance, P^{32}, which Lawrence urged Knight to try to patent together with the (d,p) process.[118] To this last proposal the examiner returned a crippling objection: the (d,p) process as described by Lawrence, Lewis, and Livingston in 1933, not the application of the process to sodium in 1934, should be the precedent; and, if Lawrence could not show that at that time he had the idea of activating the substances he bombarded, the game was up.[119] (A patent application could not then be filed on the strength of the paper of 1933, since filing had to occur within two years of publication.) This objection had at least this merit: it revealed how far the various parties were prepared to go to secure control of Lawrence's artificial radium.

The examiner held that since Na^{24} and P^{32} were produced by (d,p) in the experiments of Lawrence, Lewis, and Livington, they had been discovered then even though no one knew it. That construction of the legal mind nonplussed Lawrence. He proposed to

116. Knight to Lawrence, 8 Oct 1934, enclosing application no. 748,085; Lawrence to Knight, 9 Oct 1934, all in 35/7; Vaughn, *Patent system*, 20.

117. Knight to Lawrence, 21 Mar 1935 (35/7); Crane and Lauritsen, *PR, 44* (1933), 783–4, *45* (1934), 226–7, 430–2, and Crane, Lauritsen, and Harper, *Science, 79* (1934), 234–5.

118. Lawrence to Knight, 2 and 16 Apr, 27 Aug 1935, and to Potter, Pierce, and Scheffler, 3 Jun 1935 (35/7); examiner's response, 21 June 1935 (35/7).

119. Knight to Lawrence, 11 Sep 1935, and reply, 9 Oct 1935 (35/7).

Knight that they drop the attempt to patent radioelements. After all, he said, we have the cyclotron, "which I am sure will always be the apparatus that produces the radioactive substances." Lawrence thus reversed the estimate he had made during the patenting of the cyclotron. Then, while Poillon's Knight duelled with the examiner, Lawrence had doubted that the cyclotron had any commerical possibilities in the near term; and, when the duel concluded happily, he had questioned whether the Research Corporation should go to the additional expense of securing a Canadian patent. "The patent application on the cyclotron is very much of a gamble. It may never be of great worth," he had written Knight. "And yet," he added, "developments may come which would make it of tremendous value."[120] The developments that revalued the cyclotron were, of course, the discoveries of artificial radioactivity and neutron excitation.

Poillon declined to hide behind the cyclotron. He appealed to Lawrence's patriotism: if the generous Research Corporation were to withdraw, grasping Caltech would rush into the vacuum. "I know how repugnant it is for any right-thinking scientist to become embroiled in a discussion concerning priority of discovery....However, California Technology is quite a 'powerful Katinka' and is out for both intellectual recognition and financial return whenever proper and possible....Under these conditions might it not be possible to straighten up a little bit in your claims for patent priority?" Lawrence could scarcely deny this appeal from his benefactor for help against his rival. "It is entirely proper," he replied, "for us to look out for the commercial aspects of our work, if this can be done in a dignified and proper way." He had not counselled withdrawal from distaste for battling Caltech, but from conviction that the patented cyclotron was protection enough.[121]

Knight's strategy was to obtain an affidavit from Lawrence that he had had the idea of "irradiation" by deuterons before June 1933. That might allow patenting of (d,p); it might also require

120. Knight to Poillon, 9 Jan 1933 (35/3); Lawrence to Knight, 28 Sep 1933 (35/2). The Corporation decided to file in Canada; Knight to Poillon, 19 Oct 1933 (35/2).

121. Lawrence to Knight, 13 Sep, 9 Oct, and 23 Nov 1935; exchange with Poillon, 10 and 19 Oct 1935; all in 35/7.

Lawrence to claim discovery of artificial radioactivity. Knight's Washington correspondents arranged a meeting between Lawrence and the examiner, which they judged to be encouraging. Lawrence abandoned his application for radiosodium, substituted a claim for the (d,p) process, and swore that he had ordered the experiments of 1933.[122] The examiner rejected all claims based on the experiments and the affidavit and found a new objection: Cockcroft and Walton must have had some deuterons among their hydrogen ions; they therefore must have made something radioactive by (d,p); and consequently, by patent logic, they discovered without knowing it what Lawrence claimed as his own. The only way out seemed to be an affidavit from Lawrence's colleagues that he suggested the original bombardments with deuterons in order to search for radioactivity. This Lawrence was reluctant to seek— his colleagues might not like his running away with a patent on their joint work—but he would cooperate if necessary. He supplied a second affidavit, which the examiner again rejected as insufficient, since it did not declare (what would have been perjury) that Lawrence examined the product of the irradiation for radioactivity. Knight's correspondents judged that a sufficiently strong affidavit would win the day, but had the sense to doubt "whether upon the actual facts of the situation a fully satisfactory affidavit can be furnished."[123]

Lawrence had gone at least as far as he could. He wrote Knight: "The more I think about the matter the less enthusiasm I have for further endeavors to patent the process for producing the artificial radioactive substances." He retreated to his old position, much stronger in 1939 than it had been in 1935: "I feel that the cyclotron affords the only means of producing the radioactive materials in appreciable quantities; therefore with the cyclotron protected we have essential control of the matter." This time Poillon concurred, observing that the Research Corporation controlled

122. Knight to Potter, Pierce, and Scheffler, 12 Nov 1935; to Lawrence, 23 and 30 Dec 1935; Potter et al. to Knight, 21 Apr 1936 and 23 Sep 1937; application no. 64,411, 17 Feb 1936, replacing no. 748,085 of 1934; all in 35/7.

123. Knight to Lawrence, 27 May 1938; Lawrence to Knight, 7 Oct 1937 and 3 Sep 1938; Knight to Potter et al., 3 Oct 1938; examiner's statement, 22 Mar 1939; Potter et al. to Knight, 24 Mar 1939; all in 35/7.

not only the cyclotron, but essential features of the Van de Graaff generator too.[124]

Poillon had hoped to create and control a new industry of radio pharmaceuticals. Hence in the mid 1930s the Research Corporation invested redundantly in cyclotrons on the same principle that had built its cartel in the precipitation business: by expecting or requiring its grantees to assign improvements in the art to the Corporation. The theory is clear from the justification of a grant to Columbia to "enlarge the cyclotron...so that its field of application may be extended, and the equipment thus be made more effective for the preparation of artificial radioactive elements." In service of the same program, the Research Corporation supported work on the separation and application of biologically interesting stable isotopes, like C^{13} and O^{17}, at Columbia, and the ultracentrifuges of Jesse Beams at Virginia and of J.W. McBain at Stanford.[125] The purpose was highminded. As Poillon put it, "we do not in any way want to prevent scientists from having the free use of any discoveries that are made but if we can assess industry a reasonable sum, we will have just that much more to give to scientific research."[126] The methods, however, were those of the entrepreneur and the patent lawyer.

In 1940 it appeared likely that the Research Corporation would enjoy large royalties from its cyclotron patent. Several corporations, notably Westinghouse and American Cyanimid, deliberated building production cyclotrons for profit.[127] Radiophosphorus had shown promise in the treatment of certain sorts of leukemia. The annual cost of treating all Americans so afflicted was reckoned at between $200,000 and $500,000. Since the university cyclotrons in operation in 1940 could not supply the demand, let alone the requirements for other therapies and applications, commercial

124. Lawrence to Knight, 26 Aug 1939 (35/7); Poillon to Lawrence, 5 May 1939 (15/18); Lawrence to Poillon, 27 Apr 1939, and exchange between Poillon and Knight, 5 and 10 Jul 1939 (RC).

125. Research Corporation, "Minutes of meeting," 24 Apr 1940 (Columbia grant); cf. ibid., 25 Feb 1937 and 25 Mar 1938, and Lawrence to Poillon, 2 Jan 1938; all at RC. Infra, §8.3, for the competition between stable and radioactive isotopes.

126. Poillon to Lawrence, 4 Nov 1935 (35/7).

127. Cooksey to Aebersold, 11 Jul 1940 (1/9); W.B. Bell, American Cyanimid, to Lawrence, 27 Jul and 30 Aug 1940 (1/19); and, re Westinghouse, DuBridge to Lawrence, 28 Aug, and answer, 2 Sep 1940 (6/17).

production of radioisotopes under patents held by the Research Corporation seemed imminent and humanitarian. American Cyanimid hesitated only over the worry that universities might undersell commercial radioelements if they started to charge for the products of their cyclotrons. Both Poillon and Lawrence reassured Cyanimid that the demand would be so large, and the output of university cyclotrons available for therapy so small, that the commercial market would not be affected by the policies of university laboratories. In preparation for a windfall, the Research Corporation pushed Lawrence to patent certain cyclotron improvements.[128] The war ended these initiatives in two ways: by providing other lines of work for the interested parties and by creating, in the atomic pile, a much more efficient engine for the production of radioisotopes than the cyclotron. After the war, the Research Corporation wrote to all cyclotron laboratories to grant royalty-free licenses "for educational, scientific, experimental and research purposes." That amazed many. As the director of the Biochemical Research Foundation (Bartol) wrote in acknowledgement of this largesse: "I never knew there was such a patent."[129]

The Research Corporation had not cared to exercise its rights when the primary consumers of artificial radioelements were research teams in universities and hospitals. And, as Lewis had done with heavy water, Lawrence distributed the fruits of the cyclotron gratis throughout the world. To be sure, the product spread the fame of the machine that produced it; but the Laboratory made its gift in a true spirit of scientific cooperation. Lawrence had several reasons for not charging even the cost of production: he thereby retained the right to support only projects he thought worthwhile; he had to avoid giving the men in the Laboratory the impression that they were cogs in a business; and he wanted to repay in some measure the support he had received

128. Letters to V. Bush from J.A. Fleming, 9 Sep 1940, and J.H. Lawrence, 10 Sep 1940 (15/18); Poillon to Lawrence, 20 Sep 1940, and response, 24 Sep 1940 (15/18).

129. E.g., Joseph Barker to Sproul, 27 June 1947; E. MacDonald to Barker, 2 Jul 1947, quote (3/35); infra, §8.3, for leukemia therapy. Lawrence urged that even commerical firms be given royalty-free licenses; Lawrence to Barker, 30 Jan and 18 Feb 1946 (3/35).

from charitable foundations and public bodies.[130] The industry Poillon had envisaged did develop, after his patent expired. The first commercial cyclotrons for radioisotope production were made by the Collins Radio Company in the 1950s. In 1957 the former chief engineer at the Laboratory, William Brobeck, marketed cyclotrons for neutron therapy. By 1970 the annual sales of cyclotron-produced radioisotopes exceeded $3 million. It was but a small part of a big business—some $50 million a year—in radioisotopes for research, diagnosis, and therapy.[131] Poillon had the right idea but the wrong machine.

Other Players

Early in 1935 the chief Italian journal of physics, *Nuovo Cimento*, pointed out that Lawrence was behind Fermi's group in the discovery of Na^{24}. The Research Corporation likewise came late in the effort to patent it. On October 26, 1934, Fermi's group obtained an Italian patent covering activation by the absorption of fast or slowed neutrons and the products of the process, including radiosodium, as well. This violation of the physicist's ethos originated not with Fermi but with his patron O.M. Corbino, who had close ties to what high-tech industry then existed in Italy. "Age gave him wisdom," Mrs. Fermi writes, "[and] the boys were used to following his advice."[132]

But they, too, had been anticipated by "the inventor of all things."[133] In March 1934, a month or so after learning about artificial radioactivity and before Fermi's group had demonstrated the efficacy of neutrons, Szilard applied for a British patent on the "transmutation of chemical elements." His "invention," which he never reduced to art, had three parts: generation of neutrons to provoke reactions; separation of radioisotopes produced by the

130. Lawrence to DuBridge, 14 June 1939 (15/26A). Probably no cyclotron laboratory charged for its products before the war; J.A. Fleming to V. Bush, 6 Sep 1940 (MAT, 25/"biophys.").

131. Texas Ind. Comm., *Texas giants*, 6–7; Brobeck, interview by Seidel, June 1985 (TBL); IEEE, "1978 Conference," *IEEE trans., NS, 26* (1979), 1703–32; Highfill and Wieland, ibid., 2220–3.

132. A.P., *Nuovo cimento, 12* (1935), 123–4; Segrè, *Fermi,* 83–5; L. Fermi, *Atoms,* 101, quote; Russo, *HSPS, 16:2* (1986), 286.

133. McMillan to Mann, 3 Jan 1952 (12/31).

(n,γ) process; and utilization of the heat liberated in the transmutation. Szilard's eccentric genius is displayed to full advantage in his method of obtaining neutrons. He planned to use deuterons accelerated by a high-tension device to create neutrons in collisions with light nuclei like beryllium or deuterium (he had made good use of the indications then accumulating of the d-d reaction that had misled Lawrence); he also proposed getting his neutrons from light, via (γ,n), and sketched an apparatus for making and absorbing photoneutrons.

The productivity of the transmutation evidently depends upon the number of neutrons at work. Szilard observed that if there exists a nucleus that when struck by a neutron liberates another without capturing the first, a very rapid buildup of a free neutronic population might occur. Mixing these hypothetical neutron multipliers with the material to be transmuted would increase the efficiency of transmutation; and a large enough sample of the material capable of sustaining the chain reaction (n,2n) would make a fine explosive. For the rest, Szilard proposed to separate isotopes made by (n,γ), which are chemically identical to their parents, by exploiting a process he did take the trouble to test.[134] His scheme as of June 1934, including provision for extracting power, appears in figure 4.2.[135]

Szilard considered assigning his first British patent on isotope and energy production to the Research Corporation in return for a grant for three years to continue research on the subject.[136] Instead, he licensed it to a relative of Brasch's, a Havana importer named Isbert Adam, in return for $15,000 in research support. (Szilard thereby made more money from radioactivity than the Research Corporation, even after subtracting the $7,000 he subsequently repaid Adam to reacquire the patents in 1943.)[137] In March 1936, when Szilard obtained a second patent on chain

134. British patent application no. 7840, 12 Mar 1934, preliminary drafts, and supplements, 4 Jul and 20 Sep 1934, in Szilard, *CW, 1*, 605–28, resulting in British patent no. 440,023, granted 12 Dec 1935.

135. From supplementary application, 28 June 1934, resulting in patent no. 630,726, of 30 Mar 1936; Szilard, *CW, 1*, 650.

136. Szilard to Fermi, 13 Mar 1936 (Sz P, 17/197), and to Cockcroft, 27 May 1936, in Weart and Szilard, *Szilard*, 47–8.

137. Szilard licensed patent no. 440,023 to Adam, Dec 1936 (Sz P, 29/311).

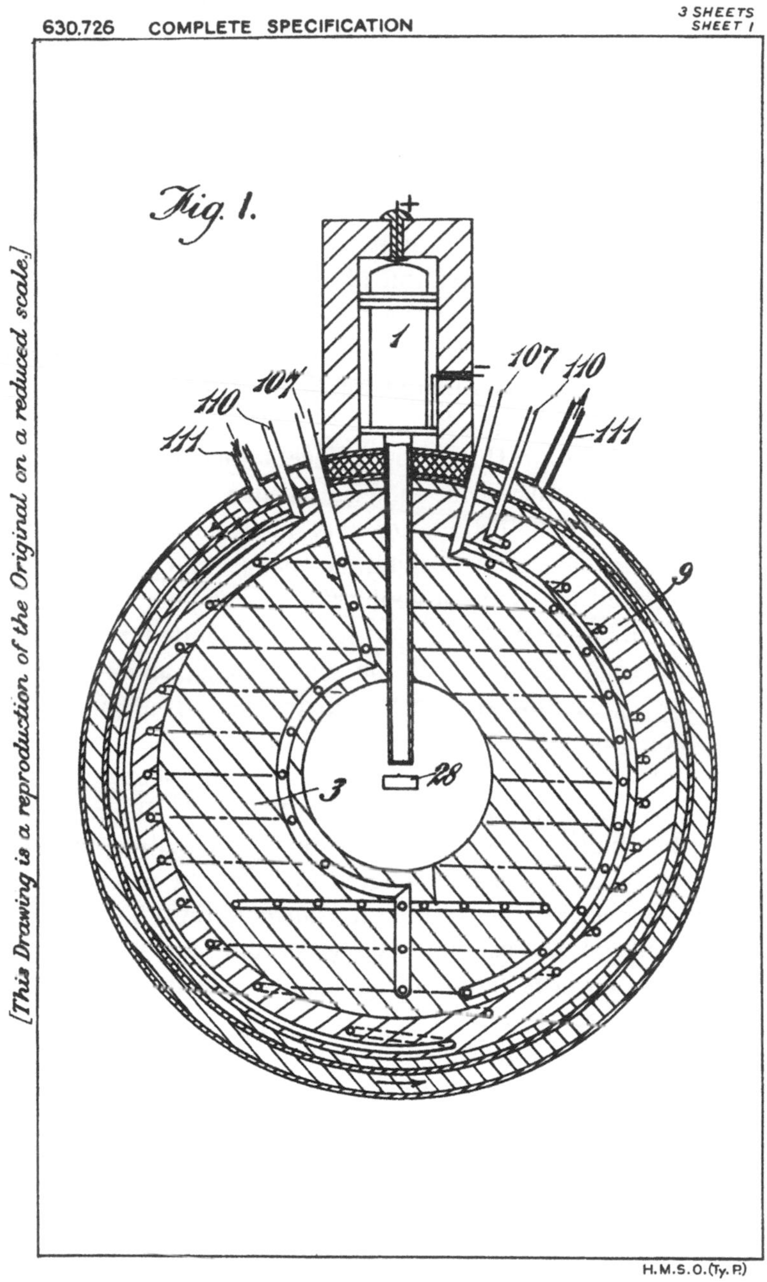

Fig. 1.

FIG. 4.2   Szilard's cornucopia of radioelements. Deuterons from the source 1 make fast neutrons from the beryllium target 28, which spread through a sphere 3 composed of all elements that might multiply neutrons and that might develop energy on absorbing neutrons. The tubes 107, 110, 111 contain a coolant that delivers the heat of the reaction to an engine not shown. Szilard, *CW, 1,* 650.

reactions, he had reason to believe that multiplication of neutrons was possible. In his applications of 1934, he had imagined three sorts of interactions: a neutron multiplication (n,2n), a neutron conversion (n,*2n*), and a neutron reduction (*2n*,n), where *2n* is a hypothetical heavy neutron with twice the mass of an ordinary one. In his definitive specification of June 1934, the basis of his patent of 1936, Szilard made the success of the chain reaction depend upon the existence of heavy neutrons, and offered indium, which in his experiments suffered an (n,*4n*) reaction, as an element of the conversion type.[138]

He mobilized fellow Hungarian refugees Eugene Wigner and Michael Polanyi to procure the material needed to initiate a chain reaction. He acquired a cylinder of beryllium, and access to a big radium bomb in a London hospital, to make photoneutrons. He convinced himself that indium could give out at least one double neutron via (n,2n) or (n,*2n*).[139] A trip to the United States provided leisure to weigh the whale he had hooked; and in March 1935 Szilard filed for an American patent on a large-scale transmutation process, similar to the earlier schemes, but with uranium and water (to slow the neutrons) as the neutron multiplier. All this was before the discovery of fission.[140] The responsibility for the chain reaction grew too heavy for Szilard to carry and in order to keep it secret he offered to assign the patent detailing it (his second British patent on transmutation) to the War Office. The official who examined the gift could see no value in it. Szilard had better luck with the Admiralty, which accepted his assignment in March 1936.[141]

As Szilard explained his actions to Fermi and to his British colleagues, he had never considered the patents to be his private property. He proposed to Fermi that they share responsibility for

138. Szilard, *CW, 1*, 643–6 (28 June 1934); memos of 13 and 28 Jul 1934 (Sz P, 29/309, 17/197); Weart and Szilard, *Szilard*, 39–40.

139. Szilard to Wigner, 7 Aug 1934; to Lange, 6 Nov 1934; to Brasch and Lange, 12 Dec 1934; to Lindemann, 3 June 1935; all in Sz P, 17/197. The important part of the letter to Lindemann is also in Weart and Szilard, *Szilard*, 41–2.

140. Szilard to Singer, 16 June 1935 (Sz P, 17/197); U.S. patent application no. 10,500, filed 11 Mar 1935, in Szilard, *CW, 1*, 654–90.

141. J. Combes to War Office, 8 Oct 1935, and director of navy contracts to Szilard, 20 Mar 1936 (Sz P, 17/197); Szilard to C.S. Wright, Admiralty, 26 Feb 1936 (Sz P, 44/476), in Szilard, *CW, 1*, 733–4.

controlling a fund secured by the promise of their patents. "It must be awkward for any scientist to have a personal interest from such patents," Szilard wrote, "while other scientists, who also could have taken out such patents, refrain from doing so." As for the research the fund might support, Szilard did not see the wisdom of the course of the Research Corporation and its Berkeley client; "I personally do not think very much of producing radioactive elements for medical purposes and I should not like to be responsible for inducing manufacturers to embark upon such an enterprise at present."[142] (In that he was not entirely free from duplicity, since he wrote by the same mail to his patron Adam that the first priority was a systematic search for long-lived elements suitable for medical purposes). Nor did Szilard think much of the cyclotron. As he wrote to encourage Adam: "The artificial production of radioactive isotopes in California that you mention depends on a principle different [from mine], which I think is not susceptible of development and will have scarcely any commercial importance."[143]

For development of his more promising scheme, Szilard thought that he could do with perhaps 1,000 pounds sterling a year and, if Fermi came in, 5,000 pounds for three years, less than a sixth of Lawrence's rate of consumption.[144] Still the sum was not easy to raise. The agent of the Italian group, G.M. Giannini, agreed that the combined patents would make a nice set and professed an interest in cooperation; Segrè liked the idea of capitalizing the patents for a research fund, and for the researchers, "which would also indirectly advance science." But nothing came of it. Szilard continued to consider himself a disinterested broker.[145] The Italians preferred to make money and hoped for a

142. M. Goldhaber to Szilard, 18 Mar 1936, in Weart and Szilard, *Szilard*, 44: "Of course, your intentions were misunderstood to be financial or otherwise unscientific." The idea of a patent pool for public purposes appears in Szilard's memo to himself of 13 Jul 1934 (Sz P, 29/309).

143. Szilard to Fermi, 13 Mar 1936, in Szilard, *CW*, *1*, 729–30; to Segrè, 1 Apr 1936, ibid., 732; to Adam, same date (Sz P, 29/310), and 14 Oct 1935 (Sz P, 44/476).

144. Szilard to Singer, 9 and 16 June 1935 (Sz P, 12/197); to Fermi, 13 Mar 1936, in Szilard, *CW*, *1*, 729–30; to Segrè, 1 Apr 1936, ibid., 731–2.

145. Giannini to Szilard, 8 Mar and 20 Apr 1936; Segrè to Szilard, 21 Mar 1936; Szilard to Giannini, 8 Aug 1936, all in Sz P, 44/476.

time to set up an industrial concern; they turned down an option on the patents offered by Metropolitan-Vickers on Allibone's recommendation and ended by selling their European rights for $3,000 to Philips of Eindhoven, which had set up a nuclear section partly as a result of a visit from Segrè.[146] They failed to interest any American corporation in their radioactive technique. Still, they obtained more than they dreamt of in the 1930s when, after much haggling, the U.S. government, which had exploited their technique during the war, paid them $400,000 as "just compensation."[147] As for Szilard, he felt obliged to explain his altruistic policies to the major British physicists and, when Fermi's group went its own way, to license Adam.[148]

There is another round to the story. In 1938 Szilard removed his headquarters from England to New York City, where he rightly expected to find greater scope for his schemes. He now concentrated on improving the neutron source. From data on neutron yields provided in papers from Berkeley and Rome, Szilard calculated that the energy that could be stored in transmuted radioactive atoms might be one hundred times the energy of the bombardment required to make the neutrons to make the transformations. Economics had given its blessing; only a little ingenuity was required to make a nuclear-powered airplane. "Perhaps we ought to think of new methods for producing really strong neutron beams." Szilard had proposed to Brasch to scale up a high-tension machine to slam electrons into metal walls at 10 MeV. The resulting x rays would sire neutrons in profusion from a beryllium target.[149]

146. Bakker to Segrè, 21 Jul 1935 (letter in Segrè's possession). Siemens of Berlin also began to take an interest in radioelements in 1935, in October, when it established a laboratory under Gustav Hertz to investigate production possibilities. Osietzki, *Technikges.*, 55 (1988), 32–3.

147. Allibone, *PRS, A282* (1964), 451, and in Hendry, *Cambridge physics*, 171–2; Segrè, *Fermi*, 84–5. The American patent covered activation by slow neutrons; the Italian group tried also to patent all neutron activation, which Lawrence thought "ridiculous." Lawrence to Poillon, 24 Sep 1940 (15/18).

148. Szilard to Cockcroft and to Rutherford, 27 May 1936, in Weart and Szilard, *Szilard*, 45–8; to Cockcroft, 21 May 1936 (Sz P, 17/197).

149. Szilard to Brasch, 3 and 10 Jul 1937 (Sz P, 29/307); cf. Szilard to Adam, 2 Oct 1935 (Sz P, 44/476), on Szilard's earlier efforts to mobilize Brasch.

For money Szilard appealed to Lewis Strauss, a Wall Street financier with an interest in radiation therapy for cancer, who thus entered on his controversial career in atomic energy. As a light inducement to cooperation, Szilard offered to give to any tax-exempt nonprofit corporation Strauss might wish to set up for producing radioelements a nonexclusive license to exploit whatever of his rights he had not sold to Adam. Strauss tried to enlist Westinghouse, General Motors, and General Electric; but all that came of it was an introduction to another supplicant, "whose friendship for the following twenty years was one of the finest experiences of my [Strauss's] life." That was Lawrence, who this time got nothing from Strauss. The future friend had decided to support Brasch and Szilard, if only a suitable place for Brasch's experiments could be found. One was. Toward the end of 1938 the ever-acquisitive Millikan offered space at Caltech, on the understanding that it would cost him nothing.[150]

Brasch intended to reach for 15 MV, which Millikan thought "exceedingly interesting and thrilling," and also expensive, over $100,000. Strauss doled out his money in droplets; Brasch complained that he could not exist on "homeopathic doses" of dollars, and raised the estimate to $200,000; by 1940 the adventure had come to an end nowhere near its goal.[151] Meanwhile fission had been discovered, and the royal road to atomic energy. Szilard kept Strauss apprised of progress by telegram, although they were almost neighbors, and asked his benefactor to find him another. Szilard had in mind Alfred Loomis, a retired investment banker and a first-rate amateur physicist, who soon became one of Lawrence's main advisors and supporters.[152] Szilard wished Loomis to help underwrite the cost of experiments he planned to try at Columbia University, where he had a guest appointment, to determine how fission might be exploited for atomic energy. To complete the circle, the new émigré professor of physics at Columbia, Enrico Fermi, was then engaged in the same line of work. We shall return to their unequal competition.

150. Szilard to Strauss, draft, n.d., and to Adam, 16 Nov 1938 (Sz P, 29/306); Strauss, *Men and decisions*, 163–5.

151. Brasch to Szilard, 27 Jan 1939, and n.d. (Sz P, 29/306); Millikan, quoted by Strauss, *Men and decisions*, 168.

152. Szilard to Strauss, 28 Feb 1939, and telegrams, Feb–Apr 1939 (Sz P, 17/198).

V
Cast of Characters

During the 1930s the Laboratory grew into the exemplar of big science. Its size, staff, and income were not so large, however, that numbers must replace people, and dollars stand for activity, in our history. Still, some systematic data may assist overviewing and understanding. We shall first look at, and then under, tables setting forth the natural and financial resources that constituted the Laboratory.

1. GIVERS

The main forces at work in the Laboratory were the technological imperatives of the accelerators, scientific interests in the fields these machines opened up, and the need to find and to satisfy financial supporters. Over the novennium 1931/40, the state, foundations, and the federal government contributed in the ratio 40:38:22 to a total of at least $390,000 for operating expenses, salaries, and supplies. In addition, the Laboratory had over $162,000 in capital investment and gifts in kind, which came from individuals and corporations, foundations, and the federal government in the ratio 48:30:22. This accounting evidently does not include the huge award of $1.15 million made by the Rockefeller Foundation in April 1940 or the matching pledge from the University of $85,000 a year for ten years. These monies were expended slowly owing to the mobilization of the Laboratory which began late in 1940.

Wild Cards

In his play for high academic stakes, Lawrence made expert use of four sets of wild cards dealt him without solicitation. The first set was calls from elsewhere. To defeat an offer from Northwestern in 1930, the Physics Department had first to win a battle on the home front against those who did not care to see mere boys promoted to full professors. Sproul took a close interest in the matter. He understood, as had his predecessor W.W. Campbell, that for Berkeley to keep and improve its place among research universities, he would have to offer inducements to innovative junior faculty that might scandalize their elders. He constituted the committee to consider Lawrence's promotion entirely of scientists, including Lewis and Birge, who shared his opinion that Lawrence was the very type of academic leader the University of California needed.[1]

To defeat Harvard in 1936, Sproul had to discover fat in a budget that the Depression and the governor had already made lean. From Harvard's large endowments and still rich friends, its president, James B. Conant, proposed to draw salaries of $10,000 or $12,000 for Lawrence, $6,000 for Oppenheimer, and also something for McMillan; provisions for other new positions; the wherewithal for a very big cyclotron; and expenses for a large staff, if Lawrence would consent to become dean of Harvard's School of Engineering and Applied Science. The possibility of the simultaneous removal of the two stars in his firmament energized Birge, who gave it as his opinion that Lawrence and Oppenheimer were, respectively, the best experimental and theoretical physicists in the country and that their loss would utterly destroy Berkeley physics. Sproul invited Lawrence to state his conditions. He did not stint: a permanent staff for the current Laboratory, a new cyclotron and laboratory for medical research, a staff for the new installation, operating expenses. Sproul returned in a week with an offer of separateness, a staff for the Radiation Laboratory, and expenses, the whole amounting to a floor of $20,000 a year in state money,

1. Sproul to George D. Louderback (a member of the promotion committee), 16 Aug 1930, and E.E. Hall to William Popper (an opponent), 23 Aug 1930 (UCPF); Lawrence to B.J. Spence, 8 and 30 Aug 1930, and to Sproul, 15 Sep 1930 (13/22), Pettit; *Twenty-eight years*, 188–9. Cf. Childs, *Genius*, 156.

and a promise to help raise funds for the medical work. This astounding response, at a time when the University's budget had not regained its level of 1931/33 and the administration had notified department chairmen that no expansion was possible, settled the matter. "I shall always be grateful to you for the honor of your confidence," Lawrence wrote Conant almost immediately after receiving Sproul's counteroffer. "It opened up possibilities for work here that a month ago I thought were quite out of the question."[2]

The generous response of the University did not yet meet Lawrence's requirements. He had a missionary zeal to establish a medical department. He believed at the time that "the new penetrating radiations discovered [!] in our laboratory" had a good chance of curing cancer. A specialist at Caltech had told him—and he was very willing to believe—that if, as John Lawrence's experiments seemed to show, neutron beams killed tumors more efficiently than x rays, "it means that it will be possible to cure more than eighty percent of cancers." The possibility, as Lawrence told Poillon, obliged him to go wherever the opportunities to continue his work were greatest.[3] Relocation at Harvard and administrative duties there would have been costly to the work. Its basis instead would be a very large cyclotron pledged by the Research Corporation and the Chemical Foundation;[4] its place, a new building on the Berkeley campus, for which Sproul promised to find the money.

Poillon suggested that Sproul try William Crocker, the old banker who had given the Sloan plant to the Medical School. After all, Poillon wrote Sproul, it would only be fair to allow Crocker to upgrade his cancer equipment from x rays to neutron beams and artificial radium. Crocker had by then given almost $300,000 to the University. Sproul went to him for more. The

2. Lawrence to Sproul, 20 Feb 1936, and Sproul to Lawrence, 28 Feb 1936 (20/19); Lawrence to Conant, 2 Mar 1936 (4/12), quote; Birge's notes for discussion with Sproul (2/32); Birge to Sproul, 12 Mar 1936 (20/19); Sproul to chairmen of departments and administrative officers, 30 Nov 1937 (23/17); Childs, *Genius*, 231–40.
3. Resp., Lawrence to Sproul, 20 Feb 1936 (20/19), and to Poillon, 22 Feb 1936 (15/17); infra, §8.3.
4. Correspondence of Lawrence and Poillon, 28 Jan–25 Feb 1936 (15/17).

pitch: "No other project [the University] could support has greater potentiality for the alleviation of human suffering....A Crocker Radiation Laboratory [would] be a lasting monument to the interest which [your] family has ever shown in public problems and the advancement of civilization."[5] Mrs. Sproul remembered the result of this appeal, even to the dollar, forty-five years later. "He [Sproul] was on the verge of losing Ernest Lawrence to Texas [!], because Ernest wanted a little laboratory. Well, my husband said he didn't know where he was going to get the money [$75,000]....Regent Crocker gave it. Lawrence was saved and the Crocker Radiation Laboratory...was born."[6]

Mrs. Sproul's slip of "Texas" for "Harvard" referred to the third raid on Lawrence that Sproul had to meet during the 1930s. The president of the University of Texas, H.P. Rainey, required a vice president and was willing to pay $14,000 a year for a big man to push Texas into the front ranks of American science. Rainey had money from oil revenues to make a cyclotron much bigger than the Crocker machine. "He almost took me off my feet," Lawrence told Sproul, and added that he had promised to go to Texas if it proved impossible to build the super cyclotron in Berkeley. Sproul knew his part in the dance. "There is no more important business before the President of the University of California at this time than the resolution of any doubts you may have about continuing as a member of our faculty." The climate for funding was much sunnier in the fall of 1939, when Texas menaced, than it had been in February 1936, when Harvard called. Lawrence had become a national figure and medical money had supported cyclotrons throughout the country. Sproul countered Texas by helping to bring in the Rockefeller grant and by arranging for the University's matching money and a building site for the fifth-generation, 184-inch cyclotron above the campus.[7]

5. Poillon to Sproul, 2 May 1936; "The proposed Crocker Radiation Laboratory," n.d., quote; Lawrence to Birge, 11 Jul, Birge to Deutsch, 20 Jul, Sproul to Lawrence, 22 Jul 1936; all in UCPF.

6. Riess, "Ida Sproul" (1981), 58 (TBL). Sproul to Daniel J. Murphy, 2 Sep, Lawrence to Crocker, 11 Sep, Underhill to Sproul, 15 Oct, Crocker to the regents, 22 Dec 1936 (UCPF). Cf. Childs, *Genius*, 245–6.

7. Lawrence to Sproul, 18 Sep 1939, and Sproul's reply, 21 Sep 1939 (17/12); Childs, *Genius*, 293–4; infra, §10.1.

The second of the four sets of wild cards was a suit of fellowship students from the United States, Canada, and Britain. Lawrence did not arrange for these fellowships himself, although sometimes he had a hand in securing their extension. Self-selected fellowship students made essential contributions to the work of the Laboratory. The third set of wild cards were important novelties found in Europe. The discoveries of artificial radioactivity and neutron activation were as much preconditions to the establishment of the Radiation Laboratory in 1936 as the Harvard offer and the contributions of students on external fellowships. The discovery of nuclear fission and the Texas offer functioned similarly in obtaining the 184-inch cyclotron and the laboratory on the hill.

The extramural prizes that certified Lawrence's status to his benefactors were the fourth set of wild cards. His first significant award, the Comstock prize of the National Academy of Sciences, came in 1937; it brought his picture to the cover of *Time* and medico-physical initiatives to the attention of the general public.[8] The award of the Nobel prize in November 1939 may well have been the key to the jackpot of the Rockefeller Foundation.

State

The financial picture is painted in more detail in table 5.1. We have rehearsed the causes of the increasing contribution of the state: the high valuation of applied science in California, Sproul's special interest in Lawrence, Birge's concern to maintain Berkeley's standing in physics, and Lawrence's rising reputation. State support jumped by a factor of four in consequence of the Harvard offer (or by a factor of six, if the average of the early years is taken as basis). Between 1936/37 and 1940/41, when the University pledged another fourfold addition to the Laboratory's regular budget to secure the Rockefeller million, its direct support increased only marginally. (The large sum for 1937/38 included $17,000 for power lines to the Crocker Laboratory.) The most important benefit on the margin was provision of adequate secretarial help in the person of Helen Griggs, an English major who

8. *Science, 86* (5 Nov 1937), 405-6, on the Comstock prize.

Table 5.1

UCRL Finances

1931/3 to 1939/40

	1931/3	1933/4	1934/5	1935/6	1936/7	1937/8	1938/9	1939/40
State								
B of Res & Pres Fund	5,170	1,374	1,580	5,630	23,000	36,100	21,200	22100
Coverage of deficit					29,000	9,800	9,800	
Federal								
WPA			1,700	4,000	8,500	8,500	>4,000	(4,000)
NYA					360	1,200	1,200	1,400
NACC						30,000[a]	600	23,000
Foundations								
RC	7,500	3,300	5,000	7,930	5,100	10,000	5,000	5,000
CF	2,500				20,000[b]	48,600[c]	48,600[c]	48,600[c]
Macy		1,000	2,200	4,500	600	7,000	3,000	
RF						15,000[a]	17,000	17,000
Others								4,000
Miscellaneous								
NRC	500	475	400					
Columbia U.	2,000		1,800					
Total	17,670	6,149	12,680	22,060	108,620	87,600	76,800	76,500

Sources: RC files; minutes of the Board of Research; president's files; budgets of the Physics Department (22/2); Lawrence's correspondence with Leuschner, Garvan, and Poillon (15/16A, 46/25R); correspondence with H.N. Shenton (Macy Foundation), 1933/5 (46/21R); and "Original gifts and loans in the establishment of the Radiation Laboratory" [1933] (20/13). The data in these sources agree well enough among themselves but conflict with some of Lawrence's retrospective estimates, e.g., Lawrence to Garvan, 6 May 1937 (15/17A).
a. Expended 1938/9.
b. Of $68,600 pledged.
c. Balance of pledge owing.

worked part-time for Lawrence in 1938 and full-time beginning in May 1939. Griggs was a blessing to the Laboratory's historians as well as to its founder. Lawrence had worked through the secretaries in the Physics Department and kept track of expenditures himself. He would run up large overdrafts in the Comptroller's Office, or, what was worse from the University's point of view, bypass the Comptroller altogether; and he did not keep track systematically of the Laboratory's staff and visitors.[9] Griggs transformed the bookkeeping. From her time, that is, from AY 1938/39, we have systematic lists of grants received and appointments made and accounts of salaries and other expenditures. In 1939/40, she handled over $70,000 worth of business, more apart from salaries than passed through the Physics Department. Lawrence calculated that she was worth more than $1,200 a year. But Sproul, who had given her initial salary without hesitation, declined to raise it, lest the precedent undermine the exploitative principles regulating secretarial life. This, and refusals of requests for campus parking permits for the Laboratory's senior non-academic staff, indicate the level on which Sproul thought it safe to deny Lawrence anything.[10]

What we would call the Laboratory's hard-money budget represented perhaps half of the state's ongoing investment in Lawrence's work. There were also the indirect costs of the building and its maintenance, and of the use of the Physics Department's shops and supplies; as well as the salaries of Lawrence, his graduate students holding teaching assistantships or university fellowships, and, in time, of assistant professors McMillan, Alvarez, and John Lawrence, who obtained a post in the University's Medical School, and Glenn Seaborg and Samuel Ruben, who had appointments in the Chemistry Department. The indirect costs elude quantification; the total in salaries ran at some $10,000 a year from 1932 to 1936 and at perhaps twice that from 1936/37 to 1939/40. To this accounting should be added the interest on the $17,000 investment in power lines; the University gradually recovered its outlay by reselling to Lawrence at 0.02 cents a kwh the power it bought from PG&E at half that amount. Much of the money Lawrence scurried to raise from outside sources went for electricity, around $10,000 annually in the late 1930s.[11] Literally and figuratively, Lawrence paid for his power.

9. M.E. Deutsch to Lawrence, 15 Jul 1935 (20/19); J.W. Barnett to Lawrence, 29 May 1941 (22/3).

10. Lawrence to Sproul, 8 June 1939 and 7 Mar 1940, and Sproul to Lawrence, 30 June 1939 (23/17); E.A. Hugill to Lawrence, 29 Jul and 26 Sep 1940 (20/21). The former Helen Griggs tells us that Lawrence augmented her salary from sources outside the University's control.

11. Lawrence to Weaver, 11 Mar 1939 (15/29). This bit of power bookkeeping amazed Weaver, "Diary," 31 Jan 1939 (RF, 1.1/205).

The state's was not the only public money that nourished the Laboratory. Both at its beginning, when it did physics and engineering, and after its entry into biomedicine, agencies of the federal government made critically important contributions. The magnet of the 27-inch cyclotron, valued at $30,000, had been paid for by the navy, although, as we know, it came as a gift from the Federal Telegraph Company. The navy further provided an electrical generator worth $500; not, perhaps, an important antici- pation of the military-academic complex, but an indication of pos- sible further benefactions. So much for the beginning. At the end, the Laboratory had two most important grants from the National Cancer Institute on the recommendation of the National Advisory Cancer Council (NACC). Lawrence was one of the institute's very first grantees.

Established by act of Congress in August 1937, the institute was authorized to support research projects that aimed at understand- ing or treating cancer. In early November a member of the NACC, Arthur Holly Compton of the University of Chicago, where two of the Laboratory's veterans were building a cyclotron, assured Lawrence (who had been lobbying him and his fellow councilman Conant) that something would be forthcoming. Lawrence notified the council's executive secretary of the "immediate and very pressing need" to complete the equipment of the Crocker Laboratory. Shielding, remote-control safety devices, and clinical furnishings would cost $18,000; the personnel to install and operate the equipment, another $12,000; in all, $30,000, which the council recommended at the end of November. It also appointed Lawrence and Compton a commit- tee of two to propose ways to spend between $50,000 and $100,000 a year on the improvement of cyclotrons. (The National Cancer Institute had about $400,000 to spend annually.) Within a week Lawrence returned suggestions for the scattering of $96,500 among the country's cyclotron laboratories.[12]

12. A.H. Compton to Lawrence, 9, 10, and 29 Nov 1937, and replies, 4 and 7 Dec 1937; Lawrence to Ludwig Hektoen, 10, 13, and 20 Nov 1937, and reply, 30 Nov 1937 (all in 13/29); Lawrence to Poillon, 14 Oct 1837 (15/17A); 75th Congress, 1st sess., 1937, *U.S. statutes at large, 50:1,* 559–62; Swain, *Science, 138* (1962), 1233-4; Harden, *Inventing the NIH,* 174. The grant ran from 1 Jul 1938; G.F. Taylor to L.A. Nichols, 15 Mar 1938 (13/33).

The $30,000 allowed completion of the medical cyclotron, but not its application to tumors. Lawrence returned to the NACC with his standard gambit—"It is almost unthinkable that the manifold new radiations and radioactive substances should not greatly extend the successful range of application of radiation therapy"— and a request for $23,000 for AY 1939/40. The sum, half of which went to salaries and a fifth to electric power, was awarded within a month of application.[13] Lawrence had smoothed the way, again by lobbying council members Compton and Conant and by mobilizing colleagues and philanthropists close to the council: Karl Compton of MIT, Warren Weaver of the Rockefeller Foundation, and Francis Carter Wood of Columbia's Crocker Institute.[14] The second NACC grant made possible the commencement of the clinical program of the Crocker Laboratory. Just as the medical application of the cyclotron was not foreseen when the 27-inch began to operate in 1932, the source of support for the clinical program was not foreseen—indeed, it did not exist—when the medical cyclotron was planned in 1936.

The federal package had two unwelcome trappings. For one, it came wrapped in red tape. The grant could not be processed at first because it was not submitted on forms acceptable to the Treasury. When money did begin to flow, it had to be metered to the dollar, every quarter, on forms that Lawrence had to certify; the press of bureaucratic fiddle-faddle brought Lawrence's secretary a secretary, and eventually the Laboratory its own business office.[15] Almost as unwelcome, at least initially, were leaks and advertisements to the press of the posssible relief from cancer by cyclotrons. Lawrence feared an onslaught of incurables and did receive pathetic letters from moribund cancer victims craving neutron irradiation as their last hope for life. The announcements also brought a rush of radiologists, eager to benefit mankind and

13. Lawrence to Compton, 19 Sep 1938 and 14 Mar 1939 (4/10), and to Hektoen, 9 Nov 1938 and 4 Mar 1939, and Hektoen to Lawrence, 4 Apr 1939 (13/29A); R.R. Spencer to Lawrence, 24 May 1939 (UCPF).

14. Letters to Lawrence from Cooksey, 29 Sep 1938 (4/21), Wood, 10 Oct 1938 (9/21), and Weaver, 6 Mar 1939 (15/29); Lawrence to Kruger, 15 Dec 1938 (10/20), Weaver, 11 Mar 1939 (15/29), A.H. Compton, 14 Mar 1939 (4/10), Conant, 24 Mar 1939 (4/12), and Cooksey, 3 Apr 1939 (4/22).

15. R.R. Spencer to L.A. Nichols, 17 Feb 1938, and to Lawrence, 1 Feb 1939 (13/33).

to share the council's money, to see how things were done in Berkeley.[16]

Between the first big magnet and the last big NACC grant, the federal government assisted the Laboratory through the Work Progress Administration (WPA) and the National Youth Authority (NYA). The first such aid came from FERA, the Federal Emergency Relief Act, in 1934/35. Lawrence asked for support for three graduate students to do odd jobs and for an electrician and cabinetmaker. In 1935/36, Lawrence asked WPA for two instrument makers, a cabinet maker, a draftsman (a graduate engineer), and an electrician (a specialist in radio), to facilitate the Laboratory's work, "which, in turn, has important applications in medicine, particularly to the problems of cancer."[17] The next year, as the Laboratory expanded its academic staff, it had a parallel increase in WPA manpower: three electricians, who rewired the Laboratory and installed additional electrical services; a carpenter by the name of House, who built a new floor, cabinets, shelves, and tables; two radio technicians, who worked on amplifiers and oscillators; a draftsman, who made all the working drawings for the apparatus and all the figures for published papers; two machinists, who provided auxiliary equipment for research; a clerk, who ordered and typed; and a young lady who autopsied rats and tested the blood of cyclotroneers. These ten free workers, who altogether cost the government some $8,500 a year if they worked halftime, constituted a very important asset, since they not only freed physicists from other work, but also assisted in experiments. Lawrence rated them "exceedingly valuable" and "urgently needed" in the Laboratory's "extended program of work, which may yield such important benefits to all of our citizens, and indeed to the whole world;" the WPA accepted the rating, and continued support at the same level for 1937/38.[18]

16. Lawrence to A.H. Compton, 17 Nov 1938 (4/10); *Time, 32:2* (28 Nov 1938), 41–2; R.D. Evans to Cooksey, 1 June 1938 (7/8); Cooksey to Lawrence, 10 and 15 June 1938 (4/21); Cooksey to Harold Ellis, Univ. News Service, 17 Dec 1941 (4/25).

17. Correspondence of Lawrence with Felix Flügel, U.C. Employment Relief Committee, 15 Aug and 13 Dec 1934 (20/21); Lawrence to Elizabeth Clark, 19 Jul 1935 (13/43). How far these desiderata were met is not indicated in the file.

18. Lawrence to WPA administrator W.C. Pomeroy, 23 Sep, 11 and 23 Nov 1936; Pomeroy to Lawrence, 27 Nov 1936, and reply, 25 Feb 1937, "urgently needed;" Lawrence to B. Hewetson, Federal Land Bank, 22 Apr 1937, other

There is no doubt that WPA help considerably advanced the work of the Laboratory. For a time, it supplied its entire shop staff, whose contributions were not only material. Birge had complained that the cyclotroneers tended to use more than their share of supplies and time in the Physics Department's shop, and although Lawrence's WPA shop did not supply all his needs, it helped to ease relations with the Department as well as research in the Laboratory.[19] An enlightening difficulty then surfaced. WPA administrators desired that Lawrence acknowledge WPA help in papers from the Laboratory. Would the Research Corporation, the Macy Foundation, and the Chemical Foundation like to be thanked along with the emblem of the New Deal, the WPA? It seemed to Lawrence that they would rather not. What he had in mind appears from the reply of the Macy Foundation, which preferred not to share the limelight even with the Research Corporation and the Chemical Foundation and advised that no acknowledgement of WPA be made until after the presidential election of 1936. "For all we know it might be interpreted...as being subversive to the spirit of the Constitution and [as an indication that] we have something to do with sinister communistic tendencies."[20]

Practical Poillon advised Lawrence to secure as much WPA help as he could and not worry further, since none of the foundations would make good the loss of WPA support. This license was what Lawrence wanted: "It seems to me entirely appropriate and in every way desirable to get as much W.P.A. assistance as possible for our work. Indeed, I believe that the Government should provide a great deal more support of scientific research." And then, an unusual disclosure: "Although I hoped and expected that Roosevelt would be reelected, I had no idea that there would be such a landslide. I think that it is really a tribute to the American public that they are not fooled and carried away by demagogu-

quotes; Lawrence to E.L. Viecz, 24 and 26 Oct 1938, and Viecz to Lawrence, 24 Sep 1938 (all in 18/33). Our estimate of half-time may be generous; cf. Birge, *History*, 4, xii, 29–30.

19. Birge, "Memorandum for the Radiation Laboratory," 21 Jul 1933 (Birge P, 33), and "Biennium report, 1936/8" (UCA, CU/68).

20. Lawrence to Poillon, 5 Oct 1936, and reply, 28 Oct 1936; Kast to Poillon, 13 Oct 1936, quote (15/17).

ery."[21] Lawrence's Democratic leaning did not survive the decline in WPA help at the Laboratory, which set in in 1939. Support for laboratory assistants and gofers through the NYA remained constant, however, at about $1,200 a year, beginning in 1937.[22]

Foundations

We approach the foundations. We have already explained the sources of income and the purposes of the Research Corporation and the Chemical Foundation. Both foundations had been hit hard by the Depression, and Lawrence deserves our admiration, as it earned him his contemporaries', for "having tapped such an apparently dry well."[23] In the grand year 1936/37, the Chemical Foundation behaved grandly and offered $68,600 for construction of the Crocker cyclotron, its accessories, and its operation. But it had overreached itself. Its German patents expired in 1937, and it could manage only $20,000; after the war it settled its outstanding obligations at 12 cents on the dollar. (Naturally Lawrence spent the entire pledge and the University had to make it good.) The Research Corporation was more careful. Beginning in 1936/37, it reduced its annual contribution to $5,000, to be used for physics research; it therefore helped importantly in keeping alive investigations of the type for which the Laboratory had been founded.[24]

Lawrence shrewdly diversified his funding base before his requirements exceeded the resources of his first foundation backers. He openly declared his motives to the two physicists he admired above all others, in order, perhaps, to obtain absolution for deviating from fundamental science. In the summer of 1935 he wrote to Rutherford that he had not experienced the difficulty he had expected in raising expenses for AY 1935/36: "the possible medical applications of the artificial radioactive substances and

21. Lawrence to Poillon, 4 Nov 1936 (15/17).
22. Cooksey to Pomeroy, 29 Jul 1939, and Pomeroy to Lawrence, 14 Dec 1939 (18/33).
23. Exner to Lawrence, 25 Jul 1933 (9/21); cf. Poillon to Lawrence, 6 Mar and 6 Apr 1933 (15/16A), admitting a difficulty in finding $500.
24. Lawrence to Sproul, 29 Sep 1936 (20/19); Underhill to Birge, 28 Nov 1938 (22/2); Board of Regents, Finance Committee, 27 Mar 1945, 3 (Neylan P, 129); Lawrence to Garvan, 6 May 1937 (3/38); Weaver to Lawrence, 6 Mar 1939 (15/29). The reneging of the Chemical Foundation made Sproul a little impatient with Lawrence's other overdrafts; Sproul to Lawrence, 30 Jan 1939 (22/2).

neutron radiation" brought what he needed. Late that year he explained the connection to father-confessor Bohr: "I must confess that one reason we have undertaken this biological work is that we thereby have been able to get financial support for all of the work in the laboratory. As you know, it is much easier to get funds for medical [than for physical] research."[25] We may back this claim with numbers. About 35 percent of the total giving of the 115 largest foundations in 1937, some $13.5 million, went to medicine and public health, including $166,000 for cancer research and treatment; the biological and physical sciences received $2.3 million in all, of which only $17,000 was reckoned as direct support for physics research.[26]

The angel of mercy and money for 1935/36 was the Josiah Macy, Jr., Foundation, established with a ledger value of $4.5 million in 1930 by Kate Macy Ladd, the daughter of the Quaker banker and philanthropist after whom she named her creation. On the advice of her doctor, Ludwig Kast, the donor expressed the wish that her foundation give special attention to medical problems that "require for their solution studies and efforts in correlated fields as well, such as biology and the social sciences."[27] Lawrence had lanced Macy's purse once before, in November 1933. He had no idea then, however, of introducing biomedical research into the Laboratory; rather, he wanted Macy's to help pay for the development of apparatus and technique to produce neutron beams as intense as x rays. To what purpose? "There is some justification," the physicist Lawrence instructed the physician Kast, "for the belief that the discovery of neutron rays is of an importance for the life sciences comparable to the discovery of x rays." Kast thought the argument plausible; the Macy Foundation gave $1,000 immediately, another grant of $2,200 in the spring of 1934, and a third (the one mentioned in the letters to Rutherford and Bohr) of $3,300 in June of 1935. All had to do with improving neutron sources. The foundation declined a further request for AY 1935/36 to support "biological studies on

25. Lawrence to Rutherford, 10 Jul 1935 (15/34), to Cockcroft, 4 Jul 1935 (4/5), and to Bohr, 27 Nov 1935 (3/3). For Bohr as confessor to physicists, see Heilbron, *Rev. hist. sci.*, 38 (1985), 223–4.
26. Twentieth Cent. Fund, *Am. found.*, 4, 33, 187, 189.
27. Ibid., 3, 25; Josiah Macy, Jr., Found., *Macy found.*, 2–6.

the effect of neutron rays."[28] After the formal establishment of the Radiation Laboratory and its embryonic medical branch under John Lawrence, Macy did grant money for biological experiments, some $7,000 in AY 1937/38 and something again in AY 1938/39 through the Research Corporation.[29]

The great neutron rays brought into existence with Macy money made possible biological experiments expensive enough to recommend recourse to the biggest spender of all the scientific philanthropies of the 1930s. The Rockefeller Foundation's Natural Sciences Division gave some $90 million during the decade in support of a single program, designed in 1932 by its director, Warren Weaver, who was trained as a physicist. The program might as well have been designed by Lawrence to meet the needs of the Radiation Laboratory after 1936: it encouraged the application of techniques and methods of physics and chemistry to the study of biology; in Weaver's classification of knowledge, cyclotron programs and isotope manufacture counted as "molecular biology." Moreover, improvement in the foundation's financial circumstances in 1936 and a new administration that favored basic science made it possible for Weaver to encourage even so demanding a supplicant as Lawrence.[30]

In May 1937 Lawrence visited Weaver, who had visited Berkeley in January and come away impressed by the possibility that neutrons might prove more effective than x rays in the treatment of cancer.[31] When asked whether the Laboratory might hope for

28. Lawrence to Kast, 11 Nov 1933, and to Poillon, 25 June 1935 (12/32); Kast to Lawrence, 25 Nov 1933, and H.N. Shenton to Lawrence, 19 May 1934 and 11 June 1935 (46/21R).

29. Lawrence to Garvan, 6 May 1937 (3/38), to Kast, 28 Aug 1937 (12/32), and to Sproul, 24 Jan 1938 (23/17); Josiah Macy, Jr., Found., *Macy found.* (1955), 156. Kast was extremely forthcoming: "We are still troubled by our reduced income [Macy could afford only $108,000 in grants in 1934]....but I am anxious to make arrangements for a renewal of the grant if you should so desire;" "will you please let us know if a small grant [for a Laboratory worker]...would be of help." Kast to Lawrence, 23 Apr 1934 (12/32) and 6 Dec 1935 (7/4); Twentieth Cent. Fund, *Am. found., 3,* 25; ibid., *4,* 22, gives Macy's book value in 1937 as $6 million.

30. Weaver, *Science of change,* 60–1, 70–2; Kohler in Reingold, *Context,* 271–9, 283. In 1937 the Rockefeller Foundation gave $9 million in grants, the Macy Foundation $211,000; Twentieth Cent. Fund, *Am. found., 4,* 26–7.

31. Weaver, "Diary," 25 Jan 1937 (RF, 1.1/205). Weaver and Lawrence had met in 1933; "Diary," 10 Feb 1933 (RF, 1.1/205).

Rockefeller support, Weaver replied that it would be an honor to help. The honor came gradually, first as stipends for postdoctoral fellows, next, in 1938/39, as a large capital grant of $30,000 to insure safety around the 60-inch cyclotron.[32] In the spring of 1939, as the big machine neared completion, Lawrence offered the Rockefeller Foundation more honor than it felt it could accept. He asked Weaver for $28,000 for the research program at the Crocker Laboratory for AY 1939/40, to be divided almost equally between the cost of cyclotron operations and the salaries of the staff. That exceeded Weaver's expectations. He knew about the application to the National Advisory Cancer Council and hesitated to pick up the larger share of a long-term project the first year of which cost over $50,000. He tendered instead a total of $50,000 over three years. Lawrence immediately recalculated that he could get by with $6,520 in salaries, $3,146 in supplies, and $7,000 in cyclotron operations, precisely $16,666, for the first year.[33] He made up some of the shortfall from the John and Mary R. Markle Foundation (established in 1927 on the fortune of a coal dealer), which was similar in purpose to, and, in 1937, about twice as rich as, the Macy Foundation; and from the Finney-Howell Foundation (established 1937), a very modest organization (ledger value $350,000) set up to give fellowships for research on cancer.[34]

Three miscellaneous sorts of contributions complete our account of the Laboratory's economy during the 1930s. The most evident of these were gifts in kind, for example, the transformer

32. Weaver, "Diary," 31 Jan 1939, and Rockefeller Foundation, minutes, 21 Jan 1938, 3801–3 (RF, 1.1/205); Lawrence to Kruger, 1 Oct 1936, and to F.B. Hansen, 28 Jul 1937 (10/20); to Weaver, 29 Nov 1937 (15/29); and to Cooksey, 12 May 1937 (4/21). Weaver rushed through a special grant of $2,000 to cover a debt on equipment that Lawrence had run up and Sproul did not wish to pay; Weaver, "Diary," 1 Jan and 24 Feb 1939 (RF, 1.1/205).

33. Lawrence to Weaver, 11 and 21 Mar, and Weaver to Lawrence, 16 Mar 1939 (15/29); Rockefeller Foundation, minutes, 5 Apr 1939, 39160–1 (RF, 1.1/205).

34. Twentieth Cent. Fund, Am. found., 4, 22, 27, 71; J.H. Lawrence to Lawrence, 6 May 1936 (11/16), and Lawrence to A.S. Woods, Markle Foundation, 9 June 1939 (12/35). The Markle Foundation originally had the grand mandate "to promote the advancement and diffusion of knowledge...[and] the general good of mankind;" in 1935 its directors decided to concentrate on medical sciences. New York Times, 4 Feb 1927, 2.

oil and radio tubes from Federal Telegraph, the generators from PG&E and General Electric, and the lead shielding from American Smelting, which Lawrence and Leuschner obtained for the 27-inch cyclotron. The total market value of these ingredients was about $2,500. It is not possible to strike a total for items given or loaned during the next several years, for the preferential prices offered the Laboratory by electrical manufacturers, or for the expert advice given gratis by engineers at General Electric, Westinghouse, Corning, and Eastman Kodak. An indication of the diversity, scale, and importance of commercial assistance was the concession given by Paramount Studios on the unused ends of cinema film for cloud chamber photography. When the concession of 1 cent a foot, which saved the Laboratory $22.50 a month under the commercial rate, ended, Lawrence mobilized Sproul, who obtained the film gratis.[35]

The least evident, smallest, and yet the most generous and useful contributor to the Laboratory's finances was Donald Cooksey, who became its associate director in the settlement of 1936. Cooksey's benefactions included subventions to needy students, fees for guest lecturers, tools for the Crocker cyclotron, and gifts to the "Lawrence Fund," a cache for all sorts of expenditures unsupervised by the Comptroller's Office, including grants-in-aid and no-interest loans to the more exploited members of the Laboratory's staff.[36]

And the most important contribution of all: the labor of the uncompensated staff. As appears from table 5.2, the Laboratory had, on the average, at least three unpaid graduate students, two or three unpaid postdocs, one or two professors on sabbatical leave, and, beginning in AY 1933/34, never fewer than three, and

35. Sproul to Emanuel Cohen, 8 Jan 1935, and V.E. Miller to Comptroller, U.C., 7–8 Mar 1935 (14/16). The file does not indicate how long Paramount supplied film at no charge.

36. Deutsch to Cooksey, 25 Jul 1938 (4/21), $200 for lectures by DuBridge, and Sproul to Cooksey, 22 Nov 1938 (4/22), up to $2,000 for tools; "Lawrence Fund," expenses (22/3), showing contributions from Cooksey of $1,190 between 9 Nov 1940 and 26 Feb 1941. For the Lawrence Fund, see Lawrence to W.B. Bell, 15 Jan 1941 (1/19); correspondence with Loomis, 1940 (46/8); letters to Cooksey from Sproul, 22 Sep 1938, and F.C. Stephens, 4 Jan 1939, and Lawrence to Stephens, 3 Sep 1940, to Corley, 16 Sep 1940, and (via Helen Griggs) to Lundberg, 19 Nov 1940 (UCPF).

Table 5.2

UCRL Staff by Means of Support,
1932/3–1939/40

	32/3	33/4	34/5	35/6	36/7	37/8	38/9	39/40
On state funds								
TA/phys. assts.	3	3	2	1	5	4	3	2
U. fellowship holders		1		1	2	2	2	3
Instructors & asst. profs.				1	1	1	3	4
Dir./asst. director	1	1	1	1	2	2	2	2
On intramural research funds								
Res. asst., assoc., fellows	3	2	4	3	7	9	13	14
Technical assistants			1	1	2	2	2	6
On external funds								
Postdoc & predoc fellows	1	4	3	5	7	4	7	4
Sabbatarians	1		1	3	3	2	1	2
WPA			2	5	7	10	10	?
NYA					3	5	5	6
On own funds								
Graduate student assts.		1	2	1	4	4	9	10
Res. asst., assoc., fellows	1	2	2	2	3	2	3	3
Total	10	14	18	24	46	47	60	56
Less WPA/NYA	10	14	16	19	36	32	45	50

as many as seven, holders of extramural fellowships. (Only people who spent more than a summer or three months during the school year at the Laboratory are counted.) A minimum estimate of their value may be obtained from the minimum cost of living in Berkeley in the mid 1930s, which was low in comparison to the East Coast: $60 or $70 a month as judged by Lawrence; $870 a year according to the more particular Miss C.S. Wu (whom Lawrence esteemed as "the ablest woman physicist that I have ever known...and altogether a decorative addition to any laboratory"); $600 a year in the opinion of the Graduate Division, whose fellowships in that amount easily provided room and board for the unfastidious ($25 a month per person double occupancy) and plenty of "rather slimy" Chinese dinners at 25 cents a head.[37]

Postdocs and sabbatical professors needed more for efficient upkeep. A room at the Faculty Club, where many regular and visiting members of the Laboratory lived, rented for $15 or $20 a month; one at the Shattuck Hotel for $17 a month. A decent dinner cost 50 or 60 cents.[38] Reckoning at $1,000 a man-year, the labor donated to the Laboratory between 1932/33 and 1939/40 came to at least $100,000. A fairer way to count would be to assign to each volunteer a stipend equal to that paid to members of the staff in the same category: graduate student, 75 cents an hour, or about $750 a year for half-time work; technical assistant, $900; postdoc, $1,200; journeyman physicist, instructor, assistant professor, $2,400; associate professor, around $3,600; professor, over $5,000.[39] Using average values for visitors ($1,800 for fellows, $4,300 for sabbatarians), we make the free-labor contribution at least $155,000. The figure is evidently too low, since it omits the unpaid help of summer visitors, some of whom, like Cooksey, Kurie, John Lawrence, and Segrè before their appointments, contributed more in a few months than a graduate research assistant might in a year; and others of whom, for example, Alexander Allen, G.K. (Ken) Green, and Roger Hickman, learned enough to be able to start cyclotrons at their home institutions.

Table 5.3 compares the total values of goods and services, exclusive of university overhead, cost of buildings, and gifts in kind, during the two four-year periods into which the Laboratory's

37. Lawrence to U.C. Nag, 29 Mar 1938 (3/23); R.E. Worley to Lawrence, 18 Apr 1939 (3/37), quoting Wu; Lawrence to Smyth, 23 Mar 1943 (18/36); Seaborg, *Jl., 1*, 40, 152, 163 (17 Jan 1935, fall 1936).

38. Cooksey to Mitchell, 15 and 19 June 1937 (13/9), and to J.A. Gray, 1 Aug 1938 and 1 Aug 1939 (14/38).

39. Lawrence, "News from Comptroller's Office," 31 Dec 1935 (22/1), Lawrence to Cooksey, 12 May 1937 (4/21), and to Weaver, 11 Mar 1939 (15/29), on graduate students; J. Brady to Lawrence, 19 Oct 1936, 27 May and 30 June 1937, 1 Mar 1939 (3/9), and Mitchell to Lawrence, 14 May 1938 (13/9), on instructors, assistant professors; Randall to Lawrence, 16 Aug 1935 (12/30), and J.R. Dunning to Lawrence, 28 Feb 1938 (4/8), on instructors; Condon to Lawrence, 17 Mar 1936 (4/15), H.B. Wells to Lawrence, 21 Jul 1937 (9/20), and Pettitt, *Twenty-eight years*, 60, on professors.

early history falls. We arrive at a grand total of around $670,000 for the Laboratory's support during the 1930s. This is roughly equal to the entire sum that went to support all of American academic physics, exclusive of new plant, in 1900; to about half of the average annual value of gifts and bequests received by the University of California during the 1930s; and to about 1 percent of the cost of an up-to-date battleship in 1940.[40]

Table 5.3

Values of Goods and Services,
1932/3–1939/40
(in thousands of dollars)

Source	1932/6	1936/40
State/university	38	243
Federal government	6	83
Res. Corp. & Chem. Fdn.	26	45
Medical foundations	8	64
Contributed labor[a]	59	96
Totals	137	531

a. Includes equivalent cost of holders of extramural
fellowships, rated as journeymen physicists.

40. Forman, Heilbron, and Weart, table I, give $475,000 for the support of American academic physics in 1900 in current dollars; Geiger, *Knowledge*, 278; *Jane's warships* (1939) gives $68 million as the cost of a battleship.

2. TAKERS

One of Lawrence's greatest strengths as institution builder was his ability to attract able students and associates. He had left Yale primarily because he had not been allowed to direct dissertations there; he had immediately found what he wanted at Berkeley, and by 1930 he had more graduate students than any other member of the Physics Department. His youth, energy, ambition, camaraderie, and solid, if not original, research projects in atomic physics enlarged his circle. In these early years he paid no stipends to his students, although they functioned as his assistants, and he had no need to keep them on after graduation to capitalize their knowledge or technique. They supported themselves or held teaching assistantships in the Department.

Movers and Shakers

The shift to long-range machine development, particularly x-ray apparatus and the cyclotron, made continuity of personnel desirable and, in the cases of Sloan and Livingston, necessary. Since in the first years of cyclotroneering Lawrence's grants did not include much if anything for academic salaries, he obtained temporary appointments for his most valuable associates in the Physics Department. On finishing his degree and the 11-inch cyclotron in 1931, Livingston became an instructor; on finishing his tenure as a Coffin Fellow, Sloan became a teaching assistant. A new pattern emerged in 1932/33. Since the Department did not reappoint instructors for a second year, Lawrence faced the grave problem in the spring of 1932 of retaining the invaluable Livingston, who could not afford to work for nothing. Sloan's progress on his commercially promising x-ray machine brought in enough to keep both him and Livingston, who received a half-time salary for setting up and running the Sloan tube at the University of California Medical School.[41] These were the first stipends Lawrence raised from outside sources. The Laboratory continued to provide for Livingston until the summer of 1934, when he transferred his art and his dependence to Cornell, and for Sloan until 1937, although

41. Lawrence to Kurie, 10 May 1933 (10/21); cf. Childs, *Genius*, 175–6.

his back injury made him a semi-invalid in 1934 and 1935.[42]

Lawrence preferred to pay as little as possible for assistance. Money for wages was the most difficult of all to raise, since the foundations had a policy—to which, fortunately for him, they did not strictly adhere—that "the universities should provide the salaries of their research men."[43] Here the Depression helped. Several of the most useful members of the Laboratory, for example, Malcolm Henderson, Jack Livingood, and Donald Cooksey, having private incomes and few prospects, paid their own ways for a time. In addition, and of utmost importance, as the reputation of the Laboratory grew, postdocs came with their own fellowship money, bringing new blood, labor, and expertise at no cost, and further advertising the Laboratory when they left it. The most frequent sources of these external rewards were the National Research Council and various agencies within the British Commonwealth, which contributed, respectively, seven and nine postdocs to the Laboratory during the 1930s, several of whom stayed for two years or more. The records for longevity belong to McMillan, who arrived in 1932 as a National Research Fellow, and Robert Thornton, who came on a Canadian fellowship. Both made their careers in the Laboratory. And there were always visitors, not only transients but residents for many months, like Frank Exner, Ryokichi Sagane, and Maurice Nahmias, who came to learn about the machines and ended by working on or with them. They, too, cost Lawrence nothing.

Appointments in the Physics Department kept many of Lawrence's people alive. Some twenty of them had teaching assistantships and/or university fellowships, generally two years of support, during their stay; and in each of the years 1936, 1937, and 1939, Lawrence achieved the feat, more notable together than the invention of the cyclotron, of placing one of his associates—Edwin McMillan, John Lawrence, and Luis Alvarez—in an assistant professorship. John Lawrence came from a junior faculty position at Yale, McMillan and Alvarez from instructorships at Berkeley, to which they rose from paid assistantships at the Laboratory, which

42. Lawrence to S. Dushman, 22 Jan 1935 (16/27 and 18/23).
43. Lawrence to Sproul, 12 Apr 1937 (20/16), quote; Birge to C.R. Jefferson, 4 Jul 1938 (Birge P, 33).

followed on a fellowship (McMillan) and on work without pay (Alvarez). A placement with similar payoff, although not engineered by Lawrence, was the appointment of Glenn Seaborg as instructor in the Chemistry Department, where he arrived after two years' service as Lewis's research assistant and Livingood's chemist.[44]

The settlement of 1936 brought several state-funded positions: a director, at $2,000 a year (Lawrence); an assistant director, at $3,000 (Cooksey); two postdoctoral research associates, who received a minimum of $1,000 each; and several technical assistants.[45] Some incumbents of these assistantships made capital contributions to the development of the machines, in particular, Charles Litton, then already the proprietor of his own engineering works, who had designed power oscillators for the Federal Telegraph Company; William Brobeck, a mechanical engineer, who took over general planning for new cyclotrons; and John A. Harvie, who came as a machinist in 1938 and ended as head of the machine shops of the postwar Laboratory. Brobeck's long association with the Laboratory began in the summer of 1937 in the ordinary way: he volunteered his services in exchange for an opportunity to learn cyclotron physics.[46] Lower down on the support ladder stood part-time technicians like Arthur Chick, who began in May 1933 assembling amplifiers, rose to odd jobs at the Laboratory and around the Sloan tube at the Medical Center, and left to succeed Coates at the Sloan tube at Columbia; Eric Lehmann, an electrician who helped around the cyclotron from 1934 or 1935 on; WPA workers; and the "charming..., smiling, witty" Helen Griggs, who spun the heads of several cyclotroneers.[47]

44. Alvarez to Lawrence, 9 Mar 1936, and phone conversation between Lawrence and C.D. Shane, 8 Nov 1945 (1/16); Chauncey Leake to Lawrence, 10 May 1936, and Lawrence to Sproul, 12 Apr 1937 (20/16), re John Lawrence; Seaborg, Jl., 1, 134–5 (2 May 1936), 257 (14 Jul 1937).

45. Birge, History, 4, xi, 23, xii, 32–3, xiii, 30; Seaborg, Jl., 1, 134–5 (2 May 1936), reporting $1,000 as the salary promised Alvarez; Underhill to Cooksey, 2 Sep 1935 and 15 Jul 1936 (4/19).

46. Lawrence to Beams, 18 Apr 1932 (12/9), re Litton; Cooksey to Lawrence, 21 May 1937, and W.J. Besler to Lawrence, 26 June 1937 (3/11), re Brobeck.

47. Lawrence to General Electric, 9 Feb 1937, Exner to Lawrence, 17 May 1937, and Chick to Lawrence, 19 Dec 1938 (3/40); documents of 1935 (4/11), Lawrence to McMillan, 11 May 1935 (11/30), and Kurie, RSI, 10 (1939), 205, re Lehmann; Aebersold to Cooksey, 14 Aug 1940 (1/9), and Seaborg, Jl., 1, 383 (20 Sep 1938), resp., re Griggs.

The trepanning of the medical purse allowed further expansion of personnel by providing stipends for the men who made isotopes and the men who improved the machines. A telling example is Livingood, who, having worked for nothing from 1932/33 to 1936/37, received a stipend from a Macy grant in 1937/38. When he immigrated to Harvard in the fall of 1938, his salary became available to Emilio Segrè, then emigrating from Palermo to Berkeley.[48] A more obvious indicator is the number and sorts of people engaged in connection with the medical cyclotron: engineer Brobeck; technician Winfield Salisbury, who returned to the University expecting to complete a graduate career interrupted by industrial employment but "worked full time on cyclotron development and was unable to take any more courses;"[49] and biophysicists Paul Aebersold and Joseph Hamilton, who came to work with John Lawrence. In all, Lawrence's grants paid for at least fifty man-years of labor during the 1930s.

A chart of the Laboratory's primary academic workers—staff, students, and long-term visitors—during the 1930s indicates their impressive growth in numbers (table 5.4). During the first two years, 1930/32, when Sloan and Livingston set the technical basis of further advance, Lawrence had about five workers a year (plate 5.1). With the setting up of "the little grey rat-trap," the "original cyclotroneers training school," "cyclotron headquarters," "the Mecca of cyclotronists," that is, the Radiation Laboratory, the number of academic workers rose to an average of fifteen during the quadrennium 1932/36.[50] Lawrence for once had more than he could handle, and briefly discouraged even volunteers from coming to Berkeley.[51] Among these fifteen were Bernard Kinsey, Franz Kurie, Edwin McMillan, Arthur Snell, Robert Thornton, and Stanley Van Voorhis, all extremely able tenants of extramural postdoctoral fellowships. With the deed of independence of the Labora-

48. Lawrence to Sproul, 31 Dec 1938 (16/14).

49. Cooksey to G.W. Stewart, 26 Dec 1945 (16/6).

50. Resp., Laslett to Cooksey, 20 Sep 1938 (10/32); Livingood to Lawrence, 16 Jul 1939 (12/11); R.D. Evans to Lawrence, 6 June 1938 (12/40); Nishina to Lawrence, 15 June 1940 (9/38), echoed in Oliphant to Lawrence, in Cockburn and Ellyard, *Oliphant*, 128.

51. Lawrence to Charles T. Zahn, 17 Nov 1933 (18/37).

Table 5.4a
Academic Staff, 1932/6

	1932/3	1933/4	1934/5	1935/6
Senior staff				
Dir./asst. dir.	Lawrence	Lawrence	Lawrence	Lawrence
Asst. profs.				
Instructors				
Res. associates	[Livingood], Livingston, [Lucci]	[Henderson, Livingood], Livingston, [Lucci]	H. White [Henderson, Livingood], McMillan, Sloan, Thornton	McMillan [Emo, Livingood], Sloan, Thornton
Postdocs				
NLF	McMillan	McMillan, Kurie	Kurie	Van Voorhis
Canadian		Thornton	Snell	Snell
British		Kinsey	Kinsey	Kinsey, Walke
UCSF				Aebersold
RF				
Medical				
Guggenheim				
Sabbatarians	Exner		Newson	Cork, Kruger, Newson
Graduate students				
Assistants		M. White [Coffin]	Laslett, Linford?, M. White [Coffin]	Cowie,[a] Laslett [Coffin?]
TA/phys. asst.		Aebersold, Paxton, H. White? Sloan, Laslett	Aebersold, Paxton	
Univ. fellowship				Paxton

Table 5.4b
Academic Staff, 1936/40

	1936/7	1937/8	1938/9	1939/40
Senior staff				
Dir./asst. dir.	Lawrence/Cooksey	Lawrence/Cooksey	Lawrence/Cooksey	Lawrence/Cooksey
Asst. profs.	McMillan	J. Lawrence, McMillan	J. Lawrence, McMillan, [Ruben?]	Alvarez, Lawrence, McMillan, [Ruben]

Instructors Res. associates	Alvarez, Emo, Kurie, [Kamen], Laslett, J. Lawrence. [Livingood], Paxton, Sloan	Alvarez [Brobeck], Hamilton, Kamen, Livingood, Lyman, Salisbury, Sloan, Snell, Van Voorhis, [Seaborg]	Alvarez [Brobeck], Cornog, Corson, [Emo], Farley, Kalbfell, Kamen, Hamilton, McNeel, Salisbury, [Seaborg], Segrè, Simmons, Thornton, Tuttle, Wilson	Seaborg [Brobeck], Corson, [Emo], Erf, Farley, Green, Kamen, Hamilton, [Langsdorf], Larkin, MacKenzie, McNeel, Salisbury, Segrè, Simmons, Tuttle, Waltman
Postdocs NRF	Van Voorhis	Langsdorf	Langsdorf, Green	
Canadian	Snell			Walker
British	Fairbrother, Hurst, Mann, Walke	Mann	Rollin	
UCSF RF	Aebersold	Aebersold Nahmias	Spear	Lewis
Medical			Aebersold, Lewis	Aebersold
Guggenheim			Erf, Marshak	Marshak
Sabbatarians	Friesen, Oldenburg, Sagane	Anslow, Sagane	Anslow, Capron	Hoag, Kruger
Graduate students Assistants	Backus, Corson, Cowie,[a] [Seaborg]	Backus, Corson, Cowie,[a] Simmons	Abelson, Backus, Condit, [Kennedy], Lyman, Mackenzie, Scott, Wright, Wu	Backus, Condit, [Kennedy], R. Livingston, Nag, Scott, [Wahl], Wright, Wu, Yockey
TA/phys. asst.	Abelson, Cornog, Kalbfell, Simmons, Wilson	Cornog, Helmholz, Kalbfell, Wilson	Lofgren (summer), Raymond	Lofgren, Raymond
Univ. fellowship	Lyman, Richardson	Lyman, Mackenzie	Mackenzie, Helmholz	Cornog, Helmholz, Wilson

Note: Brackets indicate staff whose salary or stipend did not come from the Laboratory's budget; "?" indicates status uncertain from the sources. "Staff" signifies all those who had a definite position or affiliation with the Laboratory except visitors who stayed for less than three months and students without financial support through the University on external fellowships.

a. Undergraduate.

tory, generous state financing, and bountiful external support consequent on the clinical potential of radioisotopes and neutron beams, the Laboratory's complement of academic workers more than doubled, to an average of thirty-five during the quadrennium 1936/40. And that was by no means more than Lawrence then thought he could use: "The number of men that can be kept profitably at work in connection with a [cyclotron] project is almost unlimited."[52] An account of the makeup of the Laboratory in a few representative years will put some flesh on these statistical bones.

Near the beginning of the first four-year period, in 1933/34, Lawrence had two unsupported graduate students at work on old business, photo-electricity (Cyrus Clark) and big sparks (Harvey White), and five on projects connected with accelerators. These were Leo Linford and Daniel Posin, who examined the debris from targets struck by high-speed mercury ions from Sloan's linac; Milton White, who used the 11-inch cyclotron to study proton-proton scattering; and Sloan and Hugh Paxton, who were trying to increase the intensity of the cyclotron's beams. Each of the six research associates of that year had his own project and most also collaborated in an intricate set of related researches: Kurie, a National Research Fellow, perfected his cloud chamber and studied neutron activation; McMillan, likewise a National Research Fellow, made his measurements on gamma rays, in which he was joined by Henderson, Livingston, and Lawrence; Henderson, self-supporting, and Livingston, on x-ray money, worked on deuteron disintegration; and Thornton, on a fellowship from McGill, and Bernard Kinsey, on a Commonwealth fellowship, tried to make a useful beam of lithium ions. These men brought not only the joy, but also the mess of research. "I have had to take on all the duties of a Director," Lawrence wrote his own thesis director, W.F.G. Swann, as he fought with the University to restore the janitorial services that had been discontinued as a Depression meas-

---

52. Lawrence to F.L. Hopwood, 24 Aug 1938 (9/15), estimating the number for 1938/39 at forty. Cf. Cooksey to Gray, 25 Oct 1946 (15/31), guessing at thirty-five workers in 1940, not all "members" of the Laboratory, but all involved in its "operations."

ure. "It would be manifestly impossible to get the fellows themselves to tidy up the Laboratory."[53]

In the first full year of the Laboratory's independence, 1936/37, the number of Lawrence's graduate students rose to a dozen, of external Fellows to seven, and of paid staff to ten exclusive of himself and Cooksey. The fellowship holders were Paul Aebersold in radiation biology, with support from the Medical School; Donald Hurst, on a Canadian scholarship, and Arthur Snell, an 1851 Exhibitioner; Wilfred Mann and Harold Walke, Commonwealth Fellows; Fred Fairbrother, Leverhulme Fellow; and Stanley Van Voorhis, National Research Fellow. A second National Research Fellow, Dean Wooldridge, elected to resign his award in favor of a job at Bell Labs.[54] The paid staff: Alvarez and John Lawrence, en route to assistant professorships; Dean Cowie, "an unusually ambitious and capable man and the hardest worker I have ever seen, yourself [Lawrence] included;" Count Lorenzo Emo Capodalista, who had learned his physics in Florence; Kurie, whose fellowship had expired; the perennial Sloan; L. Jackson Laslett and Paxton, now graduated, both of whom would soon go to Europe to help build cyclotrons; and the technicians Chick and Litton. There were also an unpaid Research Fellow (Livingood) and the free "Junior Research Associates," alias graduate students, Philip Abelson, David Kalbfell, Ernest Lyman, J.R. Richardson (on a University fellowship), and S.J. Simmons.[55]

By the steady-state year 1937/38, there were some forty people attached closely enough to the Laboratory to be eligible for appearance in its group photograph (plate 5.2). Lawrence could not contrive to retain the growing number of senior people on his junior staff once their fellowship or parental support had run out. In this matter of jobs he was extremely generous, even self-denying; he encouraged outside offers to his best men, whom he hoped to retain, like McMillan and Alvarez. "I shall get as many

53. Lawrence, "Report of research activities," [Spring 1934] (24/14); Lawrence to Beams, 24 Aug 1933 (2/26), and to Swann, 31 Aug 1933 (17/3).

54. Lawrence to Sproul, 2 Sep 1936 (23/17); and to Stone, 17 Jul 1935, re Aebersold (1/9).

55. T. Johnson to Lawrence, 28 June 1934 (10/1), re Cowie; Kenneth Priestley to Tuve, 3 June 1947 (5/4), from personnel records; staff listing, 1936/37 (4/21). Birge, *History, 4*, xii, 33, does not include John Lawrence among the "complete Radiation Laboratory personnel" as of 30 June 1936; but gives, ibid., xiii, 8, as the "regular staff" on 2 Sep 1936, Cooksey, Lawrence, Livingood, Kurie, and Litton.

good offers as possible," he wrote McMillan in the spring of 1935, "and allow the decision for next year to be made by [the] individuals [concerned]." And he had no trouble collecting offers; by 1935 interest in cyclotrons had become "extraordinarily gratifying (indeed amazing)." Berkeley veterans were soon building machines all over the country.[56]

Lawrence used the demand to bargain for regular rather than year-to-year appointments for his younger associates. In 1938, when ninety people applied for a single job in the physics department at the City College of New York, the Laboratory easily placed eight postdocs, each with several years' experience with the cyclotron, at leading research universities: Kurie, whom Lawrence recommended over Alvarez, went to Indiana; Langsdorf to Washington University in Saint Louis; Laslett, whom Cooksey rated over Kurie and Lawrence esteemed highly ("the outstanding graduate student [of 1936]"), to Michigan; Livingood to Harvard; Lyman to Illinois; Paxton, whom Lawrence likened to Milton White, successfully established at Princeton, to Columbia; Snell, whom Lawrence wished particularly to retain for cyclotronics, to Chicago; and Van Voorhis, "[a] storehouse of information..., a most amazing man," to Rochester.[57] "What are you going to do for help next year when all your men are leaving?" a cyclotroneer from the East asked Cooksey. It was "very distressing," Cooksey replied, since they happened to be "one's best friends;" but their leaving would not cripple the Laboratory. There remained McMillan (whom Princeton had approached), Alvarez, and Brobeck, Cooksey reminded his questioner, and half a dozen postdocs, "all of whom know the game from A to Z," and, among the graduate students, "two or three new men who are tops, [and] who

56. Lawrence to McMillan, 5 May 1935 (12/30), to M. Henderson, 3 May 1935 (9/6), and to Dayton Ulrey, 6 Nov 1935 (18/19); Cooksey to Lawrence, 29 Sep 1938 (4/21); infra, §6.3.

57. Cooksey to Lawrence, 29 Sep 1938 (4/21); Lawrence to Mitchell, 14 May 1938 (13/9), re Kurie, and to Coffin Foundation, 14 Feb 1936 (10/32), re Laslett; Cooksey to Lawrence, 29 Sep 1938 (4/21), re Laslett; Lawrence to [NRC], 27 Feb 1937 (15/22), re Paxton; Lawrence to A.W. Smith, 1 Jul, and to A.H. Compton, 22 Jul 1938 (16/33), and Kamen, *Radiant science*, 74, re Snell; DuBridge to Lawrence, 21 June 1939 (15/26A), re Van Voorhis. Snell was replaced by Thornton, who left a regular job at Michigan to return to the Laboratory; Cooksey to Allen, 5 Sep 1938 (1/14).

already know the game."[58] They, too, would be, indeed were, in demand. Lawrence invited MIT to make any offer it pleased to any of them; but no one very good, he warned, would go for anything less than an assistant professorship.[59] Weaver worried that the exodus of 1938 might injure the Laboratory. Lawrence seemed unconcerned. "[He had] so many fine younger men coming on that he does not feel badly about giving up his more experienced men."[60]

Regulation

The growth of the Laboratory, the increasing complexity of its equipment, and the continuing diversification of its activities forced a formal organization of research and development. From 1930 to the spring of 1934, Livingston did everything, and Lawrence almost everything, from machine design and construction to nuclear physics; newcomers like Henderson, Kinsey, Kurie, Livingood, McMillan, and Thornton learned how to operate, repair, and improve the cyclotron and devised and carried through experiments using its beams. To be sure, the range of these experiments was not very great: many related to deuteron disintegration and to other nuclear reactions among the lighter elements in extension of, and competition with, the work of the Cavendish Laboratory. The discovery of artificial radioactivity stimulated a new line, the making of new isotopes, and the need for chemical analysis of unweighable substances as well as the usual investigation of half-lives and radiations. At first the new line was easily assimilated into previous practice. The number of investigators did not increase rapidly, the periodic table was scoured for new activities in the old rough manner, the chemistry required stayed elementary, and the means of production—the familiar, exciting, irritating, finicky, 27-inch cyclotron—still needed constant attention and frequent repair.

58. Cooksey to "Dodie," 30 Mar, and to Lawrence, 21 [Apr] and 39 Sep 1938 (4/21); A.J. Allen (Bartol) to Cooksey, 13 June, and Cooksey to Allen, 15 June 1938 (1/14); Lawrence to A.H. Compton, 25 Jul 1938 (4/10).
59. Lawrence to Evans, 2 June 1938 (12/40).
60. Weaver, note on interview with Lawrence, 14 June 1938 (RF, 1.1/205).

The need to supply radioisotopes for biological and medical research transformed life in the Laboratory. The quantities needed rose rapidly: not only were there many enthusiastic users, but the amounts required for an experiment in biological tracing, or, especially, in medical therapy, exceeded by perhaps a thousandfold what a physical or chemical study might employ. By the spring of 1937 the cyclotron was supplying radioactive materials for the researches of two dozen physicists, half a dozen biologists, and several chemists.[61] Along with manufacture came pressure for innovation in both product and process. Besides making useful known isotopes for distribution, the staff sought systematically for new activities more suitable to the needs of the Laboratory's clients. Besides using the existing machinery, the staff worked to improve the cyclotron beam and target arrangements to make manufacture more efficient and to intensify the output of neutrons for possible clinical application.

Lawrence was not altogether happy about the reorientation of the Laboratory's work. His correspondence with Cockcroft, who stood in this case for the Cavendish and straightforward nuclear physics, took on a new tone: where before it had expressed enthusiasm and optimism, it now dwelt on dullness and routine. "Nothing surprizing has been turning up," he wrote in the spring of 1935, "simply measuring gamma ray energies and beta ray energy distributions, half lives and so forth." And again in the fall: "There is nothing particularly to mention now, excepting perhaps that we are finding that with 4.2 MV deuterons we can excite nuclear reactions over the entire periodic table....Much of the work is of almost routine character, measuring half-lives, beta-rays, gamma-rays, transmutation functions, etc., from various radioactive substances by deuteron bombardment."[62]

On November 30, 1936, Emo posted a blank sheet of paper and asked people to sign up for the shifts they preferred and to indicate the targets they wished to have irradiated for their own use. Two men constituted a shift, and six a day's "crew," under one of

61. Kurie, *GE review, 40* (June 1937), 272.
62. Lawrence to Cockcroft, 21 Mar, 2 Oct, 25 Nov 1935 (4/5). Cf. Lawrence to Kast, 6 Feb 1935 (12/32): "an unproductive period....vexed by the problem of producing more intense beams of neutrons."

its members as "captain" (the first captain was Livingood); the machine required tending from 8:30 A.M. until midnight, seven days a week.[63] Crew service, as Lawrence wrote Wooldridge, was a part of the standard "laboratory apprenticeship;" perhaps the prospect helped Wooldridge to decide on Bell over Berkeley. It could be "downright drudgery," Abelson recalled, this working thirty hours a week on a crew, when, as in his case, the indenture came on top of service as a teaching assistant and the normal studies of a graduate student.[64] Various technical improvements, which we shall describe in their place, insured that the machine did work almost daily. But little came out to delight a physicist. Lawrence wrote his conscience Cockcroft just before Christmas 1936: "The number of radioactive substances produced by deuteron bombardment is very large, and investigation of these has become almost a routine matter of little interest."[65]

In the spring of 1937, with the addition of a new oscillator, the cyclotron worked so well that even its operation had "become a monotonous job." A visitor from Joliot's laboratory, Maurice Nahmias, learned that each week's duty fell to a six-man crew, who took it in turns, two at a time, "and they work on Sundays." A crew member did not pursue his own project, but executed a series of irradiations fixed in advance. "A perfect interchangeability and solidarity." The human part of the machine, like the cyclotron itself, was adjusted for maximum output. In May 1937, Lawrence added an owl shift, from 11:00 P.M. to 3:00 A.M., to make radiophosphorus for John Lawrence among others, and on July 4 the laboratory temporarily went on 24-hour operation, with four six-hour shifts a day, to meet the demands of biomedical research.[66]

That became the recommended regime for a production cyclotron, which, Lawrence advised, should be in the care of an

63. Cooksey to Lawrence, 30 Nov 1936 (3/11); Lawrence to Cockcroft, 11 Aug 1936 (4/5); Mann to G.P. Thomson, 15 Nov 1936 (GPT).

64. Lawrence to Wooldridge, 12 Aug 1936 (18/32); Abelson, BAS, 30:4 (1974), 48.

65. Lawrence to Cockcroft, 15 Dec 1936 (4/5); the same message is in Lawrence to Cork, 24 Feb 1937 (5/1), and to T. Johnson, 9 Feb 1937 (10/1).

66. Lawrence to Segrè, 5 Apr 1937 (16/4); Nahmias to Joliot, 28 Apr 1937 (JP, F25); Kamen, Radiant science, 70–1.

experienced physicist and four graduate assistants. Without such discipline—without the dedication of graduate students to the machines—a large cyclotron could not be built or maintained. Where graduate students were not slaves to their thesis work and advisor, as at Cornell, there could be no question of "an intense nuclear program."[67] Once again Lawrence summed up the year's work for Cockcroft and once again he apologized. "Much work in both nuclear physics and biology is in progress. There is nothing particularly exciting at this moment to report." The same smooth operation, averaging about twenty hours a day, resumed in 1938 after the enlargement of the cyclotron's pole pieces to thirty-seven inches, which also brought higher energies and greater yields.[68] Typical excerpts from Lawrence's correspondence of the time: they were putting out 100 μA of 6.4 MeV deuterons, "almost night and day, most of the time manufacturing radioactive materials for biological experiments and clinical investigations;" the product consisted mainly of P^{32}, some 15 mCi a day for John Lawrence, and it was "getting awfully monotonous running the cyclotron."[69]

In addition to working on a crew, each member of the staff and every postdoc had a specialty, which Nahmias itemized as follows in the spring of 1937: Cooksey, designing and planning; Alvarez, Hurst, Kurie, and Paxton, cloud chambers; Martin Kamen, Mann, Snell, chemistry; Cowie, Laslett, Van Voorhis, electricians; Litton and Richardson, cyclotron technicalities; Emo, amplifiers; McMillan, Livingood, Walke, counters and electroscopes; Aebersold, Nurse Condit, and John Lawrence, mice. These specialists were about to turn their expertise increasingly to the construction of the medical cyclotron, which took most of their time and energy by the beginning of the new year, and, as it neared completion, curtailed the operation of the 37-inch.[70] Between the demands of the

67. Lawrence to Zirkle, 20 Feb 1940 (14/5); Livingston to Lawrence, 7 Jan 1938 (12/12), re Cornell, which he left for MIT.

68. Lawrence to Cockcroft, 1 Dec 1937 (4/5); Lawrence to Bohr, 17 Feb 1938 (3/3); Kurie, *Jl. appl. phys., 9* (Nov 1938), 700, and *RSI, 10* (1939), 205.

69. Resp., Lawrence to Bohr, 17 Feb 1938 (3/3), and to A.J. Allen, 16 Feb [1938] (1/14); McMillan to Lawrence, 27 Oct 1937 (12/30): "We hope very soon to be able to satisfy the hoards [!] of biologists that are swarming around looking for radioactive samples, and perhaps to get in a little bombardment for ourselves."

70. Nahmias to Joliot, 28 Apr 1937 (JP, F25); Lawrence to Bohr, 17 Feb 1938 (3/3); Seaborg, *Jl., 1,* 448 (19 Feb 1939).

new machine and the old manufacturing there was little time for physics. That bothered the abler men in the Laboratory.

In their autobiographies, Alvarez, Kamen, Segrè, and Thornton all comment on the inhibition to physics research presented by the Laboratory regime. Alvarez and Kamen associate this inhibition with the missing of the major discoveries in nuclear physics during the 1930s, particularly artificial radioactivity and fission; and Kamen further observes that perhaps the most interesting discovery from the chemist's point of view made among the radioactive output of the cyclotron—namely, element 43—was allowed to fall to Segrè's group of painstaking, and otherwise unoccupied, professors in Palermo. "Emphasis on keeping the machine running seemed exaggerated at times," Kamen recalls. The compulsion was particularly irksome during the scheduled sessions of neutron therapy: "The tensions created by these sessions plagued the crews badly, and they made a pill-popper out of Aebersold." Segrè remembered that most of the people at the Laboratory were more interested in the cyclotron than in the results obtained with it; Aebersold felt compelled "almost [to] eat and sleep on cyclotrons." Thornton laments that he did nothing very good in physics owing to preoccupation with machines and detectors. Robert Wilson rejected an offer to stay on at the Laboratory in favor of an instructorship at Princeton; as he later ambivalently evaluated the situation, "the sense of history being made" at Berkeley did not compensate for slavery to the cyclotron, "an activity that epitomizes team research at its worst."[71] As the technological and administrative imperative strengthened, Cooksey discovered himself to be "only a 'hear-say' physicist...., so completely busy in putting through the construction and design of our new apparatus that I can do no experiments for myself." Lawrence lost track of the work altogether: "I do not even know what substances are being bombarded or exactly what is being done."[72]

71. Alvarez, *Adventures*, 56; Kamen, *Radiant science*, 73; Segrè, *Ann. rev. nucl. sci.*, *31* (1981), 9; Aebersold to Cooksey, 14 Aug 1940 (1/9); Thornton, 10 June 1975 (TBL); Wilson in Holton, *Twentieth cent.*, 472, 478. Wilson became one of the country's most accomplished builders of accelerators.

72. Cooksey to S.W. Barnes, 19 Apr 1937 (2/24); Lawrence to Cork, 24 Feb 1937 (5/1).

Nahmias, whose sensitivity to the Laboratory regime was enhanced by culture shock, judged that most of Lawrence's men preferred measuring the half-lives of new radioisotopes to doing physics and esteemed machine making—"a mania for gadgets or a post-infantile fascination for scientific meccano games"—over both. Rather than submit to such a regime, he decided to spend most of his seven months at Berkeley in Stanford, where he could learn some physics from Fermi, then visiting professor there. Again, Lawrence understood and to some extent apologized for the situation to physicist colleagues. "To all of us working with the cyclotron there is always an urge to do what seems to be the obvious thing to make the cyclotron work better, at the expense of actual nuclear research."[73]

That was in January 1937, when Lawrence judged that half the work at the cyclotron went to improving it. Six months later, physics briefly had the upper hand when conversion to the 37-inch was slightly delayed to allow completion of promising experiments on the 27-inch. Then the balance shifted definitively in favor of the machine. Lawrence wrote Segrè: "We are all busy here, building the new medical cyclotron, and consequently some of us are not having as much time as we should like for nuclear research, but one cannot expect to do everything."[74] Most cyclotroneers were willing to accept the trade-off most of the time. Perhaps their level of modest discontent was caught best in a letter from Emo written during a year's hiatus in his attachment to the Laboratory. "I would like to be back in the midst of you all and participate, even in small measure, in the doings and developments of the lab, and put in some small research, in between times, if possible."[75]

The medical cyclotron forced a new level of organization. To facilitate its construction, Brobeck distributed the seventeen senior men into twenty interlocking committees, each with a specific and

73. Nahmias to Joliot, 12 June 1937 (JP, F25); Lawrence to Henderson, 24 Jan 1937 (9/6), quote; White to Lawrence, 18 Oct 1932 (18/21): "I have my heart set on immediately pushing the outfit up to the highest energy to which the magnet will go."

74. Cooksey to Lawrence, 1 June 1937 (4/21); Lawrence to Cork, 19 Jul 1937 (5/1); and to Segrè, 12 Apr 1938 (16/14).

75. Emo to Lawrence, 24 Oct 1937 (7/4).

detailed charge. As Kurie wrote, "In a cooperative project such as the operation of a large cyclotron no development is made without most of the staff participating." While in the organizational mood, Brobeck introduced preventive maintenance, in place of crisis management, of the cyclotron's plumbing. The first of his schedules, designed for the 37-inch cyclotron, included some twenty-four weekly checks, oil changes, and greasings, to be recorded on sheets "similar to those used in automobile service stations." When the big 60-inch came on line, the obligations of the staff enlarged as well. In June 1940, after a year's operation, Lawrence posted an announcement intended to correct the inefficiency introduced by research, thesis writing, and other academic exercises. "It is now expected that the senior men will assume full responsibility for certain phases of the work. This division of responsibility should make for a broader attack on all the problems and keep the work continually under management. Further, by placing the new members into groups under the senior members to work on particular phases of the work it is hoped that not only will the work progress in a more organized manner, but that thereby the newer members will be given a better chance to show their worth to the Laboratory."[76] When the boss was in town, the work on the machines would proceed apace. When he left, the suppressed urge to add a mite to science would out. "Professor Lawrence has gone East for a few days, and everyone seems to think that was the signal to start working on papers."[77]

This more elaborate group structure, a consequence of the forces at play within the interdisciplinary laboratory and of the need to meet external commitments, was a useful and even an inspired preparation for the role the Laboratory and its dispersed staff would take on during the war. The quasi-military organization, with its officers and cadets, suggests an extension of the analogy, already trite by 1940, between the cyclotron and an artillery piece.[78] If the cyclotron was Big Bertha, cyclotroneers were her

76. Kurie, *RSI, 10* (1939), 205; Brobeck, "Suggested maintenance operations," 25 Jul 1938 (4/21); Lawrence, announcement, June 1940, preserved by Aebersold (1/9); McMillan to Brobeck and Cooksey, 25 Apr 1939 (12/30); regime for Crocker Laboratory, divided into groups for neutron therapy, leukemia work, and tracers, in Griggs to F.C. Blake, 24 Jul 1940 (3/1).
77. Griggs to Barbara Laslett, [1939] (10/33).
78. E.g., J.K. Robertson, professor of physics, Queen's University, Ontario, en-

disciplined attendants. This analogy was drawn by the University Explorer in describing his visit to the Laboratory in 1936, even before Emo published his first schedule. The Explorer likened the 27-inch to a machine gun, as usual, and the light from the external deuteron beam to "the glow released from billions of atom citadels bursting into conflagration." He then turned his attention to the "army": to commander Lawrence, who opened the "cartridge chamber" to expel the deuteron beam, and to corporals McMillan, Kurie, Livingood, and Cooksey, who operated the controls "like a disciplined squad."[79]

Socialization

Discontent with the increasingly structured regime was minimized by close social ties, by seminars and colloquia on wider subjects, and by the satisfaction of cooperative work toward an important objective unrealizable by a single individual. Morale stayed high. In 1934, before the expansion of the Laboratory and restraint on individual action, visitors praised Lawrence for "the wonderful spirit which you have allowed to develop among the men," "the 'pep' [you instill] into your students," "the splendid and enthusiastic group of physicists you have collected;" and Alvarez, shopping for a place for postdoctoral work, contrasted lackluster Caltech with "the enthusiasm which I liked so much at Berkeley." "The human energy concentrated there is wonderful."[80]

The same and even more was said about the place by people who knew it in its organized state. Wilfrid Mann, a Commonwealth Fellow, 1936/38: "To know Ernest Lawrence is to know too why it is that the Berkeley cyclotrons give such incredible results. In the face of such irrepressible enthusiasm and such

---

titled his popular work on nuclear physics, *Atomic artillery* (1937), and the chapter on accelerators, "Bringing up the big guns;" and in 1939 Lawrence appeared with the 60-inch cyclotron, in a photograph furnished by the Laboratory and captioned, with Cooksey's approval, by the Hazard [!] Advertising Corporation, under the legend, "Gun and Chief Gunner." Hazard to Cooksey, 11 Aug 1939 (25/5).

79. "An eighty-five ton machine gun," "University Explorer," no. 103, 9 Feb 1936 (40/15).

80. Letters from T. Johnson, 28 June 1934 (10/1); J. Beams, 28 Sep and 10 Aug 1934 (2/26); and Alvarez, 15 Oct 1934 (1/16); Paxton to Phil [Abelson?], 15 Nov 1937 (14/18).

joie de vivre difficulties hardly stand a chance." Otto Oldenberg, after spending a sabbatical from Harvard at Berkeley, advised readers of the *Physical Review* of "the spirit of cooperation among [Lawrence's] collaborators which makes the work at Berkeley so pleasant and profitable." "The Laboratory represents as fine a piece of cooperative effort as exists in the annals of science."[81] Cooperativeness was the Laboratory's hallmark, and its test, "[being] capable of working well with the group," figured prominently in the recommendations for incoming postdocs.[82] For those who liked the togetherness, "Berkeley was the greatest place in the world." As Kamen wrote many years later, and after a painful break with Lawrence: "It is impossible to describe the enthusiasm and zeal for accomplishment that pervaded the Radiation Laboratory in those magical years."[83]

Tributes from short-term visitors were enough to make a strong man blush. Eugene P. Pendergast, M.D., radiologist, University of Pennsylvania: "I am writing to thank you for the greatest experience that I have ever had. A visit to your Department is just like a good tonic to an aging individual." Fred J. Hodges, M.D., roentgenologist, University of Michigan: "I have returned to Ann Arbor rested, considerably rejuvenated, and very deeply imbued with the enthusiasm and esprit de corps which pervades your entire organization." W.B. Bell, president, American Cyanimid: "The week which I spent with you, your brother, your associates and the cyclotron was one of the most thrilling in my experience." Marcus Oliphant, become professor of physics, University of Birmingham, and a tout for the cyclotron: "I know of no laboratory in the world at the present time which has so fine a spirit and so grand a tradition of hard work. While there I seemed to feel once again the spirit of the old Cavendish, and to find in you those fine qualities of a combined camaraderie and leadership which endeared Rutherford to all who worked with him. The essence of the Cavendish is now in Berkeley."[84]

81. Mann, Phys. Soc. of London, *Proc., 53* (1941), 3; Oldenberg, *PR, 53* (1938), 39, rec'd 22 Oct 1937; Birge in U.C., *1939 prize,* 27.

82. E.g., Cork to Lawrence, 31 Jan (quote) and 9 June 1939 (5/1); Wooldridge to Lawrence, 25 May 1936 (18/32); Capron, "Report," 3.

83. Emo to Lawrence, 11 Mar 1938 (7/4); Kamen, *Radiant science,* 70.

84. Letters to Lawrence from Pendergast, 13 Sep 1940 (14/22); Hodges, 3 Feb 1941 (9/9); Bell, 27 Jul 1940 (1/19); and Oliphant, 11 Jan 1939 (14/6), and *PT, 19:9* (1966), 35.

Camaraderie was promoted in many ways. The senior staff made clear the importance of the ordinary work by participating in it. Distinguished and appreciative visitors, including Niels Bohr and Arthur Compton; alumni who returned, almost by the dozen, during the summer; and organized tours, for inquisitive physicians, local schools, the National Congregationalists, the 4-H All-Stars, the Cheyenne Mountain Dancers; all this enforced the impression that the Laboratory was a special place and its members special people.[85] Visiting scientists joined regulars at a Journal Club, which met every Monday evening, from 7:30 to 9:00, to discuss the latest letters to *Nature* and papers on nuclear physics, and at a seminar on nuclear physics, which met every Thursday afternoon at 5:00. The Journal Club existed as early as 1932. At first it discussed mainly work done in other laboratories, but as more people came to work in Berkeley, reports of their activities dominated the Monday meetings and constituted the chief source of scientific information of many of the participants.[86] It was easy for Lawrence's students to conceive that the Laboratory occupied the center of the world of physics. In this connection we might construe Bethe's innuendo: "Quite generally the Rochester Ph.D.'s learn physics at the same time as cyclotronics, which distinguishes them favorably from their fellow cyclotronists."[87]

Berkeley cyclotroneers enjoyed the consciousness of braving together other dangers than ignorance. During their working hours they basked in high-power radio waves. They shared quarters with John Lawrence's rats until the stink threatened to drive them from the building. They breathed natural gas used to detect leaks and had the satisfaction, when successful, of giving their names to newly found cracks, whose locations and eponyms accompanied the old 27-inch chamber when the Laboratory gave

85. Cooksey to Lawrence, 29 Sep 1938 (4/21), on the summer's visitors; incomplete list of visiting individuals and groups (24/18).

86. M. Henderson to Lawrence, 5 Jan 1933 (9/6); Birge to Richtmeyer, 24 Oct 1933 (12/30); Mann to G.P. Thomson, 15 Nov 1936 (GPT); Alvarez to Ralph Lapp, 26 Sep 1949 (Alvarez P), and *Adventures*, 47-8; Wilson in Holton, *Twentieth cent.*, 472; Nahmias to Joliot, 28 Apr 1937 (JP, F25).

87. Bethe to Lark-Horowitz, ca. 1939 (HAB, 3).

it to a cyclotroneer at Yale. "There is one gruesome spot directly over a wheel bracket known as 'Luis's Hole' which you doubtless would never find;" it was the grand discovery of Luis Alvarez, who took in a lot of gas finding it.[88]

Flying objects made more hazards: the boss himself was surprised by pliers jumping from his pocket to the cyclotron magnet, and by a loose piece of metal, which "nipped off the end of [his] finger."[89] Then there was the constant risk of electrocution. Two Berkeley men, Coates ("[one of the] most cautious and careful workers we have ever had") and Walke, working elsewhere on Berkeley-style machines, were killed; Abelson came close to serious injury; and Jack Livingood received "a jolt that took him within an inch of oblivion," a 10,000-volt spark that jumped fourteen inches from the high-frequency generator to his head and thence, lucky for him, out his feet without crossing his chest. "It isn't possible to make the equipment foolproof when it is being rapidly altered," Lawrence said. The University ordered a thorough review, together with instruction in first aid. Gradually the Laboratory introduced interlocks and other electrical safety devices.[90]

Perhaps the most satisfactory hazard for creating a feeling of community was neutron radiation, which, by 1938, had made almost all the metal in the Laboratory radioactive.[91] Lawrence began to worry about its effects on bodies around the Laboratory in the fall of 1933; in 1934 he asked a professor of physiology to study the problem; in 1935 his brother John came to town and showed that "we have been giving ourselves undesirably great exposures to neutrons."[92] Meanwhile the cyclotron had been

88. Alvarez, *Adventures*, 41-3; Kamen, *Radiant science*, 66–8; Kurie, *RSI, 10* (1939), 205; Lawrence to John Lawrence, 24 Oct 1936 (11/16); Cooksey to Pollard, 9 Nov 1937 (14/30), quote.

89. Lawrence to Livingston, 20 Feb 1935 (12/12); *Time, 26:2* (16 Dec 1935), 32, quote; Livingood to McMillan, 5 Aug 1937 (12/11).

90. For Abelson, Lawrence to Cooksey, 6 June 1936, and for Livingood, Lawrence to Cooksey, 29 Dec 1934 (4/21); E.S. Viez to Campus Safety Committee, 12 June 1936, and Birge to same, same date (21/5); on Coates's caution, Lawrence to Johnson, 3 Apr 1937 (10/1).

91. Lawrence to DuBridge, 21 Jan 1938 (15/26).

92. Lawrence to Beams, 4 Oct 1933 and 17 Sep 1934 (2/26); to Exner, 4 Oct 1933 (9/21); to Cockcroft, 12 Sep 1935 (4/5), quote.

improved to yield 20 μA of deuterons at 4.3 MeV; they "all got the jitters," as did residents of the nearby chemistry building, and moved the cyclotron controls thirty feet from the beryllium target.[93] We shall describe later the precautions taken after 1935 and the radiological research and therapy to which study of the danger gave rise. The Laboratory had good luck. The blood counts taken now and again during the late 1930s show evidences of overexposure only in the case of Robert Cornog, who inadvertently stuck his hand in the deuteron beam. No delayed effects of radiation unambiguously attributable to exposure at the Laboratory had been identified by 1947, when Dean Cowie came down with cataracts, apparently caused by neutrons during the running in of the 60-inch cyclotron at the Carnegie Institution in 1943/44.[94] Although in 1939 Lawrence rated the Laboratory as safe as his own home, the Aetna Life Insurance Company, to whom he adressed this extraordinary statement, preferred to treat members of the Laboratory as uncommon risks.[95]

Morale was kept up by pleasure as well as by danger, by lunches, picnics, parties, joint vacations, and other collaborative amusements. The men who lived for years at the Faculty Club (Cooksey, McMillan, John Lawrence, Seaborg, Snell, and Van Voorhis) necessarily saw one another socially. Lawrence presided over a daily lunch at the club and set an example of wholesomeness as well as of accessibility by a diet of cornflakes, milk, and strawberries.[96] An annual dinner at DiBiasi's Capri Italian restaurant gave opportunity for cracks and charades at the expense of the boss in imitation, perhaps, of similar fun at the Cavendish and at Bohr's institute in Copenhagen (plate 5.3). An excerpt from

93. Lawrence to M. White, 30 Oct, and reply, 21 Nov 1935 (18/21); to Darrow, 25 Sep 1935 (6/9); Fuller to Lawrence, 30 Sep 1935 (21/5).

94. Kenneth Priestley to Tuve, 3 June 1947, and Tuve, "Memorandum regarding Mr Cowie's eyes," 18 June 1947 (5/4); infra, §8.1.

95. Lawrence to H.H. Benedict, 6 Feb 1939, and Aetna to Benedict, 15 Mar 1939 (21/5); Lawrence to Cockcroft, 16 Nov 1938 (4/5). Cf. Tuve to Aetna Life, 24 Feb 1933 (MAT, 8).

96. Lawrence to Cooksey, 29 Dec 1934, and Cooksey to Lawrence, 7 June 1938 (4/21); Alvarez to Lawrence, 15 Oct 1934 (1/6); Seaborg, *Jl.,* 2, 632, 668 (31 Aug 1940, 2 Jan 1941); Cooksey to Mitchell, 19 June 1937 (13/9); Kamen, *Radiant science,* 71–2; and Cooksey to Lawrence, 30 Sep 1932 (4/19), on the diet.

one of these entertainments will indicate the spirit, the gripes, and the affection felt by the staff:[97]

> I means intensity—our first main objective....
> M must mean mice whose smell makes us moan,
> N stands for neutrons, of moment unknown....
> S now is store-room, a creation of Jack's—
> T can't be tidyness, for this the lab lacks....
> W is for wax, which we smear on like fools;
> X hides the unknown location of tools.

The Christmas jokes were carried through the year in a special argot that distinguished cyclotroneers from the rest of mankind. An example of this "lingo (one could hardly dignify it with the term nomenclature)": "spilling soup into the can made a few mikes," which, translated, signifies that putting power into the dees caused a beam of a few microamps.[98]

Serious attachments arose with the easing of social barriers. Cooksey and Seaborg competed for Helen Griggs, who chose Seaborg; Cooksey rebounded to Millicent Sperry, a secretary in the Physics Department. Several other cyclotroneers married women in or close to the Laboratory. Lyman picked a fellow graduate student, a spectroscopist; Aebersold took "a very sweet little girl from the Purchasing Department;" Laslett ran off with Barbara Bridgeford, a graduate student in social welfare, who preceded Griggs as Lawrence's part-time secretary; and, a masterpiece of intermarriage, McMillan won Molly Lawrence's sister Elsie.[99] Perhaps the most important mechanism for the socialization of the researchers was Donald Cooksey, who not only was kindly to all on the job, lent money to people in need, and acted as gracious host to visitors, but also operated his own rest and recreation facility at his country retreat in Northern California. When he deemed a vacation to be in order, he would send the sufferer up to camp, where he also entertained visiting cyclotroneers. His surviving guest book, which begins in 1939, shows very few people

97. Poem by Laslett for dinner of Oct 1936 (4/19).
98. Griggs to Barbara Laslett, [1939] (10/33); Cooksey to Barnes, 19 Apr 1937 (2/24).
99. Cooksey to Kinsey, 23 Oct 1942 (10/18), and to W.R. Hazeltine, same date (4/25); to F.W. Loomis, 14 May 1939 (12/19); to Van Voorhis, 9 Jul 1937 (17/40) and 5 Apr 1939 (12/43); Lawrence to Randall, 28 Feb 1939 (10/33); Griggs to McMillan, 9 May 1941 (12/30).

not connected with the Laboratory. The most frequent guest was Helen Griggs; others who visited twice or more were Aebersold, Brobeck, Kamen, the Lawrence brothers, McMillan, Segrè, Thornton, and Robert Wilson.[100] A final reason for high morale, no doubt, was the excellent prospect for a permanent position outside Berkeley enjoyed by anyone who won Lawrence's support.

The camaraderie and solidarity had a negative aspect that struck that oversensitive detector Nahmias. He found that although most of the workers in the Laboratory were informative and agreeable, a few weeds, watered by the stream of visitors, had developed "a superiority complex that makes them insufferable even to some of their colleagues." They ridiculed French work and laughed at the difficulty Joliot was having in setting up a cyclotron; against which lèse-majesté Nahmias retorted that it was not a Berkeley man but Joliot, working without a cyclotron and in a contaminated laboratory, who had discovered the artificial radioactivity on which the Laboratory had come to flourish.

Furthermore—always according to Nahmias—more jealousy and animosity afflicted the staff than he had noticed in any other university in Europe or America. These unpleasant qualities are not necessarily expunged by cooperation and solidarity. Jealousy readily arises in the sacrifice of the individual to the group, which was necessary, in the opinion of the loyal Franz Kurie, to the health of the Laboratory: organization into disciplined interdisciplinary teams made possible the prompt exploitation of "byproducts of the elucidation [of the nucleus that] have shown themselves to be of untold value to other sciences."[101] But working for others and receiving little credit for it do not bring satisfaction to everyone. Team research always has the potential of arousing the sorts of feelings that Nahmias detected.

As for animosity, Nahmias perhaps responded to an undercurrent of xenophobia that may be regarded as the negative side of the feeling of pride and community for which the Laboratory was conspicuous. During the 1930s the Laboratory's reception of

100. Cooksey to Lawrence, 10 June 1938 (4/21), and to Aebersold, 11 Jul 1940 (1/9); Cooksey's guest book, 1939–41 (4/33).
101. Nahmias to Joliot, 12 June 1937 (JP, F25); Kurie, *Jl. appl. phys.*, 9 (1938), 692.

foreigners who were neither distinguished nor British was not exemplary. Here many obscure forces were at play: a wish in the nation as a whole to have nothing to do with European affairs; a strong feeling against aliens in California, which expressed itself in, among other things, a state law prohibiting the employment of noncitizens by state institutions except under special conditions;[102] and the diffuse anti-Semitism, exacerbated by job shortages during the Depression, found in many American universities in the 1930s.[103]

The University of California prided itself, rightly, on being relatively free from anti-Semitism. Its vice president, Monroe Deutsch, himself a Jew, told the campus Hillel Foundation, "There is less prejudice at this university than at any other institution in the country," and the foundation agreed. Berkeley was "a student's paradise...,[where] Jewish scholars, the world over, will always find a place for themselves." And yet, it allowed, its members found discrimination and intolerance enough.[104] There was open exclusion from some campus social groups and from organizations to which faculty belonged, like the tony Bohemian Club of San Francisco, to whose retreat Lawrence liked to take visiting dignitaries like Poillon.[105] The Laboratory was not the Bohemian Club: it had two productive and appreciated Jews on its staff, "our chemist" Kamen and Segrè; but it certainly did not seek to multiply their number. Except for Segrè, who came as a visitor, not as a refugee, the Laboratory reaped no benefit from the pool of émigré physicists, and apart from help to transients marooned by outbreak of war, it did not devote any of its substan-

102. L.A. Nichols, comptroller, U.C., to Lawrence, re Segrè, 19 Jan 1939 (16/14); infra, §10.3.

103. Among many examples concerning physicists: "It is practically impossible for us to appoint a man of Hebrew birth...in a southern institution" (re candidacy of Feenberg at North Carolina); "anti-semitism among the higher administration officials" ruled out Purdue (again re Feenberg); "there was objection only against Jewish foreigners, and not against foreigners in general" (again at Purdue); "my chief competitor is Farkas, one of the refugee Jews" (re job at Cornell). Quotes from, resp., Kevles, *Physicists*, 279–81 (both regarding Feenberg); Bethe to Lark-Horowitz, n.d. (HAB, 3); F.H. Spedding to Lewis, 8 June 1935 (Lewis P). Cf. Rigden, *Rabi*, 104.

104. Deutsch, quoted in *B'nai B'rith Hillel Call,* 1 Feb 1934, and editorial, ibid., 4 Sep 1934 (UCA).

105. Lawrence to Poillon, 8 Jul 1937 (15/17A); Childs, *Genius*, 173.

tial resources to assisting displaced European scientists. It did not "profit by the stupidity and brutality of the German government," as Deutsch had urged. In this respect the Laboratory and the departments of Physics and Chemistry behaved quite differently from the Mathematics Department and Stanford's Physics Department, where "there [was] a phantastically high density of physicists of European extraction."[106]

This is not to say that the Laboratory was openly or consciously discriminatory. On the contrary: Lawrence invited Lise Meitner to come for a visit, if she could pay her travel expenses, when the Nazis began to threaten her.[107] (She escaped via Holland and found refuge in Sweden.) That appears to be the only instance where Lawrence took the initiative. The usual situation may be illustrated by the case of Stanislaw Mrosowski, a spectroscopist from the University of Warsaw who planned to spend a sabbatical in the Laboratory learning cyclotronics. Mrosowski was en route in Chicago when the Germans invaded Poland. Anticipating that the Polish government would not be able to pay his salary, he asked for a small stipend from the Laboratory to tide him over. Lawrence and a colleague in the Physics Department, F.A. Jenkins, generously undertook to supply something for six months or so: "All of us [Lawrence wrote] are deeply sympathetic with the tragic misfortune that has befallen Poland." Meanwhile, Jenkins had telegraphed Robert Mulliken at Chicago for an "objective description" of Mrosowski. Mulliken knew perfectly well what that meant and wired back: "In doubt whether Jewish. Tall thin about 40....Nice wife not Jewish." And again, after applying to the party concerned: "Himself says he is not Jewish but entirely Slavic. Our impressions generally favorable."[108] Here is another

106. Placzek to Bethe, 22 Sep 1938 (HAB, 3); Deutsch to Sproul, 25 Jul 1933, in Rider, *HSPS, 15:1* (1984), 111; cf. ibid., 158–66. Lawrence claimed to Weaver, 24 Sep 1940 (15/30), that the Laboratory then had "several refugees," whom we have not located; in any case, none but Segrè had a paid position. Cf. Lawrence to Florence Baerwald, 20 May 1939 (2/19).

107. Lawrence to Meitner, 25 Jul 1938, and reply, from Sweden, 21 Aug 1938 (12/22).

108. Lawrence to S.W. Mrosowski, 16 Sep 1939, and Mulliken to Jenkins (telegrams), 17 and 20 Sep 1939 (13/16). Lawrence obtained $1,000 from Alfred Loomis, but Mrosowski had little of it, for he soon returned to Chicago, which had a very good record for placing émigrés; Lawrence to Sproul, 13 Jan 1940 (13/16); Rider, *HSPS, 15:1* (1984), 165–6.

subtle indicator of the general situation. A physician wishing to work with John Lawrence thought it useful to gloss his previous employment at leading Jewish hospitals. "Because of [this] experience..., I am sometimes considered Jewish. This, I can assure you, is not true."[109]

The implied bias obtruded occasionally in the correspondence of Lawrence and Cooksey, both alumni of Yale, where, as Lawrence well knew, Jews were "under a handicap." The kindly Cooksey thus recommended a cyclotroneer for election to an Eastern club: "His name would indicate that he is not Jewish....I am quite confident that he is not Jewish....I am quite confident that he would be perfectly all right at the Club." And here is Lawrence in praise of Kamen: "He is Jewish and in some quarters, of course, that would be held against him, but in his case it should not be, as he has none of the characteristics that some non-Aryans have. He is really a very nice fellow."[110] But again, as this forked evaluation shows, Lawrence did not withhold support or friendship from a Jew who had shown himself to be a very nice fellow. He liked Otto Stern, "a very jolly, pleasing personality," and also Isidor Rabi, "a very fine person," both of whom he supported for the Nobel prize in physics; he professed "the highest regard for [Kamen's collaborator Ruben], both as a scientist and a man," when advancing Ruben for a prize from the American Chemical Society; and he regarded Segrè as "an extremely good man," although not one he intended to retain at the Laboratory for any length of time.[111]

No doubt Nahmias, a Jew who did not rise to the rank of fine fellow, felt as antagonism the diffuse and subtle bias just described. There was another component to the Laboratory's xenophobia that Nahmias, who had had a good, solid European

109. L.A. Erf to J.H. Lawrence, 12 May 1939 (7/6).
110. Cooksey to Lake Placid Club, 24 Feb 1938 (4/21); Lawrence to A.J. Allen, 22 Sep 1937 (10/10), re Kamen, and to Zeleny, 5 Mar 1940 (18/44), re Rabi.
111. Lawrence to R.C. Gibbs, 15 Jan 1934 (16/49), and Crawford, Heilbron, and Ullrich, *Nobel population,* 137, for the nomination of Stern; Lawrence to Stern, 18 Jan 1932 (35/2), re Francis Simon; to Zeleny, 5 Mar 1940 (18/44), and to Rabi, 1 Nov 1941 (15/4), re Rabi; to C.L. Parsons, 5 Dec 1941 (15/33), re Ruben; to Alex Langsdorf, Jr., 24 Oct 1940 (10/30), and to Wells Surveys Inc., 27 Feb 1940 (16/14), re Segrè. Segrè was then rated by Bethe as one of the ten best physicists in the country; Bethe to LeRoy S. Brown, University of Texas, 1 Aug 1940 (HAB, 3).

education, also experienced. The culture of many members of the Laboratory was parochial. They knew no languages (the Laboratory kept a WPA worker translating articles from French and German) and little history or literature; they had small basis, apart from openness and goodwill, for sympathy with people from backgrounds much different from their own; and they were busy getting on in the world.[112]

3. THE WIDENING GYRE

As a member of the Laboratory cycled through his apprenticeship, the center of his motion was, of course, Lawrence, who ultimately assigned duties and stipends, approved research projects and the publication of research results,[113] and found jobs for journeymen cyclotroneers. His first revolutions occurred, as we have seen, in synchrony and usually in harmony with other members of the Laboratory. As his knowledge increased, he might spiral further out, occasionally colliding productively with members of Oppenheimer's group, of the College of Chemistry, or of the Medical School. Further turns might bring interactions with people from Stanford or Caltech, until—to finish with the analogy—the complete cyclotroneer would shoot forth from the Berkeley machine to build elsewhere what he had been taught to build at home. We shall examine a few turns of the spiral.

Around Mecca

In the beginning, until 1935, the disciples, or some of them, worked directly with the prophet, or, better, "the high priest."[114] But as the administrative burden and the complexity of the machine increased, Lawrence became more and more removed from the work and workers in the Laboratory. He could be very generous to his students—some parents praised him lavishly for his kindness to their offspring—but he could also be impatient and

112. The translator, Mr. Kaufmann, is mentioned in Cooksey to Viez, 19 Jan 1940 (18/33).
113. On Lawrence's prior approval of papers submitted for publication: Seaborg to Livingston, 12 Mar 1939, in Seaborg, *Jl., 1,* 45; infra, §8.2.
114. Cork to Lawrence, 14 Jan 1937 (5/1).

unreasonably demanding when the machine stopped.[115] His impatience was enhanced by his frequent colds, picked up on his travels.[116] He was often away from the Laboratory lecturing, at meetings of the National Academy of Sciences or the American Physical Society, and fund-raising (which he did to greatest effect in person in New York, where all his main supporters among foundations had their headquarters). These trips occupied more time than they need have done, since Lawrence insisted on going by train; he had worked out that flying was dangerous, that the chance of a fatal accident in a round trip from San Francisco to New York was one in a thousand, and would have nothing to do with airplanes.[117] No doubt the long journeys by rail were necessary to him for quiet relaxation and unhurried future planning. After the settlement of 1936, Lawrence was in Berkeley long enough to encourage the staff and create his family. Otherwise he left the execution of his plans to Cooksey.[118]

Lawrence retained his boyishness and ebullience for successes of the Laboratory and for approaches to his backers. It was only natural that he would come to identify with the men of wealth on whom he had come to depend and who appeared to like his company. He enjoyed the entertainments of New York bankers and brokers; he liked the style of Alfred Loomis, whose patronizing of the Laboratory, which began late in 1939, included a weekend at the plush Del Monte Hotel in Carmel for the Sprouls, the A.H. Comptons, the Vannevar Bushes, and the E.O. Lawrences. In these tastes and associations, Lawrence was followed, as in much else, by Alvarez.[119] In the same line, Lawrence set some store on

115. Letters to Lawrence from Laslett, Sr., 4 Aug 1937 (10/32); D.M. Kalbfell, 24 Jul 1939 (10/9); E. Lyman, 19 Aug 1938 (12/19).

116. Lawrence to McMillan, 11 May 1935 (12/30); to Bohr, 8 Jul 1937 (3/3); to Livingston, 23 Nov 1937 (12/12); to Archie Woods, 18 May 1940 (12/35); John Lawrence to Lawrence [1936] (11/16), and Dave Morris to Lawrence, 24 May 1940 (13/13), advising him to slow down.

117. Lawrence to E.C. Williams, Shell Development, 12 Apr 1938 (18/26), calculations based on data supplied by James Doolittle of United Air Lines; and to H.H. Benedict, 6 Feb 1939 (21/5); Childs, *Genius*, 164–5. By 1941 he had decided to take the chance; Lawrence to Cooksey, 11 Oct 1941, telegram (4/25).

118. On Cooksey: M. White to Cooksey, 13 Dec 1937 (18/21); on Lawrence's children—John (born 1934), Margaret (1936), Mary (1939), Robert (1941), Barbara (1947), and Susan (1949)—see index in Childs, *Genius*.

119. Cooksey to Charles Seymour (president of Yale), 26 Mar 1940 (18/38); Alvarez, *Adventures*, passim.

family background; whenever the ancestry of his students permitted it, he would include in his letters of recommendation the irrelevance that they came "of good family."[120] He came to dislike nonconformity (to his ideas!) and liberal causes, to hold that "science is justified only to the extent that it brings substantial riches to mankind," and to declare research scientists—among whom he enrolled himself—to be "essentially conservative people."[121] In these sentiments he disagreed altogether with his colleagues Urey, A.H. Compton, and, closer to home, Oppenheimer, whose personal style influenced the apprentice theorists at Berkeley at least as much as Lawrence's helped define the cyclotroneer.[122]

The contrast between the two men is the stuff of stories.[123] Lawrence grew up in rigid and rigorous South Dakota, worked during high school, learned to build radios, went into physics, knew little else, had no regrets, never doubted his way. Oppenheimer came from a wealthy Jewish family, grew up surrounded by books and pictures, attended the indulgent Ethical Cultural School, collected beautiful rocks, had literary ambitions, drifted into physics, suffered severe depressions. He also suffered from cultural ambitions, which caused him to learn, or affect to know, diverse languages, good food and wine, the best in music, painting, and books; he would take girls to dinner and read them Baudelaire; he sketched in charcoal and painted in oils; at Berkeley he studied with professors of Sanskrit, read Plato in Greek, and, what was worse, talked about it.[124] They moved in opposite directions

120. Letters in support of A.C. Helmholz, 16 Feb 1938 (9/4), and E. Lyman, 21 May 1939 (12/19).

121. Lawrence, "Convocation address, American College of Physicians," 29 Mar 1939 (40/22), and commencement address, Stevens Institute of Technology, 13 June 1937, *RSI, 8* (1937), 311, on science and riches; Lawrence to Urey, 20 Aug 1940 (17/40), Kamen, *Radiant science,* 179–80, and Oppenheimer in AEC, *Hearings,* 187, on conservatism ("Lawrence had a very strong objection to political activity and to left wing activity").

122. On Lawrence's repudiation of the liberal "pseudo-patriotic or philanthropic organizations" supported by Urey and Compton, see Cooksey to Poillon, 25 Jul 1940 (13/29A).

123. The most ambitious one, N.P. Davis, *Lawrence and Oppenheimer,* is too flawed by errors of fact to be useful.

124. For Oppenheimer's affectations, see Smith and Weiner, *Oppenheimer,* 16, 25, 35, 41, 44, 108, 113, 153, 155, 159, 165, 172; for his family's comfortable circumstances, ibid., 19, 34, 107, 138.

politically under the strong polarizing forces of the late 1930s. Lawrence trusted that the world would muddle through without requiring his attention. He had faith in the great powers, which, he thought, had established at Munich that international disputes would be settled by peaceful negotiations; "I am not concerned about the political atmosphere," he reassured Cockcroft, concerning a meeting in Europe planned for September 1939, "which I personally think will be all right." Oppenheimer did not allow the world to get on without his help. While Lawrence drifted complacently to the right, he rushed to the left, to the support of the Spanish Loyalists, American labor, and other pinkish causes.[125]

It might therefore appear astonishing that from Oppenheimer's first semester at Berkeley in 1929, he and Lawrence were fast friends.[126] His Jewishness bothered him more than it did Lawrence; as one of his Harvard professors once wrote, by way of recommendation, "Oppenheimer is a Jew, but entirely without the usual qualifications of his race."[127] The basis of their friendship was their fundamental honesty, kindness, and openness, qualities each would soon enough come to compromise. They also shared a liking for, and for showing off, feats of physical prowess. Oppenheimer delighted in long journeys by horseback through remote reaches of New Mexico and surprised those foolhardy enough to accompany him by his stamina. Lawrence visited the Oppenheimer family retreat near what became the Los Alamos Laboratory once or twice, but preferred to take his exercise on the tennis court or in a small boat bought with the proceeds of the Comstock prize. "He plays a good game of tennis," said the *Scientific American,* "[and] has a cruising boat on San Francisco Bay that he won't take out unless there is rough water."[128]

125. Lawrence to Cockcroft, 5 Oct 1938 and 9 Feb 1939, quote; AEC, *Hearings,* 8–9; Smith and Weiner, *Oppenheimer,* 195–7, 214, 218.

126. Among many indications, Smith and Weiner, *Oppenheimer,* 144–5, 147 (texts of 1930), 170 (1933), 174 (1934); Lawrence to DuBridge, 14 Nov 1939 (37/21); Oppenheimer in AEC, *Hearings,* 189 (events of 1940); Lawrence to Oppenheimer, 1 Nov 1940 (14/9).

127. P.W. Bridgman to Rutherford, 24 June 25, in Smith and Weiner, *Oppenheimer,* 77; ibid., 61.

128. Goodchild, *Oppenheimer,* 23, 27, Smith and Weiner, *Oppenheimer,* 184, 186, and Elsasser, *Memoirs,* 200, on Oppenheimer's machismo; Schuler, *Sci. Am., 163:2* (Aug 1940), 71, *Time, 30:2* (1 Nov 1937), 42, and Kamen, *Radiant science,* 180, on Lawrence's; Alvarez, *Adventures,* 62, and Cooksey to Van Voorhis, 21 Jul 1937 (17/43), on the boat.

At the professional level, Lawrence and Oppenheimer esteemed each other as physicists, recognized each other's ambitions, saw that they did not conflict, and possessed the tie natural to the men expected by their Department to raise Berkeley physics to national distinction. "For all his sketchiness, and the highly questionable character of what he reports," Oppenheimer wrote, "Lawrence is a marvelous physicist." "He has all along been a valued partner," Lawrence allowed in support of Oppenheimer's promotion to full professor. He might have added: "His physics [is] good, but his arithmetic awful."[129] Where Lawrence misinterpreted and mismeasured, Oppenheimer erred by factors of 100 or 1,000. When he was close. Here is a compliment from a German theorist who followed his calculations for pair production: "Oppenheimer's formula...is remarkably correct for him, apparently only the numerical factor is wrong."[130]

Oppenheimer himself did not trust the theories he elaborated. He wrote his brother in 1932: "The work is fine: not fine in the fruits but the doing....We are busy studying nuclei and neutrons and disintegrations; trying to make some place between the inadequate theory and the revolutionary experiments." And again, in 1934: "As you undoubtedly know, theoretical physics...is in a hell of a way."[131] For a time Lawrence thought that Gamow would be more helpful than Oppenheimer; but he could offer only theoretical dollars, and nothing came of his effort to bring Gamow to Berkeley.[132] Thereafter Lawrence and his Laboratory had only the theoretical advice they obtained casually from Oppenheimer's group and from an occasional short-time visitor. It is likely that Lawrence shared the views of his old friend Ernest Pollard of Yale. "To tell the truth, I don't absolutely think *any* current

129. Oppenheimer to Frank Oppenheimer, 7 Jan 1934, in Smith and Weiner, *Oppenheimer*, 171; Lawrence to C.D. Shane, 28 Jan 1936 (14/9); Serber in Brown and Hoddeson, *Birth*, 209, on Oppenheimer's arithmetic. Cf. Lamb, ibid., 313, on Oppenheimer's sloppiness in lectures.

130. Peierls to Bethe, 24 Jul 1933 (HAB/3).

131. Oppenheimer to Frank Oppenheimer, [fall 1932] and 4 June [1934], in Smith and Weiner, *Oppenheimer*, 159, 181; Serber in Brown and Hoddeson, *Birth*, 207, 209–11.

132. Gamow to Lawrence, 2 May 1934, and Lawrence to Gamow, 9 and 19 May 1934 (7/25).

nuclear theory worth much, but theoreticians regrettably have the power to divert thought from a constructive interpretation of experiment which doesn't agree with the present pervading jargons."[133]

The most productive interaction between Lawrence's and Oppenheimer's groups concerned the hypothesis of deuteron disintegration, which Oppenheimer at first accepted fully and with a certain pleasure: "that makes, as far as I can see, a hopeless obstacle to Heisenberg's pseudo q.m. [quantum mechanics] of the nucleus." But the implausibility of the consequent character of the deuteron gave him pause, and he accepted the explanation of contamination long before Lawrence did. From the shambles, as we know, he and his former doctoral student, Melba Phillips, extracted the Oppenheimer-Phillips mechanism, which did help importantly in the interpretation of disintegration experiments at the Laboratory.[134] There are a few other cases of productive interaction beyond the mixed advice Oppenheimer gave at the weekly seminars: his postdoc Robert Serber's analysis of the proton-proton scattering experiments of Lawrence's student Milton White; McMillan's measurements of absorption of γ rays, in rough confirmation of calculations by Oppenheimer and Furry; and Oppenheimer's explanation of Henderson's results on the energy dependence of the disintegration of lithium by protons.[135]

These interactions belong to the early years, when Oppenheimer felt a responsibility to help to elucidate disintegration. After 1936 the most useful of Oppenheimer's group for Lawrence's work was Sydney Dancoff, who pushed the theory of isomeric and radioactive transitions beyond what the European theorists had achieved and did not disdain to calculate the absorption of neutrons in the

133. Pollard to Cooksey, 18–20 Mar [1936] (14/30); cf. Oliphant to Lawrence, 20 Nov 1939 (14/6).

134. Oppenheimer to Frank Oppenheimer, 7 Oct 1933 and 7 Jan [1934], in Smith and Weiner, *Oppenheimer*, 165, 171; Oppenheimer to G. Uhlenbeck, ca. Mar 1934, ibid., 175–6; Oppenheimer to Lawrence, spring 1935, ibid., 193; supra, §4.1, and infra, §8.2.

135. Serber in Brown and Hoddeson, *Birth,* 210; McMillan, *PR, 46* (1934), 868–73; Oppenheimer, *PR, 43* (1933), 380; for the advice at seminars, Lawrence to C.D. Shane, 28 Jan 1936 (14/9), and Seaborg in Rabi et al., *Oppenheimer,* 48; for Oppenheimer's interest in McMillan's experiments, Oppenheimer to Lawrence, [May 1935] (14/9).

cyclotron's water shielding. This unusual interest on the part of an Oppenheimer theorist in problems central to the Laboratory inspired a still rarer thought on Lawrence's part, one perhaps not expressed since the unsuccessful approach to Gamow: the thought of adding a theorist to the Laboratory's staff. Oppenheimer did not think that a very promising employment, and Dancoff went to join Serber in Urbana.[136]

A rough measure of the closeness of the two groups may be obtained from an analysis of the composition of the thesis committees on which Lawrence served during the 1930s. Between 1931 and 1941 inclusive, thirty-seven theses were completed under these committees, twenty-one of which had Lawrence as nominal director. His associates on the thirty-seven committees divide into two classes, those with six to eight appearances and those with three or fewer, except for Leonard Loeb, who figures four times. Those in the first class: experimentalists R.T. Birge (units), R.B. Brode (cosmic rays), F.A. Jenkins and H. White (spectroscopy), theorist Oppenheimer, and chemists G.N. Lewis and Willard Libby. Perhaps the strongest showing was Libby's, who tied with Oppenheimer with eight appearances, on half of which he was the principal reader. On this showing, Lawrence's students—that is, students in experimental nuclear physics who had Lawrence as their nominal thesis director—were no closer to Oppenheimer than they were to any other active member of the Physics Department or to chemists interested in nuclear processes.

The makeup of these committees does not indicate anything about casual consultation of Oppenheimer or his students by perplexed members of the Laboratory. For graduate students, we have some measure of the extent of this consultation: six of them who had Lawrence but not Oppenheimer on their committees thank both Lawrence and Oppenheimer for advice.[137] Also,

136. Oppenheimer to F.W. Loomis, 13 May 1940, in Smith and Weiner, *Oppenheimer*, 211-2; Dancoff, *PR, 57* (1 Feb 1940), 251. On Oppenheimer's earlier sense of obligation, see Oppenheimer to Lawrence, [3 Jan 1932], in Smith and Weiner, *Oppenheimer*, 148: "If there are any minor theoretical problems to which you urgently need the answer, tell them to Carlson or Nedelsky; and if they are stymied let me have a try at them."

137. Backus, Brady, Laslett, Loveridge, Richardson, H. White; indications that these acknowledgements were not empty occur in Lawrence to Richardson, 18 Nov 1937 (15/22), and Oppenheimer to Lawrence [Apr 1937] (14/9), re Laslett.

Oppenheimer's students occasionally did a calculation or eluci-
dated a theoretical point for Lawrence's: for example, Willis
Lamb's explanation of Laslett's failure to detect decay in Na^{22} and
Eldred Nelson's collaboration with Alvarez and Carl Helmholz on
the behavior of isomeric silver nuclei.[138]

Just as many (we do not say all!) of Lawrence's "boys" adopted
his rough-and-ready approach to physics and had little time or
interest in much outside the Laboratory, so Oppenheimer's aped
his gestures, tried to acquire his tastes, went to concerts, and
talked books, art, and politics.[139] A Lawrence man rushed through
his preparation, knowing that, if he did well, he would have a
choice of eligible positions; an Oppenheimer man proceeded with
greater leisure, knowing that the art was long, and "jobs for
theorists...not too common."[140] The divergent cultures of the
theorists and experimenters, the problematic state of nuclear
theory, the peculiar fascinations of the cyclotroneers, and the
increasing importance of biomedicine in the Laboratory worked to
prevent the development there of the sort of theory-driven experiments that mark the big-machine physics of the postwar era.

Perhaps more important for the direction of nuclear science at
the Laboratory than its relations with Oppenheimer's group and
even with the entire Physics Department was its interactions with
the College of Chemistry. Besides Lewis and Libby, there was
Seaborg, who became the Laboratory's most productive chemist in
1937. As an instructor in the College of Chemistry, which he
became in 1939, after two years of easy service as Lewis's assistant, Seaborg brought excellent doctoral students to do nuclear
chemistry around the cyclotron, with consequences that were
literally earthshaking.[141] Another young chemist closely associated
with the Laboratory, Samuel Ruben, worked with Kamen and the

138. Lamb, *PR, 50* (1936), 388–9; Alvarez, Helmholz, and Nelson, *PR, 57*
(1940), 660–1, using Nelson, *PR, 57* (1940), 252, drawing on Dancoff and Morrison, *PR, 55* (1940), 122–30; infra, §8.2.

139. Serber in Rabi et al., *Oppenheimer,* 18–20, and in Smith and Weiner, *Oppenheimer,* 150, 186.

140. Oppenheimer to F.W. Loomis, 13 May 1940, in Smith and Weiner, *Oppenheimer,* 211–2.

141. Seaborg, *Jl., 1,* 8, 19, 25 (24 Aug, 12 Oct 1934, 14 Jul 1937); infra, §9.3.
Seaborg and his asociates Kennedy and Wahl are included in table 5.5.

radioisotope C^{11} on the mechanism of photosynthesis. Their discovery of a better tracer, C^{14}, was one of the high points of the Laboratory's prewar work.[142]

Spiralling further out, Lawrence's people had an opportunity, not frequently seized, of easy exchanges with their colleagues at Stanford. When Felix Bloch, a theoretical physicist with the highest European pedigree, took up an assistant professorship at Stanford in the fall of 1934, he and Oppenheimer began joint weekly seminars, three-quarters of which were held at Berkeley. This interchange resulted in collaborations between Bloch and Oppenheimer's students and between Bloch and Alvarez, and in the inspiration for an investigation by Laslett. The work of Alvarez and Bloch on the magnetic moment of the neutron was probably the most advanced piece of exact physics done at the Laboratory during the 1930s. Stanford also offered the advantage of a summer course by a visiting theorist; we do not know how many of Lawrence's people besides Nahmias took the opportunity or trouble to go to Stanford to hear Gamow, Victor Weisskopf, Rabi, John Van Vleck, or Fermi. Nor have we been able to trace much useful interaction where it might be expected, between the Berkeley cyclotroneers and the applied physicists in and around Stanford, William Hansen and the Varian brothers. These men also had an interest in accelerators, but for electrons, not ions. Their "rhumbatron" and "klystron," which did not interest Lawrence in the late 1930s, came to play a part at Berkeley during the war.[143]

In partibus

We count that the Laboratory had some fifty-four regular members from 1932/33 through 1939/40, "regular" meaning, for our purposes, persons working to prepare for a career in science.

142. Seaborg, *Jl.*, *1*, 19 (12 Oct 1934); Libby, Peterson, and Latimer, *PR, 48* (15 Sep 1935), 571–2, thanking Kurie and White for "advice and assistance;" infra, §8.3. Seaborg used a d-d source, about which he had advice from Lawrence, for his graduate work; *Jl.*, *1*, 101, 166 (12 Nov 1935, 15 Oct 1936).

143. Serber in Brown and Hoddeson, *Birth,* 211–2; Lamb, ibid., 315, and in Chodorow et al., *Bloch,* 134–5; Oppenheimer to E.A. Uehling, 12 May [1934], in Smith and Weiner, *Oppenheimer,* 179; Varney, *PT, 35:10* (1982), 28; infra, §9.1, for Alvarez and Bloch, and §10.2 for rhumbatron and klystron.

Established people, who came on sabbatical or fellowship leave, volunteers without career ambitions, and undergraduates who left without obtaining a degree, do not count as regulars. The fifty-four distribute into two groups in two different ways: (A) graduate students who obtained a Berkeley Ph.D., not necessarily by 1940; (B) people who came as fresh postdocs; (I) members of either group who left the Laboratory before the war; (II) members of either group who returned to, or remained in, the Laboratory for war work. The numbers involved appear in table 5.5. Half the Ph.D.'s and three-fifths of the postdocs, some twenty-nine people in all, made up the Laboratory's export of manpower during the 1930s. Of these, two out of three went to build or perfect cyclotrons elsewhere: they are our "cyclotroneers in partibus." They and their destinations are listed in table 5.6.

The first of the agents or disciples in partibus was Livingston, who styled himself a missionary and was expected to effect a miracle. In 1934 the Cornell physics department had just fallen under

Table 5.5

Distribution of Berkeley regulars, 1932/40

	I	II	Total
A	16	15	31
B	13	10	23
Total	29	25	54

Note: A, Berkeley's Ph.D.'s; B, postdocs from other institutions; I, regulars who left before World War II; II, regulars who did war work at the Laboratory.

Table 5.6

Cylotroneers in Partibus, 1932/9

	Date left	Destination(s)	Class
Abelson, P.H.	1939	Carnegie Inst.	A
Green, G.K.	1938	Carnegie Inst.	B
Henderson, M.C.	1935	Princeton	B
Hurst, D.G.	1937	Cambridge	B
Kinsey, B.B.	1936	Liverpool	B
Kurie, F.N.D.	1938	Indiana	B
Langsdorf, A.	1939	Washington U.	B
Laslett, L.J.	1937	Copenhagen; Indiana	A
Livingood, J.J.	1938	Harvard	B
Livingston, M.S.	1934	Cornell	A
Lyman, E.M.	1938	Illinois	A
Paxton, H.C.	1937	Paris; Columbia	A
Richardson, J.G.	1937	Michigan; Illinois	A
Simmons, S.J.	1939	Pittsburgh	A
Snell, A.M.	1938	Chicago	B
Thornton, R.L.	1936	Michigan; Wash'n U.	B
Van Voorhis, S.N.	1938	Rochester	B
Walke, H.	1937	Liverpool	B
White, M.G.	1935	Princeton	A

the control of R.C. Gibbs, an ambitious and prescient man, who decided to plant nuclear physics in Ithaca in the hope that a flourishing research tradition would grow in its shade. He brought in Livingston and, the following year, Bethe, as assistant professors to cooperate in establishing a theoretical and experimental program in nuclear physics. They did so successfully; but it proved impossible to continue development of even a sixth-scale version of the Berkeley Laboratory. The fundamental obstacle, which all cyclotroneers encountered in one degree or another, was lack of staff. Crew service had no appeal high above Cayuga's waters. Furthermore, although Cornell warmed to the cyclotron gospel, Gibbs had other interests to further as well and his even-handed division of departmental assets hampered the expansion of capital-intensive units.[144] Therefore, although Livingston built a small and effective cyclotron in good time (it gave 3 or 4 μA of 2 MeV protons within a year), and although he and Bethe worked productively together, he felt stifled as a cyclotroneer. As far as cyclotrons are concerned, as Lawrence was to say, "the larger the better." In 1938 Livingston went to MIT as associate professor of physics to build a cyclotron larger than Berkeley's 37-inch.[145]

The problem of understaffing was met in part by collecting more than one man with Berkeley experience at one place. That occurred with the second cyclotron built outside the Laboratory, at Princeton, by Milton White and Malcolm Henderson; it, too, took about a year to make, but, unlike Livingston's machine at Cornell, it was larger (36-inch poles) than its original when it came to life in 1936.[146] Other examples of double teaming:[147] the machines built or started by Berkeley pairs, as at Illinois (the second cyclotron there), Indiana, and Saint Louis; machines begun by visitors to the Laboratory and finished by regulars, as at Chicago, Harvard, Michigan, and Rochester; machines started by foolhardy types without Berkeley experience and finished with the help of one or more men from the Laboratory, as at Cambridge, Columbia, Copenhagen, Liverpool, and Paris. The mutual dependence of these young men entrusted with large and costly projects comes out eloquently in a late night letter from Thornton, then at Michigan, to Cooksey. "There are so many things under way my nerves get on edge and I am pursued by doubts. Reg [J.R. Richardson] is a great help in really discouraged moments—someone to chew with, with a proper flippant attitude towards cyclotrons."[148]

144. Livingston to Lawrence, 7 Jan 1938 (12/12); similar laments occur in Hickman (Harvard) to J.S. Foster, 15 Jan 1938 (UAV, 691.60/3), and von Friesen (Stockholm) to Cooksey, 15 Dec 1938 (17/47).

145. Bethe, "History," 1–2 (HAB); Livingston, *Particle acc.*, 37, and *RSI,* 7 (1936), 55, 61, 67; Lawrence to G.P. Harnwell, 20 Mar 1940 (14/22).

146. M. White to Lawrence, 20 Aug and 22 Oct 1936, and Henderson to Lawrence, 25 Oct 1936 (9/6); Henderson and White, *RSI, 9* (1938), 19–30.

147. See table 5.6.

148. Letter of 13 Dec 1937 (13/5).

Most cyclotroneers in partibus had to adjust their rates of performance and their levels of expectation. No one worked to Lawrence's pace. "It does take more time to get things done here in the East," Livingston discovered, pointing to a lack of "the complete stock of small things such as wires, insulation, adaptors, etc., that make things move so fast at Berkeley." Exner contrasted "the California habit of speed" with "the lethargic East." "If I haven't written you before," White wrote from Princeton, "that is because I am ashamed of the lack of progress here." The Laboratory ran almost at full tilt during summers; it came as a great surprise to Lyman, when he arrived to take up his job at Illinois in August 1938, that the physics department was closed. "The deadest place you have ever seen....They take their vacation seriously around here."[149]

The men who went to help finish cyclotrons abuilding on the Continent expressed a double culture shock. Laslett, in Copenhagen: "Here things seem to be taken with reasonable calmness and if I work at night until...say 10:00 P.M., I feel like a scab in a fink joint." Taking things calmly, Laslett went on vacation with Otto Frisch, who was expected to be the director of the Copenhagen cyclotron. It was not Frisch but Laslett who felt obliged to cut short his tour to try a new improvement on the machine. The metabolism of the Danish physicists ran at a Berkeley pace, however, in comparison with the pulse in Paris, where Paxton found his colleagues harder to move than their 30-ton magnet.[150]

The ex-Berkeley cyclotroneers formed a brotherhood, to use White's word. The European branch—Sten von Friesen, Hurst, Kinsey, Laslett, Walke—met in England, then in Denmark and Sweden, and planned a session in Lapland. The midwestern brotherhood—those at Chicago, Illinois, and Indiana—exchanged visits and provided overnight stops for Lawrence or Cooksey, hurtling East or West. The brotherhood on the Atlantic seaboard—

149. Letters to Lawrence from Livingston, 20 Sep 1934 (12/12), Exner, 26 Nov 1934 (9/21), and White, 21 Oct 1935 (18/23); Lyman to Cooksey, 7 Oct 1938 (12/19). Cf. Beams to Lawrence, 26 Oct 1936 (2/26): "Things of course are going slow with us compared to the pace you fellows set in your Radiation Laboratory."

150. Laslett to Cooksey, 3 Oct 1937 (10/32); Frisch, *What little,* 106; Paxton to Cooksey, 22 Jan 1938, and to Lawrence, 26 Sep 1938 (14/18).

those at Bartol, Columbia, Harvard, MIT, and Princeton—were the ciceroni on a standard cyclotron tour.[151] In Livingston's analogy, the cyclotron laboratories were like the California missions; located from the Midwest eastward at convenient intervals, they assured travelling cyclotroneers a welcome and a place to stay or work. When the Great Cyclotroneer himself appeared, miracles occurred: Lawrence could cure machine ills and clear up financial and personal difficulties that had refused to yield to lesser medicine. For sinners who had tried on their own and failed, the mother church had a particular indulgence. "They are like babes in the wood," Cooksey wrote of the builders of the Purdue cyclotron, who had never seen the inside of one before beginning their labor, "and need a visit."[152]

These visits promoted more than camaraderie and nostalgia. Cyclotrons multiplied in part because Lawrence wished them to. "It would please me greatly," he wrote in answer to Columbia's request for his "fatherly blessing" on their project, "if various laboratories would build cyclotrons." It was the Laboratory's policy, as Cooksey put it, to be "most interested in giving what information it can to help those who are starting in this fairly new field."[153] A large portion of the very large correspondence of Lawrence and of Cooksey during the late 1930s is devoted to what the builder of the Yale cyclotron called "the usual generosity," that is, answering questions from perplexed or would-be cyclotroneers; providing blueprints of Berkeley machines and accessories to any laboratory seriously engaged in planning or building cyclotrons; lending or giving old parts. And they could do more. To assist Bohr and Joliot, Lawrence helped to obtain fellowships from the Rockefeller Foundation for Laslett and Paxton; to help the Japanese, he and Cooksey arranged for the delivery and machining of the steel and copper for the second Tokyo cyclotron.[154] The cyclotroneers in partibus shared this ethic. Ignorant

151. Among much relevant correspondence, M. White to Cooksey, 14 Jan 1938 (18/21); Laslett to Cooksey and Abelson, [Apr 1938] (10/32); Hurst to Cooksey, 31 Jan 1938 (4/41); Capron, "Report," 2, 5; Tuve to Lawrence, 17 June 1939 (3/32).

152. Barnes to Cooksey, [1938] (2/24); Allen to Lawrence, 18 Dec 1937 (1/14); Cooksey to Lawrence, 29 Sep (4/21).

153. Lawrence to Pegram, 6 Feb 1935, answering Pegram's letter of 1 Feb (4/8); Cooksey to L.W. McKeehan (Yale), 28 June 1937 (18/38).

154. Pollard to Cooksey, 22 Apr 1937 (14/30); infra, chap. 7 (the Europeans);

of or indifferent to the patent situation, they exchanged information freely among themselves and continued to contribute to the advance of the art at Berkeley by trying new techniques on their own machines. Several major improvements came forward in this manner, in particular the capillary source introduced by Livingston and tested at Princeton and Rochester and the quarter-wave transmission line pioneered at Illinois and Columbia.[155]

Although cyclotroneers worked in many environments much different from Berkeley's, they often enough were supported by the same means as Lawrence had dispensed. The federal government assisted elsewhere as it had in Berkeley. The navy gave generators and the 500-kilowatt Poulsen arc for which Cornell and Columbia competed; had the navy had more to distribute, the union of the arcs with experts from Berkeley would no doubt have made the country (to use the elegant phrase of a Westinghouse engineer) "lousy with cyclotrons." The quantity of gifts in kind to accelerator laboratories from the navy and the War Department got a high reading on the Nahmias detector: "[They] continually distribute many pieces of equipment: generators, pumping systems, copper, electromagnets, oil, rectifiers, etc., to everyone who knows how to wangle them." Columbia's successful wangle, to meet the navy's requirement that the Poulsen magnet serve vocational education, is worthy of attention: "The magnet will be used by graduate students in connection with their training in research...; such training is strictly vocational as opposed to general cultural education."[156]

Opportunities for vocational training spread with the help of the foundations. After the National Research Council had played a brief and inexpensive part in Cornell and Illinois, the Research

---

(9/37) and Lawrence to Nagaoka, 19 Mar 1936 (9/33) and to Nishina, 18 Feb 1937, and Nishina to Lawrence, 21 Feb and 2 May 1938 (9/38) (the Japanese).

155. Some examples of interchange: Livingood to Lawrence, 3 June 1936 (12/11); Van Voorhis to Lawrence, 2 Aug 1937 (17/43), reporting from "the cyclotroneering front in the East;" Livingston to Cooksey, 3 June 1938 (12/12), sending copies of his description of the capillary source for forwarding to the "cyclotron mailing list;" Cooksey to Barnes, 29 Aug 1938 (2/24): "I guess each particular laboratory develops its own particular kinks which are useful to the others."

156. Livingston to Lawrence, 20 Sep 1934 (12/12); R.P. Jackson, Westinghouse, to Lawrence, 28 Sep 1935 (18/19); Nahmias to Joliot, 24 Mar 1937 (JP, F25); Pegram to Lawrence, 1 Oct 1935 (4/8).

Corporation, with its eye to the patent situation, made a series of grants of $2,000 or $3,000 to start or improve cyclotrons. Among its beneficiaries were Chicago, Columbia, Cornell, Purdue, Rochester, and Stanford.[157] By 1939, however, the Corporation had sensed that there was little for it in the "mad, but orderly scramble" to multiply cyclotrons within the same energy range and it concentrated its diminishing investments in accelerators on a bet on Berkeley.[158] The enlargement of opportunity and costs attendant on radioisotope manufacture brought substantial contributions from the Rockefeller Foundation to cyclotron building in Copenhagen and Paris, and lesser amounts to Rochester, Saint Louis, and Stanford.[159] Like the Research Corporation, however, the Foundation apparently decided not to support new machines within established energy regions. In 1939 it had the courage to turn down Harvard's request for "a substantial annual grant for a period of years," much to the surprise of president Conant, who had sunk money in a medical cyclotron expecting that it would bring in foundation money easily.[160]

MIT tried the National Advisory Cancer Council, without issue, and then did very well at the Markle Foundation, with $30,000. Michigan tapped its Rackham funds, an endowment given by an organizer of the Ford Motor Company, for some $25,000 a year, which made Michigan the richest American cyclotron laboratory outside Berkeley. But the brotherhood did not live by bread alone; and something in the air at Michigan drove away the two

157. For Chicago, Columbia, Cornell, Purdue, and Rochester, see the reports by Harkins (1937); Pegram, Hammett, and Dunning (1939); R.G. Gibbs and M.S. Livingston (1938); Lark-Horowitz (1937); and DuBridge (1940), respectively, in the series of bound reports at RC. Further to Rochester, DuBridge to Poillon, 9 Jul 1935 (RC); for Stanford, Bloch, Hammermesh, and Staub, *PR, 64* (1943), 49, and Staub in Chodorow et al., *Bloch*, 196–7.

158. Lawrence to Kruger, 7 Nov 1938, recommending applying for $2,000, and 16 May 1939 (10/20); Poillon to Lawrence, 17 Apr 1939 (9/20), re Indiana's application for $2,500 a year.

159. For Copenhagen and Paris, infra §7.2 and §9.3. For Rochester, Weaver, "Diary," 30 Sep 1936, 9 and 20–1 Feb 1940; Rockefeller Foundation, minutes, 18 Mar 1938 and 2 Apr 1941; Stafford Warren to A. Gregg, 23 June 1939 and 21 Aug 1940; all in RF, 1.1/200. For Saint Louis, Lawrence to Whipple, 19 Jul 1939 (15/26A). For Stanford, Bloch, Hammermesh, and Staub, *PR, 64* (1943), 49.

160. H. Shapley to Weaver, 23 Feb 1939, and reply, 15 Mar 1939, and Conant to K.T. Compton, 31 May 1939 (UAV, 691.60/3).

Berkeley cyclotroneers, Thornton and Richardson, who started their extramural careers there.[161] To finish our parallels, had Lawrence had his way, the National Advisory Cancer Council would have given all the American cyclotron laboratories in existence in 1938 save Bartol, which he thought sufficiently supported by Dupont, enough to realize their potential for radioisotope production or medical therapy. No more than the Research Corporation or the Rockefeller Foundation, however, did NACC wish to dribble away its resources to a large number of equivalent laboratories.[162]

It is not practicable to learn how much of the cyclotroneers' salaries was paid by the various contributors to cyclotron laboratories, or, indeed, the magnitude of the contributions. But we can make a rough estimate of the capital investment, inclusive of labor, in American cyclotrons operating outside Berkeley by 1940. These machines cost or represented about $250,000 exclusive of the value of the buildings that housed them. With these structures they came to about $400,000. The cost of their operation from the time of their commissioning through 1940 would perhaps raise the total expenditure on them to something like the amount spent on cyclotrons at Berkeley during the 1930s.[163]

161. R. Evans to Lawrence, 27 May 1938 (12/40), re Harvard; Lawrence to A.H. Compton, 7 Dec 1937 (13/29), and to Cottrell, 30 Mar 1934 and appended report (RC), re Rackham; Richardson to Lawrence, 11 May 1938 (37/8), on "the human atmosphere" at Michigan. Cf. S.W. Baker, *Rackham funds* (1955).

162. Lawrence to A.H. Compton, 7 Dec (13/29): $10,000 each to Chicago, Columbia, Harvard, Michigan, and Princeton; $4,000 each to Purdue and Yale; $3,000 each to Cornell and Illinois; and $2,500 to Seattle, the last little doles to provide students with cyclotron experience, not the world with radioisotopes. MIT, which had no cyclotron, was to get $25,000.

163. This number has been concocted from diverse and sometimes conflicting estimates by Lawrence, e.g., in letters to A.L. Hughes, 26 May 1938, 11 Feb and 14 Jul 1939 (18/12); to F.W. Loomis, 21 May 1938 (9/19); and to Detlev Bronk, 10 Dec 1934 (18/46); Cooksey to J.A. Gray, 24 Feb 1938 (14/38); and Hughes to Lawrence, 22 Jul 1939 (18/12). The $400,000 is estimated from Livingston, "Relative cost chart," Oct 1940 (12/12), as are also the cost of operations; it is not clear what his figures include.

VI

American Cyclotronics

Up to 1935 the 27-inch cyclotron ran in fits and starts. Then a series of improvements increased not only reliability but also energy and current: the deuteron beam rose rapidly, one μA or two at 3 MeV in 1934, which "thoroughly sold [initiates] on the cyclotron as the perfect high voltage source," to twenty μA at 6.2 MeV early in 1936, which established the machine as the preeminent producer of radioelements in the world.[1] The enlargement of the poles to 37 inches, which occurred during August of 1937, brought a beam of about twelve μA at 8 MeV, which in the following months grew to seventy-four μA, and to twice that at 6.2 MeV, the old maximum energy of the 27-inch cyclotron. The 60-inch, operational in the summer of 1939, was running in 1940 at new levels of energy, current, and dependability. Most of the instrumental changes that created the bigger beams were invented in Berkeley. In the case of two major design features of the 60-inch, however, the Laboratory adapted what other cyclotron laboratories in the United States pioneered.

1. THE ART

Berkeley's Poulsen Machines

Figure 6.1 shows the rise of beam current and energy of the cyclotrons built under Berkeley's Poulsen magnet from the onset

1. Kurie to Cooksey, 4 Mar 1934 (10/21), the yield then being 0.7 μA; infra, §6.3.

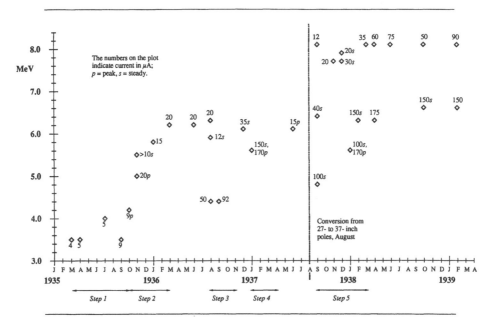

FIG. 6.1 Vital statistics of the Laboratory's 27-inch and 37-inch cyclotrons. The ordinate indicates deuteron energy in MeV; the points are labelled with currents in μA.

of substantial improvements to the start up of the 60-inch. The figure discloses that the advance occurred in five steps: (1) an increase in beam from June to October 1935; (2) a doubling of current and an increase of energy of 150 percent between November 1935 and March 1936; (3) high currents at relatively low energies, reached in the summer of 1936; (4) achievement of steadier operation; (5) a jump in the energy maximum and a doubling of current at the previous maximum between August 1937 and February 1938. The causes of these advances will indicate the range of considerations and problems the cyclotroneer had to heed.

Step 1 provides a notable instance of the principle that whatever can go wrong will. In the spring of 1935 Lawrence urged his boys to create larger deuteron beams at higher energy in order to make enough radiosodium for clinical work. His goal was modest: 5 μA at 5 MeV or 50 μA at 3.5 MeV; his eventual goal, as he told

Cockcroft, an immoderate milliampere. The boys tried many ways to a bigger beam: widening the dees, enlarging the filament of the ion source, putting up the dee voltage, and so on, all "with disappointing results." The summer was a "depressing time," "an epidemic of trouble;" whenever conditions augmented the beam, cathode rays somehow generated in the cyclotron tank would penetrate the glass insulators supporting the dees and shut down operations. It was very frustrating, this attainment of as much as 9 μA followed almost immediately by a crash. Lawrence wanted 10 mCi of radiosodium a day, which required a steady minimum current of 10 μA. "I was almost driven to distraction."[2]

He was not so distracted that he neglected to reassure his backers. "You will be interested to know," he wrote Poillon toward the end of the frustrating summer, "that we have been quite successful already in stepping up the production of radio-sodium to the point where it is now practical to begin clinical examinations." Or would be, if the machine could sustain the currents produced. "It was planned to start treating a patient at the University Hospital last Sunday with radio-sodium, but unfortunately the cyclotron broke down." The boys were working on it. "I am sure before long we will have improved the construction so that the apparatus will withstand the higher power."[3] He was right. During September supersleuthing traced the problem to a change in conductivity of the water cooling the vacuum chamber.

From the beginning, apparently, an unknown and unnoticed rectifying action had tended to establish a dc potential on the dees. Previously the water had been sufficiently conducting to prevent the accumulation of any significant charge; in the summer of 1935 it no longer carried the charge away. Following this subtle diagnosis ("Who in the world would have guessed it?"), the cure was obvious: a choke coil introduced between the accelerating system and ground destroyed the dc potential on the dees, the cathode rays stopped, and all troubles vanished immediately. That was around October 1, 1935.[4] Within a few weeks the

2. Poillon to Buffum, 9 May 1935 (RC); Lawrence to Cockcroft, 4 Jul 1935 (4/5); quotes from, resp., Lawrence to Exner, 8 Jul 1935 (9/21), to I. Fukushima, 12 Sep 1935 (9/34), to Tuve, same date (3/32), and to Kruger, 1 Oct 1935 (10/20).

3. Lawrence to Poillon, 28 Aug 1935 (RC).

4. Lawrence to Henderson (9/6), to Kruger (10/20), and to Stern (16/49), all 1

cyclotron gave 10 μA steadily at 5 MeV, all the target could take. "We can almost detect the neutron ionization here in Princeton," White wrote in admiration. The fix was durable. In the summer, with its constant shortings, the cyclotron gave "an injustly unfavorable impression of its performance;" at the end of the year it was "practically robotized" and almost as likely to work as not.[5] Still it was by no means dependable, and Lawrence could not take advantage of an offer from the Macy Foundation, engineered by Poillon, to give John Lawrence and the Mayo Clinic each $1,000 for the purchase of radiosodium for biomedical research. "We are not in a position to supply radioactive substances for clinical work....Our apparatus runs only spasmodically, and a good share of the time we have it dismantled for repair and alterations."[6]

This brings us to step 2, the increase of energy to 6.2 MeV and of current to 20 μA, obtained in January and February 1936. It was primarily Cooksey's doing. He made a new vacuum chamber that provided for, among other things, a larger final radius for the spiralling ions and many small adjustments suggested by two years' experience with the machine. Cooksey's chamber did without the notches that Lawrence had cut into the old one to alter the magnetic field in the desperate days of the previous summer. The new chamber worked almost immediately.[7] This triumph of cyclotroneering—a major intended alteration that required no fiddling, an important step toward realizing "an apparatus engineered to the point where we can depend on reliable operation to an extent justifying beginning clinical research"—may have been decisive in causing Columbia and Cambridge to proceed with the construction of cyclotrons.[8] In mid

Oct 1935; to Cockcroft, 2 Oct 1935 (4/5); to Livingston, same date (12/12); the line about guessing is from M. White to Lawrence, 21 Oct 1935 (18/23).

5. Quotes from, respectively, M. White to Lawrence, 1 Nov 1935 (18/21); Lawrence to Kruger, 1 Oct 1935 (10/20); and R. Zirkle to Lawrence, 18 Dec 1935 (18/46).

6. Poillon to Buffum, 12 and 16 Sep 1935; Poillon to Lawrence, 9 Oct 1935, and Lawrence to Poillon, 16 Oct 1935, quote (RC).

7. Lawrence to M. Henderson, 23 Dec 1935 (9/6), and to Cork, 22 Feb and 14 Aug 1936 (5/1); Cooksey, Kurie, and Lawrence, PR, 49 (15 Jan 1936), 204.

8. Lawrence to Poillon, 16 Oct 1935 (RC), quote; Pegram to Cooksey, 6 Jan 1936 (4/8), re Columbia; infra, §7.3 re Cambridge; Cooksey, "Magnetic resonance accelerator-vacuum chamber with associated parts," 20 Jan 1936 (4/20).

March the 6 MeV beam came out of Cooksey's chamber through a platinum window ten-thousandths of an inch thick. It made a splendid spectacle, a cylinder of light over ten inches long streaking through the air. This accomplishment had more to recommend it than mere display: it made possible experiments beyond the influence of the Poulsen magnet, thereby cancelling (as the Illinois cyclotroneers put it) "one of the formerly objectionable features of the cyclotron."[9]

Cooksey's can was to provide "for intense neutron sources over long periods of time and for large quantities of radioactive substances." By doing so, it showed the way to more of both; and in the summer of 1936 another chamber, with dees widened to two inches (from 1.25 inches) and a larger filament, replaced the one installed in January. This brings us to step 3: the steady current more than doubled to 50 μA, albeit at 4.3 MeV, and the peak reached over 90 μA, with no end in sight. It was at this time that (in Lewis's words) the neutron flux became a "public nuisance," disturbing experiments in the Chemistry Department 300 feet away; Lawrence reported to Poillon, with fatherly pride, the words of Mayo's Alvarez, "our baby has become a monster." Lawrence set his people to making 50 μA at the maximum energy of 6.2 MeV.[10] In December 1936 they almost complied. "For some reason, for which we are not entirely clear," the current shot up to 35 μA at just under 6 MeV with the 1.25-inch dees in place. Cooksey guessed that changing the current to the filament from dc to ac might have done it (with ac the filament could operate without bending in the magnetic field). Odder yet, the big current came out without any shimming. "This delightful mystery is unexplained. One might say unexplained either way you look at it, for no one knows why we had to use shims, or why we do not have to use shims." "There are certainly lots of things to learn about the

9. Lawrence to R.E. Zirkle, 7 Mar 1936 (18/46), and to J.E. Henderson, 14 Mar 1936 (9/5); Cooksey and Lawrence, PR, 49 (1936), 866; Lawrence and Cooksey, PR, 50 (1936), 1136; Kruger and Green, PR, 51 (1937), 57, quote.

10. Cooksey, Kurie, and Lawrence, PR, 49 (15 Jan 1936), 204; Lawrence to Cork, 1 Sep 1936 (5/1), to Cockcroft, 12 Sep 1936 (4/5), to Oppenheimer, 3 and 5 Sep 1936 (14/9), and to M. White, 5 Sep 1936 (18/21); Lawrence and Cooksey, PR, 50 (1936), 1131–6, 1139; quotes from Lewis and Alvarez in Lawrence to Poillon, 30 Sep 1936 (15/17).

cyclotron." With a little shimming and four-inch dees, it gave more than 150 μA steadily at 5.5 MeV, and 170 μA peak.[11]

In step 4, which occupied much of the first half of 1937, effort went to improving reliability of performance. There were two matters identified in 1936 that particularly needed attention, one simple and mechanical, the other complicated and electrical. The first was to do without wax. "If only we could get rid of wax joints," Lawrence sighed, "I am sure the outfit would run with very nice regularity."[12] Cooksey set to work on another chamber, which was "to be a humdinger," with so large a vacuum gap (some 7.5 inches) as to allow for dees three inches wide and very thick, and with rubber gaskets in place of wax seals. (The large gap left enough space between the wide, thick dees and the chamber walls to keep the capacitance of the system small; the thick dees resisted buckling, and hence change in capacitance, during operation.) Lawrence anticipated, rightly, that Cooksey's gaskets would be a "tremendous improvement," a great step forward in convenience and reliability.[13] The electrical matter concerned that perennial irritation, the high-frequency oscillator, which by January 1937 was on its last legs.

This oscillator worked according to Livingston's old plan, with a self-excited, tuned-grid, tuned-plate circuit. Its heart was the power tube developed by Sloan and improved with the help of Thornton and F.A. Jenkins. This water-cooled cylinder, rated at 30 kW, had a copper pipe as anode, copper wire wrapped on copper poles as a grid, and a long tungsten wire bent into a hairpin as a filament. When working it was continuously exhausted by a pump working with Apiezon oils. The Laboratory made these tubes itself, or had them made in the shop of the Physics Department, which brought their cost down to perhaps two-thirds

11. Quotes from, resp., Cooksey to A.C.G. Mitchell, 23 Dec 1936 (13/9), and to S.W. Barnes, same date (2/24); Cooksey to M. White, 10 Jan 1937 (18/21). Van Voorhis changed the filament current to ac; Dunning and Anderson, *RSI, 8* (1937), 158–9, obtained 100 μA at Columbia with a similar (and independent?) arrangement.

12. Lawrence to Newson, 18 Aug 1936 (13/43); cf. Lawrence to Cork, 14 Aug 1936 (5/1).

13. Lawrence to Sagane, 6 Jan 1937 (9/39); Cooksey to Evans, 29 Jan 1937 (7/8); Nahmias to Joliot, 28 Apr 1937 (JP, F25).

that of the equivalent commercial models.[14] The oscillator employing them had the advantage of simplicity and robustness. But it also had a serious defect.

The Livingston-Sloan oscillator drifted in frequency. To compensate the operator had constantly to adjust the magnetic field, by inspiration and experience more than by science, and thought himself successful if he could keep the beam sturdy and steady to within 10 percent. The current exciting the magnet could easily have been kept constant to within 1 percent with a feedback method, or automaton, of a type developed at the Cavendish; but such a device would only have hampered the fiddling necessitated by the drifting oscillator. There was something else: the oscillator circuit had a tendency to parasitic vibrations at unwanted frequencies, which uselessly consumed power wanted on the dees. This loss, added to consumption within the tube, reduced the effective rf power on the dees to half that of the oscillator: 12 kW to accelerate ions out of 25 kW purchased from PG&E. As Cooksey wrote Cockcroft, who had begun work on a Cambridge cyclotron: "Our methods of operation have been far from ideal and we know it."[15] To idealize them Lawrence engaged Charles Litton as the technical assistant provided for in the charter that had established the Laboratory. Litton had been the chief engineer at Federal Telegraph; he constructed in accordance with the production standards of high-tech industry, not in the experimental and jerry-rigged fashion of a physics laboratory. A 1-kW radio transmitter of tightly controlled frequency regulated the output from two 20-kW tubes. Lawrence watched in admiration. "It [Litton's oscillator] has been designed and built in the regular style of the first class engineering job that one sees in a broadcast station. It should make the operation of the cyclotron simply a matter of pushing buttons."[16]

14. Lawrence and Cooksey, *PR, 50* (1936), 1136; Sloan, Thornton, and Jenkins, *RSI, 6* (1935), 75–82. The shop time to make the grids and filament supports cost $240, a commercial tube perhaps $350; Lawrence to Fukushima, 28 Aug 1935 (9/34).

15. Lawrence and Cooksey, *PR, 50* (1936), 1136–8; Cooksey to Cockcroft, 1 June 1936 (4/19). For the artistry of the operator, Alvarez, *Adventures,* 39–40; for the automaton and its first descendents, Wynn-Williams, *PRS, A145* (1934), 250–7, and Anderson, Dunning, and Mitchell, *RSI, 8* (1937), 497–501.

16. Lawrence to M. Henderson, 24 Jan 1937 (9/6), and to T. Johnson, 9 Feb 1937 (10/1), quote.

Litton's "de luxe oscillator," as Lawrence called it, was installed in March 1937. "You can imagine the joy in tearing down the old oscillator and its associated junk." The de luxe way worked. "We now control the magnet with an automaton [Cooksey wrote the cyclotroneers at Rochester] and the beam remains constant to within ten percent for hours without attention. The boys are all complaining because the cyclotron has become so dull."[17] This is not to say that the age of heroism had passed. In late May or June the oscillator went down, leaking; the trouble was traced to the pump, and Litton summoned from Redwood City, where he had set up in business. A replacement pump arrived at 2:00 A.M. on a Sunday morning; Richardson, Oldenberg, and Van Voorhis labored day and night to revive the oscillator, but the beam would not rise above a miserable 2 μA. This time the trouble was not in the oscillator. Alvarez had placed a new cloud chamber between the cyclotron's pole faces. The iron in his equipment adversely shimmed the field. When it was put aside, the cyclotron recovered the beam and the reliability that the Laboratory had come to expect.[18]

We come to step 5. Alvarez's experiments were the last to be done on the 27-inch cycloron. During the first few weeks of August 1937 the Laboratory achieved a goal that Lawrence had been considering since February 1932 and planning since the spring of 1933, when he wrote Poillon of his desire to enlarge the poles of the Poulsen magnet to make possible the acceleration of particles to 10 MeV.[19] The $1,550 required had been available neither then nor in the summer of 1934, and subsequently improvement of the 27-inch had claimed priority. In January 1937, however, with the prospect that Cooksey's gaskets and Litton's oscillator would "approximate our dreams of a satisfactory outfit for nuclear work," it was decided to enlarge the poles to 37 inches during the summer. The goal was not only to increase the beam, as usual, but also to test a new chamber on Cooksey's

17. Lawrence to Newson, 4 Mar 1937 (13/43); Cooksey to Barnes, 19 Apr 1937 (2/24).

18. Cooksey to Lawrence, 3, 4, 8, 10, and 14 June 1937 (4/21).

19. Pelton Water Wheel Co. to Lawrence, 4 Mar 1932 and 7 June 1933 (25/2), estimating $1,550 for the work; Lawrence to Poillon, 3 June 1933 (15/16A); Lawrence and Livingston, *PR, 45* (1934), 611.

new principles for scale-up for the Crocker cyclotron. If the chamber, made from a ring of cast brass and sealed with gaskets, proved itself, Lawrence would have "no worries about funds for the new lab; but if it doesn't give fairly high output we will have to find out why as soon as possible."[20]

The new pole pieces were ordered from Moore Dry Dock Company early in June 1937 and delivered, for $490, a month later. On July 9 Cooksey's 37-inch chamber was moved into the Laboratory by block and tackle anchored by Cooksey's Packard. It leaked, but clammed up completely after liberal application of glyptol, a thick enamel for insulating wires.[21] On August 4 the 27-inch concluded its pioneering life. In eight hours the University's maintenance department installed the new pole pieces. (The business was not altogether trivial; the corresponding department at Ohio State dropped a pole during installation there and decreased its parallelism by a factor of three.) A few days later the magnet gave 14 kG. Around August 15 tests began, with new pumps, new water cooling, and new circuit components.[22] The machine was expected to run immediately on switching on and almost did so, with a little shimming by Kurie. By the end of August there was a very big beam: 100 μA at 4.6 MeV, steady, good for more than a curie of radiosodium. "Thrilling results," said Poillon.

Cooksey's new tank ran for a month without springing a leak, which, in comparison with earlier chambers, his included, amounted almost to a perpetual hermetic seal.[23] Litton's handiwork did not do so well. In order to get high currents near maximum energy, more power was required on the dees than the de luxe oscillator could deliver. The Laboratory went back to a homemade model using pumps three times as fast as Litton's.[24] A

20. Lawrence to Sagane, 6 Jan 1937 (9/39), to Cork, 23 Jan 1937 (5/1), to Allen, 1 Feb 1937 (1/14), to Buffum, 10 Aug 1937 (3/38), and to Cooksey, 12 May 1937 (4/21), quote. For earlier attempts at financing the expansion: Lawrence to Poillon, 26 Mar 1934 (15/16A), and supra, §3.3.

21. Cooksey to Lawrence, 11 June 1937 (4/21), and to Van Voorhis, 9 and 21 Jul 1937 (17/43).

22. Lawrence to Paxton, 4 Aug 1937 (14/18), and to Van Voorhis, 10 Aug 1937 (17/43); Livingood to McMillan, 5 Aug 1937 (12/11); Nahmias to Joliot, 12 June and 12 Aug 1937 (JP, F25); Green, memo to file, 14 Dec 1939 (MAT, 23/"cycl. letters"), re Ohio's poles.

23. Cooksey to Van Voorhis, 1 Sep 19 (17/43), and to Allen, 16 Sep 1937 (1/14); Lawrence to Cork, 13 Sep 1937 (5/1), and to M. White, 11 Oct 1937 (18/21); Poillon to Lawrence, 21 Sep 1937 (15/17A).

24. Lawrence to Newson, 11 Oct 1937 (13/43), and to M. White, same date (18/21).

competition to improve the creation of nothing arose between Litton and Sloan. A better pump resulted, with a new fractionating system and Litton oils rated by the inventor over Apiezon products. An estrangement resulted too. Litton thought that the Laboratory made too free with his ideas and injured his chances to patent them. It was a natural consequence of the communal ethic of the Laboratory and the vacuous patent policy of the University.[25]

During the first few months of operation of the 37-inch cyclotron, Robert Wilson was engaged in a lengthy analysis of the paths of ions between the dees. In January, shimmed to his specifications to increase the field near the center, the machine gave 170 μA at 5.5 MeV, and, a few weeks later, 150 μA at 6.2 MeV. The sensitivity of the beam to shimming may be illustrated by data collected by McMillan: Wilson's wonderworking pyramidal shims, made of ten iron disks 0.013 inches thick and 6.5 to 20 inches in diameter, cut the beam from 6 μA to 0.5 μA when placed above the chamber, and restored the 6 μA when placed below; two such pyramids, one on top and one on the bottom, made 28 μA; removal of two disks from the upper stack increased the current to 30 μA. Pyramids are full of mystery. They did not work at all at Rochester, as its chief cyclotroneer told Lawrence, in bewildered admiration of what Berkeley had achieved. "Each new figure seems more unbelievable than the last, and we wonder if there is going to be any limit to what you can extract from your machine!" All this exceeded the target's, if not the cyclotron's capacity, and the Laboratory tended to run at 100 μA or less for isotope manufacture.[26]

It also ran with unprecedented regularity, "almost night and day," as Lawrence wrote early in January. In the six months ending in September 1938, the 37-inch cyclotron operated for almost

25. Litton to Cooksey, 27 Jan, 27 Nov, and 29 Sep 1938, on Litton's improvements, and correspondence of 26–29 Sep 1938, on patents (12/9); Cooksey to Gray, 28 Sep 1938 (14/38), and to Van Voorhis, 8 Oct 1938 (17/43).

26. Wilson, *PR*, *53* (1938), 408–20, discusssed infra, §10.1; Lawrence to Crowder, 6 Jan 1938 (5/5), and to Allen, 16 Feb 1938 (1/14); McMillan to Bethe, 14 Jan 1938 (HAB, 3); Cooksey to DuBridge, 11 and 21 Jan, 5 Mar 1938 (15/26), and to Laslett, 10 Feb 1938 (10/32); McMillan, "Shimming," 30 Mar 1938 (12/30), 1–2; DuBridge to Lawrence, 25 Jan (quote) and 9 Mar 1938 (15/26); Mann, *Rep. prog. phys.*, 6 (1939), 129.

eight hours a day on average, seven days a week, and sometimes twenty-four hours a day in three eight-hour shifts. For seven or eight weeks at the end of the year, during the sojourn of an aspiring Belgian cyclotroneer, Paul Capron, the machine worked "without any serious trouble" from 8:00 A.M. to 11:00 P.M.; it attained a peak current of 300 μA of 8 MeV deuterons and made neutrons as plentifully as might 10 or 100 kilograms of radium. "In Berkeley I realized exactly all the possibilities of the cyclotron." Rather than apologize for the erratic performance of their machine, cyclotroneers could now consider it "a piece of standard equipment, a laboratory tool which can be put up by [Berkeley] experts in a short time, at reasonable expense, and without undue local development work." The weekly schedule of oilings, greasings, and other checks, introduced by Brobeck in the summer of 1938, helped to keep the machine regular.[27]

The bottom line, however, was not regularity of operations but quantity and efficiency of production. The initial routine for making P^{32}, the most readily manufactured isotope potentially useful in medicine, will indicate the process and the possibilities for improvement. One began with red phosphorus, a very disagreeable substance, which, in the system developed by Kurie, was rubbed into grooves in a water-cooled copper plate maintained in an atmosphere of helium. This elaborate target received deuterons in the usual position, outside the dees, and spread radiophosphorus everywhere. At 8 MeV it took 35 μA-hours to make 1 mCi of P^{32}. John Lawrence had set 1,520 mCi as the amount needed for experimental therapy. In the spring of 1938, when the 37-inch cyclotron delivered 60 μA of deuterons steadily at 8 MeV, it had to labor a day and a night to make one clinical dose. The case was worse for Fe^{59}, which, because it derived from the rare iron isotope Fe^{58}, could scarcely be made in quantities sufficient for tracer work—some 1 μCi per 100 μA-hr.[28]

Some creative thought about the electrical bill indicated a route

27. A.H. Compton, notes on visit to Berkeley, 16 Sep 1938 (4/10); Capron, "Report," 3; Evans to Cooksey, 1 June 1938 (7/8), quote, proposing the projected MIT cyclotron as a test; Brobeck, "Suggested maintenance operations," 29 Jul 1938 (25/1).

28. Wilson and Kamen, PR, 54 (1938), 1031–6; Kurie, RSI, 10 (1939), 199–205.

to important improvement. Experience showed that about half of the power expended in the dee circuit went to drive the ions. Hence in normal operation as much as 10 kW must have been tied up in the energy of the particles circulating within the dees. But less than a single kilowatt of beam heated the target. Wilson looked where reason pointed; toward the end of April 1938, he found that the deuteron current inside the dees was ten times what emerged from them.[29] In June, with 50 μA on the target, Wilson's probes detected 650 μA at about two-thirds of the distance from the center to the deflector. Since the presence of the probe did not diminish the target current, it appeared that the Laboratory could bake material for biologists while giving an external beam for physicists. It was a godsend. "One cyclotron may satisfy the needs of both biologists and physicists without extremely long periods of operation."[30]

For universal satisfaction, the raw material for the baking has to be available in a refractory form. Ferrous phosphide answered perfectly. By covering probes with it, cyclotroneers could make P^{32} and Fe^{59} simultaneously, much more quickly and with much higher specific activities than by Kurie's method. In developing this technique Wilson collaborated with Martin Kamen, a Ph.D. chemist from Chicago, who had arrived around Christmas 1936 in order to work with Kurie for nothing; he had ascended to salaried research assistant the following July, when Lawrence learned that he was capable of overseeing the preparation and distribution of the cyclotron's products. Wilson and Kamen's innovation raised yields by an order of magnitude. It was of the first importance for the Laboratory's functioning and financing. Kamen remembered almost fifty years later: "Bob and I were the heroes of the moment."[31]

We have operating details from the end of August to the middle of November 1939. The machine ran on most days for an average

29. Lawrence to Evans, 21 Apr 1938 (3/34), and to Chadwick, 30 Apr 1938 (3/34); Jacobsen to Bohr, 26 June 1939 (BSC).

30. Wilson and Kamen, *PR, 54* (1938), 1036.

31. Wilson, *PR, 54* (1938), 240; Wilson and Kamen, ibid., 1036; Cooksey to Lawrence, 10 June 1938 (4/21); Seaborg, *Jl., 1* (5 Jan and 1 Jul 1937), 190, 252 (re Kamen); Kamen, *Radiant science*, 65, 79–81, 117–9, quote.

of eight or ten hours; all of its internal and about 60 percent of its external bombardments went for preparation of isotopes for biomedical work. Otherwise it stopped for improvements, for example, the replacement of the old pyrex dee insulators by new Corning ware; for lack of business, as the 60-inch came on line; and for the inevitable repairs. Trouble occurred in the usual places for the usual causes: "filament burned out for no apparent reason," "oscillator failed to oscillate, fixed itself," leaks here and there, a cracked target window, blown fuses. An adjustment to the position of one dee opened holes in the chamber seal that took a day to locate and fix; then the oscillator would not oscillate; and so on.[32] There were other times when the 37-inch cyclotron acted erratically or quit altogether, and in 1940 it shut down for at least two extended periods for rewiring the controls and for rebuilding the oscillator. No doubt its later fitfulness, as compared with its earlier dependability, arose from its feeling of neglect. "If anyone cares about the old 37-inch cyclotron," Aebersold wrote, in announcing a restoration to service in 1940, "it is running fine again."[33] From early in 1938 the Laboratory had been preoccupied with assembling, testing, and operating the 60-inch cyclotron.

The Crocker Cracker

"It is a truly colossal machine." Thus Edoardo Amaldi, who passed through Berkeley in September 1939. The cyclotroneers in partibus, who knew it first from Cooksey's photographs, gasped at the size: "almost unbelievable" (Lyman); "almost too much" (Van Voorhis). Figure 6.2 shows what caused "widespread admiration and awe," what made "everybody...open mouthed" among the cognoscenti.[34] The giant's magnet weighed 220 tons; it stood 11

32. Operating results for 28 Aug–7 Oct and 30 Oct–5 Nov 1939 (25/5), and for 9–29 Oct and 6–12 Nov 1939 (25/1). Half its problems were corrected in under three hours; Tuve, memo, 24 Jan 1940 (MAT, 25/"cycl."), recording information from Abelson.

33. Cooksey to Lawrence, 30 Mar 1939 (4/22), a breakdown caused by frayed leads, discovered by McMillan by "his usual sleuth work;" Aebersold to Cooksey, 14 Aug 1940 (1/9), reopening after rewiring; Kamen to A.C.G. Mitchell, 10 Dec 1940 (13/9), first beam after a month's rebuilding of the oscillator.

34. Amaldi, *Viaggi, 6* (1940), 9; Lyman to Lawrence, 9 Mar 1939 (12/19), and Van Voorhis to Lawrence, 5 Feb 1939 (17/43).

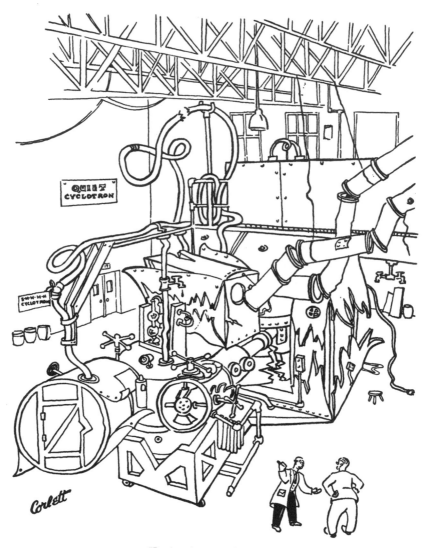

"Toughest damn atom I ever sawl"

— *The California* PELICAN

FIG. 6.2   The Crocker Cracker as portrayed in the University's student publication, *The Pelican*, 1939. LBL.

feet high; it consumed 45 kW of power. Its frame could hold the entire staff of the Laboratory (see plate 5.2). Barnes compared the ports in the vacuum chamber to manholes; Laslett admired the "amazing 153 cm beam;" Snell joked that its neutrons would reach Chicago; Aebersold worried that five feet of water did not diminish radiation enough for safety in Berkeley and that all the cyclotron's targets would be too hot to handle.[35]

Why was the Crocker cyclotron made so big? During the planning stage, Lawrence wrote Cockcroft that "the new cyclotron will be much larger than needed for any work contemplated in the near future." The rectangular design of the magnet was wasteful of material; as the American Rolling Mill Company (ARMCO), who bid on its construction, pointed out, the large spaces between the poles and the frame diverted much of the magnetic flux from the 22-inch gap containing the chamber. Lawrence acknowledged the inefficiency of the design, which he explained to the ARMCO engineers, as he later did to Chadwick: the 220 tons of metal in the magnet exceeded what was needed for the intended voltage and current "because it has been designed for the primary purposes of medical research, requiring openness and accessibility." Maximizing the flux and field between the poles had not been the goal; as usual, the Laboratory sacrificed efficiency to allowance for the possibility of future change.[36] And yet, as Lawrence made clear to a colleague wishing to make a medical cyclotron, "We recognized that for medical purposes alone such a large installation was hardly justified. On the other hand, for the sake of physics, it was important to get up to as high energies as possible." But in that case, a more efficient design would have been justified. It appears that no good reason or clear purpose except flexibility fixed the dimensions of the 60-inch cyclotron. Why then was the

Van Voorhis to Lawrence, 5 Feb 1939 (17/43).

35. Barnes to Cooksey, 14 June 1938 (2/24); Laslett to Cooksey, 21 June 1939 (10/33); Snell to Cooksey, 8 May 1939 (16/33); Aebersold to Green, 5 Oct 1939 (1/9); Cooksey to A.L. Elder, 9 Nov 1939, and to J.B. Fisk, 13 Nov 1939 (25/5).

36. Lawrence to Cockcroft, 15 Dec 1936 (4/5); ARMCO to Lawrence, 23 Dec 1936 and 4 Jan 1937 (25/3); Lawrence to C.C. Clark (ARMCO), 6 Jan and 23 Mar 1937 (25/4), and to Chadwick, 30 Apr 1938 (3/34); Cooksey to A.L. Elder, 9 Nov (25/5). In the final version, the gap held two cover plates each five inches thick, the chamber proper, ten inches in length, and spaces of an inch top and bottom for shims.

Crocker Cracker so big? "Since we can get the money for [it]."[37]

"The cyclotron has now passed beyond the stage of ordinary laboratory development; its further improvement involves problems of a primarily technological engineering character." So Lawrence wrote in his report to the Research Corporation for 1937. The builders of the 60-inch cyclotron would enjoy not only a sufficiency of funds but also the luxury of engineering from scratch. No hand-me-down magnets, no industrial discards, in short, as Nahmias put it, "no patch work." Everything had to be planned in advance: the inefficient magnet designed and tested on a scale model, bids sought and tendered in accordance with detailed blueprints, tons of metal machined and installed. The magnet and its building had to be designed to resist earthquakes. "The prospect of this job is rather appalling," Cooksey wrote shortly after starting it, "as it really seems to be engineering rather than experimentation."[38] A poignant example: Cooksey found the mold in which the 60-inch chamber was to be cast "so appalling in its complexity" that he could not understand how his own design was to be realized.[39] The Laboratory needed a first-class engineer. To its extraordinary good fortune, it got just the man it required, just when wanted, and for just what Lawrence liked to pay. William Brobeck came to work for nothing in the fall of 1937; by March 1938 he had displaced his friend Cooksey as chief cyclotroneer. "[He] puts me in the shade."[40] Brobeck reworked the original sketches, made when funding and prudence recommended 50-inch poles, into engineering plans for a 60-inch magnet based on designs by Alvarez.[41]

The Crocker Laboratory came into existence slowly because the architect of the building eluded Lawrence's pressure by dying.

37. Quotes from, resp., Lawrence to A.L. Hughes, 5 Jul 1939 (18/12), and to Yasaki, 24 Dec 1936 (9/43). Lawrence told Hughes that Garvan, the president of the Chemical Foundation, who had agreed to pay for the magnet, approved overbuilding for the sake of physics; cf. Lawrence to Garvan, 7 Oct 1937 (3/38).

38. Lawrence, "Report on progress made," 42, 1937 (RC); Nahmias to Joliot, 12 June 1937 (JP, F25); Cooksey to D.T. Uchida, 14 Mar 1938 (9/37), re earthquakes, and to Evans, 29 Jan 1937 (7/8), quote.

39. Cooksey to Lawrence, 7 June 1938, quote, to Sagane, 10 June 1938 (9/39), and to Tuve, 28 June 1938 (3/32).

40. Cooksey to "Dodie," 30 Mar 1938 (4/21).

41. Brobeck, interview, 27 Mar 1985 (TBL); Alvarez, *Adventures*, 44–5.

The delay in finding a successor left a long time for dickering for copper and steel, the two largest single items of expense. Lawrence tried to get it gratis through Buffum.[42] He did not obtain even a price concession; on the contrary, the price of steel rose and the magnet climbed over budget. Columbia Steel, a San Francisco firm, obtained the plates and disks for the frame and poles from Carnegie-Illinois Steel; Columbia's management took an interest in Berkeley's "atom buster," as they called it, and kept the price of the steel under $12,300 f.o.b. in box cars in San Francisco. Machining, by Pelton, cost $3,000. The order was placed at the end of March 1937; twelve months later the 150-ton yoke stood in place in the Crocker Laboratory.[43] The metal for the windings and the coil tanks came to about $400 less than the steel; Revere Copper and Brass supplied the copper in strips and a carefully considered plan, "rather difficult...to delineate," for winding and insulating it. The copper and the plan were turned over to a local firm, Gardner Electric, who underbid Revere for the construction of the coils. Gardner discovered that Revere had furnished some strip so rough as to threaten the paper insulation. With Gardner's vigilance, the Laboratory received carefree coils, which had taken their place under the yoke by early June 1938. The total magnet was complete in August.[44]

The cost of the Crocker magnet, including assembly, came to $31,546, very little more than the value assigned to the much smaller Poulsen magnet in 1930. The reduction in unit cost was a consequence of laminating, rather than casting, the frame and the poles, a method that ran a risk of lopsidedness. "It is a very satisfying achievement," Cooksey wrote, after determining that the poles came parallel to within 0.0014 inch, "to have achieved such parallelism out of what one might almost consider a stack of

42. Lawrence to Poillon, 16 May 1936 (15/17).

43. J.R. Gregory to Lawrence, 18 Sep 1936 (25/3), and Richard Erlin to Lawrence, 4 Feb 1937 (25/4); Lawrence to Buffum, 11 Aug 1936 and 28 Jan 1937 (3/38), to Erlin, 27 Mar, to J.H. Corley, 15 Apr, to B.M. Brock, 20 Jul 1937 (25/4), and to Cork, 26 Feb 1938 (5/1); Cooksey to David W. Mann, 17 Mar 1938 (25/5); "Orders for 60-inch," 1 Jul 1939 (25/5).

44. P.M. Mueller, Revere, to Cooksey, 7 Dec, and to Lawrence, 16 Dec 1936 (25/3); Cooksey to Mueller, 20 Mar 1937 (25/4); David G. Taylor, Gardner, to University of California, 7 and 18 Feb 1938 (25/5); Cooksey to A.J. Howell, Revere, 2 June 1938, and to Norman Grant, Moore Dry Dock, 18 Aug 1938 (25/5).

cards. It certainly shows our colleagues in other laboratories an inexpensive way to build a magnet for cyclotron use."[45]

Entrusting orders to local firms, which proved itself in the procurement of the steel and the winding of the coil, also brought an excellent product in the tricky matter of the vacuum chamber. Cyclotroneers disputed the relative disadvantages of rolling and casting the brass ring that made up the chamber walls. Cooksey favored a casting for flexibility, knowing, however, that it might be too porous to hold a vacuum. Moore Dry Dock Company of Oakland undertook to try. They poured 2,838 pounds of pure brass into the mold whose complexity befuddled Cooksey; and they made a casting full of holes. A second try produced a chamber vacuum tight but for one spot, which was plugged up satisfactorily.[46]

While awaiting the delivery of the magnet, the Laboratory organized itself into a number of interlocking groups to design and build the accelerating system and controls of the big machine. The scheme, as proposed by Brobeck in November 1937, reads like an industrial, or even military, table of organization: "The work of building each group of apparatus is given to a *committee* for which a *leader* is appointed....He may divide the work as he wishes....*Supervisors* are to be appointed for definite lines of the work and their approval must be obtained by the committee leaders for all work included in these lines. Committee leaders, supervisors and others needed are to meet periodically with the directors of the Laboratory...[as] the *directing committee*." Examples of committee assignments: magnet, vacuum chamber, vacuum pumps, oscillator. Examples of lines: mechanical design, vacuum systems, wiring, radiation. Brobeck provided for eight committees in all, as did Cooksey in a memorandum written two weeks later; their discordant schemes drove a finer division of labor, which ended in a definitive score in the summer of 1938.[47] The division is depicted in table 6.1.

45. "Orders for 60-inch," 1 Jul 1939 (25/5); Cooksey to Grant, 17 Jul 1939 (25/5).
46. Cooksey to Grant, 17 and 20 Jul 1938 (25/5), and to Lyman, 20 Oct 1938 (12/19); Brobeck, notes on meetings of 29 Jul and 24 Aug 1938 (24/13).
47. [Brobeck], "Suggested organization for the construction of the 60-inch cyclotron," 10 Nov 1937, and Cooksey, memo on committees, 22 Nov 1937 (25/4); committee assignments attached to Brobeck, "Suggested maintenance operations," 29 Jul 1938 (4/21); an identical list, undated, is in (25/5).

During the summer of 1938 the "directing committee"—initially composed of Lawrence, Cooksey, McMillan, Alvarez, Snell, and Brobeck—met at least three times at the Leamington Hotel in Oakland to decide policy at a decent distance from those who would carry it out. They constituted themselves supervisors, as in Brobeck's original proposal, in accordance with their expertise: Cooksey took charge of the chamber and ion source; McMillan, low-voltage power and wiring; Alvarez, radiation protection and electronics; Snell, controls and instruments; Brobeck, the magnet and mechanics. John Lawrence was elevated to the directorate and given all responsibility for medical matters; Winfield Salisbury remained a mere committeeman, but received authority over the oscillator and its power supply. To coordinate operation of the 37-inch cyclotron with the completion of the 60-inch, the directing committee decided to choose crew captains exclusively from their number.[48]

Snell's leaving for a position at Chicago (his assignments fell to McMillan) further restricted the pool of captains. The directors deliberated whether to enlarge it by promoting midshipmen Simmons, Abelson, and Aebersold, but decided against the dilution. The matter concerned safety as well as efficiency. Lawrence worried that the crews did not observe routine safety practices, knew nothing about first aid, and exposed themselves recklessly to electric shock. Salisbury received a nasty shock from the emission voltage of the filament, a potentially most serious mishap, which could have been prevented had he used the safety switch provided. Strong warnings were required, and sets of instructions, should time be found (none was) to write them down. Meanwhile only Alvarez, McMillan, and Brobeck would enjoy the rank of captain. Their crews would work in three watches, 8–12, 1–6, 7:30–11:30. To complete bureaucratic arrangements, Cooksey was commissioned to put signs on all the external doors of the Laboratory: "No admittance without appointment."[49]

48. Brobeck, notes on meetings of 15 and 19 Jul 1938 (24/13).
49. Brobeck, notes on meetings of 29 Jul and 24 Aug 38 (24/13); Salisbury, interview, Oct 1981.

Table 6.1

Committee Assignments for the 60-inch

	1	2	3	4	5	6	7	8	9	10	11	12	13	14	15	16	17	18	19	20
Abelson, P.		o															o			
Aebersold, P.						o	o													
Alvarez, L.			o		o	x	x	o							o					
Brobeck, W.			o		o	o		o			o	o	o				x			
Cooksey, D.	o		x				o													o
Hamilton, J.						o			o	o										
Kalbfell, D.								o		o				o				o		
Kamen, M.										x										x
Kurie, F.					x				o			o	o				o			
Lawrence, J.									o				o							
Livingood, J.				x			o		o	o		x								
Lyman, E.	x				o				o		x						o			
McMillan, E.				o	o			x	x				x		x					
Mann, W.														x				x		
Snell, A.	o	x												o					o	
Simmons, S.				o												o			o	
Van Voorhis, S.	x		o		x						o		o			x			x	

Note: 1, Medical cyclotron; 2, New oscillator; 3, 60-inch vacuum chamber; 4, Pumps; 5, Control table; 6, Protection; 7, Targets; 8, Snouting; 9, Laboratory arrangements; 10, Chemistry; 11, Automation; 12, Cooling; 13, General layout of 60-inch; 14, Deflector; 15, Generator; 16, Power for rf apparatus; 17, Deuterium and other gases; 18, Transformers; 19, Filament heating and emission; 20, Aluminum pot; o, committee member; x, committee leader.

It would not be profitable to follow in detail what went on behind these closed doors. But one feature of the development of the 60-inch deserves emphasis: the construction of cyclotrons elsewhere had eased the burden of innovation in Berkeley. In two important matters—the ion source and the dee system—the Laboratory adapted what others had introduced. In the 11-inch cyclotron, ions formed in the field of an exposed cathode, which perforce operated at the residual pressure within the "vacuum" chamber. The source for the 27-inch was only slightly more elaborate: a coiled filament insulated by glass sheathed in copper sat above (or below) the dees at the center of the diametral gap and sent its stream of electrons along the magnetic lines of force into the bottom (or lid) of the cyclotron chamber. Livingston took steps to improve the method after his translation to Cornell. He began by adapting the capillary fountain designed by Tuve and his associates for use in their high-tension apparatus. They placed an anode and a cathode opposite one another along the axis of a metal tube, narrowed the space between them into a "capillary" in which the ions formed or collected, and drew the ions out perpendicularly to the axis through a small hole. By rapid pumping outside the capillary, they could maintain the gas within it at a pressure 500 times greater than that in their accelerating tube. They managed to tease out of the hole a positive current of no less than 1,500 μA.[50]

Livingston's variation of Tuve's capillary yielded a current of positive ions of about 500 μA in December 1936. Of these perhaps a fifth were protons; it remained only to conduct them from the capillary hole to the collection cup or target. Collecting proved taxing and at first Livingston and his colleagues at Princeton and Purdue, who tried to follow his lead, got but a feeble current. Profiting from this informal testing service, Lawrence and Cooksey stayed with what they knew to work in the conversion to the 37-inch cyclotron, although Cooksey did not like to recommend their flat filament to others. Lengthy tinkering—paid

50. Lawrence and Cooksey, *PR, 50* (1936), 1131–4; Tuve et al., *PR, 46* (1934), 1027–8, and *PR, 48* (1935), 242–51. At first the Cornell cyclotron used the exposed incandescent cathode (Livingston, *RSI, 7* (1936), 64), as did the Princeton cyclotron (Henderson and White, *RSI, 9* (1938), 25).

for by the Research Corporation—eventually multiplied the deuteron beam at Cornell twelvefold, from 3 to 35 μA, and then doubled that. In September 1938 the Laboratory was planning a big capillary. "It should give 4 m.a.," Cooksey wrote Lawrence, "Oh dear." Livingston's version of 1938 is shown in figure 6.3: from the capillary's hole at the cyclotron's center issues a stream of positive ions created in the plasma around the hole by electrons shot from the hot cathode at the top of the source. The gas to be ionized enters the upper alembic-shaped vessel through the pipe that also brings current to the filament. The little circles on the surface of the alembics indicate cooling tubes. The hooks or "feelers" extending from the left dee toward the hole supply the electric force to extract the positive ions. With this arrangement, Livinston and his associates brought 70 percent of the 100 μA of protons issuing from their source to the collector and the cyclo-

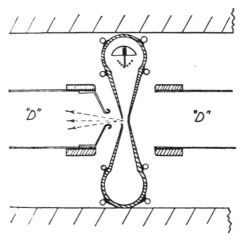

FIG. 6.3 Livingston's capillary source. Livingston, Holloway, and Baker, *RSI, 10,* (1939), 65.

troneers at Purdue managed to obtain five or six times the current to target they had from a hot tungsten filament.[51]

As Livingston observed, his handiwork had important advantages in addition to increasing the current. The beam was not only bigger but also narrower and better focused owing to the small size of the exit hole and its placement at the center of the cyclotron. At Berkeley the circulating current was ten times as large as the beam to target, whereas Livingston brought home 70 percent of the protons he started with. Because of the better definition of the beam, the wide dees in use at Berkeley to accommodate the ions from the extended exposed filament were no longer required. The vacuum and magnet gaps could be narrowed, reducing power consumption and increasing the field and therewith the maximum available energy.[52] When McMillan and Salisbury finished their other chores for the 60-inch cyclotron, they adapted Livingston's paragon in the manner shown in figure 6.4. Here a truncated metal cone, which Livingston had mentioned as a possibility, serves as anode to a hot cathode at its base. The exit hole at the top of the cone occupies a place just off center in the cyclotron's median plane. After some experimentation with the size and orientation of the hole and the feelers, McMillan and Salisbury obtained 90 μA of deuterons to the target. Under the same conditions, the old filament source delivered 14 μA.[53] To take advantage of the reduction in the vacuum gap made possible by the capillary source, one put iron disks ("filler plates") into the chamber or, as at the Carnegie Institution, made them integral with thick cover plates.[54]

51. Livingston to Lawrence, 2 Dec 1936, and reply, 15 Dec 1936 (12/12); Cooksey to Barnes, 28 Jan 1937 (2/24), and to Lawrence, 29 Sep 1938 (4/21); Lawrence to Livingston, 21 Feb 1939 (12/12); Livingston, Holloway, and Baker, RSI, 10 (1939), 63, 65–6; Henderson, King, and Risser, PR, 55 (1939), 1110. A preliminary version of the paper by Livingston, Holloway, and Baker, signed by Livingston alone and probably dating from 1938, is in 12/12.

52. Livingston, "A capillary source," 7–8 (12/12); Livingston et al., RSI, 10 (1939), 63, 67.

53. McMillan and Salisbury, PR, 56 (1939), 836, letter of 2 Oct. The Berkeley design was often reprinted, e.g., in Livingston, RMP, 18 (1946), 293–9, and Fremlin and Gooden, Rep. prog. phys., 13 (1950), 258–9; cf. Mann, Rep. prog phys., 6 (1939), 131–2, and Livingston and Blewett, 84–6, 159–61.

54. J.A. Fleming to ARMCO, 30 Sep 1939 (MAT, 23/"cycl. letters").

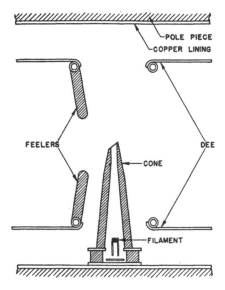

FIG. 6.4   The Laboratory's modification of Livingston's capillary source. McMillan and Salisbury, *PR, 56* (1939), 836.

The old dogs learned new tricks also in the design of the dee system. In the standard Berkeley arrangement, which most early cyclotron laboratories followed, the oscillator fed power to the dees by inductive coupling, in the manner of a transformer, and the system was tuned by adjusting the inductance of the coil on the secondary or dee side. Stiff copper rods supported the dees and sat on pyrex tubes carried by the chamber walls. These pyrex tubes were the insulators whose failure gave so much trouble with the 27-inch cyclotron. Lawrence and Cooksey realized that the stopgaps they employed to block stray cathode rays and to cool the glass could not be extended to much higher voltages. When necessary, they supposed, they might adapt Sloan's system of vacuum insulation, in which the secondary coil hangs from a point on its length where the voltage induced in it has a node. But they did not say how they would proceed.[55] Insulation was not the only

55. Lawrence and Cooksey, *PR, 50* (1936), 1134–5; Sloan, *PR, 47* (1935), 62–71. Examples of the system of pyrex insulators: Lawrence and Cooksey, *PR, 50* (1936), 1136; Henderson and White, *RSI, 9* (1938), 27; Nahmias, *Machines*, 48–9, and plate VI (Berkeley, Cambridge), 196–8.

problem in the old dee system. As we know, it transferred only about 50 percent of the power developed by the oscillator into the dee circuit and it returned an unwanted, destabilizing feedback.[56]

Transmission lines resolved or moderated these problems. Green and Kruger at Illinois improved stability and annihilated feedback by coupling a coaxial line between the oscillator and the dee circuit. The innovation had the additional advantage of distancing the bulky oscillator from the work area around the cyclotron. But it did not bring more power from the oscillator to the dees and did not relieve the stress on the insulators.[57] The solution, as developed by Dunning and Anderson at Columbia, was to introduce a second set of transmission lines and to make the dees part of them. The scheme may be clear from figure 6.5, in which the coaxial lines coupled to the oscillator as in the Illinois plan are in turn coupled inductively to straight conductors or stems that open out to become the dees. The stems, copper pipes a few inches in diameter, are fixed coaxially in much larger steel cylinders by movable metallic disks. Plate 6.1 shows the cylinders and the dees whose stems they carry, for the MIT cyclotron built by Livingston. Each pipe-and-cylinder pair constitutes a transmission line in which a standing electrical wave may be set up. (The whole may be likened to a transmission line with infinite load: the dees and the stems make up the central wire and the chamber walls and cylinders the outer sheath.) By a clever choice of dimensions and electrical parameters, the cyclotroneer can arrange that the voltage maximum comes at the dees and no voltage exists where the disk or "spider" grips the stem. The cylinder may then be maintained at ground potential. Evidently the distance from the dee tip to the voltage node must be approximately one-quarter of the wavelength at which the cyclotron resonates. Hence the designation of the pipe-and-cylinder as a "$\lambda/4$ transmission line."[58]

56. Nahmias, *Machines*, 50.

57. Kruger and Green, *PR, 51* (1937), 57–8 (Nov 1936); Henderson and White, *RSI, 9* (1938), 19–30.

58. Dunning and Anderson, *PR, 53* (15 Feb 1938), 334, presented at APS meeting, 28–30 Dec 1937; Nahmias, *Machines*, 51–2, 60; Livingston and Blewett, *Particle acc.*, 168–76.

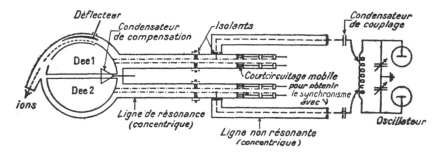

FIG. 6.5 Transmission-line coupling of the oscillator to the dee stems. The nonresonant line from the oscillator taps into the quarter-wave line ending in the dees. Nahmias, *Machines atomiques*, 53.

Dunning and Anderson recommended their double application of transmission-line technology for its greater efficiency, higher voltages, and higher frequencies; its elimination of insulators, shielding of cables, and ease of adjustment; in short, at the bottom line of cyclotroneering, for bringing a "considerable increase in energy and intensity of the ion beam." In the summer of 1938 the Laboratory's directing committee in conclave cyclotron.[59]

The astounding result—dimensions cyclotroneers could "hardly swallow"—appears in plates 6.2 and 6.3.[60] In plate 6.2 McMillan sits on the external surface of the left dee support while Alvarez sprawls on the upper pole of the magnet; the large segmented cylinders entering the dee supports are the untuned transmission lines coupling the oscillator to the dee system. In plate 6.3 Green stands with one hand on the mechanism that adjusts the position of the left dee stem while Cooksey looks on from a bower of tubes and cables. The untuned transmission lines now run into a metal housing resting on the external dee supports; their other ends enter a large cabinet on the mezzanine that contains the oscillator. The long black hose penetrating the enlarged cylindrical attachment to the dee system above Green's head provides the power for the beam deflector.

59. Meetings of 29 Jul and 24 Aug 1938 (24/13); Cooksey to Pollard, 19 Sep 1938 (14/30); cf. Lawrence to Paxton, 13 May 1938 (14/8).
60. Paxton to Lawrence, 12 June 1938 (14/18), quote.

The complexity of the transmission lines and dee supports, the tedious determination by calculation and trial of its dimensions and electrical constants, was perhaps the most difficult challenge for the builders of the Crocker cyclotron. In their success they had the satisfaction of outdoing Harvard, which required Salisbury's services after its radio frequency men had failed to couple their oscillator to their quarter-wave dee supports.[61] The dee system was but one of many headaches. There were the oscillator itself, Salisbury's larger and stabler version of the 37-inch system, a self-excited oscillator tapped into the dee stems through choke coils at nodes located by cut-and-try;[62] the vacuum pumps for the oscillator and the chamber, improved by Sloan to a capacity of 5,000 liters/second;[63] and 10,000 details of wiring, cooling, controlling, and protecting the machine.

A cyclotron is more than the sum of its parts. Although the major components of the 60-inch had proved themselves separately, a beam did not issue from the whole merely by turning on the switches. Beam hunting began at least as early as February 1939, when "most everybody in the laboratory...[was] working day and night on the medical cyclotron." All the planning promised an early success, and Lawrence arranged to announce the start up over CBS on April 15. But parasitics plagued operation. Once Salisbury had suppressed them, the transmission system showed itself unable to put sufficient power on the dees. No resonance had occurred by April 4. Cooksey cancelled the broadcast. Alterations in the lengths and positions of the lines improved their behavior and disclosed malfunctions in the deflector system and oscillator circuit. After four months of coaxing there was still no beam.[64]

61. Correspondence between R.D. Evans and Lawrence (7/8).

62. Salisbury to Livingston, 24 Feb 1940 (16/6), a letter of fifteen pages, esp. 9, 12–15, explaining how to find the nodes and itemizing the advantages of a self-excited design; Cooksey to Barnes, 7 June 1938 (2/24).

63. Cooksey to Tuve, 28 June 1938 (3/32), to Livingston, 29 Oct and 14 Nov 1938 (12/12), and to A.L. Elder, 9 Nov 1939 (25/5).

64. Lawrence to Whipple, 21 Jan 1939 (15/26A), quote; Cooksey to Lawrence, 29 Mar and 2 Apr 1939 (4/22), to Van Voorhis, 19 Apr 1939 (17/43), and to Snell, 10 May 1939 (16/33); Lawrence to Cooksey, 26 Apr and 2 May 1939 (4/22).

Toward the middle of May, a Geiger counter detected a rudimentary proton stream. With much shimming and trimming they made it to the target, a few μA strong, at 7.5 MeV, with only 10 kW on the oscillator. The transmission lines behaved beautifully, delivering 200 kV across the dees with an input of 80 kW. (Ultimately they helped the 60-inch to consume less power than the 37-inch for similar currents: for 70 μA of 8 MeV deuterons the smaller machine needed 45 kW to the magnet and 70 kW to the oscillator; for 80 μA of 16 MeV deuterons, the larger also needed 45 kW to the magnet but only 40 kW on the oscillator.)[65] Probes showed a milliamp of positive ions within the chamber. Lawrence was much gratified: "We will obtain really prodigious circulating currents," he boasted justifiably to Cockcroft.[66] In June the first deuterons arrived, at 19 MeV. The reactions of the proprietors were characteristic. Cooksey: "They stick out in the air 153 centimeters. Oh boy!!" Lawrence: "The only difficulty in the way of going on up to a hundred million volts is the financial one." The intent to build bigger surfaced in the announcement of the beam in the *Physical Review.*[67]

During July and August the new machine was tuned to higher currents: from 5 μA of 16 MeV of deuterons at the end of June to 15 μA on July 26, with a filament source; and from 50 μA a month later to 100 μA or more in early October, with a capillary. The neutron yield per μA was five times as great at 16 MeV than at the old maximum of 8 MeV.[68] During the first three weeks of September, a shield of water tanks was installed for protection against neutrons and bombardments began with 15.8 MeV deuterons. The local press was admitted and amazed. The staff, who

65. Lawrence to Weaver, 21 Feb 1940 (15/30).

66. Lawrence to Pollard, 16 May 1939 (14/30), to J.A. Fleming, 1 June 1939 (3/32), and to Cockcroft, 2 June 1939 (3/32); Lawrence et al., *PR, 56* (1939), 124, letter of 12 June.

67. Cooksey to Laslett, 12 June 1939 (10/33); Lawrence to Bush, 8 June 1929 (3/13), quote, and, same day, same words, to Weaver (15/29); Lawrence et al., *PR, 56* (1939), 124. The deuterons emerged on June 7 (Seaborg, *Jl., 1*, 495), following protons around June 1 (Lawrence to DuBridge, 1 June 1939 (15/26A)).

68. John Lawrence to Cockcroft, 29 June 1939 (4/5); Lawrence to Hopewood, 26 Jul 1939 (9/15), and to Hevesy, 4 Oct (9/7); Cooksey to Sagane, 29 Aug 1939 (9/39); Amaldi, *Viaggi, 6* (1940), 9; Lawrence, *Science, 90* (1939), 407–8, paper of 23 Oct.

had "all decided," according to Cooksey, to bring the crew system to the Crocker, organized to squash all remaining bugs and to build a treatment room before neutron therapy began. Altogether in its twelve weeks' running in, the new cyclotron operated for 155 hours (about half for physicists and half for biologists), stopped for construction for 300 hours (primarily for shielding and building the treatment room), and shut down for repairs and adjustments (mainly to the ion source, electrical controls, and oscillator) for 260 hours.[69]

Although the medical cyclotron had shown what it could do, it had not contrived to do it steadily by the end of the year. Around Thanksgiving, Kamen had to advise a supplicant for Fe^{59} that "the new machine still continues to be a creature of chance and in its present condition cannot be expected to run continuously for another month." Nor did it then. It did not work steadily enough to bake probes and could not be depended on for neutron therapy. It suffered from what its operators liked to call "morning sickness," little difficulties with obscure symptoms, like stuck relays and failed heaters, that neophytes could not diagnose. It also had major problems, like shorts in the magnet coils, that kept it in bed for weeks.[70]

From its resumption of service in mid February, it performed brilliantly, with "almost unbelievably large currents," as much as 200 μA, far more than the targets could take; it ran "with amazing smoothness and stability," seldom missing a scheduled bombardment.[71] As Lawrence wrote to J.S. Foster, who was trying to secure money to build a cyclotron at McGill, the consequence of this steady, high-energy, high-current bombardment was an output

69. Cooksey to Lawrence, 29 Mar 1939 (4/22); operating reports on 60-inch, 28 Aug–12 Nov 1939 (25/5).

70. Kamen to P.F. Hahn, Rochester, 25 Oct and 21 Nov 1939 (15/26A); Lawrence to Whipple, 30 Jan 1940 (15/27); Abelson to Evans, 19 Feb 1940 (7/8); Abelson, memo, 24 Jan 1940 (MAT, 25/"cycl."), on morning sickness. From 19 Sep 1939 to 30 Apr 1940, the 60-inch cyclotron required 3,665 hours of shop time, the 37-inch 285 (25/6).

71. Quotes from Lawrence to DuBridge, 14 Mar 1940 (6/17), to Whipple, 20 Feb 1940 (15/27), and to Kruger, 14 Mar 1940 (10/20), resp.; Lawrence to Weaver, 12 Mar 1940 (15/30). By July the 60-inch could give 305 μA; Lawrence had to quell rumors that he had half a milliamp (Lawrence to Pollard, 2 July 1940 (14/30)).

of neutrons and radioactive materials "no less than prodigious."[72] The figures in table 6.2 will assist in the estimation of the prodigy.

The intensity of the source increases with the duration of the bombardment until the specimen has been irradiated for a time approximately equal to its period of decay. The figures in the last two columns of table 6.2 usually represent a "day's" bombardment, which might mean anything from eight to twenty-four hours. (Table 6.3 gives more precise yields, in $\mu Ci/\mu Ah$, for the 60-inch in 1941.) The expense of running the machines was reckoned as follows. The direct cost of operation of the 60-inch averaged out to a little over 10 cents a microamp hour. Now the total capital cost of the 60-inch cyclotron through September 1939 was $63,500; of the labor by then lavished on it, $20,750; and of the Crocker building, $75,000. We have in all about $160,000.[73] Depreciating this investment over eight years, as the Laboratory did, and reckoning 2,400 hours of running time each year, we arrive at an amortization of $8.33 an hour. The costs—power, supplies, target preparation—for an hour's operation at 100 μA was something over 100 times 10 cents, say $10.50. Maintenance and repairs added another $6 an hour. The real cost of neutron therapy or isotope manufacture at the 60-inch therefore came to about $25.00 an hour, some sixty times the salaries of its attendants.[74]

2. BREEDING AT HOME

From Physics to Physic

The multiplication of cyclotrons in the United States began in 1934, with Livingston's little machine at Cornell (plate 6.4). No fewer than eight were commissioned the following year at Bartol,

72. Lawrence to Foster, 26 Mar 1940 (7/18); Thomas, *HSPS, 14:2* (1984), 363–7.

73. Brobeck to Cooksey, 27 Dec 1940; orders prior to 1 July 1939, and Cooksey to F.A. Denhard, 22 Sep 1939 (25/5).

74. "Operating costs," 14 Dec 1940 (22/3). Recalculation altered various costs but kept the total, $25 an hr, the same; ibid., and "Operating costs," 19 Dec 1940 (22/3).

Table 6.2

Representative Yields from Berkeley Cyclotrons

	d energy (MeV)	d current (μA)	n flux/sec ($\cdot 10^{-8}$) (from d on Be)	Ra-Be equiv. (kg)	Ra equiv. Na^{24} (mg)	γ equiv. Na^{24} (mC)
27" cyclotron						
May–June 34[a]	2.6	~1	5	0.02	>0.2	
Jul–Sep 35[b]	3.0	10	100	0.4	50	
Aug–Sep 36[c]	5.0	20			200	
June–Aug 37[d]	5.5	25	25000	100		1
37" cyclotron						
Apr–Jul 38[e]	8.0	50	>25000	>100	500	~16
Oct 39–Feb 40[f]	8.5	100			1000	10
60" cyclotron						
Nov 39–Feb 40[g]	16.0	100	>2500	>100		~80

a. Lawrence to Cooksey, 21 May 34 (4/19), and to Oliphant, 5 June 34 (14/6); Livingston et al., NAS, *Proc.*, *20* (1934), 470–5; Lawrence, *PR*, *47* (1935), 17.

b. Lawrence to Exner, 17 Jul 35 (9/21), and to Cockcroft, 4 Jul 35 (4/5); Rasetti, *Viaggi*, 77–8.

c. Lawrence to Randall, 17 Jul 3 (13/5).

d. Lawrence to Cockcroft, 16 Apr 37 (4/5); Kurie, *GE review*, *40* (June 1937), 2⁴0; Cooksey to Kamen, 12 Jul 37 (10/10); Lawrence to Paxton, 4 Aug 37 (14/18); Livingood to McMillan, 5 Aug 37 (12/11). The values of flux and Rn-Be equivalent were confirmed by Fermi.

e. Lawrence to Chadwick, 30 Apr 38 (3/34); Cooksey to Laslett, 10 Feb 38 (10/32).

f. Gentner, *Nwn*, *28* (1940), 394.

g. Birge, *The 1939 prize*, 21; Lawrence to Weaver, 21 Feb 40 (15/30).

Table 6.3

Yield in $\mu Ci/\mu Ah$ at the 60-inch

Element	Half-life	Yield
H^3	20 y	0.1
C^{14}	> 1000 y	$5 \cdot 10^{-5}$
Na^{24}	14.8 h	10^4
P^{32}	14.3 d	200
Fe^{59}	47 d	0.03
I^{131}	8.0 d	20

Source: Hamilton, *Jl. appl. phys.*, 12
(June 1941), 459.

Columbia, Illinois, Michigan, Princeton, Purdue, Rochester, and Washington. They came in two almost standard sizes: poles around 16 inches in diameter (Cornell, Illinois, Rochester, Washington) and poles around 35 inches in diameter (Bartol, Columbia, Princeton, Purdue). Michigan exceeded them all, and Berkeley too, before the 60-inch, with poles of 42 inches, became the standard for the cyclotrons of the later 1930s (table 6.4). As we know, Lawrence cooperated fully with the builders of these machines, most of whom were his students and associates. Neither he nor the Research Corporation had any interest in restricting the spread of their patented technology.

These first American cyclotrons were made for physics. An indication of their orientation is that Livingston designed the Cornell cyclotron for acceleration of protons and deuterons without giving priority to either; he had separate chambers available for each from the time his cyclotron began to operate in 1935. Although deuterons, the particles of preference at Berkeley, are particularly effective in making radioisotopes and generating neutron beams,[75] protons are better for close examination and

75. Lawrence to Barton, 2 Aug 1935 (2/25): "We are experimenting with deu-

Table 6.4

U.S. Cyclotrons by Year of Commissioning
and Size

	Machine(s) (location)	Pole size(s) (inches)
1934	Cornell	16
1935	Illinois-1, Rochester, Washington	16, 20, 13
1935	Bartol, Columbia, Princeton, Purdue	38, 35, 35, 37
1935	Michigan	42
1936	Chicago-1	32
1937	Yale	27
1937	Harvard, Ohio State	42, 42
1938	Illinois-2, Indiana, MIT, Chicago-2	42, 45, 42, 45
1939	Stanford	27
1939	Pittsburgh, Saint Louis	44, 42
1939	Carnegie	60

interpretation of nuclear reactions and forces. Moreover deuterons, which have a lower resonant frequency in the cyclotron than protons, can be accelerated to relatively higher energies with relatively less bother about the always painful oscillating circuit. As medical and biological applications of cyclotrons became paramount in acquiring financial support for their construction, parameters were set to favor the production of deuterons. Most of the large cyclotrons in operation in 1940 were not suitable for the acceleration of protons. In this circumstance Hans Bethe saw cause for regret and even alarm.[76] But with his colleague Living-

terons exclusively;" and to R.P. Jackson (Westinghouse), 22 Oct 1935 (18/19): "deuterons of energy above 3 MeV will certainly be used [preferentially] for the manufacture of radioactive substances."

76. Bethe to C.G. Suits (General Electric), 21 Jul 1945 (HAB, 20). Cf. J.A. Fleming to McMillan, 30 Oct 1939, and reply, 6 Nov 1939 (3/32); Gentner, *Ergebnisse, 19* (1940), 159.

ston's cyclotron, which delivered 3 or 4 μA of 2.0 MeV protons (or 1.35 MeV deuterons), Bethe could test the predictions of yields of nuclear reactions in light elements computed by himself and his students on the basis of Bohr's theory of nuclear structure.

The only applications envisaged by Livingston in his first description of the Cornell cyclotron concerned nuclear problems, viz., the study of "all types of reactions" caused by "all available bombarding particles." At first Cornell used deuterons in the reaction Be(d,α)Li in order to infer the states of the beryllium nucleus from the energies of its ejected alpha particles; they subsequently exploited the reverse transformation Li(d,n)Be for neutrons for scattering experiments; and they ended, in the late 1930s, with careful measurements of deuteron-induced disintegrations of nitrogen, carbon, and oxygen.[77] Their allegiance to the original purpose of their machine was secured by its size: it could not make radioisotopes in sufficient quantities for biomedical studies and foundation support.[78] The same things are true—enforced devotion to physics—of the Illinois cyclotron. Like its twin at Cornell, it was used to generate secondary beams that themselves became the instrument or object of study, for example, gamma rays and neutrons.[79] That was not enough for cyclotroneers, however, and, as we know, both Green and Kruger at Illinois and Livingston at Cornell made important improvements in the equipment of cyclotrons.[80]

The first of the larger cyclotrons to be completed outside Berkeley was Princeton's. It was more powerful (several μA of 10 MeV deuterons) and more costly ($12,000 inclusive of motor generator) than Cornell's. It was adaptable to isotope production. Henderson and White emphasized magnet design and explicitly preferred deuterons; their machine, although financed by the university and

77. Livingston, *RSI*, *7* (1936), 57–8, 67, quote. On Be excitation, Gibbs and Livingston, "Report of progress...with the support of a grant of $3,000" (RC); on neutron scattering, Bacher and Swanson, *PR, 53* (1938), 676, 922, and Bacher, *PR, 55* (1939), 679–80; on disintegrations, Holloway, *PR, 57* (1939), 347–8, and Holloway and Moore, *PR, 57* (1940), 1086, and *PR, 58* (1940), 847–60.

78. Gibbs to Poillon, 30 June 1938 (RC).

79. Kruger and Green, *PR, 51* (1937), 57, 699–705; dissertations by Green on the cyclotron and gamma rays (1937) and by R.Z. Watson on the doings of neutrons from the d-d reaction (1938).

80. Supra, §6.1; Kruger, Green, and Stallman, *PR, 51* (1937), 291.

not by outside medical interests and dedicated, according to their first description of it, to nuclear physics, proved a useful manufacturer of radioelements, to which it came to devote much of its time. Like Cornell's cyclotron, Princeton's began to run within a year or so of its commissioning.[81]

Rochester's cyclotron also took but a year to build. Its promoter, Lee DuBridge, "one of the earliest to see a great future for the cyclotron," had been a National Research Fellow at Berkeley, though not in the Radiation Laboratory. He built with medical support and the help of industry and the Research Corporation.[82] His 20-inch machine was dedicated at first to proton work in the unexplored region above the reach of high-tension apparatus and the Cornell and Illinois cyclotrons. Working at 2 or 3 μA, he and his colleagues reaped a good harvest of the known reactions—p capture and (p,α) disintegrations—in the "fast" proton range of 1–4 MeV, and they discovered a new process, (p,n), which sets in at about 3 MeV. With this success, they decided to enlarge their magnet's poles to 26 inches in the expectation of having protons of over 7 MeV. They reached 7.2 MeV and followed (p,n) reactions in at least thirty-five elements. As their experimental work continued Cornell's, so their theorist, Victor Weisskopf, continued Bethe's. As at Cornell, theory went hand in hand with experiment.[83]

The enlargement of the poles, which occurred in 1937, made possible the production of 4.5 MeV deuterons. They sufficed to make radioisotopes in sufficient quantities for tracer work by physicians and biologists at Rochester, and for an investment of

81. Lawrence to Henderson, 3 May 1935 (9/6), and to Chadwick, 31 Jan 1936 (3/34); Henderson and White, *RSI, 9* (Jan 1938), 19, 20, 28, 30; White to Lawrence, 22 Aug and 22 Oct 1936 (18/21), and 22 Sep 1936 (3/33); Henderson to Lawrence, 17 Dec 1934, 31 Mar and 26 Oct 1936 (9/6). Lawrence later estimated the costs with improvements at around $15,000; Lawrence to Hughes, 26 May 1938 (18/12).

82. Lawrence, "Cycl. inst," 3, quote; DuBridge to Poillon, 9 Jul 1935 (RC); DuBridge and Barnes, *PR, 49* (1 June 1936), 865; Barnes to Cooksey, 19 Aug 1936 (2/24).

83. DuBridge to Lawrence, 21 Dec 1936 (15/26); DuBridge and Barnes, *PR, 49* (1936), 865; Barnes, DuBridge, et al., *PR, 51* (1937), 775; DuBridge to Barnes, 25 Jul 1938 (6/17); DuBridge, "Nuclear physics research program," 21 Feb 1940 (RF, 1.2/200/161/1978); Weisskopf, *Selected essays,* 13–4.

$30,000 by the Rockefeller Foundation, to be split about equally between the physicists and the doctors. The physicists used their share to attract Van Voorhis, who later rearranged the radio frequency system to double the power reaching the dees. The proton bombardments and the physics program retained the ascendency at Rochester, and only one day a month was devoted to deuteron irradiation of materials for biomedical experiments.[84]

The balance fell out quite differently for cyclotrons of greater power than Rochester's. Columbia and Michigan mark the transition. Columbia began in February 1935, asking Lawrence's opinion about the conversion of a general purpose 14-inch magnet. He advised a magnet and machine of the type—and hence for the purposes—that Livingston was then pursuing at Cornell. An opportunity then presented itself to build bigger, in the form of the navy's 500-kW Poulsen arc, not Berkeley's twin, but big enough to support pole pieces 36 inches in diameter, which NYU decided it could not use. That was in the summer of 1935.[85] By the time the magnet with its 36-inch poles was in place at the end of 1936, the biomedical economics of cyclotrons had been discovered; Columbia University, in consideration of medical applications, put up money for the completion of the machine and the chairman of the physics department and a colleague from chemistry went to the Research Corporation for more, "to produce very usable quantities of artificial radioactive substances." They got what they wanted, some $1,850.[86] By the winter of 1938/39, having shimmed away the bad asymmetry of the poles and perfected the quarter-wave transmission line, the Columbia group, which by then included Paxton, was ready to make P^{32} and Na^{24} for biological tracing. Eventually they had a neutron flux equivalent to that from 50 kg of Rn-Be and could detect on the

84. DuBridge, "Nuclear physics" (RF); DuBridge to Gregg, 21 Aug 1940, and Warren to Gregg, 23 June 1939 (RF, 1.1/200/161/1977–8); DuBridge to Lawrence, 4 Apr and 18 May 1938 (15/26); Barnes to Cooksey, 25 Feb 1938 and n.d. (2/24); Kamen to Hahn, 12 Sep 1940 (15/27).

85. Pegram to Lawrence, 1 Feb and 2 Oct 1935 (4/8); Lawrence to Pegram, 6 Feb 1935 (4/8), and to Livingston, 22 Jul 1935 (12/12); supra §3.4.

86. Pegram to Cooksey, 10 Mar 1936 (4/8); Sagane to Lawrence, 14 Nov 1936 (3/39); Pegram and L.P. Hammett to Poillon, 20 Jan 1937, and "Report on chemical and biological investigations with radioactive indicators," 1937 (RC).

roof of their fourteen-story physics building neutrons that had passed through their colleagues on the way up from the cyclotron in the basement. "The cyclotron really operates very beautifully now."[87] Between the decision to make a cyclotron, taken in February 1935, and its completion in 1938, the purpose the machine was to serve, the basis of its financing, and the magnitude of its hazards had been transformed.

The Michigan cyclotron was very well supported from the beginning—from the summer of 1935—by the university's Rackham funds. The chairman of the physics department, H.M. Randall, and the machine's main builder, James Cork, who had spent 1935/36 on sabbatical in Berkeley, apparently had indicated to the university's medical school, whose collaboration they obtained, that the large cyclotron they planned would yield radioisotopes for clinical use. That was more than Berkeley had by then achieved, and it was doubtless with relief that Randall learned from Lawrence in the summer of 1936 that the 27-inch could make samples of Na^{24} equivalent to 50 mg of radium. "You can proceed with the construction of a cyclotron with the definite assurance that you will be able to produce enough radiosodium and other radioactive substances for medical investigations."[88] The Michigan cyclotron put out its first beam (several μA of 5 MeV deuterons) in August 1936, about a year after the physics department had assigned it first priority. Lawrence, who happened to be in town at start up, "got a great [and well-deserved] kick out of seeing the second large cyclotron in the world go into operation." A new tank installed with Thornton's help in the winter of 1937/38 gave an intense beam that reached to 200 μA in the summer of 1939. That more than met the debt to the medical people, who had been satisfied with 15 μA.[89]

87. Pegram to Joliot, 23 June 1938 (JP, F28); Cooksey to Laslett, 18 Aug 1938 (10/32); Paxton to Joliot, 14 Nov 1938 (JP, F28); Darrow to Lawrence, 27 Jan 1939 (6/9); Dunning to Tuve, 29 Sep 1939 (MAT, 23/"cycl. letters"), quote; Pegram, Hammett, and Dunning, "Combined report on cyclotron and connected researches," 1939 (RC); Amaldi, *Viaggi, 6* (1940), 9.

88. Randall to Lawrence, 16 Aug 1935 (12/30); Lawrence to M. White, 18 Aug 1936 (18/21); Lawrence, "The work," 7–8; Lawrence to Randall, 17 Jul 1935 (13/5), quote.

89. Lawrence to Poillon, 12 Aug 1936 (15/17), quote, and to Cork, 14 Aug and 1 Sep 1936, and Cork to Lawrence, [Aug 1936] (5/1); Sagane to Lawrence, 14 Nov

Apart from the 27-inch cyclotrons begun at Yale in 1937 and at Stanford in 1940, built on the cheap for experiments in physics,[90] all the machines commissioned from 1936 on were at least as large as Michigan's and, like it, dedicated largely to biomedical work. Physics became a secondary or tertiary goal of cyclotron builders, according to their reviews of their discipline. Kurie, Chadwick, and Mann, for example, gave pride of place to radiobiology, chemical tracing, and clinical treatments in their accounts of uses of the cyclotron; and Mann, writing in *Nature* in April 1939, offered Kamen and Wilson's internal target, which multiplied the harvest of therapeutically relevant isotopes, as "the most important recent" improvement in the cyclotron art and a grand "stimulus to those who are in favour of the immediate construction of a cyclotron for medical purposes in Great Britain."[91] Lawrence was also a stimulus in this cause. His recommendation to the British that their medical cyclotron have 50-inch poles ("[the 60-inch] is probably larger than necessary"), might suggest that machines of the Michigan class were not big enough to practice medicine. It appears that 50 inches was an arithmetical compromise between the excessive 60-inch and the adequate Michigan type: as Lawrence wrote Livingood, "a 40-inch cyclotron really fills the bill for most medical purposes." Or, to put the point backWard, since the 37-inch was adequate for therapy and isotope manufacture, a 50-inch would do.[92]

1936 (9/39); F.J. Hodges to Lawrence, 3 Oct 1938 (9/9); Thornton to Cooksey, 13 Dec 1937 (13/5); Cork to Lawrence, 11 Aug 1939 (5/1).

90. By using hand-me-down parts and making most of the apparatus in the laboratory (including the magnet at Yale), the cost of these two instruments was kept to around $5000 each exclusive of labor. The Yale cyclotron made gamma rays for study via proton bombardment; the Stanford cyclotron made neutrons for study via deuteron bombardment. Pollard to Cooksey, 28 June, 22 Aug, 22 Nov, and 7 Dec 1937, 29 Mar 1939 (14/30), and to Lawrence, 20 May 1937 (4/21); Cooksey to J.A. Gray, 24 Feb 1938 (14/38); Lawrence to F.W. Loomis, 21 May 1938 (9/19), and to Hughes, 26 May 1938 (18/12); Bloch, Hamermesh, and Staub, *PR, 64* (1943), 49; Staub in Chodorow et al., *Bloch,* 196–7.

91. Kurie, *GE review, 40* (1937), 70–2, and *Jl. appl. phys., 9* (1938), 699–701; Chadwick, *Nature, 142* (8 Oct 1938), 632–3; Mann, *Nature, 143* (1939), 583, and *Rep. prog. phys., 6* (1939), 132–3.

92. Lawrence to Frank L. Hopwood, St Bartholomew's Hospital, London, 24 Aug 1938 (9/15); Lawrence to Livingood, 20 Jul 1939 (12/11), and to Hughes, 5 Jul 1939 (18/12).

Lawrence's optimum medical cyclotron, although costly, came to considerably less than the Crocker machine. Late in 1938 someone at the Laboratory took the trouble to estimate the differences in price: for the magnet, metal and installation, $40,000 for the 60-inch, $18,000 for the 50-inch; for the remaining parts, $40,000 versus $32,000; for labor, $8,000 and $6,000. Since something could be saved by making some parts in departmental shops and by gaining concessions from manufacturers, Lawrence quoted $50,000 in all as the ante "to do it right from the standpoint of clinical work." But if only $20,000 were available, it might just do.[93]

The largest cyclotron started in the United States after the Crocker was almost its identical twin, erected by, of all people, Lawrence's again friendly rival Merle Tuve. The Carnegie Institution of Washington, with its high investment in high tension, could scarcely remain aloof from the new forces and opportunities in high-energy physics. "If you want an equipment for producing artificial activity [as opposed to one for exact physics], there is nothing that even faintly begins to compare with a cyclotron, nor has there ever been." So Tuve answered an enquiry from Bell Labs in the spring of 1940. By then he had been cyclotroneering for a year. During the spring of 1939, Brobeck, Cooksey, and John Lawrence joined interested parties in the Washington area to discuss the desirability of building a big cyclotron at Carnegie for service to local institutions. The consensus held that Carnegie could do no better than to build a 60-inch machine. As the director of Tuve's unit explained to the president of the Carnegie Institution, the machine could do physics for the Department of Terrestrial Magnetism and chemistry and biology for other units of the institution, local universities (Johns Hopkins, George Washington, Catholic) and the federal government (the Department of Agriculture, the National Cancer Institute).[94] It would not practice medicine. "Medical research will be given no preference whatsoever and, in fact, will be undertaken only in the sense of

93. Memo of 1 Nov 1938 (25/5); Lawrence to Hughes, 26 May 1938 (18/12).
94. Tuve to W. Schockley, 13 May 1940 (MAT, 19/"Tuve letters"), quote, and report, 16 May 1939 (MAT, 21/"extra copies"); J.A. Fleming to V. Bush, 6 Sep 1940 (MAT, 25/"biophys.").

fundamental work in physiology and biochemistry." Physics would have a third of the machine time.[95] The Carnegie philanthropies had their own resources and freedom to determine their own programs.

Once commissioned, the Carnegie cyclotron came into existence in the manner then standard. Tuve and his main associate, Richard Roberts, received an "enormous roll of blue prints" from Berkeley, "almost lost our mind[s]" reading through them, learned that a visit to Berkeley and Berkeley cyclotroneers was indispensable, hired two (Green and Abelson), and started shopping for steel.[96] They differed from Berkeley in contracting with ARMCO for the magnet ($16,025), part of which was cast (as at Chicago, Columbia (pole tips only), Harvard, MIT, and Ohio State, among others), and with GE for water-cooled coils to run it (as at Columbia, Harvard, MIT, and Rochester). Nonetheless, the whole would resemble the Crocker magnet very closely, in weight as well as in shape, the coils like "the turret[s] of a battle ship."[97] The analogy was not idle. Like all other cyclotron builders in 1939 and 1940, the Carnegie Institution faced the free-market competition of an arms race. "War orders have made loads of extra work in getting contracts that are within starting distances of estimates based on Berkeley costs."[98]

Cyclorama

There were twenty-two cyclotrons completed or under construction in the United States in 1940. They came in four sizes, as indicated in table 6.5, which also contains information about their coming-to-be. As the table shows, the quantity of steel and copper in the magnet increases much more rapidly than the diameter of its poles, somewhere between the cube and the square. Since the

95. Fleming to Cork, 22 Nov 1939 (MAT, 23/"cycl. letters").
96. Tuve to Lawrence, 17 June 1939, quotes, and Lawrence to Fleming, 1 June 1939 (3/32); Tuve to Arthur Hemmingdinger, 20 Nov 1939, and Cooksey to Tuve, 10 June 1939 (MAT, 24/"cycl. reports").
97. Fleming to C.C. Clark, ARMCO, 5 and 30 Sep 1939, Tuve to Dunning, 5 Oct 1939, and Fleming to Lawrence, 18 Oct 1939 (MAT, 23/"cycl. letters"); Fleming to Robert B. Nation, International Nickel, 26 Feb 1940 (MAT, 25/"cycl. letters"); A.B. Hendricks, Jr., GE, to R.W. Hickmann, 1 May 1940 (UAV, 691.60/3).
98. Tuve to DuBridge, 17 Feb 1940 (MAT, 19/"Tuve letters").

magnet's metal was the single largest expense in the construction of a cyclotron, its unit cost became a matter of concern as prices rose owing to inflation in the worldwide buildup of arms. In 1935 mild steel boiler plate, which Cook and Henderson regarded as almost the equivalent of ARMCO iron, cost 1.6 cents a pound. With iron so cheap, Lawrence wrote, a good magnet could be made from scratch more cheaply than refurbishing an old navy arc, and he urged DuBridge to build bigger—the Rochester cyclotron initially was to have 14-inch poles—since the raw ingredients came for a song.[99] In 1937 the prices of steel and copper were about twice what they had been in 1935, and in 1938 they increased another 150 percent. The implicit and symbolic competition between cyclotrons and armaments for strategic materials in the late 1930s became explicit and realistic in the early 1940s, when several incomplete machines secured high-priority allocations of iron and copper in the national interest.[100] Only medicine could cope with the rapidly inflating capital requirements of cyclotroneers.

The second set of information in table 6.5 reveals that, with a few exceptions easily explained away, a baby cyclotron took about a year to build, a small cyclotron perhaps a year and a half, a medium one two years or a little more. The exceptions: Washington, a three-year birth in the baby class, was almost entirely the work of one man; Yale, two years acoming in the small class, also had a very small crew and assembled its own magnet; in the middle class, Columbia was delayed by navy bureaucracy and by its own innovativeness, while Purdue suffered from lack of resources, and both had the disadvantage of doing almost entirely without a man from Berkeley.[101] But a physicist or two who knew what they were about, did not hanker after novelties, and had the help of a graduate student, a competent shop, and enough money for their project, could bring a 90-ton cyclotron from the drawing boards to first beam in two years or less. The record of Berkeley men in partibus may be read from table 6.5.

99. Lawrence to Tuve, 12 Sep 1935 (3/32), to Livingston, 13 Aug 1935 (12/12), and to DuBridge, 12 Sep and 16 Oct 1935, and reply, 20 Sep 1935 (15/26).

100. Lawrence to Hughes, 26 May 1938 (18/12); Livingston to Cooksey, 28 Jul 1938 (12/12); U.S. Steel to Lawrence, 2 Sep 1938 (14/38); J.A. Gray to Lawrence, 21 Nov 1938 (14/38).

101. John D. Howe to Cooksey, 6 Jul 1937 (25/1), re Purdue.

Table 6.5

U.S. Cyclotrons by Size, 1940

Size[a]	poles (inch)	Magnet Fe (ton)	Cu (ton)	Commission dates Plan	Magnet	Beam	Builders
Baby (1-2 MeV)							
Cornell	16	3.5	0.5	F 34		Jul 35	Livingston[c]
Illinois-1	16	3.5	0.5	Feb 35	Sep 35	Jul 36	Kruger,[b] Green
Washington	13			W 35/6		May 38	Loughbridge
Small (3-7 MeV)							
Rochester	20	15	2	W 35/6	Apr 36	Aug 36	DuBridge,[b] Barnes
Rochester	27					Feb 38	DuBridge,[b] Barnes, Van Voorhis[b]
Stanford	27				W 39/40	F41	Bloch, Staub
Yale	27	17	2	Jun 37	Aug 37	May 39	Pollard
Medium (8-12 MeV)							
Bartol	38	62	10	W 35	Aug 36	Jan 38	Allen
Berkeley	37						Cooksey et al.
Chicago	41	60	10.5	Sp 36	Mar 37	Nov 38	Newson;[b] Snell[b]
Columbia	35	65	7	Feb 35	Sp 36	Aug 38	Dunning, Anderson, Paxton[c]
Harvard	42	70	16	Sp 34	Nov 37	Oct 39	Hickman, Evans, Livingood[b]
Illinois-2	42			W 38/9			Kruger,[b] Lyman,[c] Richardson[c]
Indiana	45	70	10	F 38	May 39	Sp41	Kurie,[b] Laslett[c]
MIT	42	70	16	Sp 38	Feb 39	Sp40	Livingston[c]
Michigan	42	80	15	Aug 35	Mar 36	Aug 36	Cork,[b]Thornton[b]
Ohio State	42			Dec 37	Jun 38		Smith, Pool
Pittsburgh	47			F 39			Allen, Simmons[c]
Princeton	35	40	8	F 35	Mar 36	Oct 36	White,[c] Henderson[b]
Purdue	37	48	8	F 35	Nov 36	Jan 39	W.J. Henderson
Saint Louis	42			F 39			Thornton,[b] Langsdorf[b]
Large (16 MeV)							
Berkeley	60	196	22	Sp 36		Jun 39	Brobeck, Cooksey, et al.
Stanford	27				W 39/40	F41	Bloch, Staub
Carnegie	60	~11	~11	Sp 39	Sep 39	1944	Tuve, Roberts, Green,[b] Abelson[b]

Note: F = fall, W = winter, Sp = spring.
a. Livingston, "Cost estimates," ca. 1 Nov 1940 (12/12), established the size categories.
b. Postdoctoral experience at the Laboratory.
c. Berkeley Ph.D.
d. F = fall, W = winter, Sp = spring

How many cyclotrons did the United States require? Of what sizes and capacities? In May 1938 Lawrence had an easy recipe: "There should be a cyclotron laboratory in every university center which will provide ammunition for unending work in nuclear physics and biology as well as in clinical medicine." Cooksey explained the situation to Karl Darrow, an industrial physicist and physics popularizer, who calculated that "the country may need a thousand cyclotrons." In October 1940 Urey judged that the country had as many as it needed, or could afford. "There are cyclotrons and Van de Graaff machines in most of the universities of the United States....Some institutes have one or two of each."[102] The natural limit to their reproduction might well have been sighted. Their costs were increasing as the square or cube of their size and their number perhaps linearly in time, while the national funding base showed no prospect of rapid enlargement. Indeed, there were signs of retrenchment: the National Advisory Cancer Council declined to distribute $100,000 for the capital improvement of cyclotrons, as Lawrence counselled in 1938 and 1939; the Research Corporation, with a reduced income, cut off support to leading cyclotron laboratories like Rochester in 1939 and 1940; the Rockefeller Foundation turned down Harvard and looked askance at all applications in support of new cyclotrons that did not go beyond the reach of Berkeley's 60-inch. The sovereign remedy for weak finances—"getting the phosphorus up in millicuries will bring you the [needed] backing, and support"—had, by repetition, made foundations resistant.[103]

In this Malthusian situation, the American Institute of Physics and the cyclotroneers at Harvard and MIT called a conference on applied nuclear physics that met in Cambridge from October 28 to November 2, 1940. For it Livingston drew up the chart reproduced as table 6.6. Apart from the baby cyclotrons, installation costs, including labor, did increase with at least the square of the pole size, whereas the operating expense of small and medium

102. Lawrence to A.L. Hughes, 26 May 1938 (18/12); Cooksey to Lawrence, re Darrow, 29 Apr 1938 (4/21); Urey to Hevesy, 25 Oct 1940 (Urey P, 2).
103. For NACC and RF, supra, §5.1; for RC's cutback, DuBridge to Lawrence, 25 May 1939 (15/26A); Cooksey to Snell, 10 May 1939 (16/33).

Table 6.6

Livingston's Classification of Cyclotrons, 1940

Class	Large	Medium	Small	Baby
Deuteron energy	16 MeV	8–12 MeV	3–7 MeV	1–2 MeV
Typical installation	U. Calif.	Harvard/MIT	Rochester	Cornell
Size (pole diam.)	60 inches	42 inches	27 inches	16 inches
Operations Crew	15	5	7	2
Energy (deuterons) MeV	16	11.5	4.5	1.4
Installation cost	$182,000	$60,000	$25,000	$2,500
Installation time (yrs)	2	2	1.5	1
Operating costs[a]	$60,500	$23,400	$20,000	$3,000
Running time (hr/d, hr/yr)	7/2400	7/2400	7/2400	3/1000
Operating costs[a]/hr	$25.20	$9.75	$8.30	$3.00
Beam (μA)	200	20/100[b]	2/50[b]	40
μA hr/day	1400	140/700[b]	14/350[b]	120
Operating costs[a]/μA hr	$.125	$.487/$.0975[b]	$4.15/$.166[b]	$.075
μA hr/mC of P^{32}	5	10	~ 100	
Costs[a]/mC of P^{32}	$.625	$4.87/$.975[b]	$415/$16.60[b]	
Neutron equiv (kg Rn-Be)	1,200	60/300[b]		~ .25
Neutron equiv/ μA Rn-Be	6	3		~ .006
Flux @ 100 cm (n units/m)	7.5	.38/1.9[b]		.0032
Costs[a]/n unit	$.056	$.44/$.082[b]	$15.60	
Flux @ 100 cm (r units/m)	30	3/15[b]		
Therapy costs/100 r units	$1.40	$10.80/$2.16[b]		

Note: Prepared for "Conference on Applied Nuclear Physics," MIT, Oct 1940 (12/12).
a. Includes amortization and overhead.
b. Numbers left of the slash indicate current, those right of the slash projected performance.

cyclotrons came to about the same.[104] The last represented a true gain in efficiency: the typical medium machine needed less tending than the smaller ones and consumed no more power. For a long time the price of power was the most worrisome part of

104. Installation costs include replacement value of apparatus, cost of space, technical services, and directors' salaries; operating expenses include direct costs, salaries of crew, 4 percent interest on the installation price, 20 percent obsolescence for the apparatus, 10 percent depreciation of the building. Direct costs for the larger machines fell out considerably under Livingston's estimate for installation, e.g., $35,000 for the 42-inch cyclotron at Saint Louis (18/12).

Lawrence's budget: the 27-inch ate up around $1,500 a year in 1933/34 and 1934/35 and almost twice that in 1935/36; after its enlargement to 37 inches, it required as much as 50 kW for the magnet, and (for 100 μA of deuterons) around 40 kW for the oscillator, which, at the Laboratory's cost of 2 cents/kWh and at Livingston's figure of 2,400 operating hours a year, amounted to almost $4,500 per annum.[105] The Harvard and MIT cyclotrons—made identical in size to "short circuit [covetousness]"—required only $1,500 a year for power, about a third that of the 37-inch. Their more compact magnets of ARMCO iron, transmission lines, and lower unit costs (around 1.5 cents/kWh) compassed the reduction.[106]

The most significant figures in Livingston's table concern output. From the beam currents (numbers to the right of the slash are Livingston's estimates of the expected eventual performance of the Harvard and MIT machines), the assumed operation of seven hours a day, and the operating costs, the unit price of the common tracer P^{32} is readily deduced. The advantage of the medium machines leaps to the eye: when in stride, the Cambridge 42-inchers would produce P^{32} at about $1/mCi, very much cheaper than the Rochester cyclotron could do and not much more than the cost at the Crocker. (These figures must be taken as approximations; from operational data on the Berkeley machines, the Carnegie Institution deduced that a mCi would cost $6.50 at the 37-inch cyclotron and $2.25 at the 60-inch.)[107] A similar story emerges from Livingston's figures on neutron production. From the neutron intensity in equivalents of Rn-Be follows the neutron effect in "n units" per minute at one meter from the cyclotron's

105. Crew size: Lawrence to Hughes, 26 May 1938 (18/12), minimum of three; proposed budget of Harvard cyclotron laboratory, 24 May 1939 (UAV, 691.60/3), crew of five. Power costs: unsigned memo, 3 Dec 1934, and Leuschner to Lawrence, 16 Jan 1935 (25/1); Time, 30:2 (1 Nov 1937), 40 ($1.50/hr); Lawrence to J.A. Gray, 14 Feb 1938 (14/38), and to Oliphant, 2 Aug 1938 (14/6); "Operating costs," 14 Dec 1940 (22/3).

106. Evans to Cooksey and Lawrence, 28 May 1938 (12/40), quote; Livingston, Buck, and Evans, PR, 55 (1939), 1110; budget of 24 May 1939, attached to Hickman to Evans, 23 Jan 1940, and R.B. Johnson to Hickman, 19 Nov 1940 (UAV, 691.60/3).

107. Fleming to Bush, 9 Sep 1940 (MAT, 25/"biophys."). The figures for the 37-inch appear quite reasonable.

beryllium target; and from the *n* units and the hypothesis that energy delivered by neutrons to living tissue is four times as destructive as the same quantity of energy delivered by x or gamma rays, Livingston arrived at the penultimate line of his chart. The bottom line, the cost of therapy per 100 roentgens/minute, shows that price did not limit the effective treatment of cancer by neutrons.[108]

To bring out clearly the relative excellence of the cyclotron as a factory for radioisotopes and therapeutic neutrons, Livingston rated the electrostatic generators of 1940 as shown in table 6.7. A comparison of the three classes of generators with the largest classes of cyclotrons (table 6.6) shows that although the cyclotrons cost more than the corresponding Van de Graaffs in both capital investment and operating expenses, they enjoyed so great an advantage in beam and energy that they manufactured radioisotopes and neutron doses at much lower unit prices. The standard moderate cyclotron made—or could make, after improvement—a millicurie of radiophosphorus in six minutes for less than a dollar; the top-of-the-line generator, Tuve's 19-foot pressurized Van de Graaff, also after perfection, would need about twenty hours—and $166—to do the same. It was not that Van de Graaff and his associates had been idle or ill-financed. Between 1936 and 1940 both the cyclotron and the generator increased their effective beam energies by a factor of four and beam currents by a factor of ten or more: from 4 to 16 MeV and 20 to 200 μA at Berkeley, from 0.9 to 3.5 MeV at the Carnegie Institution, and from the Carnegie's 10 μA to MIT's 100 μA and more.[109] But an electrostatic generator that could hold 3.5 MV was a technological freak—Westinghouse was trying for 5 MV but could scarcely reach 3—and MIT's Van de Graaff represented the effective energy limit in 1940 if a respectable current were desired.[110] Clinical doses of radioisotopes lay beyond its capabilities.

108. For definition of the *n* unit and the comparative biological effects of x and neutron rays see infra, §8.3.

109. Supra, table 6.7; Hafstad and Tuve, *PR, 48* (15 Aug 1935), 306–8; Elsasser to Joliot, 13 Sep 1936 (JP, F28); Livingston to Lawrence, 5 Feb 1939 (12/12).

110. Amaldi, *Viaggi, 6* (1940), 9, reporting that after two years the Westinghouse engineers had got only 2.9 of their hoped-for 5 MV.

Table 6.7

Livingston's Classification of Electrostatic Generators

Type	Large	Medium	Small
Energy	3–5 MeV	2–3 MeV	1–2 MeV
Typical installation	Carnegie	M.I.T.	Carnegie
Size (diam/pressure)	18 ft/50 lbs	15 ft/atmos.	8 ft/atmos.
Operations Crew	3	3	3
Energy (deuterons)	3.5	2.5	1.2
Installation cost	$75,000	$45,000	$7,500
Installation time (yrs)	2	1.5	1
Operating costs[a]	$20,000	$17,000	$3,800
Running time (hr/d, hr/yr)	7/2400	7/2400	7/2400
Operating costs[a]/hr	$8.30	$7.10	$1.60
Beam (μA)	15/50[b]	3/300[b]	10
μA hr/day	105/350[b]	21/	70
Operating costs[a]/μA hr	$.55/$1.66[b]	$2.36/$.024[b]	$.16
μA hr/mC of P^{32}	1000		
Costs[a]/mC of P^{32}	$555/$166[b]		
Neutron equiv (gm Rn-Be)	1500		70
Neutron equiv/ μA (gm Rn-Be)	100		7
Flux @ 100 cm (n units/m)	.02		.0008
Costs[a]/n unit	$6.90		$40.00

Notes: Prepared for "Conference on Applied Nuclear Physics," MIT, Oct 1940 (12/12).
a. Includes amortization and overhead.
b. Numbers left of the slash indicate current, those right of the slash, projected performance.

It is not easy to state the goals of the leaders of cyclotron laboratories in 1940. On the one hand, their instruments had opened up vast fields of biological and medical research, held promise of discoveries in nuclear chemistry and physics, and constantly challenged and enticed *Homo faber*. Theirs was an exciting and progressive line of work. On the other hand, the cyclotron brought slavery to physicians and to the chase for money, regimentation of laboratory work, and no long-range research project.

It was a tool constantly in need of improvement lest it condemn itself and its attendants to routine manufacture in the service of others. After consulting with Lawrence and Conant, Karl Compton summed up the situation: "To maintain an active program and a well rounded staff has required more aggressive salesmanship than the scientific profession relishes...., an abnormal competitive element which is unfortunate."[111] The cyclotroneers escaped the logic of their situation—an increasingly competitive struggle for large sums in an increasingly inelastic market, a growing disparity between builder-physicists and operator-technicians, a tightening tension between service to others and science for oneself—by going off to war.

111. Compton to M.C. Winternitz, 24 Nov 1941 (4/12). Lawrence approved Compton's assessment as "excellent in every way;" Lawrence to Compton, 29 Nov 1941 (4/12).

VII
Technology Transfer

1. AN EASY CLONE

The first working cyclotron outside the United States came to life not in a great center of nuclear physics in Europe, but at the Institute for Physical and Chemical Research (Riken) in Tokyo. This precociousness was born of a conjunction of forces characteristically Japanese: a conviction on the part of government and industry that excellence in Western science was essential to Japan's place in the sun; an ability to assimilate foreign designs; and no vested interest in any machinery for splitting atoms. Just as the great substantive discoveries and instrumental improvements in experimental nuclear physics were being made in Europe and the United States, the Japanese, pulling themselves out of the Depression by military adventure and economic aggression, found the money and motive for multiplying particle accelerators in Tokyo. In 1932 Riken examined the leading Western alternatives and decided on a Van de Graaff; they next added a Cockcroft-Walton; and in 1935 they undertook to make a cyclotron.[1]

They stayed close to their prototype. They procured a Poulsen arc magnet as a gift from the Japan Wireless Telegraph Company and gave it symmetric poles. They raised other capital costs from a foundation (the Mitsui) then recently established by industry and got their oscillator free from the Tokyo Electric Radio Com-

---

1. Hirosige in Nakayama et al., *Science and society in Japan,* 202–3, 206–7, 213.

pany. They succeeded, where Lawrence had failed, in obtaining operating expenses from the utility that supplied their power, the Tokyo Electric Light Company.[2] For technical guidance, Riken sent a man to Berkeley. Lawrence had warned that otherwise they would never make a cyclotron in a reasonable length of time. "It is rather ticklish in operation, and a certain amount of experience is necessary to get it to work properly." The emissary was Ryokichi Sagane, who, despite his name, was the son of Hantaro Nagaoka, the elder statesman of Japanese physics and an important voice in the allocation of research funds. Sagane arrived in Berkeley in the fall of 1935 and remained for over a year. He did his job if anything too well: following his instructions, his compatriots copied Berkeley mistakes as well as successes and Lawrence had to instruct them to fill up holes they had drilled in the tank plates in emulation of a measure once tried and immediately discarded at the Laboratory to improve the magnetic field.[3] The machine, with 27-inch poles on the Poulsen magnet, wax seals and glass insulators in the chamber, and an oscillator arranged in the Berkeley manner, took just over a year to build. When it started up in April 1937, its builders again aped Lawrence and called in the press. The reporters celebrated Riken's "large and fantastic laboratory for the atomic nucleus" and ranked it second only to Berkeley's.[4]

The main responsibility for the construction of the cyclotron rested on Tameichi Yasaki, who had the benefit of only a brief sojourn in the Laboratory. Not being proficient in the high art of shimming, he could not coax more than a few μA from the machine. With Sagane's trained touch, Tokyo gained a very respectable deuteron beam of 30 μA (later 47 μA) at 3 MeV. The cyclotron ran well, eventually for as long as thirty hours at a stretch, at least as reliably as Berkeley's 27-inch.[5] Its anticipated

2. Ibid., 217; Weiner, ICHS, XIV, *Proceedings, 2,* 354–7; Nishina et al., Inst. Phys. Chem. Res., *Sci. papers, 34:854* (1938), 1664–8.

3. Lawrence to Iwao Fukushima, 22 Jan 1934, and Fukushima to Lawrence, 8 Sep 1935 (9/34); Lawrence to Yasaki, 24 Dec 1936 (9/43).

4. Nishina et al., Inst. Phys. Chem. Res., *Sci. papers, 34:854* (1938), 1658–68; Hirosige in Nakayama et al., *Science and society in Japan,* 214.

5. Nishina to Lawrence, 1 June and 13 Jul 1937, 21 Feb 1938 (9/28), and to Bohr, 28 Aug 1937 (BSC); Sagane to Lawrence, 10 Mar 1938, and to Cooksey, 15 May 1938 (9/39); Lawrence to Cork, 29 May 1937 (5/1).

success caused the Japanese to adopt still another of Lawrence's practices. Long before they got the first beam from their 26-inch cyclotron, they had determined to make a bigger machine, indeed the biggest. With extraordinary confidence and deferential effrontery, they asked to copy the Crocker cyclotron while it was under construction. Lawrence met the request with his usual generosity and enlightened self-interest. He helped the Japanese order the necessary metal in the United States, where it was cheaper than in Japan, and had Brobeck and Cooksey help Sagane oversee the machining of the iron, the coiling of the copper, the procurement of the coil tanks, and so on. Why was he so forthcoming? No doubt the flattery implied by the copying, the tug of the exotic East, and appreciation that Riken had pioneered the cyclotron abroad inspired his actions; so, too, did the hope that two 60-inch machines could be built at less than twice the price of one. "I should be glad to help the Japanese in this way," he wrote Revere Copper and Brass, "because, in addition to helping out the scientific work in Japan, it should be possible for us to get somewhat better price quotations if we place double orders for the material and equipment." When it became clear that there was no economy of scale, Lawrence made a virtue of necessity. He wrote Columbia Steel: "My position is only that of one who desires to be of help in furthering important scientific work in this county and abroad."[6]

The metal for the Japanese Crocker arrived in Tokyo in the spring of 1938. Its completion was delayed by an impulse to innovate and by a new sort of trouble with oscillators. Here the cyclotron made its first acquaintance with war. Riken required for its oscillator precisely the same sort of tubes that the Japanese Army used in its broadcasting stations in China; none could be bought in Japan and import restrictions inhibited procurement abroad. Cooksey suggested to Sagane that they make their own tubes, in accordance with blueprints he provided, and Lawrence offered to supply any further information required "about our design and

6. Nishina to Lawrence, 30 Jul 1936, and reply, 18 Aug 1936, in Weiner, ICHS, XIV, *Proceedings, 2,* 355–6; Lawrence to Paul M. Mueller, Revere, 10 Feb 1937, and to Richard Erlin, Columbia, 27 Mar 1937, and Mueller to Lawrence, 16 Feb 1937, all in (25/4); correspondence between the Laboratory and Japanese purchasing agents, Mar 1937–Apr 1938 (9/37).

technique."[7] In 1940, having decided that their innovations were not improvements, the Japanese applied to the Laboratory for all blueprints of the perfected 60-inch cyclotron. Here again they ran into war. The Laboratory, previously so cooperative, now would give nothing. Cooksey wrote in 1943: "When the Japs left this country [after a visit in 1940], they were making every effort to obtain from us the final details about our construction and methods. We have consistently refused them any information for approximately the last three years." Why details of the cyclotron became confidential before Pearl Harbor will be explained in due course.[8]

The two Tokyo cyclotrons bracketed the prewar foreign implementation of Lawrence's prepotent invention. The installations that had achieved a beam by 1940 are listed, with some vital statistics, in table 7.1. As appears from the table, five cyclotrons were operating in Europe in 1939. The stories of four of them—we neglect Stockholm's, since it came on only briefly, literally on the eve of the war—instance the disparity between Lawrence's Laboratory and the physics institutes of the Old World.

Conversion rates will assist evaluation of the sums to be mentioned. Between 1934 and 1939, the Danish krone held steady at 22 cents; the French franc declined gently but gravely from 6 cents to 3 cents; the German Reichsmark held at 40 cents; the Italian lire fell from 8 cents to 5 cents; and the British pound, having hovered around $4.95 since 1934, dropped to $4.43 in 1939.

2. RESISTANCE

A Preference for High Tension

The major laboratories for nuclear physics in France and Britain had declined the opportunity to clone the 27-inch cyclotron from the obsolescent Poulsen-arc magnets decommissioned by the

7. Sagane to Lawrence, 24 Jul 1939, and Cooksey and Lawrence to Sagane, 29 Aug 1939 (9/39); Weiner, intervention in Brown and Hoddeson, Birth, 283.
8. Yasaki to Cooksey, 29 Nov 1940, and reply, 2 Jan 1941 (9/43); Cooksey to Livingston, draft for Lawrence's signature, 18 Nov 1943, passage removed from letter sent (12/12); infra, §10.2.

Table 7.1

Foreign Cyclotrons by Size, 1940

| | Magnet | | | Commission dates | | | |
Size[a]	poles (inch)	Fe (ton)	Cu (ton)	Plan	Magnet	Beam	Builders
Small (3-7 MeV)							
Leningrad-1	24					Sep 37	Rukavichnikov
Tokyo-1	26	23		Sep 35	Sp 36	Apr 37	Yasaki,[b] Sagane,[b] Watanabe
Medium (8-12 MeV)							
Cambridge	36	46	8	Sp 36	1936	Aug 38	Cockcroft, Hurst[b]
Copenhagen	36	35	3	W 35-6		Nov 38	Frisch, Jacobsen, Laslett[c]
Heidelberg	40	80	80	1937	Sp 38	Dec 43	Gentner[b]
Leningrad-2	40			1937?		1946	Kurchatov, Alikhanov
Liverpool	36	46	8	Sp 36	1936	Mid 39	Chadwick, Kinsey[b], Walke[b]
Osaka	40					1939	Kikuchi
Paris	32	30?		1936	Nov 36	Mar 39	Joliot, Nahmias, Paxton[c]
Stockholm	35			1937	Sep 38	Aug 39	von Friesen[b]
Large (16 MeV)							
Tokyo-2	60	196	22	1936		> 1941	Yasaki,[b] Sagane[b]

a. Livingston, "Cost estimates," Ca. 1 Nov 1940 (12/12), established the size categories.

b. Postdoctoral experience at the Laboratory.

c. Berkeley Ph.D.

French radio service in 1932. Their disinclination had both negative and positive causes. On the minus side, the cyclotron of 1932 ran unsteadily, could not accelerate electrons, and when operating delivered only a fraction of a μA to the target. It retained its reputation for unreliability after it had attained the ability to work an eight-hour day, perhaps because visitors who came when it was under repair advertised their disappointment.[9] As late as the winter of 1935/36, the cyclotroneers themselves acknowledged that their machines did not function most of the time. At Princeton they expected to operate only two hundred hours a year. Making his usual virtue of necessity, Lawrence reassured Chadwick that he need not worry much about the power bill for his planned cyclotron since it would run so infrequently. We may recall Lawrence's inability to supply $2,000 worth of radiosodium to clients of the Macy Foundation in 1935.[10]

By the winter of 1937/38 a well-made cyclotron could work steadily if not driven at its maximum capabilities. In the spring of 1938 the 37-inch ran at an average of eight hours a day, seven days a week, and Princeton's cyclotron went on for three months without serious mishap. Nonetheless the impression remained abroad that, as Lawrence wrote to an English physician trying to promote a medical cyclotron in Britain, "the cyclotron is still a very unreliable apparatus;" and occasionally he had to reassure American inquirers that "the cyclotron is no longer a capricious laboratory device....[but] an efficient and thoroughly rugged and reliable apparatus." And yet, between the desire to improve performance and the need to repair and replace parts, cyclotroneers often had their machines apart. In July 1938, in a tour of all the cyclotron laboratories in the East, Livingston saw nothing to see. "Don't let this get out," he wrote Lawrence, "but I did not find a single cyclotron operating."[11]

9. Rasetti, *Viaggi, 3* (1936), 78; Goudsmit to Bohr, 31 Jan 1938 (Frisch P); Lawrence to Cockcroft, 12 Sep 1935 (4/5).
10. Henderson to Lawrence, 17 Dec 1934 (9/6); Lawrence to Chadwick, 31 Jan 1936 (3/34); supra, §6.1.
11. Lawrence to Hopwood, 24 Aug 1938 (9/15); M. White to Lawrence, 10 June 1938 (18/21); quotes from, resp., Lawrence to Hopwood, 24 Aug 1938 (9/15), and to Hughes, 26 May 1938 (18/12), and Livingston to Lawrence, 28 Jul 1938 (12/12).

In addition to their unreliability, the early Berkeley machines were depreciated in Europe for their finicky and "empirical" character. The Europeans had the "general impression [Lawrence acknowledged] that the cyclotron is a very tricky and difficult apparatus to operate." As for empiricism, the missionary cyclotroneers freely admitted the charge, and even gloried in what elevated a possible science into an actual art. In describing the Cornell cyclotron, Livingston pointed to the size of the gap between pole faces, the height of the dee aperture, the position of the source filament, and the shimming of the magnet as parameters that could only be fixed by "experimental maneuvers;" at Princeton, Henderson and White admitted to the method of cut and try in setting the dimensions of their magnet, the position of their filament, and so on, and to their inability to justify many design details, "except to say that [they are] known to work."[12] Nor did the mother church affect to know the principles of its practice. In describing the definitive version of the 27-inch cyclotron late in 1936, Lawrence and Cooksey recommended maximizing the beam by ad hoc adjustments to, among other things, the deflecting potential, the shimmed magnetic field, and the positions of the dees, deflecting plate, and source filament.[13]

All this summed to *trop d'empirismes*, too much tinkering, according to Nahmias, who thus depreciated the cyclotron after a visit to Princeton. The criticism was almost out of date when Nahmias made it in the spring of 1937; cyclotrons then being built would operate with fewer *empirismes* as well as with greater regularity. Here one story is worth a thousand words. Cooksey and Kenneth Bainbridge of Harvard visited the Bartol cyclotron in April 1938. Its creator, Alexander Allen, threw the switches; a beam came immediately, without fiddling. Bainbridge, who had never seen a cyclotron work, cried in astonishment, "Why, he just turned it on!"[14]

12. Lawrence to Chadwick, 31 Jan 1936 (3/34); Livingston, *RSI, 7* (Jan 1936), 57–9 (quote), 64, 66; Henderson and White, *RSI, 9* (Jan 1938), 21 (quotes), 25, 28; W.J. Henderson et al., *JFI, 228* (1939), 574–5.

13. Lawrence and Cooksey, *PR, 50* (1936), 1139.

14. Nahmias to Joliot, 24 Mar 1937 (JP, F25); Cooksey to Lawrence, 29 Apr 1938 (4/21).

In comparison with cyclotrons, again according to Nahmias's survey of March 1937, Tuve's improved two-meter Van de Graaff generator, which operated with great reliability at 1,000 kV and even at 1,200 kV, "under good conditions, compounded of low humidity, good fortune, and infinite other ingredients," had the advantage of few empirical adjustments to achieve a strong, homogeneous beam. High-tension apparatus not only produced better beams for exact work, but it did so by scaling up devices familiar to physicists. The cyclotron could not make headway in Europe until it could demonstrate advantages so decisive that physicists there would undertake to master the alien field of radio technology. After a visit home to the Cavendish in 1934, Bernard Kinsey succinctly explained his countrymen's reluctance to make cyclotrons: "They are all scared stiff at the thought of setting up an oscillator."[15] Lawrence himself pointed to the radio engineering as the most challenging part of his operation: "The difficulties encountered [in making a new cyclotron] are similar to those found when a new radio broadcasting station of design and power that's never been used before is first constructed." The high-frequency oscillator for Bohr's cyclotron was to have an energy and to present a difficulty larger than the shortwave transmitter of the Danish state radio.[16] Why trouble with a finicky machine and unfamiliar technology when other sorts of accelerators seemed capable of doing the same or similar jobs?

Tuve's group had advertised their technique by veiled reference to the superiority of the point of view of the investigator in a high-tension laboratory to his counterpart in a (or rather in *the*) cyclotron laboratory. Their call for detailed exploration of reactions initiated by homogeneous beams before subliming to high voltages met with widespread agreement outside Berkeley.[17] A clear conscience and a pure beam did not exhaust the advantages

15. Quotes from, resp., Tuve, Hafstad, and Dahl, *PR, 48* (15 Aug 1935), 332–3, 325 (quote), 336–7, and Kinsey to Lawrence, 4 Oct [1934] (10/18).

16. Lawrence, CBS broadcast, "Adventures in science," 15 Apr 1939, 4 (40/15); Bohr in *Politiken,* 11 Nov 1938 (clipping in 3/3).

17. Tuve, Hafstad, and Dahl, *PR, 48* (15 May 1935), 316; Breit, *RSI, 9* (1938), 63–74; Wells, *Jl. appl. phys., 9* (1938), 682; Gentner, *Ergebnisse, 19* (1940), 197; Jacobsen, *Fys. tidsskr., 39* (1941), 50; Cockcroft, Br. Assn. Adv. Sci., *Report,* 1938, 382.

of high-tension over magnetic-resonance acceleration. As Tuve told Nahmias, a Van de Graaff version of high-tension accelerators could work at continuously variable voltages, could push electrons as well as positive ions, could make x rays, might achieve as much as 10 MeV, and would do it all at less expense than any other model. Hence, he said, Westinghouse was building a Van de Graaff with two concentric spheres, the larger 10 meters in diameter, large enough to enclose an ordinary laboratory space, and capable of maintaining a pressure of 10 atmospheres. If all went well, and, as we know, it did not, Westinghouse would get 10 MeV for $15,000.[18]

Decisions taken at the Cavendish in 1935 and 1936, when the laboratory recognized the need to go to higher energies, indicate the considerations then at work against cyclotrons. In May 1935 its newly appointed building committee for a high-tension installation approved a report drawn up by Cockcroft and Oliphant, who stressed the "immediate importance to develop apparatus for accelerating charged particles by at least two million volts." They recommended "an extension of the [Cockcroft-Walton] method of producing high steady potentials which has been in successful operation for two [more accurately three] years." In reaching this conservative conclusion they had the assistance of Philips of Eindhoven, which they had visited the preceding January. The Philips laboratories were "an eyeopener" to Oliphant, so he wrote Rutherford, "and in the opinion of Cockcroft...far superior to any in America." Oliphant eyed a million-volt high-tension set and desired to build a similar one, at twice the voltage, at the Cavendish.[19] In explaining their decision to Lawrence, Cockcroft pointed to the Cavendish's interest in x rays as well as positive ions and to the existence of "plenty of [American] laboratories who would be capable of using the cyclotron method."[20]

Cockcroft and Oliphant calculated that their laboratory would require a building sixty feet long, forty feet high, and forty feet wide, with concrete walls a foot and a half thick and a mobile

18. Nahmias to Joliot, 24 Mar 1937 (JP, F25).
19. Cockcroft and Oliphant, "Memorandum on the construction of a high voltage laboratory" (UA Cav, 3/2/i); Oliphant to Rutherford, 7 Jan [1935] (ER), and 24 June 1935, in Cockburn and Ellyard, *Oliphant*, 61–2, resp.
20. Cockcroft to Lawrence, 18 Jul 1935 (5/4).

crane to install and repair the large tubes, transformers, and generators. They estimated the cost of the building at 4,000 pounds and that of the apparatus at under 3,000 pounds, over twice the price of the cyclotron then nearing completion at Princeton. They underestimated by much more than a factor of two.[21] "The approximate cost of 15,000 pounds [for the building] staggered me, as I imagine it will you," Oliphant wrote Rutherford. "I can see the new laboratory receding into the distance if we are not careful....It is a thing we need urgently, and not in some distant future when all the cream has been scooped off by folks whose results we dare not trust too deeply."[22] From which it appears that Oliphant had in mind to do physics and further battle with Berkeley with his machine.

By July 1935 Rutherford had accepted these objectives and promoted Oliphant and the building. The former he made assistant director of research in place of Chadwick, who left the Cavendish for a professorship at Liverpool. The latter he lobbied for so effectively before the council of the senate of Cambridge University that it recommended proceeding at once with university funds to be repaid by proceeds from an outside appeal for the 250,000 pounds the laboratory deemed necessary to meet all its research needs. Rutherford's reluctance to ask for money has been overestimated.[23]

The great cost of the building, which was completed in 1937,[24] and technical difficulties precluded going directly to 2 MeV. Again the Cavendish had an opportunity to consider a cyclotron and a more modest establishment. Again Oliphant went to Philips and again returned inspired. He decided to buy a copy of the 1.2 MV installation he saw at work. "After a stiff fight with some of

21. Cockcroft and Oliphant, "Memo" (UA Cav, 3/2/i); Cockcroft to H.C. Marshall, 21 May 1935 (UA Cav, 3/3/iii).

22. Oliphant to Rutherford, 25 Aug 1935 (ER), in Cockburn and Ellyard, *Oliphant*, 61–2.

23. Cockcroft (?) to Appelton, 10 Nov 1937, enclosing a memorandum on needs drawn up for Rutherford in 1935 (UA Cav, 4/2/vii); Rutherford to Lawrence, 18 Jul 1935 (15/34); Council of Senate, "Report on a high-tension laboratory," *Camb. Univ. rep.*, *66* (1935), 232; Oliphant's appointment, ibid., *65* (1934/5), 982–3, 1098; Rutherford's reluctance to appeal, Crowther, *Cavendish*, 186.

24. Cockcroft to Lawrence, 16 Nov 1936 (4/5); Rutherford to Lawrence, 21 Dec 1935 (ER); *Camb. Univ. rep.*, *68* (27 June 1938), 1156.

my colleagues and with Philips themselves over the price [just under 6,000 pounds] I have persuaded the Prof. [Rutherford] to invest in one of these sets." This trouble-free, ready-made instrument arrived in its impressive building ("a cinema outside, a cathedral within") after Christmas 1936 and worked as advertised.[25] By then its promoters had forsaken it. Oliphant had been appointed professor in Birmingham the previous June, effective October 1937. Cockcroft had at last pronounced in favor of a cyclotron, and returned from a visit to Berkeley in 1937 intending to recommend selling Cambridge's high-tension equipment. That pleased Lawrence immensely. "The Cavendish Laboratory has expended large sums of money in installing high voltage equipment," he wrote in 1937, in a puff of Berkeley. "Although the Cavendish Laboratory pioneered with high voltage methods the distinguished scientists there have come to the conclusion that the cyclotron is superior, and are adopting it."[26] The Philips accelerator, which had seemed essential to the Cavendish's place in nuclear physics, fell eventually under the management of a Swiss physical chemist, Egon Bretscher, who used it as a neutron source to make radioisotopes for chemists and biologists.[27]

Several other British institutions followed Cambridge in preferring high tension to the cyclotron in the second generation of particle accelerators. At Bristol, for example, two members of the staff each tried to make a Van de Graaff generator, while a third, Cecil Powell, built a Cockcroft-Walton machine. In a few years, when they wished for a cyclotron, they judged that they could not acquire one on their own resources.[28] At Oxford, the New Clarendon was designed for a Van de Graaff or a Cockcroft-Walton, although the professor, F.A. Lindemann, doubted that his university would put up the money for such expensive apparatus. As for

25. Oliphant to E. Bretscher, 9 Jul and 30 Aug 1936 (Mrs. Bretscher), quote; *Nature, 143* (17 Sep 1938), suppl., 522; Emo to Lawrence, 24 Oct 1937 (7/4); Peierls to Bethe, 13 Nov 1937 (HAB/3), quote.

26. Cockburn and Ellyard, *Oliphant*, 63; Lawrence, "The work," 1937, 11; Cockcroft's intention as reported by officials of the Rockefeller Foundation, "F.B.H." and "E.B.," "Diary," 9 May 1937 (RF, 713D).

27. Bretscher and Feather, "Experiments involving element 94," report to MAUD Committee, 19 Dec 1940, 2, and Oliphant to Bretscher, 20 Apr 1936 (Mrs. Bretscher); Cockburn and Ellyard, *Oliphant*, 68–9.

28. Tyndall, "History," 28–9; Powell, "Development" (1943), 3, 6.

the cyclotron, "an instrument most popular abroad," it had the disadvantages, according to Lindemann, of not being cheap either and of not accelerating electrons. When in 1938 he decided that he wanted a cyclotron and proposed that the Ministry of Health pay for it, the ministry declined on the ground that physicists had not yet "decided on the scientific basis of the cyclotron." Lindemann rejected this opinion, which he supposed to derive from a prominent British physicist, as the nonsense it then was: "There is no possible doubt about the scientific basis of the cyclotron in Oxford and still less in California....I have no doubt that if one liked to spend the money one could buy a ready-made cyclotron from America which would function quite satisfactorily."[29] One did not like to spend the money, and Oxford got no cyclotron.

High-tension machines were also the preferred second-generation accelerators on the Continent. In Germany, where university physics had been all but destroyed by the implementation of Nazi racial laws, the leading centers of experimental nuclear physics sheltered in the institutes of the quasi-independent Kaiser-Wilhelm-Gesellschaft. In 1934 Walther Bothe, whose work had set up the discovery of the neutron, became head of the physics department of the Kaiser-Wilhelm-Institut für Medizinische Forschung in Heidelberg. He had considerable experience in high-voltage technique. With the help of Wolfgang Gentner, who had worked with Joliot and who was to become the first German cyclotroneer, he set up a Van de Graaff at 950 kV, which began to operate in 1937. Despite the name of their institute, Bothe and Geiger's work with high tension centered on basic physics, for example, photoinduced nuclear reactions, work that in Joliot's judgment was for a time the most productive "in radioactivity and nuclear physics."[30] That made a curious reversal of the situation of cyclotron laboratories in physics departments in the United

29. Lindemann, "Report," 4–5 (FAL, B13), and letter to I.O. Griffith, 7 Nov 1937 (FAL, B18); Lindemann to Registrar, Oxford Univ., 18 Dec 1938 (FAL, B24), quote. Lindemann's student James Tuck was trying to build a betatron; he had the help of Szilard's prodding from a distance, but did not finish it before the war. Lindemann to Ralph Glynn, 29 Nov 1938, and to Chadwick, 19 Apr 1944 (FAL, D245).

30. Brunetti, *Viaggi, 4* (1938), 19; Gentner to Joliot, 3 May 1937, in Goldsmith, *Joliot-Curie*, 61–2.

States, which then were beginning to produce radioactive materials for biological work.

Also in 1937 the Kaiser-Wilhelm-Institut für Physik, a brand new institute built with a gift from the Rockefeller Foundation, officially opened its doors. Designed to emphasize nuclear physics, it had a tower fifteen meters in diameter and fifteen meters high to house a cascade generator of 2 MV. The Kaiser-Wilhelm-Gesellschaft's two high-tension installations were the only machines in Germany in 1937 capable of furnishing particle beams of a million volts or more. Not until 1938 did Bothe put forward a proposal to erect a small cyclotron, at a cost of 20,000 RM, which Gentner was to complete during the war.[31]

Hedged Bets

In Copenhagen and Paris they planned cyclotrons along with the more familiar high-tension apparatus. In both cases the decisions were influenced, if not inspired, by the policies of the Rockefeller Foundation, which, as we know, in 1932 adopted Weaver's program excluding support for basic physics. Weaver did not wait for supplicants. In May 1933 he discussed opportunities with Niels Bohr, then shopping at the Rockefeller Foundation's headquarters in New York; Bohr's Institute for Theoretical Physics, which had been extended in the 1920s with some $45,000 of Rockefeller money, seemed poised for a push toward biology. Bohr himself had been lecturing in a vague philosophical way on the connection between his interpretation of quantum physics and the limits of biological research. More encouraging, no doubt, to Weaver was the possible collaboration of Georg von Hevesy, who was considering resigning from his institute for physical chemistry at the University of Freiburg (which, to complete the circle, had been built with the help of $25,000 from the Rockefeller Foundation). In the summer of 1933 Hevesy decided to leave Nazi Germany; in January 1934 the Rockefeller Foundation granted $6,000 to establish him in Bohr's institute for three years.

31. Richter, *Forschungs.*, 52; Bothe to Reichsforschungsrat, 7 Apr 1938, ibid.; W.E. Tisdale (RF) to file, 4 Oct 1935 (RF, 717); Debye, *Nwn, 25* (1937), 257–60; Heilbron, *Dilemmas,* 175–9.

Hevesy was the world's expert in the use of naturally radioactive substances as tracers in chemical research. If he and Bohr could be teamed with August Krogh, professor of physiology at the University of Copenhagen, whose institute, yes, the Foundation had helped to build, the sort of group that Weaver wished to create would come into being and the Foundation would have the satisfaction of integrating and capitalizing its earlier investments. By the spring of 1934 Bohr had become actively interested in biology, "aided undoubtedly," according to W.E. Tisdale, the Rockefeller man who spoke with him, "by the ramifications involved in the purification of the German race." Bohr's plans, then still "rather vague and philosophical," firmed during the next six months, under the impact of the neutron activation work of Fermi's group: he would collaborate with Hevesy and Krogh, beginning with heavy-water physiology and continuing with artificial radioactive substances to be produced by a high-tension set.[32] In the spring of 1935, the three put forward a proposal, which the Rockefeller Foundation immediately funded, for $54,000 over five years, of which $15,000 was payable on request for apparatus and the balance provided an annual grant for research expenses. With a pledge from the Danish Carlsberg Foundation to pick up Hevesy's salary when his Rockefeller stipend expired, the project had the guarantee of a good long life.[33]

The $15,000 was to go for "the installation of an apparatus for the production of radio-active materials, patterned after the equipment of Lawrence at Berkeley." The rationale for the installation, according to Tisdale, advised by Hevesy, was to make possible preparation of radioisotopes not obtainable from high-tension apparatus operating at 1 MV. "Von Hevesy assures me that this project, involving as it does so much physics, is completely orien-

32. Robertson, *Early years*, 91–3, 106–8; "University of Copenhagen—Biophysics," [1940] (RF, 713D); Tisdale and D.P. O'Brien, "Log," 10 Apr 1934, and Tisdale, "Diary," 10 Apr 1934 (ibid.); Aaserud, *PT, 38:10* (1985), 41–3; Heilbron, *Rev. gen. sciences, 35* (1985), 212–19.

33. RF, "Docket," 17 Apr 1935, and Tisdale, "Diary," 30 Oct 1934 and 22 May 1935 (RF, 713D). The $54,000 brought the RF's benefactions to Bohr's enterprises to $109,000 (it had given $10,000 for a liquid-air machine in 1933/4), more than the total Lawrence had by then raised from external sources.

ted toward bio-physical problems and is in no wise an attempt on Bohr's part to obtain equipment to permit him in any wise to compete with the Rutherfords, the Lawrences, and others who are working in the field of pure [!] physics." Nonetheless, as Tisdale could not help but realize, a particle accelerator "would not be limited in its usefulness to the single purpose of preparing radioactive materials for the cooperative problem, but would also permit of studies in nuclear physics from the physics point of view." It was an awkward matter, this possible application to physical science of a machine built for biological research, but the eminence of Bohr, as Weaver had earlier remarked, would "probably protect us from anything that would reach real embarrassment" in so blatant a compromise of the Foundation's policy against supporting pure physics.[34]

Although the cyclotron had been reviewed favorably in Denmark in 1934 as the machine of choice above a million volts, it was not the preference of the Copenhagen group. In November 1934 they considered the merits of Cockcroft-Walton, Van de Graaff, and Lawrence machines, and decided on a high-tension set designed for 2 MV. Having obtained 150,000 kroner for his primary objective from the Carlsberg Foundation, Bohr asked the Rockefeller Foundation for a cyclotron as a secondary piece of apparatus.[35]

Hevesy did not wait for either machine. He procured a source for his experiments by raising enough Danish money to buy 600 grams of radium for Bohr's fiftieth birthday in October 1935. The cost of the gift, $21,000, exceeded the estimate for the cyclotron; but it could be used immediately, mixed with beryllium, to yield the neutrons to convert sulphur into radiophosphorus for Hevesy to feed to rats. The foundations of the hall to house the high-tension apparatus were going down as the radium came to hand.[36]

34. Tisdale to Weaver, 27 Feb 1935 and 16 Nov 1934, and Weaver to Tisdale, 5 June 1934 (RF, 713D).

35. Ambrosen, *Fys. tidsskr., 32* (1934), 143; Tisdale, "Diary," 30 Oct 1934, and Tisdale to Weaver, 16 Nov 1934 and 27 Feb 1935 (RF, 717D); Jacobsen to Bohr, 13 Feb 1935 (BSC), and Frisch to Placzek, 24 Mar 1935 (Frisch P). Cf. Weiner, ICHS, XIV, *Proceedings, 2* (1975), 356–7.

36. E. Rasmussen to Bohr, 30 Oct 1935 (BSC); Frisch, *What little,* 98–9, and in Rozental, *Bohr,* 140; Aaserud, *PT, 38:10* (1985), 45.

The apparatus itself consumed much more than its budget—it swallowed the Carlsberg Foundation's 150,000 kroner and the Rockefeller Foundation's $15,000—and gave back much less. Unable to get beyond the region already well explored at Cambridge and elsewhere, it offered no incentive to applications to basic physics and so served its purpose, as an Italian physicist disdainfully reported, "of biological research." Early in 1937 Bohr returned to the Rockefeller Foundation for $12,500 to complete his cyclotron, for which he had also to raise additional support in Denmark. That came primarily from the Thrige Foundation of Odense, a charity run from the profits of a large electrical concern, which gave the magnet and generators.[37]

In Paris Joliot did Bohr one or two better. The discovery of artificial radioactivity in 1934; the Nobel prize for chemistry in 1935; and, not least, the coming to power of the Popular Front, which appreciated science, in 1936, and which Irène Curie served for a time as undersecretary of state for scientific research; all this catapulted Joliot from *chargé de recherches* and consort of Marie Curie's daughter to a professorship at the Collège de France and the leadership of French nuclear physics.[38] Early in 1935, before acquiring the prize or the professorship, Joliot had turned to the Rockefeller Foundation for help in converting his research to biophysics, since, according to Tisdale, "he had no hope of competing with the Rutherfords, the Lawrences, etc., who seem to have a great deal of capital behind them." As for himself, Joliot said, he had access to a small Van de Graaff at an engineering school outside Paris near Arcueil, where he wished to build two more accelerators, both high-tension machines, that is, transformers in series (to reach 2 MV) and an impulse generator (3 MV). All would be used to make isotopes for biological research; three were required so that at least one would always be working to meet the expected high demand. Not a word about cyclotrons.

37. Weaver to Tisdale, 23 Feb 1937, and RF, "Docket," 19 Mar 1937 (RF, 713D); Frisch to Knauer, 21 Jan 1936 (Frisch P); Wick in *Viaggi, 5* (1939), 26 (visit of Sep–Oct 1938); Bohr in *Politiken,* 16 Nov 1938 (clipping in 3/3); Bjerge et al., Dansk. Vidensk. Selsk., *Math.-fys. meddelser, 18:1* (1940). The grant of $12,500 on March 9, 1937, brought Bohr's total take from the Rockefeller Foundation since 1923 to $122,300 (memo, 12 Apr 1937, RF, 713D).

38. Goldsmith, *Joliot-Curie,* 40–1; Weart, *Scientists in power,* 46–7.

Joliot had collected promises for the land and for operating expenses calculated at 120,000 francs; all he needed was the capital investment, some two million francs for erecting and equipping the laboratory, from the Rockefeller Foundation. "We know the great profit that science and humanity have drawn from the judicious help that the Rockefeller Foundation has given to similar undertakings."[39]

The Foundation's initial response was not favorable. Weaver doubted that Joliot could compete successfully with the Cavendish, Berkeley, Caltech, and the Carnegie Institution, and he doubted that Joliot's conversion to biology was sincere. "I suppose that Joliot and his associates have been only human in their desire to give a biological slant to their proposal." But Joliot's professions convinced the men in the field. There was no alternative to making big complicated tools of physics to create biologically useful radioisotopes and no way to make up-to-date tools without research into their basis in physics. The magnitude of Joliot's proposal forced the Foundation to reconsider its policy of closing out support for physics per se. "The proposal as a whole is indicative of the limitations that are inherent in our saying that 'pure physics' has gone far enough—now let's give no more support to it but only to its applications. That attitude runs into a ditch every time."[40] The Foundation eventually did decide to help Joliot, but not before he had helped himself.

By the end of 1935 Joliot had added the impulse generator at the laboratory of the Companie générale electrocéramique at Ivry to his arsenal. This device, rated at 3 MV, proved a very poor source of radioisotopes but a cornucopia of x rays, of which it gave the equivalent of the gamma radiation from more than a kilogram of radium. That impressed the Rockefeller Foundation. In 1937, after Rutherford had advised a major British center for cancer treatment to invest in 20 grams of radium for a single bomb, Tisdale informed the same center that Joliot could give in 19 seconds an irradiation of high-voltage x rays equivalent to six

39. Tisdale to Weaver, 15 Feb 1935, and Joliot, "Proposal," 9 and 22 Feb 1935 (RF, 500D), and "Projet," Jan 1935 (ibid.).

40. Weaver to Tisdale, 21 Jan and 5 Feb 1935 (quote); Tisdale to Weaver, 15 Feb 1935; Alan Gregg to O'Brien, 28 Feb 1935 (quote); all in RF, 500D.

hours' exposure to a gram of radium. Such apparent progress; lavish support from the French government, which put up well over two million francs to buy and furnish a "Laboratoire de synthèse atomique;"[41] strong collaboration with French biologists; and "the ability and expertise which merited the award of the Nobel prize" brought Rockefeller support to Joliot's nuclear enterprises, some $20,684 for the study of life, disease, and death, in December 1937.[42] Joliot stuck to his word: neither he nor his wife did any of their research in nuclear physics or chemistry during the 1930s with beams from Joliot's accelerators. Instead, they used their standard, old-fashioned source, polonium, which gave enough alpha particles to use either directly or, via beryllium, as a source of neutrons.

When the committee planning expansion of the Cavendish went to Paris during Easter 1937, it inventoried Joliot's equipment as follows: the impulse generator at Ivry; a small Van de Graaff at the Collège de France; two more Van de Graaffs brought from Arcueil for exhibit in Paris; and, again at the Collège, "in a room surrounded by paraffin and borax solutions," a cyclotron under construction. It was the stepchild of Joliot's family of accelerators. Arno Brasch visited the family in the spring of 1938. In his report, he made no mention of the cyclotron, which was not yet working, but lavished praise on the complex of high-tension generators, which had no equal anywhere in his experience.[43]

41. Joliot, "Laboratoire," Sep 1937, and "Rapport," 1938. The total cost of Joliot's laboratory by the end of 1939 ran to seven million francs exclusive of salaries; Joliot, "Rapport," 1939. All in RF, 500D.

42. Tisdale, "Diary," 22–26 June 1937, on conversation with Sir Edward Mellanby; "October docket," 24 Sep 1937; and "Trustees' confidential bulletin," Dec 1937; both in RF, 500D. Cf. Joliot to Tisdale, 17 Jul 1937 (JP, F35), asking for $2,500 a year for an unspecified number of years.

43. "Visit of laboratory subcommittee," Easter 1937 (UA, Cav 3/1, item 31); Joliot to J.A. Ratcliffe, 13 Apr 1937 (JP, F28); Tisdale, "Diary," 5 Apr 1938, reporting a conversation with Brasch (RF); Langevin, "Rapport" on Joliot, 1943, 15 (Langevin P); Goldsmith, *Joliot-Curie*, 63.

3. RECEPTION

The leaders of British nuclear physics began to warm to the cyclotron in 1935. No doubt the increasing strength of its currents and yields played a major part in their revaluation. So did a worry that they might be left behind. "Everyone in America is building cyclotrons!" So Fowler overestimated the situation, while making clear his understanding of the reason for the stampede: Lawrence was getting "something like a beam," namely a microamp of 11 MeV alpha particles. Chadwick saw the cyclotron as an engine for strengthening his position at Liverpool while indulging his aesthetic sense: the "magnetic resonance accelerator," he wrote its inventor, "ranks with the expansion [cloud] chamber as the most beautiful piece of apparatus I know....I must have a cyclotron apparatus. When I look at your cloud chamber photographs and see the enormous number of recoil tracks I realize what I am missing." Cockcroft declared his desire to see a cyclotron built in Britain, but did not yet—in the summer of 1935—know how or whether to commit the Cavendish to it. "The medical applications would probably provide an excuse."[44]

The Cavendish soon committed itself. The initiative came from an unlikely source, the Soviet Union, which, by detaining its citizen Peter Kapitza during his visit home in 1934 had upset Cambridge physics. Kapitza held a special professorship, financed by the Royal Society, to preside over a laboratory for high magnetic fields and low temperatures established within the Cavendish through the generosity of the industrial chemist Ludwig Mond. To coax Kapitza to cooperate, the Soviet government proposed to reproduce the Mond Laboratory in Moscow. In the late fall of 1935, Cambridge agreed to accept 30,000 pounds from the USSR for the Mond's magnetic equipment, for duplicates of the liquefaction plants, and for other apparatus; and by the end of the following year Kapitza had his stuff and the Cavendish its cash.[45]

44. Fowler to Bohr, 3 Jul [1935] (BSC); Chadwick to Lawrence, 29 Dec 1935 and 11 Mar 1936 (3/34); Cockcroft to Lawrence, 31 Aug 1935 (4/5). Traces of earlier interest: Walton to Lawrence, 13 Nov 1933 (18/1), and Lawrence to Cockcroft, 26 Jan 1935 (4/5).

45. Badash, *Kapitza*, 1, 101–2; Cockcroft to Lawrence, 16 Nov 1936 (4/5); *Camb. Univ. rep.*, 67 (9 Mar 1937), 749.

Rutherford decided not to replace the big generator for producing the very high magnetic fields that were Kapitza's specialty, but inclined to buy a large electromagnet, for—among other things—accelerating ions. Cockcroft consulted Lawrence on the dimensions of the putative all-purpose magnet. The reply—"the bigger the magnet the better"— did not cause the Cavendish to commission construction. Lingering doubts may have been resolved by notification by Lawrence of the fine initial performance of Cooksey's first vacuum chamber. "If you are undertaking the construction of [a cyclotron], whoever is directly in charge of the work will probably derive some comfort from [Cooksey's experience], because in many quarters it is not realized that it is possible to build an accelerator with a predictably satisfactory performance." This reassurance was dated February 3, 1936; on February 22, Rutherford wrote that he had decided to proceed with the magnet, but not to dedicate it to resonance acceleration. It would be available "for general purposes, and also probably for use as a cyclotron." Cockcroft provided details: they thought to have a magnet with pole pieces 100 cm in diameter, capable of 17.5 kG, the pole faces mounted vertically rather than horizontally as at Berkeley in order that the instrument be adaptable to cosmic-ray work.[46]

Meanwhile Chadwick was busy raising money for a Liverpudlian cyclotron. He approached A.P.M. Fleming, director of research of Metropolitan-Vickers, who went to inspect Berkeley's 27-inch in November 1935 and agreed to create something similar for 5,000 pounds. That was just twice what Princeton expected to pay and over twice the 2,000 pounds that Chadwick had in hand. Lawrence sent the advice of experience: commission Metro-Vick with the 2,000 pounds and goad or embarrass Fleming into providing the rest gratis. Fleming was interested in cancer, as Lawrence had learned during Fleming's visit to Berkeley. He would therefore be interested in the recent discovery at Berkeley that a certain mouse tumor is more strongly affected by neutron

46. Cockcroft to Lawrence, 16 Oct 1935 and 25 Feb 1936, and Lawrence to Cockcroft, 25 Nov 1935 (4/5); Lawrence to Rutherford, 3 Feb 1936, and Rutherford to Lawrence, 22 Feb 1936 (ER). Lawrence also sent news of Cooksey's reliable chamber to Chadwick, 31 Jan 1936 (3/34).

beams than by x rays. "If malignant tumors in general are correspondingly more sensitive to neutron radiation, neutrons will supersede x rays in the treatment of cancer....You might tell Dr Fleming that in view of this important possibility, we are definitely planning to go forward with the treatment of human cancer with our cyclotron."[47]

Fleming decided to be generous—eventually he gave Chadwick oscillator tubes and other essential parts—and he, Chadwick, and Cockcroft joined forces to design magnets for three cyclotrons. Fleming wanted a small one, for a 1 MV or 2 MV machine; the Cavendish, rich in rubles, wanted "a very large magnet indeed;" "for my part [Chadwick wrote] I must build for ten million volts." All wanted to put the poles vertical for magnetic work and cosmic-ray studies. Lawrence discouraged their uprightness. The vacuum chamber had to be removable and accessible, and mounted horizontally on wheels to roll out on tracks for servicing and repair. When Lawrence rejected vertical poles and implied that a cyclotron required a dedicated magnet, the Cavendish had just come into additional wealth that allowed it to multiply magnets outside the Mond laboratory. On May 1, 1936, the vice chancellor announced that the automobile manufacturer Lord Austin had given the 250,000 pounds sought for the Cavendish Laboratory. That put an end to Rutherford's penny-pinching. "It amazed me [Pollard, of Yale] to see the free way in which money is passed around. The Cavendish...seems to be wallowing in cash." Cockcroft got rid of his Morris car and collected plans for magnets with horizontal poles.[48]

No doubt the main force that drove the creation of cyclotrons at Cambridge and Liverpool was the desire of the directors of

47. Lawrence to Chadwick, 27 Nov 1935 and 24 Mar 1936, and Chadwick to Lawrence, 11 Mar 1936 (3/34); Allibone to Lawrence, 27 Sep 1935 (1/15).

48. Chadwick to Lawrence, 11 May, and Lawrence to Chadwick, 26 May 1936 (3/34); *Cambr. Univ. rep.*, 67 (9 Mar 1937), 748; Cockcroft to Lawrence, 28 Apr 1936 (4/5); New Buildings Committee, "Minutes," 23 Jul 1936 (UA Cav, 4/2/1); Pollard to Cooksey, 22 Aug [1937] (14/30). Oliphant's recollection, in a letter to Chadwick, 20 Mar 1967 (Cockburn and Ellyard, *Oliphant*, 62), that Rutherford "blew up, becoming red in the face, and shaking his pipe," when Oliphant urged the expenditure of Austin money on a cyclotron, does not fit the facts; it is more likely that Oliphant remembered Rutherford's reaction when he announced his decision to go to Birmingham.

both laboratories to have the complete equipment for nuclear research. But that did not drive Metro-Vick, whose attitude was decisive for cyclotroneering in Britain. In September 1935 an official of Metro-Vick, George McKerrow, had asked Cockcroft whether they should try to acquire rights under Fermi's patents or whether they should work up their own process.[49] Cockcroft answered with Berkeley's latest neutron yields. They excited McKerrow greatly: "The business is just on the edge of practical possibility." Since Philips intended to work Fermi's patents via neutrons from high-tension machines, Metro-Vick's best bet appeared to be cyclotron production. At the end of February, just after the Cavendish had decided on a cyclotron, McKerrow asked Cockcroft for drawings of Lawrence's machine.[50] After more careful consideration, the industrial possibilities, already circumscribed by the Research Corporation's patent on the cyclotron and Fermi's patents on nuclear activation, must have seemed less promising. Oliphant was no doubt right in placing the blame for the delay and inefficiency of the first British cyclotrons on the growing indifference of Metro-Vick.[51]

Hardware

You cannot always get what you can pay for. Take the magnet, for example. Everyone at Berkeley knew the importance of having poles absolutely symmetrical and the gap between them big enough to admit shims, accommodate a proton chamber, and allow easy service. Lawrence suggested a gap of over 7 inches for magnets of pole pieces 36 inches (90 cm) in diameter, the size chosen at Liverpool, Cambridge, and Copenhagen; Joliot began with the hope of 100 cm, but the cost reduced him quickly to 80. The British adopted a wide gap (8 inches), as at Berkeley, despite the resultant sacrifice in magnetic intensity (a maximum of 18 kG); but the Continentals, reaching for 20 kG and not consulting Lawrence, narrowed the gap to 3.5 inches (9 cm) at Copenhagen

49. McKerrow to Cockcroft, 14 Sep 1935 (CKFT, 20/62).

50. McKerrow to Cockcroft, 3 Nov 1935, 11 and 26 Feb 1936, and Cockcroft's replies, 17 Sep 1935, 6 Feb 1936 (CKFT, 20/62); Cooksey to S.W. Barnes, 27 May 1938 (2/24), quoting the opinion of a Philips expert "that a cyclotron will never yield protons of energy greater than 6 MeV or so;" supra, §4.3.

51. Oliphant, *PT, 19:10* (1966), 47, in Weart and Phillips, *History*, 188–9.

and Paris and landed in trouble.[52]

All the magnets were built by large industrial contractors, in England by Brown-Firth of Sheffield and Metro-Vick; in Denmark by the Thrige works; in France—or rather in Switzerland, where Joliot had his magnet made by the only firm in Europe he thought capable of it—by Oerlikon of Zurich.[53] In every case the construction of the mammoths—forty-six tons of steel and eight of copper for the robust British, thirty-five tons of steel and three tons of copper for the Danes, thirty tons for the delicate French—went slowly. In the spring of 1937 the Cavendish discovered that Metro-Vick was winding its magnet at the rate of one coil a week and would finish in eight months, well over a year after commissioning, if something were not done. And the Cavendish magnet had precedence over Liverpool's. After appeal to Metro-Vick, they expected delivery in August, then in October; but neither Cambridge nor Liverpool had its magnet for Christmas. As they realized their orders competed with commissions for armaments. This difficulty also stymied magnet making in Sweden, whose noble neutrality and Nobel industry earned it an enviable trade in arms.[54] In Copenhagen they looked forward to having their donated magnet around Easter 1936, but it had not arrived by September; when it came it exceeded all expectations as to the magnitude of its field, but failed in homogeneity, which could not be corrected by shimming because of the narrow gap. That made a most awkward situation, since Thrige, which had built and donated the magnet, did not like to admit and rectify its

52. Lawrence to Cockcroft, 25 Nov 1935, and Cockcroft to Lawrence, 15 Nov 1936 (4/5); Chadwick, *Nature, 142* (1938), 632; Rasmussen to Bohr, 30 Oct 1935 (BSC); Frisch to Meitner, 12 Nov 1935 (Frisch P); Jacobsen, *Fys. tidssk., 39* (1941), 37; Secretary, Collège de France, to Joliot, 30 Jul 1936, and Oerlikon to Joliot, 9 Nov 1936 (JP, F25); cf. Cooksey to Oliphant, 10 Sep 1938 (14/6). Lawrence volunteered advice and blueprints to Joliot in Sep 1936, but by then the dimensions of the magnet had been set; Lawrence to Joliot, 12 Sep 1936, and Joliot to Lawrence, 13 Oct 1936 (JP, F25).

53. Joliot to Directeur général des douanes, 20 Jan 1938 (JP, F25).

54. Sagane to Lawrence, 27 June 1937 (9/39); Cockcroft to Lawrence, 2 Aug and 13 Nov 1937 (4/5); planning meeting at Metro-Vick, "Minutes," 13 May and 4 June 1937 (UA Cav, 3/1/35–36); Chadwick to Lawrence, 7 Aug 1937 (3/34); Emo to Lawrence, 24 Oct 1937 (7/4); Chadwick, *Nature, 142* (1938), 632; Jacobsen, Dansk. Vidensk. Selsk., *Math.-fys. meddelser, 19:2* (1941), 4; Friesen to Cooksey, 8 Aug 1937 (17/47).

mistake.[55] The deficiency had to be corrected by cutting slits in the pole stems to admit shims and by reducing the acceleration chamber to a brass ring sealed directly onto the pole faces.[56]

The precise Swiss precisely calculated their delay in advance: seven and a half months, counting from November 1, 1936. They finished the following May and tested in June; the magnet proved "eminently satisfactory," *magnifique*, "beyond expectation," "much better [according to Wolfgang Gentner, who saw it at Oerlikon] than the American ones." It could give 22 kG, though, to be sure, at a great expenditure of power, 130 kW. Then it occurred to Joliot that the 9-cm gap that came with the big field might not accommodate the dees. He asked Lawrence. The answer—that the large capacitance between the dees and the tank consequent on their proximity would certainly make many difficulties in the construction of the oscillator circuit—was not reassuring.[57]

The narrowness of the gap exactly matched the place that Joliot planned to put his cyclotron, a sub-basement at the Collège de France beneath the foundations of the chemistry building, enlarged from a cellar for storage of dangerous materials, a place condemned by government architects for lack of heat and ventilation. Into this hole Joliot expected to fit all the auxiliary equipment for his cyclotron and to hook it up to the vacuum chamber crammed with little working space into the 9-cm gap between the pole pieces of his magnificent magnet. An American visiting in 1939 saw a machine "built in a strangely cramped way by a Swiss firm and...stowed away in a strangely narrow subterranean chamber where working conditions...are not only uncomfortable but positively dangerous." Another inspector of Joliot's cyclotron cave judged "the French attack on nuclear physics [to be] about as adequate as their preparations to repel Hitler."[58]

55. Frisch to Meitner, 27 Jan, 8 Sep 1936, 6 May 1938, and to Weisskopf, 2 May 1938 (Frisch P); Weaver, "University of Copenhagen—Biophysics," ca. June 1938 (RF, 713D).

56. Bohr to Scherrer, 25 Aug 1939 (BSC).

57. Oerlikon to Joliot, 9 Nov 1936 (JP, F25); quotes from, resp., Paxton to Joliot, 16 Jul 1937, and Joliot to Nahmias, 19 Jul 1937 (JP, F25), and Gentner to Joliot, 23 June 1937, in Goldsmith, *Joliot-Curie*, 62; Joliot to Nahmias, 31 May 1937, and Nahmias (then in Berkeley) to Joliot, 12 June 1937 (JP, F25).

58. A. Guilbert, architect, reports of May and of 13 June 1936, and "Note sur les défauts du batiment de chemie nucléaire," Jul 1937 (JP, F34); Paxton to

The oscillator provided as fine an opportunity as the magnet for delay and frustration. Joliot entrusted his difficult radiofrequency system to Culmann et compagnie, specialists in a French specialty, electric furnaces, who offered in June 1936 to install a modification of Livingston's original design, adjustable down to a wavelength of 17 meters, for 45,000 francs without the tubes. The plan was scrapped a year later in favor of an ambitious design with several stages and regulators. The price rose faster than the power, reaching 359,282 francs by June 1938, some two years after the original contract; and that omitted 449,800 francs for equipment that Joliot, who had ordered economies on wiring, regulators, and safety devices, decided he did not need. But with all this the furnace maker never got his oscillators going properly and Joliot had eventually to ask the Rockefeller Foundation for salaries for two specialists in high-frequency construction to correct the installation.[59] The other would-be cyclotron laboratories would not meet the price of commercially built oscillating systems (30,000 DKr from Philips, too much from Metro-Vick) and decided to make them.[60] This decision did not result in disaster because each laboratory had acquired the most important instrument for the efficient construction of a cyclotron: a man trained in Berkeley.

Software

Already in the winter of 1935/36 Lawrence was offering students as well as advice and blueprints to Chadwick and Cockcroft.[61] He recommended Bernard Kinsey, who in three years at Berkeley as a Commonwealth Fellow had become "a thorough master of the art of high frequency oscillators," to the Cavendish. But it was Chadwick who acquired him and made him responsible

Lawrence, 8 Jul 1937 (14/18); quotes from, respectively, Darrow to Lawrence, 27 Jul 1939 (6/9), and L.E. Akeley to Lawrence, 21 Jul 1941 (1/12).

59. Culmann to Joliot, 25 June 1936, 23 Sep 1937, 21 Jan, 5 Feb, and 27 June 1938 (JP, F25); Joliot to H.M. Miller, 28 Nov 1939 (JP, F34).

60. Frisch to Meitner, 27 Jan 1937 (Frisch P), and Chadwick to Lawrence, 7 Aug 1937 (3/34). Bohr's group had the advice of the son of P.O. Pedersen, who had perfected the Poulsen system; Rasmussen to Bohr, 16 Nov 1935 (BSC).

61. On distribution of Berkeley blues: Cooksey to Cockcroft, 19 Apr 1937 and 1 Aug 1939 (4/5); Joliot to Cockcroft, 2 Aug 1936, and Bothe to Cockcroft, 9 Sep 1938 (CKFT, 20/4, 14); Nahmias to Joliot, Apr–Aug 1937 (JP, F25).

not only for the oscillators, which he completed in the summer of
1937, but for the entire installation of the Liverpool cyclotron. A
visitor with Berkeley experience, James Cork, judged the oscillators to be "stupendous." Cambridge chose something less spectacular, a simplification of an oscillator circuit devised by the BBC.[62]
In the spring of 1937 Lawrence proposed to Chadwick and to
Cockcroft that they find a stipend for Harold Walke, another
Commonwealth Fellow nearing the end of his tenure, who wanted
only "enough to barely live on..., to go on with research for a
while." The Cavendish did not supply the pittance, since they
were getting a Berkeley man for free, Donald Hurst, who had an
1851 Exhibition fellowship. With his help ("Hurst is a great
acquisition"), Cambridge got the first faint evidence of a beam in
August 1938. Unfortunately for Walke, Chadwick found him a
place, which he took up in October 1937. Walke and Kinsey succeeded in getting the Liverpool machine going in 1939. Later in
the year, while replacing the grid resistance of Kinsey's stupendous
oscillator, Walke touched a 230-volt line, which was enough to kill
him.[63]

Early in 1937 Joliot realized that he needed Berkeley experience. He obtained a fellowship from the Rockefeller Foundation
for Nahmias to go to the United States and arranged with Bohr to
divide the services of one of Lawrence's students for a year, again
with Rockefeller money. The grant for a travelling student was
made, but Lawrence declined to supply one on these terms: Joliot
and Bohr must each have a man for a year. As Tisdale reported
Lawrence's argument, "the job of designing, installing, adjusting,
and testing a cyclotron is considerably longer, more arduous, more
tricky [!], and more difficult than other people expect." As Nahmias misreported it, all the cyclotroneers building outside Berkeley
had had "experience of at least two years with the monster." The

62. Lawrence to Cockcroft, 25 Nov 1935, quote, 21 Mar and 11 Aug 1936 (4/5);
Chadwick to Lawrence, 25 Feb 1936, 7 Aug 1937, and 16 Apr 1938 (3/34); Cork to
Lawrence, n.d. [Aug 1938] (5/1); Cockcroft, *Jl. sci. instr., 16* (1939), 40–1.

63. Lawrence to Cockcroft, 7 Jul 1937, quote, and 5 Oct 1938, and Cockcroft to
Lawrence, 24 Oct 1938 (4/5); Emo to Lawrence, 24 Oct 1937 (7/4); *Nature, 142*
(Sep 1938), 522; Lawrence to Rutherford, 28 May 1937 (ER); Lawrence to
Chadwick, 7 Jul 1937, and Chadwick to Lawrence, 7 Aug 1937 and 24 Dec 1939
(3/34); the testimonial to Hurst is in Cockcroft to Lawrence, 13 Nov 1934 (4/5).

Rockefeller Foundation doubled its grant accordingly. The Berkeley experts—Paxton (Paris) and Laslett (Copenhagen)—arrived at their stations in July and September 1937, respectively.[64]

Laslett needed all his talents to bring about the reconstruction of the Copenhagen magnet and to design and help make the thousands of components deemed too expensive to buy. He built and wired the switchboard that controlled the measurement instruments and safety equipment. By October he had proved himself to be of great use; in the end he was "indispensable," and Bohr arranged to keep him on with a grant from the Rask-Oersted Foundation, which recirculated dollars from Denmark's sale of the Virgin Islands to the United States in 1917.[65] That the Copenhagen machine could give a tentative beam in November 1938 (or, rather, a few flashes on a fluorescent screen, "which can hardly have been anything but positive ions"), and so became the second European cyclotron to operate, was "to a large extent due to his efforts."[66]

Paxton shouldered a more formidable task. Joliot had left his magnet in Zurich while awaiting completion of its subterranean den. It was buried there in January 1938 with a vacuum chamber designed by Paxton and built by Oerlikon between its teeth. Swiss efficiency again threw British bumbling into relief: it took Metro-Vick over a year to finish Liverpool's vacuum tank and perhaps as long to do Cambridge's, although both were built almost exactly to plans furnished by Lawrence.[67] (Cooksey had designed and built

64. Memo to file, Fall 1937 (RF, 713D); Tisdale to Weaver, 8 Jan 1937, to diary, 10 Mar and 12 Apr 1937, and to file, 5 May 1937, quote, all in RF, 713D; Nahmias to Joliot, 28 Apr 1937, and Lawrence to Joliot, 25 May 1937 (JP, F25); Lawrence to Bohr, 8 Jul 1937 (3/3).

65. "F.H.B" and "E.B.," "Diary," 9 May 1937, and "University of Copenhagen—Biophysics," [1938] (RF, 713D); Frisch to Meitner, 7 Oct 1937, 6 May and 8 Aug 1938, and to Ellen and V.F. Weisskopf, 10 Dec 1937 (Frisch P); Jacobsen, Dansk. Vidensk. Selsk., *Math.-fys. meddelser*, *19:2* (1941), 4; Lassen, *Fys. tidssk.*, *60* (1962), 92.

66. Jacobsen to Bohr, 25 and 30 Nov 1938, first quote (BSC); Hevesy to Lawrence, 5 Dec 1938 (9/7), quote. Cf. Lassen, *Fys. tidssk.*, *60* (1962), 92–4; Frisch, *What little*, 106.

67. Paxton to Schnetzler (Oerlikon), 9 Nov 1937, and Oerlikon to Joliot, 14 Jan 1938 (JP, F25); Nahmias, *Machines*, 35; Joliot to Guilbert, 5 Oct 1937 (JP, F34); planning meeting at Metro-Vick, "Minutes," 13 May 1937 (UA Cav, 3/i/35); Cockcroft to Lawrence, 13 Nov 1937 (4/5); Chadwick to Lawrence, 16 Apr 1938 (3/34).

his tank in a month or two.) But Joliot's high-frequency system would not work, or put enough voltage on the dees if it did; leaks and sparks still plagued the vacuum system and insulators when Paxton left Paris in September 1938 to help start up the Columbia cyclotron. By the end of January 1939, Nahmias, who had returned with such Berkeley experience as he had permitted himself to acquire, had cured some of these ills and sought a better vacuum, more dee voltage, and a beam. He was at last requited, on March 3, 1939. He immediately called Tisdale. "Nothing will do but that I must come over and see the phenomenon—which I did. Needless to say that there is great elation."[68]

4. THE NEW WORLD AND THE OLD

Running In

None of the European cyclotrons was in regular operation in 1938. The Cavendish had only "evidence" of a deuteron beam at the end of August (they had been expecting since March), but nothing on target; they switched to protons, coaxed forth 0.02 μA, and started shimming. Lawrence advised more voltage on the dees; hit the ions hard enough, he said, and "the beam will come through without any attention to shims at all." That worked: before Christmas Cockcroft had 12 μA of 5 MeV protons. But in trying to reach the design energy for deuterons, some 11 MeV, he ran into trouble from parasites, faulty insulators, and a badly machined chamber. At the end of May 1939, the deuteron beam amounted to only 3 μA at 9 MeV; the following month it had risen to 5 μA steadily on target and as much as 15 μA at peak performance.[69]

68. L.A. Turner to Frisch, 23 Aug 1938 (Frisch P); Pegram to Joliot, 23 June 1938, Paxton to Joliot, 14 Nov 1938 and 4 Mar 1939, to "Ignace," 29 Jan 1939, and to Nahmias, 12 Feb 1939 (JP, F25); Tisdale, "Diary," 4 Mar 1939 (RF, 500D).

69. Hurst to Cooksey, 31 Jan 1938 (4/21); Cockcroft to Lawrence, 24 Oct 1938, 21 Jan 1939, and [June] 1939, Lawrence to Cockcroft, 16 Nov 1938, and Cockcroft to Cooksey, 29 May 1939 (4/5); Cockcroft, *Jl. sci. instr., 16* (1939), 41-2.

That was enough to do experiments and almost to impress visitors. J.G. Spear, of the Strangeways Research Hospital, Cambridge: "The Cambridge cyclotron is now beginning to function and is running at about 5 micro-amps—sufficient to make biological experiments possible but not up to the Berkeley standard yet." Bohr judged Cockcroft's cyclotron to be as capable as his own. Lawrence applauded the onset of steady operation and the 15 μA "so early in the game." It had taken three years from the commissioning of the magnet, experts and blueprints from Berkeley, the largest British electrical manufacturer, and the resources of the Cavendish to accomplish the feat. That was the fastest English pace; they still sought a stable beam in Liverpool. "With your cyclotron working so nicely [Lawrence wrote Cockcroft], Chadwick and Kinsey should feel much better, for I am sure that with all their trouble they must have doubted that a cyclotron could be made to work satisfactorily [in Britain]." Both British cyclotrons operated satisfactorily and at substantial currents during the early war years, when they provided information for guiding speculations about the possibility of nuclear explosives.[70]

A similar story can be told of the running in of the Copenhagen cyclotron. It, too, was designed to produce deuterons at over 10 MeV. In December 1938 its builders had about 1 μA of 4 MeV deuterons, with which they drove a beam of neutrons equivalent to the yield from a kilogram of radium mixed with beryllium. The Danish press, which believed Bohr could do anything, advertised that he had made a kilogram of radium. Frisch wanted to use this fine source for physics; Bohr insisted that the machine be adjusted to give bigger currents at higher energies, in order to make in quantity the radioisotopes for which the Rockefeller Foundation had paid.[71] Instead, it broke down.

Repairs brought back the microamp of 4 MeV deuterons, "a real thrill, and a great relief," but the machine did not yet run well.[72] The main difficulty was the same as Cambridge's, too little

70. Spear to Cornog, 28 June 1939 (16/41); Bohr to F. Paneth, July 1939 (BSC); Gowing, *Britain and atomic energy*, 61, 403; infra, §10.2.
71. Frisch to Meitner, 26 May 1938, to Weisskopf, 12 Dec 1938, and to Placzek, 18 Dec 1938 (Frisch P); Bohr to Lawrence, 20 Dec 1937 (expecting a beam "within a few months"), and 11 Nov 1938 ("working well") (3/3).
72. T. Bjerge to C. Lauritsen, 20 Jan 1939 (BSC), quote; Frisch to Meitner, 20 Jan, and to Bohr, 22 Jan 1939 (Frisch P).

voltage on the dees; not until the fall of 1939 was Bohr's cyclotron "brought to the stage of producing an efficient beam of high-speed particles." During November the Copenhagen group caught a glimpse of 9 MeV and a grant from the Thrige Foundation to develop the electrotechnical part of the cyclotron "to the utmost efficiency." They achieved a steady deuteron beam at 9.5 MeV in the spring of 1940. During 1939 and 1940, the cyclotron manufactured isotopes for Hevesy as planned. It ran until March 1941, when it was idled for improvements and to conserve electrical power.[73]

Joliot was getting a fair yield of deuterons in June 1939, but the oscillating system still did not perform satisfactorily. He had the machine pulled apart for modifications, which, however, could not be completed before the mobilization of the laboratory. Joliot succeeded in recalling Nahmias from the French Army late in 1939 and in acquiring from the Rockefeller Foundation, then still in business in France, 60,000 francs for stipends for an expert and a helper in high-frequency electronics. The machine had not been returned to working order before the occupation of Paris. "I am thankful for that," Nahmias wrote from the temporary safety of Marseilles, where he had found a job in a cancer clinic, "because being in such a messy state it may look unworthy of a German lab."[74]

The fall of France closed off the Rockefeller Foundation's subvention to Joliot on orders from the U.S. government and brought the German military into the sub-basement of the Collège de France. The first officers to contemplate the broken, unkempt cyclotron wished to confiscate it for its copper and other strategic materials. The proposal came to the attention of the head of research for German Army Ordnance (Herereswaffenamt), Eric Schumann, who had been alerted to the possibility of nuclear explosives by Paul Harteck. Schumann immediately flew to Paris

73. Jacobsen, Dansk. Vidensk. Selsk., *Math.-fys. meddelser, 19:2* (1941), 4, 23, 27, 30; Bohr to Tisdale, 25 Nov 1939, and Weaver to file, 3 Jan 1940 (RF, 713D); S.H. Jensen to Frisch, 25 Nov 1939 (Frisch P); Hevesy to Lawrence, 11 Sep 1939 (9/7); Hevesy to Urey, 21 Jan and 10 Nov 1941 (Urey P, 2).

74. Nahmias to Hansen, 18 Oct 1940, Joliot to H.M. Miller, Jr., 28 Nov 1939 and 28 Mar 1940, Miller to Joliot, 6 Dec 1939, and to file, 25 Nov 1939, and Nahmias to Miller, 10 Dec 1940 (RF, 500D).

and saved Joliot's cyclotron. An agreement was then concluded according to which Joliot would continue his research and admit three Germans to work in his laboratory.[75]

According to the field representative of the Rockefeller Foundation, Joliot thought the Germans were sincere in their expressed wish "to follow the edict that science was international, that scientific work should go on, and that, as Joliot represented a distinguished member of the scientific world, his laboratory should be approved of and supported by the Germans in every way possible." This highly implausible reading of the situation turned out to be realized. The Germans put Joliot's former collaborator Gentner, then in the service of the Heereswaffenamt, in charge of their presence in the laboratory. Gentner managed to discourage his superiors from removing the cyclotron—it would be too costly and dangerous to rip it from its cellar—and to protect it from entrepreneurs like Manfred von Ardenne, who was trying to interest Nazi agencies in building accelerators to assist in research on the exploitation of nuclear energy.[76] With the help of Gentner and German experts in radio technology and of Oerlikon, which made up the steel and copper Joliot procured into new parts for the electromagnet, the Paris cyclotron was at last set to going reliably, with an output of alpha particles equal in number and energy to those from 100 kilograms of radium.[77]

It is said that the cyclotron faltered when Bothe wanted to use it to study the fission of uranium. The French operating crew then sabotaged it in subtle ways that gave Bothe the impression that he had mishandled the controls. Gentner understood and overlooked the maneuver. He protected Joliot, whom he knew to be a leader of the Resistance, and helped French physicists elude the Gestapo. The German authorities did not approve of Gentner's style of supervision and in 1942 returned him to

75. D.P. O'Brien to Joliot, 31 Oct 1940, and Joliot to R. Letort, 20 Oct 1941 (JP, F34); O'Brien, "Diary," 12 Sep 1940 (RF, 500D); Irving, *German atomic bomb,* 36, 40–1.

76. O'Brien, "Diary," 12 Sep 1940, and Nahmias to Hansen, 18 Oct 1940 (RF, 500D); Ardenne to Joliot, 27 Nov 1940 (JP, F28); Irving, *German atomic bomb,* 76–8, 89; Ardenne, *Mein Leben,* 157–9, describing activities little in keeping with the humanitarian intentions he proclaims, ibid., 149, 268.

77. Langevin, "Rapport" on Joliot, 1943, 15 (Langevin P); Oerlikon to Joliot, 27 Sep 1941, 15 and 22 Jan 1942, 26 Jan 1948 (JP, F25).

Heidelberg, where he made the first cyclotron in Germany, give an indication of a beam in December 1943. This machine had been planned since 1937, together with one for the University of Leipzig, both of which were ordered from Siemens in 1939. Siemens worked on them and on a larger one for the Heereswaffenamt during the war, while Krupp tended to two others, for Manfred von Ardenne and for the Research Institute of the German Post Office. The very sizable expenditures in strategic material and trained mechanics this manufacture required were intended by the physicists and the manufacturers as investments for a postwar competition in useful radioisotopes. It was also, according to a consensus of industrialists and Nazi officials, "a matter of prestige for Germany, which must be pursued even during the war, although cyclotrons have no decisive military importance." These industrialists also invested in betatrons, "exclusively," according to Steenbeck, "for business purposes. If the Americans were working on it, we must hurry, so that after the war, no matter how it turns out, we can bring our apparatus into the market place as soon as possible, if necessary with the American firm we license."[78]

Two cyclotrons were operational when the war ended: Bothe and Gentner's in Heidelberg, and the Post Office's in Miersdorf. Neither enhanced its builders' market value. The Russians stole the Miersdorf machine. As for Gentner, he had no need of out-of-date cyclotronics to enhance his merits. The French remembered his friendship and courage in the matter of Joliot's cyclotron and made him an officer of the Légion d'honneur.[79]

Running Over

It is scarcely an exaggeration to say that a week's sweating over a Berkeley cyclotron was worth six months' immersion in its blueprints. In just a few days in the spring of 1937 Cockcroft saw enough to "feel...that the uncertainties in my mind about

78. Osietzki, *Technikgesch.*, 55 (1988), 36; Steenbeck, *Impulse*, 124.
79. Weisskopf in MPG, *Gedankfeier*, 24–5; Gentner, ibid., 41–4; Weart, *Scientists in power*, 156–60; Goldsmith, *Joliot-Curie*, 62, 99–101; Joliot, "Autobiographie," in Joliot, *Textes choisis*, 91–2; Gentner in FIAT, *Nucl. phys.*, 2, 28–31; Salow in FIAT, *Nucl. phys.*, 2, 32–3.

cyclotron operation have been completely removed." His visit followed immediately after one by Bohr, who found in the working machine reassurance for himself and his patron: "The decisive importance of the new grant for our work [he wrote the Rockefeller Foundation] has become still more clear to me during my stay in Berkeley, where I have been most impressed by the ingenuity with which Professor Lawrence and his group in the Radiation Laboratory ha[ve] developed his wonderful cyclotron into an ever more efficient but of course an ever more complicated apparatus."[80] Also in the spring of 1937 Nahmias arrived from Paris and Sten von Friesen from Stockholm, presaging, so Lawrence fancied, a "world wide epidemic of cyclotron construction."[81]

The epidemic continued in the winter of 1938/39 with Oliphant from Birmingham and Bothe and Gentner from Heidelberg. Their letters point not only to the value of the information they acquired but also to a quality present in an exaggerated degree in Berkeley and rapidly dwindling in Europe. Bothe: "I am especially impressed by the atmosphere of enthusiasm and comradeship ruling in your laboratory." And well he might be, since by then the institutes of the Kaiser-Wilhelm-Gesellschaft were riddled with Nazis. Among the "noteworthy" items he mentioned in his official report on his trip to the United States was "the model camaraderie among the 10 or 15 members of the cyclotron crews."[82] Gentner had not expected the great hospitality and generosity he experienced in Berkeley, where no one seemed to mind that he represented (though he did not approve) a totalitarian state. He remained for seven weeks in the cyclotroneers' Mecca: "For here [as he explained his long sojourn to his sponsors] is the center of cyclotron construction, and all other installations are more or less close imitations of the fundamental work of Professor Lawrence."[83]

80. Mann, *Nature, 143* (8 Apr 1939), 585; Cockcroft to Lawrence, 13 Apr 1937 (4/5); Bohr to Weaver, 4 Apr 1937 (RF, 713D).
81. Lawrence to W. Buffum, 5 Apr 1937 (3/38).
82. Bothe to Lawrence, 20 May and 30 June 1939 (3/6); Bothe, "Bericht über eine Vortrags- und Studienreise nach USA," 19 Aug 1939 (MPG, Bothe NL/30).
83. Gentner to Cockcroft, 28 Feb 1939 (CKFT, 20/11); Gentner, "Bericht über die Reise nach Nordamerika," attached to Bothe to Generalverwaltung, Kaiser-Wilhelm-Gesellschaft, 25 May 1939 (MPG, KWG Akten/1063).

As for Oliphant, he was swept of his feet: "Many things about the cyclotron are now clear, which formerly were hazy....I return with a greater confidence and a greater belief in the cyclotron, in physics, and in mankind."[84] Oliphant had plenty of money—some 60,000 pounds from Lord Nuffield, the magnate of Morris Motors, more than enough to outdo Austin's cyclotron at the Cavendish— and plenty of enthusiasm. Cambridge copied the 37-inch; Birmingham would exceed the 60-inch. Oliphant expected to be finished by Christmas 1939 and did manage to erect his magnet, "of phantastic dimensions," according to Frisch, who saw it in August. But Oliphant's new confidence in mankind was misplaced; war stopped construction, and Lord Nuffield's cyclotron was obsolete when it started up in 1950.[85]

The culture shock experienced by some Europeans who spent time in American accelerator laboratories makes the same point in reverse. The lust after machinery, the squandering of time on mere technical improvements, offended them as uncivilized and unscientific. "Americans are mostly coarse types, very good workers but without many ideas in their heads....Their number is impressive, it is true, but one should not worry too much about their technical facilities. It will be a long time before they get from them what they can." So wrote Walter Elsasser, a Göttingen Ph.D., who had worked in Germany and in France and was to make his career in the United States. He excepted Lauritsen from his indictment: "Everything in his laboratory is built with great simplicity and without the technical elegance that Americans love so much....Lauritsen is a European by birth as well as in spirit."[86] Likewise Emilio Segrè remarked on the want of subtlety of the machine makers at the Radiation Laboratory. Segrè trained all over Europe—in Rome with Fermi, in Hamburg with Stern, and in Amsterdam, where he continued his studies of spectroscopy in the laboratory of the old master, Pieter Zeeman. This itinerary provided a perspective quite different from Berkeley's: Fermi and

84. Oliphant to Lawrence, 11 Jan 1939 (14/6).
85. Oliphant to Lawrence, 19 Jul and 20 Aug 1938 (14/6); Frisch to "Franz," 16 May 1939, and to J. Koch, 28 Dec 1939 (Frisch P); Cockburn and Ellyard, *Oliphant,* 73–4, 78, 136.
86. Elsasser to Joliot, 13 Sep 1936 (JP, F28). This sentiment does not recur in Elsasser, *Memoirs,* chap. 8, "Passage to the New World."

Stern had command of deep theory as well as of experimental technique, and Zeeman had made a career of accuracy in measurement. As Franz Kurie wrote of himself and his fellow cyclotroneers: "One feels quite the blundering caveman beside one's spectroscopic brothers."[87]

We already know some of Nahmias's ideas about American techno-physics. Here is his summary: "I've observed here [Tuve's lab] and at Van de Graaff's a certain rush to realize projects immediately after we first discuss them. [Americans] work quickly and in groups, but [he reassured Joliot] you should see their alarm when one talks about future European installations." The coordination needed to realize the projects oppressed him so heavily that he decided not to do experimental work in Berkeley. "I occupy my time better in reading nuclear physics, biology, and electro-technology than in hypnotising myself in front of an electroscope with the nth new period [of radioactive decay]."[88] When Lawrence suggested that Joliot ask the Rockefeller Foundation to extend Nahmias's stay to enable him to learn by helping to assemble the 37-inch, Joliot declined, thinking that his emissary had learned and suffered enough.[89]

The fascination with hardware and the subordination of the individual to the group that characterized Berkeley by the late 1930s were to spread from accelerator laboratories to other parts of physics and from the United States to the rest of the world. Ryokichi Sagane may serve as a weather vane. After a year at Berkeley and a return there, he toured laboratories in the United States and then visited the Cavendish. "I was rather disappointed and also astonished," he wrote Lawrence. Although he judged that some pieces of native apparatus showed some ingenuity, it was clear to him that the British like the Japanese would have to derive their methods from the Americans. "So far as the experimental techniques are concerned, America has surpassed very far the England."[90] The award to Lawrence of the Nobel prize in phy-

87. Segrè, *Ann. rev. nucl. sci., 31* (1981), 1–18; Kurie, *Jl. appl. phys., 9* (1938), 692.

88. Resp., Nahmias to Joliot, 24 Mar and 12 June 1937 (JP, F25).

89. Nahmias to Joliot, 2 Aug 1937, Lawrence to Joliot, 25 May 1937, Ragonot to Nahmias, 9 Dec 1937, and Joliot to H.M. Miller, 5 Jul 1937 (JP, F25); Tisdale, "Diary," 12 Jul 1937 (RF, 500D).

90. Sagane to Lawrence, 14 Nov 1938 (9/39).

physics for 1939—an event of great importance for our history—was at once an emblem of this dominance and the certification of the cyclotron at the international level. The American style of physics established a beachhead in Europe before the war. Lawrence's machine was the landing craft.

PLATE 5.1   The early staff of the Radiation Laboratory in the transition year 1932/3 assembled around the 27-inch cyclotron. Left to right: Jack Livingood, Frank Exner, M.S. Livingston, David Sloan, Lawrence, Milton White, Wesley Coates, L. Jackson Laslett, and Commander T. Lucci. LBL.

PLATE 5.2   The staff of the Radiation Laboratory in 1938 assembled under the yoke of the 60-inch cyclotron magnet. Top, left to right: Alex Langsdorf, S.J. Simmons, Joseph Hamilton, David Sloan, Robert Oppenheimer, William Brobeck, Robert Cornog, Robert Wilson, Eugene Viez, J.J. Livingood. Center, left to right: John Backus, Wilfred Mann, Paul Aebersold, Edwin M. McMillan, Ernest Lyman, Martin Kamen, D.C. Kalbfell, William Salisbury.   Bottom, left to right: John Lawrence, Robert Serber, Franz Kurie, R.T. Birge, Ernest Lawrence, Donald Cooksey, Arthur Snell, Luis Alvarez, and Philip Abelson.

PLATE 5.3   A Rad Lab party at Di Biasi's restaurant in Albany. From left to right, standing: Robert Cornog, Ernest Lawrence, Luis Alvarez, Molly Lawrence, Emilio Segrè; second row: Gerry Alvarez (seated), Betty Thornton, Paul Aebersold (standing), Iva Dee Hiatt, Edwin McMillan, Bill Farley; first row: Donald Cooksey, Robert Thornton, and Bob Sihlis. LBL.

PLATE 5.4   Oppenheimer and Lawrence flanking Fermi, probably early 1940. Courtesy of LBL.

PLATE 6.1   The dee system at the MIT cyclotron.   The large cylinders
are the quarter-wave lines.   Livingston and Blewett, 158.

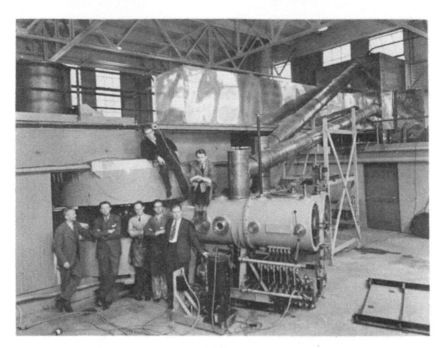

PLATE 6.2   Laboratory staff lolling around the poles and dee supports of the 60-inch cyclotron. Left to right above: Alvarez, McMillan; left to right below: Cooksey, Lawrence, Thornton, Backus, Salisbury. LBL.

PLATE 6.3   Donald Cooksey, G.K. Green, and the mechanism of the dee stem. LBL.

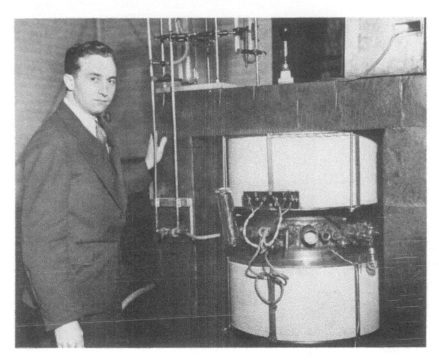

PLATE 6.4   Livingston at the Cornell cyclotron. Courtesy of Lois Livingston.

PLATE 8.1   Water shielding of the 37-inch cyclotron. LBL.

PLATE 8.2   Livingood and Seaborg after a successful hunt. They are hurrying through Sather Gate (the south entrance to the Campus) to the post office to send their latest findings to the *Physical review*. Note the dress for the occasion.   Courtesy of G.T. Seaborg.

PLATE 8.3 Emilio Segrè's ionization chamber, modeled on one used in Rome and in great demand at the Laboratory where precision instrumentation was in short supply.

PLATE 8.4   Robert Stone and John Lawrence treating Robert Penney at the 60-inch neutron port. LBL.

PLATE 8.5   Radioautograph of a tomato leaf at the top of a growing tomato plant thirty-six hours after the plant had absorbed a solution containing P^{32}. Arnon, Stout, and Sipos, *Am. jl botany, 27* (1940), 794.

PLATE 9.1   Luis Alvarez at work in the Rad Lab. LBL.

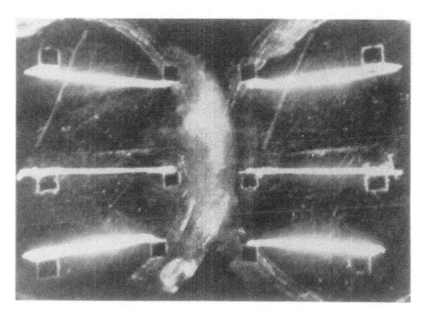

PLATE 9.2   Tracks of fission products. The upper thick bright line is the record of the uranium film; the tracks of the oppositely recoiling fragments run from the film at about 60°. Corson and Thornton, *PR, 55* (1939), 509.

PLATE 9.3   Edwin McMillan at about the time of the discovery of neptunium. Courtesy of the *Oakland Tribune*.

PLATE 10.1 Lawrence, the Comptons, Bush, Conant, and Loomis dis-
cuss the proposal for the 184-inch cyclotron in Berkeley in March 1940.
Left to right: Lawrence, Arthur Compton, Vannevar Bush, James B.
Conant, Karl Compton, and Alfred Loomis. LBL.

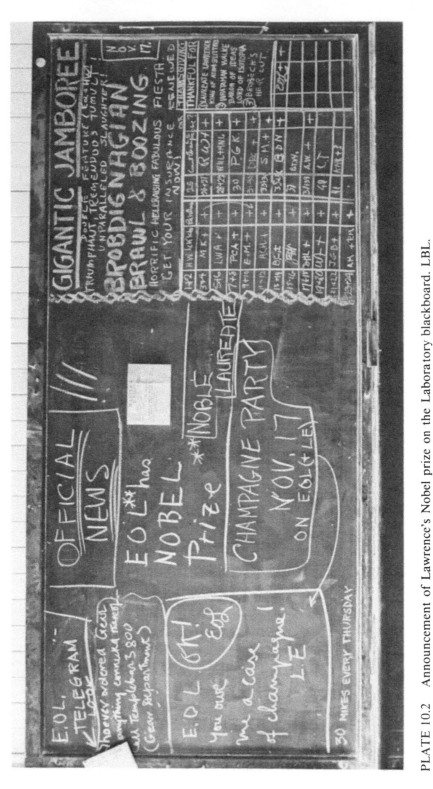

PLATE 10.2   Announcement of Lawrence's Nobel prize on the Laboratory blackboard. LBL.

VIII
New Lines

1. PREVIEW AND OVERVIEW

Beginning in 1934 and 1935 the Laboratory pioneered two sorts of interdisciplinary work centered on products of the cyclotron. The earlier, radiochemistry, enlarged its domain as the energy of the deuterons it chiefly employed increased. Up to the summer of 1935, when the cyclotron first bombarded at 4 MeV, direct activation by deuterons did not penetrate much beyond potassium, although for a time Lawrence thought he had transmuted platinum.[1] In late 1935, with the energy at 5 MeV, the activation extended to antimony; in late 1937, when the 37-inch operated at 8 MeV, and recovered for Berkeley the lead in energy usurped by Michigan a year earlier, it reached uranium.[2] Here the Laboratory had a monopoly.

Neutron activation could be practiced anywhere, with modest means, and throughout the periodic table. Here the cyclotron conferred the advantage of intense neutron beams obtained from deuterons incident on targets of lithium or beryllium. The range and complexity of neutron activation were much increased in October 1934, when Fermi inserted a block of wax for one of lead in a measurement of absorption. (He had in mind, perhaps, the finding of his collaborators that the intensity of activation by

1. Lawrence to Pegram, 6 Feb 1935 (4/8); Cork and Lawrence, *PR, 49* (1936), 205 (talk of 20–21 Dec 1935).
2. Lawrence to Cooksey, 1 Sep 1936 (5/1); supra, fig. 6.1; Kurie, *Jl. appl. phys., 9* (1938), 700–1.

neutrons depended in some cases on the material on which the irradiated substances stood, and the observation by Joliot and Curie that neutrons directed at a layer of paraffin made a particularly strong impression on an ionization chamber.) The silver thimble whose induced activity served Fermi as a measure of neutron flux responded more enthusiastically to neutrons passed through wax than to neutrons striking it directly.[3] He explained that nuclei susceptible to (n,γ) reactions capture slow neutrons more easily than fast ones. Between 1934 and 1936 Fermi's group identified more than 80 radioelements—about half the radioelements known in 1937—made by neutrons fast and slow.[4] The Rome work created both an opportunity and a complication at Berkeley. It indicated that everything could be activated by neutrons; and it required that experimenters learn to discriminate between the effects of fast and slow ones.

The study of the reactions induced by deuterons or neutrons at Berkeley produced more than new entries, some of great importance, in the lengthening list of activities. The Laboratory had to its credit the first artificial (d,p) and (d,n) reactions, which, however, it had to share with Caltech; the first experimental demonstrations of K-electron capture, whereby a nucleus decays by swallowing one of its nearest satellite electrons; a full demonstration of the Oppenheimer-Phillips mechanism of (d,p) transformations; and early examples of isomeric nuclei.

Usually teams of two or three worked at the experiments, but even those who labored alone had important help from other members of the Laboratory. Everyone depended upon the crew on duty to keep the cyclotron going. Many turned to Kurie and his collaborators and successors, for example, J.G. Richardson and Ernest Lyman, to determine in their cloud chamber the sign of the beta particles given off in decays under study. Since each staff member became an expert on a few elements—Kamen recalled

3. Amaldi et al., *Ric. sci.* 5:2 (1934), 282–3, letter of 22 Oct, and *PRS, A149* (1935), 522–58, rec'd 15 Feb 1935, in Fermi, *CP, 1*, 757–8, 765–94; Curie and Joliot, *CR, 194* (1932), 273–5, in Joliot and Curie, *Oeuvres*, 359–60. We follow the reconstruction in Dragoni, *Physis, 18* (1976), 139–60. Cf. the recollections of Amaldi in Weiner, *History of physics*, 309–15; Segrè in Fermi, *CP, 1*, 639–44, in ICHS, X (1964), *Acts, 1* (1964), 149–54 , and in *Fermi*, 77–83.

4. Rasetti, *Nuovo cim., 14* (1937), 377.

that when he started work in 1937 only bismuth and tellurium were unassigned—an explorer had to consult experts on neighboring elements, or McMillan, who became the authority on them all, when hacking through the tangle of reactions in the middle reaches of the periodic table.[5] And also in the middle reaches, particularly among the transition metals of the fourth and fifth period and the platinum metals of the sixth, the chemistry easily acquired by physicists no longer sufficed for clean separation of the elements of an activated target. Chemists became essential collaborators.

The first chemist to have an ongoing connection with the Laboratory, Glenn Seaborg, began in the spring of 1936, when Livingood, perplexed by the mess of activities in a tin target irradiated with deuterons, sought his help.[6] Their ongoing collaboration—Seaborg at the sink, Livingood at the electroscope—produced the largest quantity of information about nuclear reactions obtained by any group in the Laboratory. In the fall of 1938, when Livingood went to Harvard, Seaborg continued the collaboration by mail, struck up a new one with Segrè, and brought his students from Chemistry to help in the increasingly difficult chemical separations.

The collaboration of Livingood and Seaborg instanced not only the assimilation of chemists into the Laboratory, but also a fundamental change in the character of its nuclear chemistry. Their first try with tin was made in the old hit-and-run manner: Livingood bombarded with deuterons for three hours; Seaborg separated for a few more; Livingood observed the decaying fragments; and the two enriched the *Physical Review* with rough indications of three activities in indium (if they were not owing to tin contamination), two in tin, and two in antimony. There were perhaps twenty-four hours of experiment in all and three months from irradiation to manuscript. The business had, of course, to be repeated.[7] Gradu-

5. Kamen, *Radiant science*, 66; Cooksey to S.W. Barnes, 19 Apr 1937 (2/24); cf. McMillan to Barnes, 18 Apr 1937 (2/24), and McMillan's compilation, "Atomic masses derived from disintegration data," mentioned in Lawrence to Evans, 4 Feb 1936 (7/8).

6. Seaborg, *Jl.*, *1*, 128 (3 Apr 1936).

7. Livingood and Seaborg, *PR*, *50* (1 Sep 1936), 435, rec'd 29 June; Seaborg *Jl.*, *1*, 137, 196–8, 438 (19 May 1936, 30 Jan 1937, 2 Feb 1939).

ally Livingood and Seaborg's experiments became more elaborate and reliable. Between February and June 1937, for example, they bombarded seven elements from chromium to antimony, some many times, with deuterons or neutrons; Seaborg split the activities into a great many samples, which Livingood filed and observed for as long as they showed life, sometimes for two years and more. Among the products disclosed that busy spring was Fe^{59}, which became the standard, if rare, tracer for iron in studies of the blood.[8]

Like radiochemistry, the second pioneering line, radiobiology, danced to the tempo set by the machine. In the fall of 1933, the cyclotron's beam of 3 MeV deuterons drove out so great a flux of neutrons from beryllium that, Lawrence wrote Poillon, "we are already worried about the physiological effects on us." Lawrence was then about to set off for the Solvay Congress in Brussels; en route he planned to consult Wood at Columbia about the physiological effects of neutron irradiation, which, he thought, might have a satisfactory side, "of great medical importance."[9] Lawrence supposed that neutrons might destroy cancers more effectively than x rays, because, as Chadwick had shown, neutrons pass more readily through dense materials like lead and bone than through light hydrogenous matter like paraffin and body tissue. This differential absorption would make images taken by neutron rays the inverse of those by x rays—what appears light in the one being dark in the other—and perhaps also confer some special therapeutic benefit. "There is [therefore] some justification for the belief that the discovery of neutron rays is of an importance for the life sciences comparable to the discovery of x rays."[10]

Losing no time, Lawrence discussed the potential of big neutron beams with Dave Morris, treasurer of the Research Corporation, vice president of the Macy Foundation, and, in 1933, Roosevelt's new ambassador to Belgium.[11] At Morris's suggestion, between

8. Seaborg, Jl., 1, 200–51 (6 Feb–4 June 1937), 314 (12 Feb 1938), 486, 500 (13 May and 22 June 1939); Livingood, Fairbrother, and Seaborg, PR, 52 (1937), 135; Livingood and Seaborg, PR, 55 (15 Feb 1939), 414–15; infra, §10.3.

9. Lawrence to Poillon, 4 Oct 1933 (15/16A), quote, and to Exner, same date (9/16).

10. Lawrence to Kast, 18 Nov 1933 (2/29). The argument recurs in Lawrence, Radiology, 29 (1937), 313.

11. Current biography, 1944, 483, for Morris; Josiah Macy, Jr., Found., Review (1937), 53–67.

Brussels and Berkeley Lawrence talked over his expectations with Macy's president, Ludwig Kast, who saw enough in them to grant the Laboratory $1,000 by the time Lawrence's train arrived in San Francisco. Two other grants of over twice that amount followed progress in the aggrandizement of the beam. In January 1934 Berkeley's neutron ray, "more penetrating than either x rays or radium," was intense enough to penetrate the *New York Times*.[12] In February Lawrence talked about it with the University Explorer over NBC. But by then his energies were fully engaged in proving his unstable deuteron and he found it more comfortable to hint at atomic power than at cures for cancer.[13] The collapse of the Berkeley teachings about deuterons and the discovery of artificial radioactivity then shelved development of neutron sources and concern with neutron hazards.

The steady increase in energy and current of the deuteron beam reopened the question of safety, and the "discovery" of radiosodium put a premium on pushing intensity and danger further. In April 1935 H.F. Blum, of the University's Department of Physiology, who wanted to try the effects (which he expected to be catastrophic) of neutrons on tissue containing lithium, agreed to look into safety.[14] During the summer and early fall of 1935, John Lawrence, then very much concerned with safety, was in Berkeley recovering from an automobile accident. He followed up Blum's experiments and learned that, as his brother put it, "neutron rays are considerably more lethal biologically than x rays." "We are getting harmful doses of neutrons in a few minutes when we stand near the magnet."[15]

12. Lawrence to Poillon, 15 and 26 Mar 1934 (15/16); M. Churchill to Lawrence, 5 Jul 1934 (UCPF); supra, table 5.1; "New neutron ray more powerful than either the x ray or radium," *New York Times*, 17 Jan 1934, 1. Cf. ibid., 18 Jan 1934, 30, and 21 Jan 1934, §IX, 5:3, and Sproul's report for 1932–4, in Birge, *History*, 4, xi, 22.

13. "University Explorer," no. 38, aired 5 Feb 1934, and John F. Royal, NBC, to Lawrence, 19 Feb 1934 (40/15).

14. Lawrence to Leuschner, 12 Apr 1935 (20/13); Lawrence to Kast, 8 Mar, and to H.N. Skenton, 25 June 1935 (2/32), re the lithium experiments.

15. Quotes from, resp., Lawrence to Rutherford, 13 Aug 1935 (ER), and to Tuve, 12 Sep 1935 (3/32). A detector responsive to the gamma rays produced by (n,γ) on cadmium registered 1,000 counts/m 75 feet from the target; Lawrence to Cockcroft, 2 Oct 1935 (4/5).

Exposure of laboratory personnel to penetrating radiation had long worried directors of laboratories with high-voltage x-ray equipment, especially Tuve, who began in 1929 to press for studies of the long-term biological effects of the rays from his Tesla coil. As he wrote Aetna Life, who were doubtless pleased to read it, he took "extreme precautions," setting the tolerable dose at under 0.1 roentgen a day and the total integrated exposure to less than 1 percent of what it took to kill a rat.[16] A roentgen, or "r unit," is the quantity of x rays that makes about a billion ion pairs (1 esu to be exact) when passing through a cubic centimeter of air. The second International Congress of Radiology, meeting in Stockholm in 1928, took the roentgen as its unit and during the 1930s most workers in the field adopted a dose of 0.1 r of x rays as the tolerable daily allotment.[17] What should be the limit for the more deadly neutron rays? John Lawrence first suggested no more than a quarter of the x-ray limit, in accordance with his preliminary measurements on the relative lethality of the two radiations to mice. Berkeley tended to be more generous than other places. MIT's Evans recommended a limit of 0.001 n unit/day (an "n" unit being the quantity of neutron radiation registering 1 r in a certain standard detector), Aebersold ten times that, McMillan four times Aebersold's limit. No national or international level of tolerance to neutron rays was established before the war.[18]

The discovery of the uncertain danger of the neutron background brought prudence to the Laboratory. In 1935 the dose four feet from the beryllium target was 0.2 r an hour, five times the allowable x-ray rate per day; a water barrier cut it down "quite a lot," but not enough to stop the Laboratory's neutrons from spoiling experiments in the Chemistry building.[19] The 37-inch had a more extensive shield, a wall of water three feet thick; Fermi thought this extravagant, but the National Advisory Cancer

16. Tuve to Fleming, 9 Dec 1929 (MAT, 4); to Captain C.M. De Vain, Naval Medical Officer, Philadelphia Division, 1 Apr 1932, and to Walter S. Paise, Aetna, 24 Feb 1933, both in MAT, 8; Tuve, "Report on high-voltage work," 3 Dec 1929 (MAT, 9/"lab. file").

17. Hacker, Dragon's tail, 15–8.

18. J.H. Lawrence and Tennant, Jl. exptl. med., 66 (1937), 687; Aebersold to Evans, 9 Nov 1940 (7/8); McMillan, "Shimming," 23.

19. Lawrence to F.C. Wood, 31 Jul 1935 (9/21); to Zirkle, 4 May 1937 (18/46); and to Johnson, 9 Feb 1937 (10/1).

Council, which authorized $5,000 for it, did not. The three feet of water reduced the radiation at the control desk by a factor of three; the addition of a layer of water 1.5 feet deep above the cyclotron drove it down by a factor of ten (plate 8.1).[20] As for the 60-inch, with four feet of water around the entire machine, the dose at the controls fell to an insignificant 0.001 r/d, or so the heads of the Laboratory claimed.[21] Aebersold calculated that a five-foot shield would be necessary to keep the exposure at the controls, some forty feet from the target, to 0.01 n/d (n units per day). The wisdom in the field was that the background activity in Berkeley exceeded norms in other cyclotron laboratories.[22] Further to prudence, the cyclotroneers carried ionization gauges in their pockets so as to meter what *Time* called the "new lethal death ray hurled by magnet[s]" and *Science Service* represented as a "deadly danger for young researchers." They also had their blood drawn "every so often."[23] Documentation of the seriousness of the neutron hazard returned Lawrence to his idea of neutron therapy.[24]

In 1936 the Laboratory began to acquire biologists and physicians just as it did chemists. John Lawrence returned for the summer and came back for good in 1937. Macy money and Rockefeller riches allowed him to build up a sizable group, which, by 1940, numbered himself as director; five doctors of different sorts, including a visiting fellow, but not counting Stone, who supervised neutron therapy on behalf of the Medical School; a nurse and laboratory assistant; and four operators of the 60-inch cyclotron. The biomedical staff distributed itself into three

20. Lawrence to Rutherford, 24 Feb 1937 (ER), in Oliphant, *PT, 19:9* (1966) in Weart and Phillips, *History*, 190; Seaborg, *Jl., 1*, 201, 399 (8 Feb 1937, 22 Oct 1938); Kurie, *GE review, 40* (June 1937), 271; Lawrence to Hughes, 5 Jul 1939 (18/12), to Hektoen (NACC), 20 Nov 1937 (13/29), and to Harnwell, 20 Mar 1940 (14/22); J.A. Fleming to R.D. Evans, 9 Oct 1939 (MAT, 23/"cycl. letters").

21. Cooksey to Moore Dry Dock Co., 3 Aug 1939 (25/5), and to Harnwell, 20 Mar 1940 (14/22).

22. Aebersold to Evans, 9 Feb 1940 (7/8); Cowie to Cooksey, 24 Sep 1940 (5/4), reporting the opinion of Newson, Snell, and Thornton.

23. Lawrence to Cockcroft, 12 Sep 1935 (4/5); *Time*, 16 Dec 1935, 32; *Science service*, 27 Feb 1936, and similar reports in the *New York World Telegraph*, 4 May 1936, and the *New York Times*, 14 June 1936 (11/16); John Lawrence to Lawrence, 24 Mar [1936] (11/16), on bloodletting, quote; Tuve, "Memorandum concerning Cowie's eyes," 1947 (5/4).

24. Lawrence, "The work" (Aug 1937), 18, and *Radiology, 29* (1937), 315–6.

research groups, one for neutron therapy, another for work on leukemia, and a third for biological tracers.[25] The enduring members, besides John Lawrence, were Paul Aebersold, who obtained his Ph.D. in 1939, after several years at the Laboratory on fellowships in radiological studies from the Medical School, for work on collimating neutron beams for radiation therapy; and Joseph Hamilton, who worked in radiology at the Medical School after receiving his M.D. there in 1936, came to the Laboratory as a Finney-Howell Fellow, graduated to research associate on NACC funds, and ended, after the war, as director of the Crocker Laboratory.

Neutron therapy began at the 37-inch cyclotron in September 1938. It seemed at first to offer some advantage over treatment by x rays. So did the ingestion of P^{32} in cases of chronic leukemia and polycythemia vera, and radioiodine for diseases of the thyroid. Radiosodium, in which so much was invested—cured nothing.[26] The 60-inch cyclotron, which could treat more patients than the 37-inch, improved statistics. It appeared that the neutron ray was a cruel disappointment, but that radiophosphorus and radioiodine afforded many sufferers true benefits. The tracer research also had notable successes, particularly in elucidating steps in photosynthesis. And, as in the case of technetium in radiochemistry, discoveries of importance in radiobiology were made outside the Laboratory by people using radioactive preparations made in Berkeley.

2. RADIOCHEMISTRY

Gluttony at the Periodic Table

The Laboratory's earliest radiochemistry, apart from radiosodium, concerned nitrogen and oxygen. These elements lent themselves to experiment: they could be obtained very pure and deployed without fear of surface contamination; they have few

25. Lawrence to A.H. Compton, 14 Mar 1939 (4/10), and to F.C. Blake, 12 Jul 1940 (3/1).

26. Lawrence, "Report to the National Advisory Cancer Council," 12 Oct 1939 (13/29A); Aebersold to Zirkle, 19 Aug 1939 (1/9).

naturally occurring isotopes to confuse analysis; and they are easily excited by deuterons at 2 or 3 MeV. The experimental setup, which remained standard for gases, is indicated in figure 8.1. The first to work it were Lawrence, Henderson, and McMillan. They attacked the air and detected three groups of alpha particles and two of protons, which they assigned to reactions of nitrogen, but found no radioelements and did no chemistry. The ranges they measured for the alpha particles and protons did not agree with more careful determinations by Cockcroft; once again Lawrence had to remeasure and retract. McMillan persisted, substituted nitrogen for air and Livingston for Lawrence and Henderson, and uncovered a positron activity that lasted about two minutes. A little chemistry showed that the active substance formed water; a little reasoning ascribed the activity to a new radioelement, O^{15}, half-life 126 seconds, and to the reactions $N^{14}(d,n)O^{15}$, $O^{15} \rightarrow N^{15} + e^{+}$.[27] In a parallel investigation, Henry Newson, who came from and returned to the Chemistry Department at the University of Chicago, found that F^{17} ($\tau = 1.16$ m)

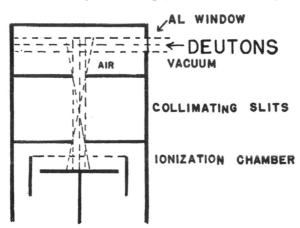

FIG. 8.1 Experimental arrangement for irradiation of gases. Lawrence, McMillan, and Henderson, *PR, 47* (1935), 276.

27. Lawrence, McMillan, and Henderson, *PR, 47* (1935), 273–7, rec'd 2 Jan; Cockcroft to Lawrence, 28 Sep 1935, and answer, 22 Oct 1935 (4/5); McMillan and Livingston, *PR, 47* (1935), 452–7, rec'd 21 Jan, adumbrated in Livingston and McMillan, *PR, 46* (Sep 1934), 437–8.

could be made by (d,n) on O^{16}. He thereby recovered a known activity made by (α,n) on N^{14} and raised his reputation in the Laboratory. (Lawrence had judged him to lack the pushiness needed to accomplish anything there.)[28]

Meanwhile McMillan and Lawrence worked on aluminum, which has but one natural isotope, and Henderson attacked magnesium, which has three. They used an apparatus similar to that of figure 8.1 with the target mounted in the beam. Protons, positrons, alpha particles, and neutrons came off aluminum, which could provide them all via the reactions $Al^{27}(d,p)Al^{28} \rightarrow Si^{28} + e^+$, $Al^{27}(d,\alpha)Mg^{25}$, and $Al^{27}(d,n)Si^{28}$. The activity of Al^{28} (τ = 156 sec), the only one studied, was scarcely fresh, having been prepared in France the natural way, by (α,n) on phosphorus, and in Italy the Italian way, by (n,γ) on aluminum. By placing a series of foils in a line, McMillan and Lawrence measured the "excitation function," the yield of radioaluminum as a function of the energy of the incident deuterons. That brought nothing new either: the excitation function agreed with Gamow's theory.[29]

With a little more energy—3MeV—Lawrence and McMillan, now joined by Thornton, got results that diverged from theory. Oppenheimer became interested, calculated, and concluded that disintegration via (d,p) followed the Oppenheimer-Phillips process at the higher bombarding energies. Meanwhile Henderson was getting two different radioactive products from the heaviest magnesium isotope, Mg^{26}, that is, Mg^{27} (τ = 10 m) via (d,p) and Lawrence's Na^{24} via (d,α). Although both activities were known (Fermi had made Mg^{27} by neutron capture), Henderson could claim the first case in which two different products resulted from the bombardment of a single isotope by the same charged particle. He determined the excitation functions for both, and consulted Oppenheimer. The sage authorized the conclusion that the Na^{24} came into existence by Gamow capture and the Mg^{27} by the mechanism of Oppenheimer and Phillips.[30]

28. Newson, PR, 48 (1935), 790–6, rec'd 3 Sep; Lawrence to W.D. Harkins, 14 Oct 1935 (13/43).

29. McMillan and Lawrence, PR, 47 (1935), 343–8, rec'd 12 Jan; Seaborg, Jl., 1, 282–3 (18–19 Oct 1937).

30. Lawrence, McMillan, and Thornton, PR, 48 (1935), 493–9, rec'd 1 Jul; Henderson, ibid., 855–61, rec'd 16 Sep.

The Laboratory's appetite at its first sitting at the periodic table appears from the menu of the meeting of the American Physical Society held in Berkeley at the end of December 1935. Cyclotroneers gave twelve talks, only one of which concerned machinery. Otherwise the subjects were elements excited by deuterons: copper, nitrogen, and oxygen, whose excitation functions Newson followed to energies above the nuclear potential barrier, where the Gamow curve no longer holds; phosphorus, argon, nickel, cobalt, zinc, and arsenic, made radioactive by Paxton, Snell, Thornton, and Livingood; nitrogen, fluorine, sodium, aluminum, silicon, phosphorus, chlorine, argon, and potassium, whose beta and gamma emissions gave employment to Kurie, Richardson, Paxton, Cork, and their cloud chamber. For most of this work, deuteron energies ran about 3.5 MeV and the elements studied were no heavier than arsenic (atomic number, Z, = 33). This was a little tame and routine for the boss. Lawrence's name appeared on two papers at the APS meeting. In one, with Cooksey and Kurie, he described improvements in the cyclotron that resulted in 6 MeV deuterons; in the other, with James Cork, he announced the discovery that platinum nuclei (Z = 78!) "resonated" when hit by such rapid particles. This response from the tough platinum nucleus, which he thought he had "transmuted" to gold, was most gratifying. As Lawrence wrote the Macy Foundation in October 1935, six months earlier he would not have thought such alchemy possible with energies attainable in the Laboratory.[31]

During 1936 the Laboratory worked its way forward from iron, using the faster deuterons then available and trusting in the efficiency of the Oppenheimer-Phillips process apparently so potent in aluminum. (In fact, as Bethe later showed in an elaborate, but approximate, calculation, the Oppenheimer-Phillips process would not have been detectable below Z = 30.)[32] At the

31. Abstracts of papers at APS, 20–21 Dec 1935, in *PR, 49* (15 Jan 1936), 203–9; Lawrence to Rutherford, 10 Jul 1935 (ER); to Shenton, 10 Oct 1935 (12/32); to Cockcroft, 2 Oct and 25 Nov 1935 (4/5), and to Chadwick, 27 Nov 1935 (3/34). Newson's definitive results were published after he left Berkeley, in *PR, 51* (1937), 620–3.

32. Bethe, *PR, 53* (1938), 42, 47–9, previewed in Bethe, *RMP, 9* (1937), 201–4. Oppenheimer's calculation had presupposed a deuteron more weakly bound than

meetings of the spring and early summer of 1936, Van Voorhis introduced a duplicitous copper, which, having imbibed a neutron perhaps in the manner of Oppenheimer-Phillips, decays by either a positron to nickel or an electron to zinc; he thereby found much, but missed more, since the predominant mode of decay of Cu^{64} is via a process, K-electron capture, then undetected. Livingood reported on the unseparated messes he made with 5 MeV deuterons on several metals and also on his attempt, fleetingly successful, to make the first artificial-natural radioelement (Bi^{210}, alias RaE) via the Oppenheimer-Phillips process $Bi^{209}(d,p)Bi^{210}$. He thought he glimpsed the faint beta decay of RaE, and also alpha particles of the right range to arise from RaE's descendent Po^{210}.[33] The man who first identified RaE, Rutherford, was delighted to know that Lawrence could make a, or perhaps any, link of a naturally occurring radioactive series. "[It is] a great triumph for your apparatus."[34]

Another triumph seemed in the offing. Cork had continued with the experiments on platinum. Together he and Lawrence identified four activities, two arising (according to them) from platinum isotopes excited by an unknown process more powerful than Oppenheimer-Phillips and two from iridium isotopes produced by an unlikely (d,α) reaction. They disclosed further that the excitation function of platinum did not increase monotonically with energy, but showed several bumps or resonances.[35] Oppenheimer developed a new theory to account for it all. At this point, in April 1937, Niels Bohr passed through Berkeley on his way to Japan.

Bohr had just put the finishing touches on his theory likening the nucleus to a liquid drop with modes of excitation incompa-

the best measurements suggested and a Coulomb repulsion effective to vanishingly small distances; both presuppositions favor Oppenheimer-Phillips.

33. Van Voorhis, *PR, 49* (1 June 1936), 876, and Livingood, ibid., APS meeting of 30 Apr–2 May, in Washington; Livingood, *PR, 50* (15 Aug 1936), 385, 391, APS meeting of 17–19 June, in Seattle; Livingood, *PR, 50* (1936), 432–4, rec'd 29 June.

34. Rutherford to Lawrence, 22 Feb 1936 (15/24). Cf. McMillan, *PT, 12:10* (1959), and Oliphant, *PT, 19:10* (1966), in Weart and Phillips, *History*, 264, 186–7, resp.

35. Cork and Lawrence, *PR, 49* (15 Jan 1936), 205, and ibid., 788–92, rec'd 7 Apr.

tible with Lawrence's platinum resonances. At a seminar arranged especially to discuss the latest curiosity of Berkeley experiment and theory, Lawrence announced that his measurements conflicted with Bohr's ideas, but agreed perfectly with Oppenheimer's, and Oppenheimer gave what Kamen remembered as "a typically stupefyingly brilliant exposition of its theoretical consequences." Bohr declared that the data had to be wrong. When he left, the resonances became sharper, the experiments more convincing.[36]

Lawrence did not wish to repeat the saga of the disintegrating deuteron. He summoned McMillan, who had once traced an apparent activity of platinum under deuteron bombardment to radioactive nitrogen driven by recoil into the surface of the metal. McMillan realized that he needed chemical advice (the separations on which Lawrence and Cork had relied were hurriedly done by Newson during his last days at the Laboratory). He called on Kamen and a new graduate student in chemistry, Samuel Ruben. It took them over three months of strenuous chemical work—eighteen hours at a time—to separate the activities and to trace the exotic "resonances" to just plain dirt. By rubbing Laboratory grime into platinum foils before bombarding them, Kamen was able to reproduce most of Cork's measurements.[37]

Products

In the summer of 1936, Segrè, then newly appointed professor of physics at the University of Palermo, visited Berkeley, to see the cyclotron, to escape the heat at his main place of sojourn, Columbia, and to survey possibilities of escaping from the heat in Italy should war threaten. He returned to Palermo "still dreaming of the cyclotron" and carrying some bits of copper strip that he had scavenged from the chamber of the 27-inch. He and his associates separated radioisotopes of copper, zinc, and perhaps manganese from the scrap. 'Twas but antipasto. "We would like very

36. Nahmias to Joliot, 28 Apr 1937 (JP, F25); Kamen, *Radiant science*, 76–8; Lawrence to Segrè, 5 Apr 1937 (16/14), and to Nishina, 13 Apr 1937 (9/38); McMillan and Livingston, *PR, 47* (1935), 454.

37. Lawrence to Oppenheimer, 26 Apr 1937 (14/9), to Segrè, 28 May 1937 (16/4), and to Bohr, 8 Jul 1937 (3/3).

much to have more copper," Segrè wrote. "I think you can send any substance in a letter."[38] Meanwhile the cyclotron had been opened for repairs. Lawrence salvaged more copper and the molybdenum strip that protected the dee edge at the exit slot. He had it all cut up and sent in several letters. It was an act of head-strong generosity. Lawrence suspected that the molybdenum con-tained an activity of long life but of too little promise to add its investigation to the rigorous Berkeley routine. "We are all very busy here, but there is nothing very exciting at the time to report."[39] There would have been some excitement had they kept the hot molybdenum.

If the long activity arose via (d,n) on molybdenum, it belonged to element 43. Number 43, alias davyum, lucium, nipponium, and masurium, had been nondiscovered several times. No trace of it had turned up in the surveys of x-ray spectra by H.G.J. Moseley and his successors; the only evidence for its existence when Lawrence sent Segrè his second installment of scrap was three faint x-ray lines observed by Walter and Ida Noddack in 1925 in the course of their successful detection of element 75 (rhenium). Segrè and Carlo Perrier, Palermo's professor of mineralogy and an accomplished analytical chemist, took the Laboratory's molybdenum apart. They separated a large amount of radiophosphorus, which they found to contaminate everything from Berkeley, and handed it to colleagues in physiology to administer to rats.[40] They tried to carry the residual activity on molybdenum and the elements immediately below it, niobium and zirconium, but to no avail; its chemistry was closer to that of rhenium, the heavy homologue of "masurium." In April Segrè notified Lawrence: "All the activity is due to some substances which have all chemical characters one would expect to find in the element 43." Perrier and Segrè saw indications of three different radioisotopes of 43, to which the Palermo group soon attached half-lives of 90, 50, and 80 days, in order of relative abundance. As for stable isotopes of "masurium," Perrier and Segrè declared

38. Segrè to Lawrence, 9 Sep, 2 Nov, 18 Dec 1936 (16/14), and *Mezzo secolo*, 23–4, 58–9.

39. Lawrence to Segrè, 6 Jan 1937 (16/14); McMillan, *PT, 12:10* (1959), in Weart and Phillips, *History*, 265; Segrè, *Ann. rev. nucl. sci., 31* (1981), 8.

40. *Nature, 139* (1937), 836–7, 1105–6; Kirby, in *Gmelins Handbuch*, 2–3.

them to be "absent," indeed, nonexistent; no known element could be invoked to act as a carrier for the new activity.[41] Consequently, just after the war, when weighable amounts of element 43 were created in nuclear reactors, Perrier and Segrè annihilated "masurium" and named their element, the first made by man before its discovery in nature, "technetium."[42]

"The cyclotron evidently proves to be a sort of hen laying golden eggs." So Segrè wrote Lawrence at the start of the analysis of the molybdenum strip. He publicly acknowledged his gratitude not only by the usual professional thanks, but also, what Lawrence no doubt preferred, by advertising the cyclotron. Perrier and Segrè concluded their presentation of the radiochemistry of element 43: "We hope also that this research carried on months after the end of the irradiation and thousands of miles from the cyclotron may help to show the tremendous possibilities of this instrument." Segrè asked for more active long-lived material and sent some purified uranium oxide for irradiation by slow neutrons in the hope of making more unnatural elements, perhaps alpha emitters from transurania. "I think that when you are producing neutrons every point near to the cyclotron gets a stronger irradiation than with the most powerful sources available in Europe."[43]

Lawrence at first was more impressed by the salvage and application of P^{32} than by the news of element 43: "We are all amazed here to to see the amount of good work you have done with such a trivial amount of radiophosphorus." As for claims based on complex radiochemical analysis, he had reason to be wary: "Of course there are difficulties."[44] By the fall of 1937, when sending P^{32} for Segrè's colleagues and more scrap for Segrè himself, he wrote, with his usual optimism: "We are only too glad to send you

41. Segrè to Lawrence, 28 Apr 1937 (16/14); Perrier and Segrè, Nature, 140 (1937), 193–4, letter of June 13, and Jl. chem. phys., 5 (1937), 713–4, rec'd 30 June, confirmed ibid., 7 (1939), 155; Cacciapuoti and Segrè, PR, 52 (1937), 1252–3, letter of 17 Nov; Segrè in Nicolini et al., Technetium, 5–6.

42. Kirby, in Gmelins Handbuch, 7; Perrier and Segrè, Nature, 159 (1947), 24. The first traces of naturally occurring technetium, arising from the spontaneous fission of U^{238}, were found in 1961; Alleluia and Keller, in Gmelins Handbuch, 12–3.

43. Segrè to Lawrence, 7 Feb, 13 June, and 1 Jul 1937 (16/14); Perrier and Segrè, Jl. chem. phys., 5 (1937), 716.

44. Lawrence to Segrè, 5 Apr and 28 May 1937 (16/14).

material because you have accomplished so much with what little we have furnished." Indeed, although it was not known at the time, Segrè had accomplished precisely what Lawrence had prepared: the creation of new materials interesting in themselves and applicable to medicine. The element has been in clinical use since 1963. One isotope, Tc^{99}, which gives a useful gamma ray, is an important agent in visualizing tumors and abscesses of the liver, in imaging the living skeleton, and in brain scanning. Its production by accelerators became the basis of a multi-million dollar industry.[45]

As Aristotle said, the road from Athens to Corinth runs also from Corinth to Athens. In the case of technetium, Europeans detected a new element in material made in Berkeley. In the case of the isobars of mass three, the Laboratory made a discovery prepared in Europe. Soon after Rutherford and his collaborators had identified the d-d reactions, the Cavendish and also Fritz Paneth and G.P. Thomson at Imperial College, London, sought to produce enough H^3 and He^3 in Cockcroft-Walton accelerators to investigate their physical and chemical properties. All failed. Rutherford supposed that the elusive isobars combined with stray protons and neutrons to form alpha particles, of which, however, no trace could be found.[46] Efforts to detect He^3 by the spectroscope also failed, except in passing and at Princeton, where physicists saw and then did not see lines in the spectra of the products of d-d reactions ascribable to a light helium isotope.[47] By 1935 the case of the uncollectible isobars was attracting attention on both sides of the Atlantic.

The most promising route appeared to be the isolation of "triterium," as Rutherford called it, from heavy water. This quest rested on the assumption that H^3 is stable. Weighing in its favor was a report by Tuve's group of the presence of tritium in the

45. Lawrence to Segrè, 20 Oct, 23 Nov, 23 and 27 Dec 1937 (16/14); Kirby, in *Gmelins Handbuch*, 8; A. Seidel, in ibid., 308–9; NRC, *Phys. persp., 1*, 138.

46. Oliphant, Harteck, and Rutherford, *PRS, A144* (1934), in Rutherford, *CP, 3*, 395; Rutherford, *Nature, 140* (1937), in *CP, 3*, 427–8 (1937); Paneth and Thomson, *Nature, 136* (1935), 334.

47. Harnwell, Smyth, and Urey, *PR, 46* (1934), 437; Smyth et al., *PR, 47* (1935), 800–1. Seaborg also planned a try; *Jl., 1*, 46 (29 Jan 1935). For a time Lewis thought he had H^3; Lawrence to Stern, 20 Nov 1933 (16/49).

heavy water made by Urey and F.G. Brickwedde and the apparent slight excess in mass of lightest helium over heaviest hydrogen.[48] But Tuve's result had not been duplicated and the argument from the masses was far from secure. Although the Cavendish's values of the energies and masses entering into the reaction $H^2(d,p)H^3$ were regarded as so reliable that the mass of H^3 determined from them stood as one of the most certain of nuclear constants, the mass of He^3, deduced from the reaction $H^2(d,n)He^3$, stumbled over the difficulty of measuring the kinetic energy of the liberated neutron. Rutherford's group made He^3 just slightly heavier than H^3; analysis by the meticulous T.W. Bonner and W.M. Brubaker, who also used the reaction $Li^6(d,\alpha)He^3$, made the masses the same within experimental error; and Hans Bethe and R.F. Bacher, scrutinizing it all in the spring of 1936, awarded He^3 the tenuous excess of about two ten-thousandths of a mass unit.[49] Excess in the atom, as in the human, may be a sign of instability. Already in 1934, the Cavendish physicists conjectured that He^3 might transmute into stable H^3 by emission of a positron, in the manner then just made fashionable by the studies of Joliot and Curie.[50]

The first in the field were Walker Bleakney and his associates at Princeton, who had electrolyzed 75 tons of ordinary water down to its heaviest cubic centimeter before learning of the Cavendish evidence for the existence of H^3. They set their precious material free in Bleakney's mass-spectrometer, found traces of particles of mass five that they declared to be molecules of H^2H^3, and inferred that H^3 constitutes about one part in a billion of ordinary water. Their colleagues at Princeton, G.P. Harnwell and H.D. Smyth, corroborated their finding by running the product of a gas discharge in deuterium into the spectrograph; and, in a further dividend, the Princeton mass-spectroscopists identified He^3 in the d-d product as well. A year later Tuve's group again reported

48. Tuve, Hafstad, and Dahl, *PR, 45* (APS, Apr 1934), 840–1.
49. Oliphant, Kempton, and Rutherford, *PRS, A149* (1935), in Rutherford, *CP, 3*, 401–3; Bonner to Bethe, 10 Dec 1935 (HAB/3); Bonner and Brubaker, *PR, 49* (1936), 19–21; Bethe and Bacher, *RMP, 8* (Apr 1936), 87, 147, 197 (the 0.0002 given on 197 disagrees with the 0.0004 on the preceding pages).
50. Cockcroft to Lawrence, 14 Nov 1934 (5/4).

"stable hydrogen atoms of mass 3 in numerous electrolytic deuterium samples" put through their magnetic analyzer.[51]

The British were unable to duplicate the feats at Princeton and Washington. From Norsk Hydro, which by the mid 1930s had become the world's largest producer of deuterium, the Cavendish received 11 grams of the heaviest and most expensive remains of the electrolysis of 13,000 tons of water. (In 1935 Norsk Hydro-Elektrisk Kvaelstofaktieselskab sold almost pure heavy water for $1.25/g in quantities over 50 grams; in 1938, the price had declined to 75 cents/g for lots of 25 grams.)[52] Aston could not find a drop of heaviest water in the Norwegian stock. Meanwhile, Mark Oliphant and Fritz Paneth and G.P. Thomson had failed to confirm Harnwell and Smyth, and guessed that the Princetonians, who had begun to doubt themselves, had been misled by the release of helium that had been dissolved into the glass walls of their discharge tube.[53] By mid 1937 a consensus of sorts had been reached. "The claims of the Americans...were ill-founded," Thomson wrote Rutherford after visiting Princeton. "I am glad that you are coming to a similar conclusion." The next move appeared to Thomson to be to persuade Lawrence to devote a large amount of cyclotron time to irradiating a bucket of heavy water with deuterons in order to search for the spectrum of He^3.[54] Nothing seems to have come of his proposal. The Laboratory had a more certain manufacture than the elusive isobars of mass three.

In July 1939 owing to the idleness of the 60-inch cyclotron as it awaited adequate shielding, the matter was reopened in Berkeley. Luis Alvarez thought to fuse deuterons in the 37-inch and feed the product into the 60-inch, which he would use as a giant mass-

51. Lozier, Smith, and Bleakney, *PR, 45* (1934), 655; Harnwell, Smyth, et al., ibid., 655–6 (both letters dated 21 Apr), and *PR, 46* (1934), 437; Bleakney et al., *PR, 46* (1934), 81–2; Tuve, Hafstad, and Dahl, *PR, 48* (15 Aug 1935), 337; Hafstad and Tuve, *PR, 47* (1935), 506.

52. Lawrence to Urey, 21 Mar 1935 (17/40), and to J.E. Henderson, 27 Aug 1935 (9/5); Cooksey to Snell, 21 Nov 1938 (16/33); Joliot to Norsk Hydro, 11 Mar 1938 (JP, F25).

53. Paneth and Thomson, *Nature, 136* (1935), 334; Smyth et al., *PR, 47* (1935), 800.

54. Thomson to Rutherford, 27 May 1937, and Mann to Thomson, 7 June 1937 (GPT); Rutherford, *Nature, 140* (21 Aug 1937), 303–5, in Rutherford, *CP, 3,* 424–8.

spectrograph. He apparently accepted Bethe's revaluation of the mass data in 1938, which corrected in the wrong direction made He^3 definitely heavier than H^3 and allotted light helium a period of 5,000 years. "This would mean that He^3 cannot be found in nature."[55] This was to ignore the recomputation made by Bethe and Livingston in 1937, using revised values of the relevant parameters, which lowered the mass difference by a factor of ten; and also the reconsiderations of Bonner, who found that neutrons carried off more energy in $H^2(d,n)He^3$ than he had thought, and lowered the mass of He^3 below that of H^3.[56] Rutherford thought it safest to assume the stability of both isobars. Holding with Bethe, Alvarez was alert to any anomaly that indicated that He^3 does not have a period of five millennia.[57]

When Alvarez and a graduate student, Robert Cornog, began their search, the magnetic field of the 60-inch cyclotron was set to accelerate alpha particles. As a preliminary check on the background radiation through the machine, Alvarez watched an oscilloscope that monitored the current to the target while the operating crew decreased the field. The current through the cyclotron fell to zero, as expected, when the field no longer held the alpha particles in phase with the radio frequency potential on the dees. At the end of the test, the crew turned off the field after readjusting it for accelerating alpha particles. As the rapidly changing field passed through the setting for mass three, a sudden burst of particles registered on the oscilloscope. The general phenomenon—a momentary spike in a cyclotron beam as the magnet current sweeps rapidly through the resonance point—had been noticed by M.C. Henderson and Milton White, who explained that the changing magnetic flux set up eddy currents in the pole faces that acted like shims.

Alvarez did not expect to see a spike. It made his day ("one of the finest moments of my scientific life"). It indicated particles of mass three in the cyclotron's source. The source was natural

---

55. Bethe and Bacher, *RMP, 8* (Apr 1936), 197; Bethe, *PR, 53* (1938), 313–4.

56. Bonner, *PR, 53* (1938), 711–3, challenged by Rumbaugh, Roberts, and Hafstad, *PR, 54* (1938), 675–80, on the basis of Bethe's new numbers.

57. Livingston and Bethe, *RMP, 9* (Jul 1937), 324, 331, 373, 378, 379; Rutherford, *Nature, 140* (1937), in *CP, 3,* 424; cf. Alvarez, *PT, 35:1* (1982), 26.

helium from a deep well in Texas, where it had lain for geologic ages. Evidently He^3 had a half-life greater than 5,000 years. In fact, as Alvarez proclaimed, it is stable.[58] He and Cornog shimmed the 60-inch to give a He^3 beam and estimated the relative abundance of the light and heavy isotopes as 10^{-7} for atmospheric, and 10^{-8} for well helium. (These numbers are low by a factor of ten; the higher percentage in the atmosphere arises from creation of He^3 by cosmic rays.) That solved half the old problem. They then looked for *radioactive* H^3 in the product of d-d reactions passed into an ionization chamber. A new long activity with chemical properties of hydrogen rewarded their search.[59]

The most obvious next step was to determine the half-life of the radioactive hydrogen. A first estimate, made by Thanksgiving day, 1939, was 230 days, later diminished to 150 days, and then raised to perhaps ten years. The lower numbers may stand as a warning to unwary experimenters: as Cornog discovered to his chagrin, they measured the rate, not of decay, but of the leak of hydrogen through a rubber tube used in the apparatus.[60] This last number came from McMillan, who had measured the half-life of H^3 without discovering it. He had guessed that an activity he had noticed in 1936 and reckoned at ten years belonged to a supposititious Be^{10} made by (d,p) along with the B^{10} made by (d,n) in the usual cyclotron irradiation of beryllium. After the disclosure of the activity of H^3, physicists at the University of Chicago got a nice radioactive gas on dissolving a specimen similar to McMillan's. They supposed they dealt with the product of the reaction $Be^9(d,H^3)Be^8$ and that McMillan's half-life characterized tritium.[61] The period is, in fact, 12.6 years. Tritium is constantly

58. Alvarez and Cornog, *PR, 56* (1939), 379, letter of 31 July; Henderson and White, *RSI, 9* (Jan 1938), 29–30; Cooksey to Allen, 2 Aug 1939 (1/14); Alvarez, "Adventures," 16–21, *PT, 35:1* (1982), 27–8, and *Adventures*, 68–71, quote.

59. Alvarez and Cornog, *PR, 56* (1939), 613, letter of 29 Aug.

60. Alvarez and Cornog, abstract of paper for APS meeting, 22 Nov 1939 (2/2), *PR, 57* (1940), 249, and *PR, 58* (1940), 197; Cornog in Trower, *Discovering Alvarez*, 26–8.

61. McMillan, *PR, 49* (1936), 875; O'Neal and Goldhaber, *PR, 57* (1940), 1086–7. McMillan's sample was lost before it could be tested for H^3; Ruben and Kamen, *PR, 59* (1941), 349.

made in the atmosphere by cosmic rays. The heavy Norwegian rainwater examined by the Cavendish consequently contained radioactive hydrogen, which lived long enough to reveal itself to a Geiger counter after the war.[62]

Lawrence was delighted with the identification of the isobars of mass three. He gave it pride of place in his report to the Research Corporation on the Laboratory's work for 1939. When the report was submitted early in 1940, negotiations with the Rockefeller Foundation over the 184-inch cyclotron had reached their critical phase. Lawrence gave the result a gloss that he doubtless expected Poillon to pass on to the Foundation. "Radioactively labelled hydrogen opens up a tremendously wide and fruitful field of investigation in all biology and chemistry."[63]

Processes

Until the late spring of 1937, experimentalists knew only one way for artificial radioelements to decay: by the emission from their nuclei of a positive or a negative electron. Theorists had observed, however, that a nucleus liable to produce a positron might also transmute by capturing one of the two atomic electrons—the so-called K electrons—closest to it. The heavier the nucleus, the stronger the pull on the K electrons and the greater the likelihood of capturing one of them. The possibility was first aired by that inventive interpreter of Fermi's theories, Gian Carlo Wick.[64] A means of detecting the process, should it occur, lay close to hand. An atom containing the stable nucleus created by K-electron capture would lack an electron in its innermost shell. An electron from the next shell is likely to fall into the hole and to emit an x ray, called a K_α ray, in the process. K-electron capture by an unstable nucleus of charge Z betrays itself by a K_α ray characteristic of element $Z-1$. The higher the Z and the longer the life of a positron emitter, the more likely the competing K process is to occur. None of the naturally occurring radioelements, all of which have high Z, decay by releasing positrons. Hence the K-prospector looked as though it were among the longest-lived

62. Johnston, Wolfgang, and Libby, *Science, 113* (1951), 1–2.
63. Lawrence, "Report [for 1939]," n.d. (15/18).
64. Wick, Acc. naz. lincei, *Atti, 19* (1934), 319–24 (4 Mar 1934).

emitters of positive electrons he could make as far up the periodic table as his means allowed.

Fermi's theory modelled beta decay and the competing capture process in analogy to the theory of electromagnetic radiation: just as an atomic electron can release or absorb a photon in leaving or reaching an excited state, so a proton can give rise to a positron or capture an electron while turning into a neutron. The analogy breaks down in that the energy of the beta particle created in the process is not equal to E_{max}, the difference in energy of the nucleus before and after the creation; rather, the energy may take any value up to E_{max}, as indicated in figure 8.2. To save the principle of energy conservation, physicists had supposed that a second particle is created in the beta decay, a "neutrino," whose lack of charge and vanishingly small mass (necessary to produce the asymmetry in figure 8.2) protected it from observation. With a neutrino mass of zero, the shape of the theoretical beta curve is determined chiefly by the value of E_{max}. Fermi's calculation, which took the strength of the interaction between electron and neutrino to be proportional to the amplitude of their fields, did not give quite the degree of asymmetry observed. Theorists at the University of Michigan, E.J. Konopinski and his professor, G.E. Uhlenbeck, came closer in the cases of P^{32} and Al^{28} by making the interaction proportional to the product of the amplitude of the electron field and the gradient of the neutrino field.[65]

In 1936 Christian Møller in Bohr's institute compared the predictions of Fermi and of Konopinski and Uhlenbeck (K-U) for the total probabilities per second of positron emission (λ_+) and K-electron capture (λ_K) in a hypothetical susceptible nucleus of high Z. (λ is the inverse of the period of the activity.) His result: on both theories λ_K is much larger than λ_+, the disparity being the greater the smaller E_{max}. Then he specialized to the curious results of Cork and Lawrence, whose "platinum" decayed with a period of 49 minutes by emission of positrons of $E_{max} = 2.1$ MeV. Neither Fermi nor K-U could give enough positrons to fit these data. Møller supposed that K-electron capture must have occur-

65. Konopinski and Uhlenbeck, *PR, 48* (1935), 7–12. Cf. Rasetti, *Elements*, 193–200, and (for Al^{28}) Cork, Richardson, and Kurie, *PR, 49* (15 Jan 1936), 208.

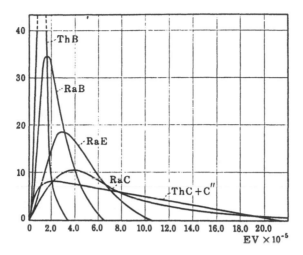

FIG. 8.2 The beta-ray spectra of some naturally occurring radioisotopes. The ordinate indicates the percentage of the total activity at the energy of the abscissa to which it corresponds. Rasetti, *Elements*, 146.

red in the Berkeley experiments 9 times or 47 times as often as positron emission depending on whether events followed Fermi or K-U. He advised looking for x rays from the "iridium" formed from the "platinum" decay. Lawrence could not find the x rays and, as usual, thought he had caught out the theorists: "It looks to be a serious difficulty for the Fermi theory." Since Bohr's institute had no machine for making radioplatinum, experimentalists followed up Møller's lead by seeking K_α from the decay product of the heaviest available positron emitter. This was Sc^{43} ($Z = 21$, $\tau = 4$ hours), made by alpha particles from radon on (ancestor) calcium via (α, p). A search for K_α from (descendent) calcium by J.C. Jacobsen failed. Calculations indicated that for Sc^{43}, $\lambda_K/\lambda_+ = 5$ according to K-U and 0.1 according to Fermi. Calculation and measurement in Copenhagen therefore favored Fermi.[66]

Berkeley had by then plumped for K-U on the basis of its apparently better fit to measurements of beta decay. Here the primary instrument of research was Kurie's cloud chamber and the

66. Møller, *PR, 51* (1937), 84–5, rec'd 9 Nov 1936; Cork and Lawrence, *PR, 49* (1936), 788–92; Jacobsen, *Nature, 139* (1937), 879–80, letter of 6 Apr; Lawrence to Segrè, 5 Apr 1937 (16/14).

primary researchers himself, J.R. Richardson, and Hugh Paxton. By using hydrogen, in which slow particles have a better chance to show their presence than in oxygen, then the usual medium in the chamber, they obtained close agreement with K-U for N^{13}, F^{17}, Na^{24}, and P^{32}. The cloud chamber men concluded that "[the K-U theory] completely describes the process of emission of a beta particle."[67]

As they drifted further along the periodic table, however, their conviction dissipated. Active chlorine, argon, and potassium could not be fitted to K-U unless each contained two unresolved activities. And nothing fit unless E_{max} were put higher than the limit to which, as judged by the eye, the experimental curve tended.[68] (They could not measure all the way to the maximum because they could not register enough of the very few fastest particles.) Ernest Lyman, another graduate student associated with the Kurie group, confirmed the disconfirming of K-U in the cases of P^{32} and RaE. As he observed, however, P^{32} and RaE have unusually long periods (14 and 5 days respectively), and might die in ways not dreamed of in the competing theories.[69]

Fermi himself had pointed out that substances like RaE might escape his theory. He called attention to their position on the so-called Sargent curves, a plot of $\log E_{max}$ against $\log\lambda$ for the naturally radioactive substances. Its author, B.W. Sargent, who took up the project while at the Cavendish, divided the empirical points into two classes, each of which fell roughly along a straight line (fig. 8.3). For a given E_{max}, an element of class II, which included RaE, has a much longer life than an element of class I. Their decay apparently required an inhibiting change of nuclear spin. At a meeting of the American Physical Society in Seattle in June

67. Kurie, Richardson, and Paxton, *PR*, *48* (1935), 167, and *PR*, *49* (1936), 368–81, rec'd 7 Jan 1936, 372, quote. Konopinski had worried about N^{13} and the difficulty of putting the theories to the test; Konopinski to Bethe, 10 May 1935 (HAB, 3).

68. Kurie, Richardson, and Paxton, *PR*, *49* (15 June 1936), 203 (APS meeting, 20–1 Dec 1935); Newson, *PR*, *51* (1937), 624–7, rec'd 16 Feb.

69. Lyman, *PR*, *50* (1936), 385, and *PR*, *51* (1 Jan 1937), 1–7, rec'd 27 Oct 1936. By 1939, experiment favored a mixture of Fermi and K-U (Walke, *Rep. prog. phys.*, *6* (1939), 20–1), and theory gave little support to either (Breit, *RSI*, *9* (1938), 64).

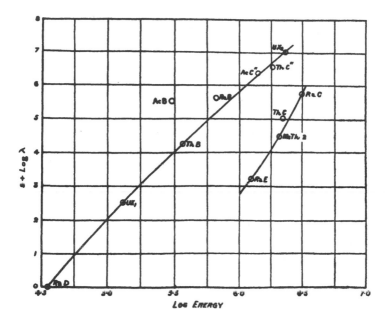

FIG. 8.3 The Sargent curves relating the maximum energy of the electrons emitted by natural radioelements to the decay periods. Sargent, *PRS, A139* (1933), 671.

1936, Lawrence introduced Laslett to talk about a singularly long-lived positron emitter, Na^{22} (τ = 3 years), first made by Otto Frisch by (γ,n) on F^{19}, then in comparative plenty by Laslett by (d,α) on Mg^{24}. Its exceeding longevity interested Willis Lamb, who was brought forth at the same meeting by his mentor, Oppenheimer. Lamb reported that his calculations showed that if Na^{22} did require a change in nuclear spin, it would be twice as likely to decay by K-electron capture as by positron emission according to Fermi's theory, and thirty times as likely according to K-U. He proposed as a test not looking for the K_α line of neon but counting the relative numbers of alpha particles (one for each atom of Na^{22} created) and positrons (one for each positron decay) in the process. K-electron capture was not, and has not been, observed in Na^{22}.[70]

70. Sargent, *PRS, A139* (1933), 671, and in Shea, *Otto Hahn,* 227–9; Frisch, *Nature, 136* (1935), 220; Lamb, *PR, 50* (1936), 388–9; Laslett, ibid., 388, and *PR, 52* (1937), 529–30, rec'd 25 June, and Laslett to Lawrence, 26 June 1936 (10/32).

While Laslett wrote up his results, Alvarez entered the game and picked up the chips. He noticed that Sc^{43} lies on the first Sargent curve and looked around for a neighboring element with a radioisotope on the second. The cyclotroneers had been exploring the region. By the end of 1935 they had reached zinc and had studied at least one element, argon, carefully; but they did not stop to examine the transition elements below zinc closely enough to find the eligible positron emitters they had activated in them.[71] Nor did Livingood's surveys of 1936 or a direct search among the activities of copper pick them up.[72]

Harold Walke, who was so unsure of himself that he had to be careful, then looked closely at the activities of the first few elements in the fourth period. He found a strong positron activity induced on titanium; chemical separation pointed to an isotope of vanadium (later identified as V^{48}) produced by (d,n). The period of decay, 16 days, placed the new activity on the second Sargent curve. That was the combination desired: a Z high enough, a life long enough. Alvarez attacked radiovanadium with McMillan close behind, opening loopholes, "following the job." Alvarez did not attend the Laboratory picnic on Sunday, June 20. That day he detected rays from the decaying vanadium with a penetrating power appropriate to the K rays of titanium. K-electron capture, long expected in theory, thus materialized in the laboratory. Alvarez made the ratio λ_K/λ_+ about 1, closer to Fermi's theory than to K-U. Lawrence praised the work as "especially significant."[73]

71. Livingood, *PR, 49* (15 Jan 1936), 206 (Zn); Thornton, ibid., 207 (As, Ni, Co); Snell, ibid., 207, and *PR, 49* (15 Apr 1936), 555–60 (A).

72. Livingood, *PR, 50* (1 Sep 1936), 425–32; Lawrence to Segrè, 5 Apr 1937 (16/14). Alvarez, *PR, 54* (1 Oct 1938), 486, miscredits Yukawa and Sakata, Phys.-Math. Soc. Japan, *Proc., 17* (1935), 467–79, and *18* (1936), 128–30, rather than Wick with the first suggestion that K-electron capture might compete with positron emission. Cf. Rasetti, *Elements* (1936), 202–3, and Segrè in Trower, *Discovering Alvarez*, 11–2.

73. Walke, *PR, 51* (1937), 1011, talk at APS, Washington, ca. 1 May; Alvarez, *PR, 52* (1937), 134–5, letter of 21 June; Walke, *PR, 51* (1937), 439, Hurst and Walke, ibid., 1033, and Walke, *PR, 52* (1937), 663, letter of 6 Aug (K, Ca); Allen and Alvarez, *RSI, 6* (1935), 329; Cooksey to Lawrence, 17 and 21 June 1937, and Lawrence to Cooksey, 25 June 1937 (4/21).

Once the process had been seen everyone saw it. Livingood, now disciplined in collaboration with Seaborg and another chemist, Fred Fairbrother, a Leverhulme Fellow from the University of Manchester, found a positron activity in manganese. Later Livingood and Seaborg showed that this isotope (or rather isotopes: Mn^{52} and Mn^{54}) decays by K-electron capture and that Zn^{65}, which Livingood had examined earlier, does so too. Then Otto Oldenburg, on sabbatical from Harvard, found a K process without positron competition in tantalum excited by neutrons (Ta^{180}, $\tau = 8.2$ hours).[74] In all this there was a difficulty, however, which McMillan pressed on Alvarez. Perhaps the K electron does not jump into the nucleus but out into the world, driven by a gamma ray originating from an excited state of the stable final nucleus? If the probability for "internal conversion" (the release of an atomic electron that absorbs the gamma ray) were sufficiently high, only the x rays and the converted electrons would appear in the radiations. To decide the question, the experimenter must determine the element from which the K ray emerges: if from element Z (Z being the atomic number of the radioelement), then internal conversion; if from element $Z - 1$, then K-electron capture.

Alvarez took up this problem with the positron emitter Ga^{67}, which Wilfred Mann had made by (d,n) on zinc. The radiation from Ga^{67} consists of electrons, gamma rays, and x rays characteristic of zinc. Ernest Lyman and another graduate student showed that all the electrons had about the same energy. Alvarez explained: a nucleus of Ga^{67} swallows a K electron and ends in an excited state of Zn^{67}, which emits a monoenergetic gamma ray that has a moderate possibility of internal conversion; homogeneous electrons demonstrate the conversion and the zinc x rays the filling of the holes in the zinc atom's electronic structure. Walke found a better demonstration with long-lived V^{49} ($\tau = 600$ days), which decays only by K-electron capture and only into the ground state of titanium. No gamma rays or ionizing radiations

74. Livingood, Fairbrother, and Seaborg, PR, 52 (1937), 135, letter of 30 June; Livingood and Seaborg, PR, 54 (1938), 239, 391; Seaborg, Jl., 1, 355–6 (23 June 1938); Oldenburg, PR, 53 (1938), 35–9, rec'd 22 Oct 1937.

complicate the picture. Like the V^{48} in which Alvarez had made his discovery, V^{49} appeared to die out more closely to Fermi's than to K-U's specifications. By 1939 K-electron capture had been recognized in some twenty isotopes, including two of element 43.[75] K-electron capture proved to be as common as theorists expected. Among other consequences of its ubiquity, it ruled out the possibility that platinum could be the source of the positrons seen by Cork and Lawrence.[76]

Some pieces of the platinum puzzle fit well with the study of another nuclear process, in which, like the detection of K-electron capture, the Laboratory pioneered. This was the behavior of isomers, forms of the same unstable nucleus differing in internal energy. A pair of isomers can decay in several ways: each might emit a beta particle, or the more energetic isomer may relax into the lower by throwing off a gamma ray, or both processes might occur together. Isomerism first came to light in 1921, when Otto Hahn deduced that the third member of the radioactive chain descending from uranium UX_2 (Pa^{234}) consists of two beta emitters, both of which, he thought, arose directly from $UX_1(Th^{234})$. No other instance was found. It took theorists some time to devise an explanation. In 1934 the inventive Gamow thought to trace the difference between Hahn's isomers UX_2 and UZ to the presence in one of them of a hypothetical antiproton-proton pair in place of two neutrons. Another idea, put forward by a student of Heisenberg's, C.F. von Weizsäcker, in 1936, preserved the upper isomer long enough to emit a beta ray by supposing that a big difference in spin discouraged it from dropping immediately to the lower. Still, the matter was neither clear nor persuasive; Hahn's partner Lise Meitner expressed skepticism about isomerism and Bethe, though accepting the phenomenon, hedged over whether the pair UX_2 and UZ was an example of it. The Cavendish's Norman Feather and Egon Bretscher cleared the matter up early

75. Oldenburg, *PR, 53* (1938), 35–9, rec'd 22 Oct 1937; Alvarez, *PR, 53* (1938), 606, letter of 15 Mar, and *PR, 54* (1 Oct 1938), 486–97, followed up by Helmholz, *PR, 57* (1940), 248; Walke, Williams, and Evans, *PRS, A171* (1939), 360; Walke, *Rep. prog. phys.,* 6 (1939), 22–3. The vanadium isotopes were misidentified as V^{47} and V^{48}.

76. Lawrence recognized the problem; Lawrence to Bohr, 8 Jul 1937 (3/3).

in 1938. They made UX_2 the excited metastable state and the only direct descendent of UX_1, and UZ the rare result of UX_2 nuclei that could not restrain their gamma radiation. By an appropriate attribution of the complex beta rays, they showed that UZ belonged on the first, and UX_2 on the second Sargent curve.[77]

Feather and Bretscher had the encouragement of the first unequivocal example of isomerism among artificially active elements. That was the work of Arthur Snell, the most complete and exact bit of radiochemistry accomplished at the Laboratory to that time (August 1937). He was inspired by the apparent existence of too many active bromines. Fermi's group had found two, with periods of 18 minutes and 4.5 hours, which they supposed to arise by slow neutron capture in the two stable bromine isotopes, Br^{79} and Br^{81}. Then a Soviet physicist, I.V. Kurchatov, found a third activity ($\tau = 36$ hr) in bromine hit by neutrons from a Rn-Be source, for which there was no obvious available antecedent in natural bromine. Kurchatov proposed that whereas Fermi reactions were of the ordinary type (n,γ), his occurred via the then still unestablished route (n,2n), giving rise to a suppositious active Br^{78}.[78]

Snell checked these results by trying to make the Italian radioisotopes Br^{80} and Br^{82} and the Soviet Br^{78} in other ways than by neutron bombardment of natural bromine. He examined no fewer than twenty-eight different reactions involving As, Se, Br, Kr, and Rb activated by deuterons, alpha particles, and neutrons. Among his most significant results: Br^{78} does exist—he made it by (α,n) on the single arsenic isotope As^{75} and by (d,n) on Se^{77}—but its period (6 min) and its decay mode (positron emission) exculpated it from responsibility for Kurchatov's activity; Br^{83}, hitherto

77. Gamow, PR, 45 (1934), 728–9; von Weizsäcker, Nwn, 24 (1936), 813–4; Meitner, in Bretscher, Kernphysik, 41; Bethe, RMP, 9 (1937), 225–6; Feather and Bretscher, PRS, A165 (1932), 530–1, 545–50. Cf. Flammersfeld in Frisch, Trends, 71–6.

78. Amaldi et al., PRS, A149 (1935), 522–58, and Amaldi, Phys. rep., 111 (1984), 128ff.; Kurchatov et al., CR, 200 (1935), 1201–3; Livingston and Bethe, RMP, 9 (1937), 348–50. According to Golovin, Kurchatov, 25–6, Kurchatov proposed the existence of bromine isomers to the Soviet Academy of Sciences in March 1936.

entirely unknown, also exists (τ = 2.5 hr); and the three previously known activities had to be shared among Br^{80} and Br^{82}. From his inventory of twenty-eight reactions, Snell could show that both Fermi activities belong to Br^{80}. And, to complete his happiness, he created Kurchatov's activity (Br^{82}) by deuterons on selenium. With the advice of Bohr—this result dates back to April 1937 or earlier—Snell pinned the reaction on Se^{82}, since Se^{81}, which could have given rise to Br^{82} by the familiar (d,n) reaction, does not exist naturally. On this explanation, Snell gave the first example of a (d,2n) reaction.[79] The excitement of the discoverers of such arcana may be hard for outsiders to share. But in the breast of the nuclear physicist, they inspired "great joy."[80]

The joy came also to McMillan, Kamen, and Ruben, who were still laboring on Lawrence's dirty platinum when Snell completed his work. Platinum appeared to have three periods, all activated by slow neutrons and decaying by fast electrons, and gold and iridium behaved similarly. It seemed too much of a good thing. "The results of this work so far [McMillan's group wrote] do not seem to be capable of any simple explanation without the introduction of a fantastic number of isomeric nuclei."[81] Candidate isomeric nuclei turned up everywhere after the summer of 1937, and frequently for the first time in Berkeley; of the seventeen pairs of artificial isomers established by 1939, eleven were discovered or first confirmed by members of the Laboratory.[82]

Livingood and Seaborg were the most successful hunters (plate 8.2). Earlier investigators had not carried their separations of zinc activated by neutrons very far. Seaborg went further. He and

79. Snell, *PR*, 52 (1937), 1007–22, sent 3 Aug; Livingston and Bethe, *RMP*, 9 (1937), 326, are agnostic about (d,2n); Livingood and Seaborg, *RMP*, 12 (1940), 38, accept it. Cf. Segrè and Helmholz, *RMP*, 21 (1949), 271–2, 291.

80. Walke, as quoted in Cooksey to Lawrence, 17 June 1937 (4/21); cf. Seaborg, *Jl.*, *1*, 296 (26 Nov 1937), on the excellence of Snell's work. The original three bromine acitivities were confirmed by Bothe and Gentner, *Nwn*, 25 (1937), 284, letter of 19 Apr, by photoexcitation.

81. McMillan, Kamen, and Ruben, *PR*, 52 (1937), 375–7, rec'd 19 Jul.

82. Walke, *Rep. prog. phys.*, 6 (1939), 26, and Walke, Williams, and Evans, *PRS, A171* (1939), 360–82. Some thirty-four isomeric pairs were known in 1940, and over seventy-five pairs.by 1949. Frisch, *Ann. rep. prog. chem.*, 36 (1940), 16–21; Segrè and Helmholz, *RMP, 21* (1949), 255–9.

Livingood in consequence added two new nickels to their treasury of isotopes, cleaned up earlier misattributions by Livingood and by Robert Thornton, and identified isomers of Zn^{69}. In all there were but three zincs, which Livingood and Seaborg prepared in a total of eleven different ways. Their attributions have stood.[83] Then there were iron and its neighbors. With separated fractions from fifteen bombardments of iron, four of chromium, and two of manganese, all by deuterons, and several irradiations of the same with alpha particles and neutrons, Livingood and Seaborg straightened out a great many reactions and uncovered a pair of isomers in Mn^{52}.[84] The most interesting and complex of the isomers came to light during a lengthy study of activated tellurium, which attracted Livingood and Seaborg not only as another radiochemical puzzle, but also as a possible quarry for a useful radioactive iodine.

Their first irradiation of tellurium with deuterons took place on March 26, 1938. They moved on to iodine, then back to tellurium, glimpsing, losing, and finally establishing the existence of an iodine with the biologically useful period of 8 days. That brought them through the summer. In September Livingood left for Harvard; he set up an electroscope in his kitchen to examine samples mailed him by the indefatigable Seaborg.[85] There were too many telluriums. Joseph Kennedy, Seaborg's graduate student, helped to untangle them. He soon confirmed his collaborators' conjecture that the 8-day iodine, I^{131}, descended from not one but two telluriums, isomers of Te^{131} with periods of 1 hour and 1.2 days. But on further examination, the hour period shrank to 45 minutes and then split into two, of 30 and 55 minutes; and, in addition to the three telluriums that had replaced Kennedy's two, there were three more, of periods of 10 hours, 1 month, and several months. After

83. Livingood and Seaborg, PR, 53 (1938), 765, and PR, 55 (1939), 457–63, rec'd 15 Dec 1938; Thornton, PR, 53 (1938), 326, and Livingood, PR, 50 (1936), 425; Seaborg, Jl., 1, 412, 416 (3 and 15 Dec 1938). Alvarez, PR, 53 (1938), 606, and Mann, PR, 54 (1938), 649–52, also helped to purge Thornton's "zincs."

84. Livingood and Seaborg, PR, 54 (1938), 51–5, rec'd 10 May, and PR, 54 (1938), 391, rec'd 11 Jul; Seaborg, Jl., 1, 360–1 (3 Jul 1938).

85. Seaborg, Jl., 1, 333, 339–43, 347–8, 353, 359–61, 367, 373, 377–8 (26 Mar–6 Sep 1938); Livingood and Seaborg, PR, 54 (1938), 775–82, rec'd 7 Sep 1938, which itemizes five radioiodines, three of which were new.

three weeks of hard work in January 1939, Kennedy and Seaborg proved that the 30-minute (corrected to 25-minute) activity was a parent of I^{131} and the lower of a pair of isomers whose upper level was the activity of 1.2 days.[86]

To shorten a story already sufficiently long, Seaborg, Livingood, and Kennedy labored on the tellurium system until the end of December 1939, when they declared on the basis of sixteen different reconfirmed reactions that there exist precisely four radiotelluriums, three of which come in two isomers each. With a clever chemical technique, soon to be described, they showed which isomeric state was the lower; and, with the help of two graduate students, Carl Helmholz and David Kalbfell, who examined the specimens in a beta-ray spectrograph, they identified conversion electrons knocked out by the gamma rays emitted in transitions from upper to lower isomeric states.[87]

After Livingood, Seaborg took up with an isomer hunter of even greater resourcefulness, Segrè, who brought the experience of identifying isomeric Cu^{65} in cyclotron scrap.[88] Segrè required something in addition to Seaborg before he would begin their planned search for isomers of element 43: a better detector than Livingood and the Laboratory's other hunters of new activities employed. This detector consisted of an ionization chamber of a type used in Rome connected to a dc amplifier built by Lee DuBridge during a summer at Berkeley.[89] Segrè's electrometer (plate 8.3) later served in the detection of H^3, C^{14}, and plutonium. It began by registering a beta ray, no gamma ray, and an x ray from activated molybdenum, which Seaborg and Segrè diagnosed as an electron converted from a gamma ray with almost 100 percent efficiency and the associated K radiation of element 43. They sent a letter to the *Physical Review* announcing these results; but

86. Seaborg, *Jl.*, *1*, 364, 394, 403, 406–7, 424, 427–30, 432–6, 439 (28 Jul, 11 Oct, and 8 Nov 1938, 11–30 Jan and 2 Feb 1939); Seaborg and Kennedy, *PR*, *55* (1939), 410, letter of 31 Jan.

87. Seaborg, Livingood, and Kennedy, *PR*, *55* (1939), 794, letter of 31 Mar.

88. Seaborg, *Jl.*, *1*, 361 (9 Jul 1938); Perrier, Santangelo, and Segrè, *PR*, *53* (1938), 104–5. Cf. Segrè in Nicolini et al., *Technetium*, 8.

89. Seaborg, *Jl.*, *1*, 361–4 (9–18 Jul 1938); Segrè to Lawrence, 7 Jan 1938, and reply, 7 Feb 1938 (16/14). The standard armamentarium of detectors in 1940 is described in Seaborg, *Chem. rev.*, *27* (1940), 207–10.

their pleasure suffered an interruption when Lawrence told them that Oppenheimer thought so high a degree of conversion impossible. Lawrence asked them to withdraw their report. The frequency with which the Laboratory had had to retract published results had declined since Lawrence had gone full time into fundraising and administration, and he did not wish to risk a throwback. Seaborg and "a rather agitated Segrè," who had lately been the master of his own ship, complied. Ten days after this contretemps, the *Physical Review* published a letter from Bruno Pontecorvo describing a similarly high conversion in rhodium. Lawrence conceded that Segrè and Seaborg should resubmit.[90]

They soon confirmed their observations. They asked a graduate student, Philip Abelson, who had built a good spectrograph, to determine whether the x radiation they had found belonged to element 43 (it did), and they established that the lower isomer associated with the 6-hour activity had a life of at least forty years. These are the isomers of Tc^{99} (the lower has a life of almost a million years), the clinically important species of technetium.[91] The collaboration was close and demanding. Segrè participated in the chemical separations and Seaborg in the physical measurements, and they wrote up their results together.[92]

Since atoms of isomeric nuclei possess precisely the same chemical properties, it might not seem advisable to think about ways to separate isomers chemically. Segrè thought it could be done, however, by modifying the Szilard-Chalmers process, in which nuclei rendered active by an (n,γ) process (e.g., radioiodine) and knocked out of a compound (e.g., ethyl iodide) by the neutrons they absorb, are collected by combining them with suitable molecules (e.g., in a precipitate of silver iodide). His thought was at first received unenthusiastically because the recoil from the isomeric

90. Seaborg, *Jl.*, *1*, 367–8, 376, 381, 383, 390, 393, 396 (26 Jul, 1 and 30 Aug, 14 and 20 Sep, 8 and 14 Oct 1938); Segrè and Seaborg, *PR*, *54* (1938), 772, letter of 14 Oct (originally 14 Sep); Kalbfell, ibid., 543, determined the energy of the conversion electron.

91. Lawrence to Sproul, 31 Dec 1938 (16/14); Seaborg, *Jl.*, *1*, 399, 405, 412–5 (23 Oct, 12–14 Nov, 3–10 Dec 1938); Seaborg and Segrè, *PR*, *55* (1939), 808–14; Segrè, *Ann. rev. nucl. sci.*, *31* (1981), 9.

92. Seaborg, *Jl.*, *1*, 366ff., 418 (20 Dec 1938), 424 (10 Jan 1939), 437–8 (1 Feb 1939).

transition seemed insufficient to free an atom from its chemical bonds. Ralph Halford, an instructor in the Chemistry Department, with whom Seaborg discussed the matter, thought that he might be able to effect a chemical separation. Segrè and Seaborg irradiated a liter of ethyl bromide, which, after treatment by Halford, gave a hydrobromic acid enriched in the lower isomer of Br^{80}. This isomer thus stood revealed as the 18-minute activity observed by Fermi's group, by Kurchatov, and by Snell.[93] Seaborg and Kennedy immediately applied the scheme successfully to tellurium, confirming the double origin of 8-day iodine; and then, together with Segrè, tried hard, but with few positive results, to separate isomers of several other metals, from manganese to platinum.[94]

The Groaning Board

The periodic table of the elements began as a sort of Ouija board, arranged on arbitrary principles and operating with mysterious powers. The concept of natural isotopes and the theory of the nuclear atom diminished the mystery by explaining that the principle of arrangement by weight was a happy approximation to the true regulation by atomic number. Why the natural isotopes have the weights they do remained a mystery. The discovery of means to split atoms and make natural isotopes radioactive made possible determination of the exact binding energies of nuclei and an understanding, or the beginning of one, of the genetic inter-relationships among the elements.

The Laboratory made very material contributions to the new knowledge. The cyclotron plugged two holes in the periodic table, the spaces at 43 (technetium) and 85 "eka-iodine," later astatine. The first, we know, was found by Segrè and Perrier in

93. Szilard and Chalmers, *Nature, 134* (1934), 462 in Szilard, *CW, 1* 143–4; Szilard to Hopwood, 28 Aug 1934 (Sz P, 17/197); Seaborg, *Jl., 1,* 422–3 (4–7 Jan 1939); Segrè, Halford, and Seaborg, *PR, 55* (1939), 321–2, letter of 13 Jan. DeVault and Libby, ibid., 322, followed up Segrè's suggestion with different chemistry; Seaborg had long been interested in ways to separate radioelements without recourse to carriers, e.g., Grahame and Seaborg, *PR, 54* (1938), 240–1, and Seaborg and Livingood, *JACS, 60* (1938), 1784–6.

94. Seaborg and Kennedy, *PR, 55* (1939), 410, letter of 31 Jan; Seaborg, *Jl., 1,* 465–6, 469–82, 478–84 (24 Mar–5 May 1939).

molybdenum irradiated by deuterons in the 27-inch; the second, found in an experiment that Segrè proposed, came from bombardment of bismuth by alpha particles in the 60-inch. At Hamilton's bizarre suggestion, astatine was fed to guinea pigs, who obligingly concentrated it in their thyroids to demonstrate its similarity to iodine; the Laboratory was nothing if not interdisciplinary.[95]

The total record appears in figure 8.4. It indicates the percentage of reactions known in May 1935, July 1937, and December 1939 that the Laboratory either discovered or helped to clarify. The comparison with Cambridge is particularly striking: in 1935 Berkeley matched the Cavendish only in reactions initiated by deuterons; by 1937 the Laboratory dominated study of deuteron interactions and was extending itself vigorously into neutron and proton work; by 1939 it had risen to dominance in alpha- as well as deuteron-induced reactions, and to hegemony in neutron work, while Cambridge, under the management of W.L. Bragg, had lost its position in nuclear chemistry. Other laboratories, for example, Fermi's in Rome, Bothe and Gentner's in Heidelberg, and DuBridge's at Rochester, specialized in one sort of reaction only.[96] A cruder, though perhaps more impressive index to Berkeley's contribution appears from comparison of the percentage of all reactions known at a given time for which Berkeley was credited or co-credited: in 1935, one reaction in ten; in 1937, one reaction in four; in 1939, almost one reaction in two.[97]

95. Corson and Mackenzie, *PR*, *57* (1940), 250; Corson, Mackenzie, and Segrè, *PR*, *57* (1940), 459, letter of 16 Feb, and *PR*, *58* (1940), 672–8, rec'd 16 Jul; Hamilton, *Jl. appl. phys.*, *12* (1941), 454–5; Cornog in Trower, *Discovering Alvarez*, 26.

96. These generalizations can be documented in other reviews than those used for table 6.2, e.g., Grégoire, *Jl. phys. rad.*, *9* (1938), 419–27, and the running score in Diebner and Grassmann, *Phys. Zs.*, *37* (1936), 359–83, *38* (1937), 406–25, *39* (1938), 469–501, *40* (1939), 297–314, and *41* (1940), 157–94.

97. Birge, in U.C., *1939 prize*, 23, credits the Laboratory with the discovery of over one half of all isotopes (223, according to him) "discovered" by cyclotrons.

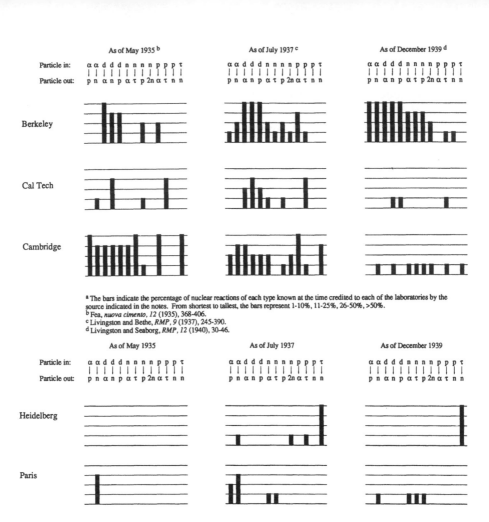

a The bars indicate the percentage of nuclear reactions of each type known at the time credited to each of the laboratories by the source indicated in the notes. From shortest to tallest, the bars represent 1-10%, 11-25%, 26-50%, >50%.
b Fea, *nuova cimento*, 12 (1935), 368-406.
c Livingston and Bethe, *RMP*, 9 (1937), 245-390.
d Livingston and Seaborg, *RMP*, 12 (1940), 30-46.

FIG. 8.4   The Laboratory's contribution to the inventory of artificial radioactive substances during the 1930s.

3. RADIOBIOLOGY

Mice and Men

By the time the danger from neutron rays was appreciated, high-energy x rays no longer held promise for cancer therapy. The result obtained by Lauritsen and Packard of Columbia's Institute of Cancer Research in 1931—that 550 kV rays had no more deleterious effect on Drosophila eggs or the common mouse tumor "sarcoma 180" than an equal quantity of 50 kV rays—had been substantiated and extended. Lawrence had conceded to Wood that the 1 MV Sloan plants probably would have no greater curative properties than standard x-ray apparatus, and experience at the University's Medical School fully confirmed the concession.[98] Hence the possible medical value of neutron therapy held unusual interest for Lawrence both for itself and as a replacement for a played-out technology. A principal objective of the very first experiments with neutron irradiation was to compare its biological effects with those of x rays.

John Lawrence and his technical advisor, Ernest Lawrence, who in July 1935 became "consulting physicist" to the Medical School, exposed $120 worth of rats near the beryllium target of the 27-inch cyclotron and at the Sloan machine in San Francisco. The neutrons appeared to be about ten times as effective as x rays per roentgen in altering the makeup of rodent blood, or five times as effective per unit of ionization since (they estimated) a roentgen of neutrons made twice the ionization in rat tissue that a roentgen of x rays did. Since the standard tolerable limit of x rays was 0.1 r/day, they recommended prudently that the maximum for n rays be 0.01 r/day.[99] While the Lawrences zapped rats, Aebersold and Raymond E. Zirkle, a medical physicist visiting from the

98. Packard and Lauritsen, *Science, 73* (1931), 321–2; J.H. Lawrence to Ridenour, 6 Feb 1940 (14/5). This was to go beyond Tuve, *Radiology, 20* (1933), 289–33, talk of Dec 1931, who showed that the penetrating power of 2,000 kV x rays in water should not much exceed that of 200 kV rays, but would draw "no biologic conclusions."

99. Lawrence to Chauncey Leake, 10 Jul 1935, and reply, 12 Jul (20/6); Lawrence and Lawrence, NAS, *Proc., 22* (Feb 1936), 126–7, 133; Lawrence to Poillon, 2 Mar 1936 (15/17), itemizing the costs of John Lawrence's summer experiments: $120 for mice, $150 for a technician, $312 for travel and expenses.

University of Pennsylvania, tried the effects of the radiations on delaying the growth of wheat seedlings. Here one roentgen of neutrons did the damage of 20 r of x rays. Would neutrons prove ten or twenty times as effective as x rays in other biological contexts? "The general question is of more than theoretical interest, for it bears directly on the possibility of using very fast neutrons in the treatment of tumors."[100]

John Lawrence returned to Berkeley early in February 1936 to take up the general question. With the help of Aebersold, the Lawrence brothers cooked some mice with 84 r/m of neutrons, and other mice with 32 r/m of x rays; neutrons killed with a third the dose (as measured in roentgens) needed for death by x radiation. The numbers fell out differently for sarcoma 180. About four times as large a dose of x rays as n rays was required to prevent pieces of tumor irradiated apart from the mouse from taking after implantation. Call the quantity of x rays needed to kill the tumor (mouse) X_t (X_m) and the corresponding quantities for neutrons N_t (N_m). Then $N_t = (3/4)N_m(X_t/X_m)$. That is an exciting equation. It says that although the x radiation required to kill a tumor might exceed that required to kill its host, the lethal neutron dose for the tumor might not. The quantity X_t/X_m has only to be less than 1.3. Unfortunately, according to the measurements of Aebersold and the Lawrences, $X_t/X_m = 3.5$. Still, neutrons appeared to hurt tumors more, and the body less, than x rays.[101]

The doctor paid his next visit in the summer of 1936. "Things are humming," his brother wrote the Chemical Foundation, then eager for news relevant to their big commitment to the medical cyclotron. "The [27-inch] cyclotron is in operation daily, hundreds of mice and hundreds of tumors are being killed by neutron rays." John Lawrence and Aebersold repeated the work on

---

100. Zirkle and Aebersold, NAS, *Proc.*, 22 (Feb 1936), 136–7; cf. Zirkle, Aebersold, and Dempster, *Am. jl. cancer*, 29 (1937), 556–62, on the edge of neutrons over x rays in the destruction of Drosophila eggs; Axelrod, Aebersold, and Lawrence, SEBM, *Proc.*, 48 (1941), 252, give a table of differential effects of neutron and x rays on various biological systems.

101. Lawrence, Aebersold, and Lawrence, NAS, *Proc.*, 22 (1936), 543–57, rec'd 23 Jul 1936, amplified in J.H. Lawrence and Robert Tennant, *Jl. exptl. med.*, 66 (1937), 667–88; Lawrence to Exner, 4 Feb 1936 (9/21), and to Cork, 22 Feb 1936 (5/1).

sarcoma 180 with another mouse tumor, obtained from Yale. It took 3,600 r of x rays, or 700 r of neutrons, to kill half the tumor particles before implantation; on healthy mice, 400 r of x rays had the same lethality as 120 r of neutrons. Hence neutrons killed Yale tumors 3,600/700 times, and Berkeley mice 400/120 times, as effectively as x rays. In this case, $N_t = (2/3)N_m(X_t/X_m)$.[102] Numbers were moving in the right direction. The addressee was William Crocker. John Lawrence's preliminary, but "highly significant," results had figured in Ernest Lawrence's declaration to Sproul in late February 1936 of the need for a clinical arrangement like Lauritsen's to test the efficacy of neutrons on human cancers. Now John Lawrence's firm comparative data about mouse tumors helped to convince Crocker to give the clinic. At the beginning of September, Sproul pitched effectively, as follows: "The newly [!] discovered neutron ray...seems to provide a means of overcoming the handicap which now limits the effectiveness of the x ray in the treatment of cancer. It appears that it can be used to increase the destruction of cancerous tissue without increasing the damage to the normal tissue." Greatly overplaying John Lawrence's results, Sproul suggested that neutrons might be three times more effective—wreak thrice the havoc to tumors for the same damage to the body—as x rays.[103]

With Crocker's gift secured, the problem of staff for the clinic demanded solution. During the spring of 1936, the dean of the Medical School, Langley Porter, pressed by R.S. Stone to tighten ties with the Laboratory, met with Sproul and Ernest Lawrence for dinner at the Bohemian Club. Subsequently, Porter's assistant, Chauncey Leake, sought appointments for John Lawrence and Paul Aebersold in the Medical School. The business went slowly. As Poillon, himself a physician, had warned, "Medical men are extremely jealous of their prerogatives and...even to have a physicist suggest what they might do is received with anything but acclaim." By the time John Lawrence's appointment came through, he had decided to return to Yale; he kept up his work at

102. Lawrence to Buffum, 11 Aug 1936 (3/38), to Exner, 5 Aug 1936 (9/21), and to Cork, 14 Aug 1936 (5/1); Lawrence, *Radiology, 29* (1937), 316–8.

103. Lawrence to Sproul, 20 Feb 1936, addendum; Sproul to D.J. Murphy, 2 Sep 1936; both in UCPF. Lawrence had to press his more cautious brother to publish his more complete data; Lawrence to John Lawrence, 20 May 1936 (11/16).

Berkeley on visits that totalled about six months during the academic year 1936/37.[104] Neither he nor Stone wished to postpone neutron therapy until the Crocker cyclotron started up. Ernest Lawrence, concerned to show quick progress, agreed. The Lawrence brothers may have been especially, though irrelevantly inspired by the dramatic improvement of their mother under the rays of the Sloan tube. She had long complained of abdominal pain. In November 1937, the Mayo Clinic discovered an inoperable uterine tumor and gave her three months to live. John Lawrence brought her to San Francisco; Stone irradiated her several times with supervoltage x rays; the tumor melted away. Although the swelling may not have been a tumor at all, its erasure by x rays could not but have encouraged the Lawrences to press to make available an agent they had reason to believe would be still more powerful and beneficial.[105]

As a preliminary to its clinical use, the diffuse neutron beam from the beryllium target of the 37-inch cyclotron had to be collimated and directed to a treatment port. Aebersold had the job, which he discharged by rearranging the water tanks of the cyclotron shielding and by lining the beam channel with lead (fig. 8.5). That kept the intensity within the channel almost twenty times that outside it at the port 70 cm from the beryllium target, where the patient received about 12 r/m. Carpenters transmuted a window of the Laboratory into a door opening into a demountable treatment room entirely screening the cyclotron; "the patients will hardly know they are next to such a monster."[106] A parade of physicians, including one from the National Advisory Cancer Council, trooped through the Laboratory during the summer and gave their

104. Stone to Porter, 30 Mar, and Porter to Sproul, 31 Mar 1936 (UCPF, 393/"physics"); Lawrence to Sproul, 12 Apr, and Leake to Lawrence, 13 Apr 1936 (20/16); Poillon to Lawrence, 25 Feb 1936, and Lawrence to Poillon, 5 Oct 1936 (15/17).

105. Lawrence to Sproul, 9 Sep 1938 (UCPF, 446/416), and to E.R. Crowther, 6 Jan and 4 Feb 1938 (5/5); Childs, *Genius*, 198–9, 278–9; Lawrence, "Faculty research lecture," 20 Mar 1938 (40/20): "It is evident that...the neutron rays...have important medical applications in therapy." Cf. John Lawrence to Lawrence, [Mar 1936] (11/16).

106. Cooksey to Allen, 5 Sep 1938 (1/14); Lawrence to Exner, 22 Sep 1938 (9/21), quote, to Hopwood, 5 Oct 1938 (9/15), and to Cockcroft, same date (4/5); Aebersold, *PR, 56* (1939), 714–27; Compton, notes on visit, 16 Sep 1938 (13/29).

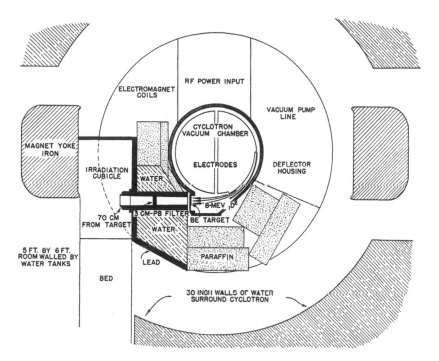

FIG. 8.5 Aebersold's arrangement for neutron therapy. The treatment room is at the left, within the magnet yoke. Not all the water tanks surrounding the cyclotron are shown. Aebersold, *PR, 56* (1939), 717.

blessings to the general work, if not to the therapeutic intiative. Their good judgment pleased and surprised Cooksey. "My opinion of doctors as a whole has risen tremendously." The first patients were exposed on September 26, 1938, just in time for a visit by the NACC's Arthur Compton. Their skin showed effects no worse than those caused by 12 r/m of x rays. Lawrence informed Sproul. "It gives me great pleasure to report an event of historic interest....I personally believe, and my views are shared by my medical colleagues, that this will be the beginning of a new method of cancer therapy which in a few years will be as widespread as that of x rays and radium." What better time to ask for money? $2,000 for power and supplies, $1,400 for furnishings? "We could, of course, slow down our activities, including bringing to a halt the clinical therapy, but in view of the great immediate importance of this pioneering work it would be no less than tragic to do so."[107]

107. Cooksey to Lawrence, 10, 14, 15 and 18 (quote) June, 1938 (4/21); Lawrence to Sproul, 12 Oct 1938 (13/29 and UCPF), quote, and to Crowder, 6 Oct

After five months' experience with therapy at the 37-inch, Stone judged that the results of single erythema doses—doses sufficient to redden the skin—gave encouragement for "a complete course of therapy" and hope for "better results than are now being obtained." It would have been difficult to reach a different conclusion as the cyclotroneers began to fish for a beam in the 60-inch cyclotron. The first patient to absorb neutrons from the Crocker cracker, Mr. Robert Penney, received treatment on November 20, 1939 (plate 8.4). A regular clinical program did not begin until the end of January 1940. Then a few tumors vanished. "Dr Stone and John are very enthusiastic about the results." Thus brother Ernest. But John himself would not go beyond the meaningless formulation of the weather forecaster: "[There is] better than a fifty-fifty chance that neutrons are going to be of great value in therapy."[108] He was therefore just better than half wrong when Stone evaluated the program in 1948. Only one of the 24 patients treated at the 37-inch cyclotron in 1938 and 1939 was then alive, and only 17 of the 226 treated at the 60-inch between 1939 and 1943. All but one of the 250 had been considered incurable. The survivors suffered what Stone described as "distressing late effects" that might not have occurred had they undergone x-ray rather than neutron therapy. He judged that he and John Lawrence had overexposed their patients. "Neutron therapy as administered by us has resulted in such bad late sequelae in proportion to the few good results that it should not be continued." Stone's negative evaluation put an end to fast-neutron therapy for two decades.[109]

After much experimentation, treatment of human cancers by neutron rays recommenced around 1970, at, among other places, the Hammersmith Hospital in London, which used a cyclotron constructed in 1952. By 1978 over 3,000 patients had been treated at eleven centers in Europe, Japan, and the United States. The therapy proved effective against advanced, superficially placed

---

1938 (5/5); Lawrence, "Report to the Research Corporation," 3 Jan 1939 (15/17A).

108. Lawrence to Frank Hinman, 12 Jan (24/18), to Poillon, 31 Jan (15/8), and to Kruger, 14 Mar 1940 (10/20), quote; John Lawrence to Ridenour, 6 Feb 1940 (14/22), quote.

109. Stone, *Am. jl. roent.,* 59 (June 1948), 771, 775–6, 784. Cf. Sheline et al., ibid., *111* (1971), 31–41.

tumors that could be irradiated with little damage to neighboring normal tissue. The experience at Hammersmith through 1984 was that 70 percent of such tumors regressed after treatment with 7.5 MeV neutrons in comparison with 35 percent after treatment with x rays.[110] Stone and the Lawrences had the right idea but the wrong dosage.

Grand Elixirs

Lawrence had brought the idea of clinical use of Na^{24} to Poillon (who marked it "basic, with important commercial applications") in his report for 1934. He had in mind, apparently, that the gamma rays of radiosodium might replace radium's in the general treatment of cancer. In the spring of 1935, about the time that he began to worry about excessive exposure of cyclotroneers to neutrons, he also began to plan provisionally for clinical tests of Na^{24} and P^{32}. He was confirmed in this intention by a visit in April of Charles Sheard, head of biophysical research at the Mayo Foundation; and also by the Board of Regents, who responded to Sproul's report about the discovery of Na^{24} by declaring their interest in "the far reaching biological aspects of the discoveries and the new possibilities of radio therapy."[111] In May, Lawrence lectured the board of the Research Corporation on the power of the rays from radiosodiuim; in July, he could make 50 mg of the stuff, enough, he thought, for clinical tests; in August, treatment began, or would have, had the cyclotron not broken down.[112]

Radiosodium did not answer expectations. As advertised in the Research Corporation's patent, it did not cause harmful side effects, to dogs at least, even when given in large doses; but then neither did it do outstanding damage to tumorous tissue. Two leukemia patients received doses of radiosodium in the spring of 1936; one had 147 mCi in all, the largest quantity of active

110. Field, *Curr. top. rad. res. quart.*, *11* (1976), 1–85; Raju, *Heavy particle therapy*, 78–169; and Catterel in Acc. dei XL, *Memorie, 8:2* (1984), 251–4.

111. Lawrence, "Report [for 1934]" (RC); Lawrence to Leuschner, 12 Apr 1935 (20/13), and to Rutherford, 17 Apr 1935 (ER); Sproul to Lawrence, 31 May 1935 (20/19).

112. RC, Board of Directors, "Minutes," 23 May 1935, 1061 (RC); Lawrence to Beams, 20 Jul 1935 (2/26), and to Poillon, 28 Aug 1935 (RC).

material administered that year. Neither benefitted or suffered. Hamilton inferred that he might practice safely on normal people. He fed his subjects from 80 μCi to 200 μCi of patent-pending Na^{24} while they sat with one hand (and its arm) in a lead cylinder grasping a Geiger counter (fig. 8.6). The counter indicated that active material reached the hand within a few minutes of ingestion. In subsequent, more careful experiments with the same setup, radioisotopes of sodium, chlorine, bromine, and iodine made it from mouth to hand in from three to six minutes.[113] Injected in one arm, the tell-tale tracers arrived in the other in about twenty seconds. Radiosodium accordingly had some employment in studies of circulation and water balance in the body. It became a diagnostic tool for vascular disorders; for a time it stood literally at the cutting edge of research, as an indicator of the best site for amputation of impaired parts.[114]

The most useful of the harmless tracers of Hamilton's experiment was radioiodine because it concentrates very strongly in a particular organ, the thyroid, which has at least as much iodine as all the rest of the body. The first in the field were a group in Cambridge, led by Saul Hertz of the Massachusetts General Hospital and including Robley Evans of MIT. They gave I^{128} made with neutrons from a Rn-Be source to rabbits, which, when "finely minced," disclosed that they had deposited radioiodine in their thyroids very soon after eating it—which was fortunate, since I^{128} has the inconveniently short half-life of 25 minutes. Hamilton wanted something with a week's demi-duration, and so informed Seaborg. "I then told him that I would try." A month later, Livingood and Seaborg presented their 8-day iodine, I^{131}, untangled from the results of irradiating tellurium with deuterons.[115] With this I^{131}, Hamilton and Myron Soley of the Medical School

113. Hamilton and Stone, SEBM, *Proc.,* *35* (1937), 595–8, on leukemia; Hamilton, NAS, *Proc.,* *29* (1937), 521–7, and *Am. jl. physiol.,* *124* (1938), 667–78, on radiosodium and radiohalides; Hamilton and Alles, *Am. jl. physiol.,* *125* (1939), 410–3, on dosing dogs with radioalkalis; Lawrence to Hopwood, 28 Jul 1938 (9/15).

114. A.H. Compton, notes on visit to Berkeley, 16 Sep 1938 (13/29); Hamilton in Wisconsin, Univ., *Symposium,* 339–40.

115. Hertz, Roberts, and Evans, SEMB, *Proc.,* *38* (1938), 510–3; Seaborg, *Jl.,* *1,* 332, 340 (30 Mar and 2 May 1938).

FIG. 8.6 Hamilton's arrangement for detecting the circulation of radioactive salts. The hand in the lead cylinder grasps a Geiger counter. Hamilton, *Jl appl. phys.*, *12* (1941), 449.

showed that uptake of iodine by patients suffering from toxic goiter or overactive thyroid exceeded tenfold the uptake by normal persons, and that parts of thyroids invaded or destroyed by cancer cells could not fix iodine at all. (This last information came from autoradiography: sufferers about to have their defective thyroids cut out ingested radioiodine; sections of the removed thyroids were laid against x ray film; and the portions of the sections containing the active material photographed themselves.) It appeared that the rate of manufacture of the hormone issuing from the thyroid, thyroxin, depended upon the organ's ability to take up iodine and that radioiodine would not be a good weapon against thyroid cancer.[116]

116. Hamilton and Soley, *Am. jl. physiol.*, *127* (1939), 557–72, and *131* (1940),

Research and development then split. L.I. Chaikoff of the University's Department of Physiology and several collaborators used cyclotron-produced I^{131} to study the general biochemistry of iodine; by 1942 they had made important progress in elucidating the process of its fixation in the thyroid.[117] Hamilton and Lawrence, and also Hertz and Roberts, moved on to therapy. The treatment of noncancerous hyperthyroidism began at Berkeley in 1940 and about the same time at the Massachusetts General Hospital, which received I^{131} from the Laboratory for the purpose. Both groups had much the same experience. In four of five cases treated, the cyclotron-produced I^{131} (or, rather, combination of I^{131} with the 12.7-hour I^{130}, also discovered by Livingood and Seaborg), diminished the goiters, relieved symptoms (some of many years' standing), and increased emotional stability within a week or two of administration. Doses ranged from 5 to 28 mCi. No deleterious side effects occurred within six years of commencement of treatment. Still, a cautious physician could not rule out development of thyroid cancer or damage to kidneys or bone marrow. Chemical and surgical treatment improved; and the best judgment when the first clinical experiences were evaluated limited radioiodine to patients intolerant of the new drugs or unable to undergo therapy. Effective treatment of thyroid cancer by I^{131} dates from after the war.[118]

The work with radioactive species of alkalis and halogens, however promising and stimulating, was but side play in the Laboratory's great drama of radiomedicine. The protagonist there was radiophosphorus—easy to make, of convenient half-life (about 14 days), with a good strong beta ray (maximum of 1.7 MeV), and known to concentrate in bones. In 1935, Hevesy and a colleague

---

135–43; Hamilton, Soley, and Eichorn, Univ. Calif., *Publ. pharm., 1* (1940), 339–67; cf. Hamilton, *Jl. appl. phys., 12* (1941), 448–53.

117. Hamilton and Soley, *Am. jl. physiol., 127* (1939), 557–72; Hamilton, *Radiology, 39* (1942), 553–63; Chaikoff and Taurog in Wisconsin, Univ., *Symposium,* 308–24.

118. John Lawrence to G. Failla, 23 Jan 1940 (5/7); Hertz and Roberts, *Jl. clin. inv., 21* (1942), 624, and Hertz, Roberts, Means, and Evans, *Am. jl. physiol., 128* (1940), 565, 575; Hamilton and Lawrence, *Jl. clin. inv., 21* (1942), 624; Hamilton and Soley, *Am. jl. physiol., 131* (1940), 135–43; Hertz in Wisconsin, Univ., *Symposium,* 379–80, 387–9.

at Bohr's institute (for theoretical physics!) fed rats P^{32} made from Rn-Be neutrons on S^{32}, measured the hot phosphorus in the feces, destroyed the animals, and found, among other things, that bone is a dynamic tissue. Further inquiry disclosed that young rats concentrated phosphorus very quickly and substantially in their growing bones.[119] The inquiry continued in San Francisco at the University's Medical School with samples of P^{32} having activities 100 or 1,000 times Hevesy's and with chicks in place of rats. The researchers, S.F. Cook, K.G. Scott, and Philip Abelson, confirmed that new phosphorus goes primarily to the bones, and also to the musculature; and they found that between 4 and 60 days after ingestion, the labelled phosphorus migrated from muscle and small intestine to bone and bone marrow. The spleen enjoyed high deposits throughout.[120]

These results encouraged the speculation that P^{32} might help control blood diseases such as polycythemia vera (a multiplication of red blood cells causing nosebleeds and an enlarged spleen) and leukemia. John Lawrence treated a lady suffering from polycythemia vera in 1936; her symptoms remitted (permanently, as it happened), and plans for proceeding to leukemia were made. Early in 1938 John Lawrence took the entire output of P^{32} from the 37-inch cyclotron. Part he fed to cancerous mice, part, in therapeutic doses, to a leukemic human. The mice confirmed the supposition grounding the therapy: they concentrated the active phosphorus in their fast-growing tumors, especially in tissue invaded by leukemic cells, and did so at the expense of deposition in their bones. Hence the indication: P^{32} as a weapon against cancers of the bone and bone marrow.[121] The human suffered from chronic leukemia of the bone marrow (myelogenous leukemia). He got 70 mCi of P^{32} over two months, which made his blood picture normal; a great triumph, which, as Lawrence

119. Chievitz and Hevesy, *Nature*, *136* (1935), 754–5, and Dansk. Vidensk. Selsk., *Biol. meddelser*, *13* (1937), in Hevesy, *Selected papers*, 60–2, 63–78.

120. Cook, Scott, and Abelson, NAS, *Proc.*, *23* (1937), 528–33.

121. J.H. Lawrence, *Jl. nucl. med.*, *20* (1979), 563; Lawrence to DuBridge, 5 Mar 1938 (15/26); J.H. Lawrence, "Present status of the biological investigations," 25 Aug 1938 (13/29); J.H. Lawrence and Scott, SEBM, *Proc.*, *40* (Apr 1939), 694–6.

rightly cautioned in telling Chadwick, "should not be mentioned in public." But in private, he advised, it might be just the thing to mention to people considering building a medical cyclotron. That was certainly the message that Karl Compton took home from his visit to the Laboratory during April 1938, when the patient's blood appeared so nearly normal that it did not allow firm diagnosis of his condition. As MIT's Robley Evans reported Compton's reaction: "Immediately on arriving home from his visit to your laboratory, Compton called me over to describe with unbounded enthusiasm your leukemia work....I am sure Compton's visit to 'cyclotron headquarters' did much to kindle his already enthusiastic support [for a cyclotron for MIT]."[122]

This was to compound Lawrence's optimism with physicists' ignorance of medical matters. John Lawrence was much more circumspect than his brother. He wrote Evans that the remission accorded his patient might have been accomplished by x rays, a point that Ernest Lawrence then kept in mind. In a report on the treatment of two cases of myelogenous leukemia published in a biomedical journal, John Lawrence and his collegues limited themselves to facts about uptake of P^{32} in the blood and retention in the body, and made no clinical inferences.[123] By June 1939, John Lawrence was treating a dozen patients, who took a total of 20 or 25 mCi a year, and had in consequence a life expectancy that he judged to be similar to that procured with x rays, some two or three years. In July he left for Europe, to bring tidings of the Laboratory's progress in radiobiology and radiomedicine to the British Association for the Advancement of Science. He mentioned Hamilton's work with radioactive alkalis and radioiodine, the indications for leukemic therapy, and the occurrence of remissions under treatment.[124]

122. Lawrence to Chadwick, 30 Apr 1938 (3/34); Evans to Lawrence, 15 Apr 1938 (7/8); Lawrence to A.H. Compton, 19 Sep 1938 (4/10), same news, same message.

123. J.H. Lawrence to Evans, 9 and 12 Apr 1938 (7/8); J.H. Lawrence to Hektoen, 22 Sep 1938 (13/39); Tuttle, Scott, and J.H. Lawrence, SEBM, *Proc., 41* (May 1939), 20–5.

124. J.H. Lawrence to L.A. Erf, 2 June 1939 (7/6); Lawrence, *Nature, 145* (27 Jan 1940), 125–7; Tuve, "Present technical status of the use of radioactive tracers," 10 May 1939 (MAT, 25/"Hamilton Club"), 4.

The meeting of the British Association ended as Germany invaded Poland. Two days later, on September 3, 1939, Britain declared war. John Lawrence then set sail in the *Athenia*, which the Germans promptly torpedoed. The physician saved himself, bravely, after tending the wounded. Ernest Lawrence pulled what strings he could to obtain a berth for John on the first American ship sailing from Britain to the United States. The most useful of these strings involved leukemic politics. Early in August the Laboratory had received a visit from F.C. Walcott, an influential Republican senator from Connecticut until the Democratic landslide in 1936, who had come West as a guest of Herbert Hoover. He had a son, Alex, suffering from leukemia. The senator placed his hopes in the cyclotron and P^{32} and asked that John Lawrence stop to see Alex in New York on his return from Europe. When John seemed stuck after the sinking of the *Athenia*, Walcott pulled his strings, attached to the American ambassador to Britain and the president of the merchant marine. Within a week John had a berth on the *Nieuwamsterdam*.[125] He was restored to his leukemia patients, who now included Alex Walcott, in October. Results were mixed. P^{32} did nothing for Alex. Evaluating the situation in February 1940, a few months before Alex died, John Lawrence could not rate P^{32} more effective than other therapeutic agents in the control of leukemia. His best hope was that "possibly it will turn out to be slightly better."[126]

The continuing study of the metabolism of phosphorus in mice supported the hope. The physiologists at Berkeley showed that the generation of phospholipids in tumor cells took place as vigorously as in the most active organs, the liver, kidney, and small intestine, and that, in contrast to the normal active organs, tumors retain their P^{32} for long periods of time.[127] Lawrence and his associates

125. Childs, *Genius*, 293; Walcott to Lawrence, 12 Aug 1939, Lawrence to J.H. Lawrence, 19 Sep 1939, and telegrams among the Lawrences and Walcott, Sep 1939 (18/6). In the event, John did not need his brother's influence; he sailed not on the *Nieuwamsterdam* but on an American ship chartered to bring home the survivors.

126. J.H. Lawrence to Ridenour, 6 Feb 1940 (14/22); infra, §10.1.

127. Jones, Chaikoff, and Lawrence, *Jl. biol. chem.*, *128* (1939), 631–44, ibid., *133* (1940), 319–27; and *Am. jl. cancer*, *40* (1940), 235, 241–2, 250; Lawrence to M.W. Schramm, International Cancer Research Foundation, 29 Jul and 18 Aug 1938 (9/22), requesting $2,100 for this work.

at the Crocker Laboratory confirmed the phosphorophyllic tastes of leukemic tissue.[128] In this respect humans behaved like mice. Lawrence and Lowell Erf, a physician expert in hematology and supported on fellowships, gave tracer doses of P^{32} to seven patients about to die of various cancers. In their last moments, as autopsy disclosed, they had put as much phosphorus per gram wet weight in their malignant tissues, or in tissues infiltrated by malignant cells, as in their most active organs. These results, according to John Lawrence, had constituted the rationale for the therapeutic use of radiophosphorus. The argument, however, was indirect. The first direct trials against cancerous growth in mice were, in Erf's words, "very disappointing."[129]

On May 18, 1940, Arthur Compton and James Murphy of the National Advisory Cancer Council looked in to see what the council supported at Berkeley. Ernest Lawrence, on John's advice, had gone to Sonoma to recover from a sore throat; but he came down to the Laboratory to meet men so important for his future work. The day did not go well. Murphy, an expert on animal experimentation, graded the Laboratory's procedures and facilities poor or worse. Cooksey later visited Murphy's institute. "[I] was tremendously impressed with the facilities....It was obvious that our set up was terrible in his eyes." The conversation on May 18 switched to politics. Compton complained that an article he had written for a newspaper outlining the duties of scientists toward science and the nation had been misinterpreted. He planned to reply. Lawrence and Cooksey told him, politely no doubt, that he should have kept his mouth shut in the first place. Berkeley reserved its radicalness for technology and medicine. The report of the site visitors helped the NACC decide not to underwrite the expansion that John Lawrence suggested to it: investigation of differential effects of n rays and x rays on animals; metabolic studies of iron, sodium, potassium, sulphur, and iodine on hundreds of animals; methods to improve uptake of potassium and boron in

128. J.H. Lawrence et al., *Jl. clin. inv.*, *19* (1940), 263–71; Tuttle, Erf, and Lawrence, ibid., *20* (1941), 57–61; Erf and Lawrence, ibid., 567, 575.

129. Erf and J.H. Lawrence, SEBM, *46* (Apr 1941), 694–5; J.H. Lawrence, *Nature, 145* (27 Jan 1940), 127; Erf to Theodore B. Wallace, Smith Kline and French, 6 Mar 1940 (7/6).

tumor tissue. On first consideration, NACC declined support for animal experimentation and cut back that for Stone's clinical application of neutron rays.[130]

Ernest Lawrence was not accustomed to such rebuffs. He accepted that the council might not care to support experiments with animals to refine neutron therapy or clinical use of radioiosotopes. "But it is totally incomprehensible to me that there should be any suggestion of curtailing the clinical program with neutron rays....It seems to me a primary obligation on the part of all of us to see that the program of exploring these possibilities be carried forward full steam ahead....The cancer program simply must go forward as Dr Stone and my brother have planned it." Compton, the recipient of this appeal and demand, allowed that Stone's program would most probably be funded in full.[131] It was. But the council stood its ground on animal experimentation.

The clinical program with radioelements rested on a financial and technical base quite different from that of neutron therapy. Treatment did not require immediate access to a cyclotron or special facilities at the Laboratory. The chief therapeutic agent, P^{32}, came so plentifully that, in John Lawrence's estimate, he could treat all the chronic leukemia in California without interfering with other obligations of the cyclotron.[132] The cost of machine time for making P^{32} could be passed on to the patient. The positive signs of phosphorus therapy outweighed the negative and opened the brief flurry of interest in commercial production of radioisotopes by cyclotrons arrested by the war. The psychology of support appears plainly from the initiative of Hans Zimmer, professor at the Harvard Medical School, dying of leukemia, who believed that he had obtained some benefit from P^{32}. It bothered him that the amount of radiophosphorus available fell far short of the nation's, if not of California's needs. He appealed to the Carnegie Foundation for $12,000 for a cyclotron for his Medical School.[133] Although John Lawrence advised against rushing

130. Cooksey to Poillon, 25 Jul 1940, and J.H. Lawrence to Dr. Smith, NACC, 1 June 1940 (13/29A). On John Lawrence as Laboratory doctor: Cooksey to Lawrence, 29 Mar 1939 (4/22).

131. Lawrence to A.H. Compton, 21 Jul, and reply, 26 Jul 1940 (4/10).

132. J.H. Lawrence to V. Bush, 10 Sep 1940 (15/18).

133. Zimmer to E.P. Keppel, 31 Aug 1940 (15/18);

commercial production, Poillon pushed it, confident in and counting on the Research Corporation's patents on the cyclotron and the Van de Graaff.[134] Despite the attitude of the NACC, the Laboratory had no trouble expanding the basis of support of its biomedicine in 1940/41. The clinical program in radioisotopes had the support of the Jane Coffin Childs Fund for Medical Research, the Darian Foundation, Merck and Company, the Columbia Foundation, and the Donner Foundation.[135]

In 1948 Byron Hall of the Mayo Clinic evaluated the results of the clinical use of P^{32} in Berkeley, the Medical School of the University of Rochester, and his own institution. "It is too early [he wrote] to make a final evaluation of this form of therapy." But he allowed that it had brought relief to sufferers from polycythemia vera and the chronic forms of leukemia. The clinic had by then treated 154 cases of the one and 33 of the other. Symptoms in almost all of the 124 cases of polycythemia vera for which adequate follow-up data existed improved or disappeared. In 85 percent of the cases, the blood picture remitted satisfactorily and risk of hemorrhage and thrombosis decreased. Remissions of up to five years were achieved, but most lasted under two years. Complications—drop of white-blood-cell count, severe anemia, acute leukemia—occurred in a total of 30 percent of the cases. In general, P^{32} raised the life expectancy of sufferers from polycythemia vera as much as vitamin B_{12} did that of victims of pernicious anemia. Twenty patients with chronic myelogenous leukemia were followed. Many obtained the same sort of remission they would have acquired from x rays. Eleven of the patients died within six years of treatment. Chronic lymphatic leukemia proved more receptive. Six of six patients continued in remission from ten to twenty-six months after treatment. Radiophosphorus conferred no benefit on patients with acute leukemia, Hodgkin's disease of the bone marrow, or multiple myeloma.[136]

134. Poillon's correspondence with American Cyanamid, General Electric, and Westinghouse, Apr–Oct 1940 (RC, "cyclotron licensing"); Poillon to Lawrence, 26 Sep 1940 (15/26A); RC, Board of Directors, minutes, 29 Oct 1941, 1496.

135. Correspondence between the donors and recipients, 1940/41 (UCPF, 515/400 and 541/312).

136. Hall in Wisconsin, Univ., *Symposium*, 353–76; J.H. Lawrence, Manowitz, and Loeb, *Radioisotopes and radiation*, 50.

Rosetta Stones

On May 25, 1939, the "University Explorer," reaching deep to advertise the Laboratory, announced the arrival of radioactive fertilizer in Hawaii. A professor at the University there had imported P^{32} from Berkeley by Pan American clipper to test its power on pineapples. An explorer of the consequences of the innovations of the Laboratory had (and has!) his work cut out: "The influence of the cyclotron has been felt in so many different fields of science that no one can predict its ultimate value to mankind."[137]

The influence began to spread at home, in 1937/38, among members of Berkeley's Chemistry Department. The first practitioners of radioactive tracing were Libby, Seaborg, and, above all, Ruben, through whose efforts University biologists took up the technique. Purely chemical applications, which were not irrelevant to the biologist, included the vast fields of exchange reactions and reaction mechanisms; the nature of the inquiry may be indicated by studies by Libby's students of exchanges between various valence forms of sulphur and of photochemical processes in solution, and by Libby himself of the reactions of recoil nuclei activated by neutrons. Another conspicuous line, detection of impurities by their radioactivity, which had been an unwelcome and misleading annoyance, was practiced for a time by Seaborg and Livingood, who found copper in nickel, iron in cobalt, and phosphorus and sulphur everywhere.[138]

Ruben and P^{32} inspired a major direction of research in the Physiology Department on the metabolism of phospholipids or phosphatides, which occur in all living tissues in connection with fatty deposits. Ruben, Chaikoff, and their students and colleagues, who included the chemist I. Perlman and, occasionally, John Lawrence, ground up rats and birds fed radiophosphorus under various regimes—fasting, normal diets, fatty diets—to learn the loci of the creation and destruction of phospholipids. Following John

137. "University Explorer," 25 May 1939 (40/15).
138. Kamen, *Science, 140* (1963), 587, re Ruben; Seaborg, *Chem. rev., 27* (1940), 250–73, esp. 256, 263, 266; Voge and Libby, *JACS, 59* (1937), 2474; Rollefson and Libby, *Jl. chem. phys., 5* (1937), 569–71; Seaborg and Livingood, *JACS, 60* (1938), 1784–6.

Lawrence's interests, they also examined cancerous mice. They extended the findings of Segré's associates in Palermo, who, with an Italian touch, had fed olive oil and Berkeley phosphorus to their rats; which, when anatomized, disclosed that the liver is the fastest metabolizer of phospholipids. The Berkeley group further showed that excised bits of liver, kidney, and intestine continued to work at phospholipid metabolism in vitro, and that the rate of its metabolism in tumors transplanted from one animal to another was characteristic of the tumor, not the animal.[139]

Two other sustained programs in radioactive tracing that thrived on the Laboratory's output deserve mention. One grew in the Biochemistry Department around David Greenberg's ongoing investigations of mineral metabolism. Inspired, he said, by the "revolutionary nature and potential importance" of radiotracing, Greenberg and his associates began as others did, feeding radiophosphorus to rats and examining deposits in tissue and feces. They went on to the phosphorus metabolism of rachitic animals. They were the first to use radiocalcium (Ca^{45}), a commodity more costly than Fe^{59}, as a tracer. In 1940 and 1941 they published information on the metabolism, deposition, and elimination of manganese, iron, and cobalt, the last, as they found, a possible cause of polycythemia.[140] The second program, conducted by S.C. Brooks at the Medical School, studied the transport of ions into and within plant cells. Here radioactive indicators brought light where darkness had long prevailed. According to an authoritative reviewer, the results of Brooks and his associates released biologists from "the necessity of postulating mysterious properties for cellular membranes and protoplasm."[141] Further

139. Artom et al., *Nature, 139* (15 May 1937), 836–7, and ibid. (26 June 1937), 1105–6; Perlman, Ruben, Chaikoff, and others, in various combinations, *Jl. biol. chem., 122* (1937), 169–82, *123* (1938), 587–93, *124* (1938), 795–802, *126* (1938), 493–500, *127* (1938), 211–20, *128* (1939), 631–44, 735–42. This and similar work from other laboratories is reviewed in Greenberg, *Ann. rev. biochem., 8* (1939), 276–7; Hevesy, ibid., *9* (1940), 649–54; and Hamilton, *Jl. appl. phys., 12* (1941), 445, and *Radiology, 39* (1942), 545–8.

140. Greenberg reviewed his own work in *Ann. rev. biochem., 8* (1939), 269–70, quote, and in Wisconsin, Univ., *Symposium* (1948), 263–9, 270–3, 279–81. Cf. Kohler, *From medical chemistry*, 329; Greenberg, "Recollections" (UCA); Hamilton, *Radiology, 39* (1942), 566–7.

141. Loofbourow, *RMP, 12* (1940), 275, with references to the literature.

attacks on these mysteries occurred at the School of Agriculture, where Perry Stout and his co-workers made the uptake of alkali and halide salts by growing trees and shrubs their subject of study. They settled the vexed question whether salts rise through the bark as well as through the xylem (the answer is no) and demonstrated by persuasive autoradiography the deposit of P^{32} in the leaves and fruit of growing plants (plate 8.5).[142]

It did not take much, apart from the material, to set up with radioactive tracers and to get quick, publishable results. There were so many animals and plants, so many elements, so many permutations of experimental circumstances. Hence the Laboratory received many requests for its products, which, if they did not save lives, might improve careers. Lawrence honored requests from outside the University on the basis of merit, other factors being equal. His most generous support went to workers in Rochester and in Copenhagen, the two places where, in the opinion of Merle Tuve, no fan of the Laboratory, work with tracers second only to Berkeley's was being done.[143] In Copenhagen, Hevesy's group continued work with P^{32}, following it throughout the body and its products, into blood, eggs, milk, across cell walls, to and from the liver, and into the brain. Like Chaikoff's group, they paid much attention to phospholipid metabolism; and they worked out other biochemical pathways, notably the role of phosphorus compounds in the enzymatic breakdown of carbohydrates (glycolisis). Lawrence encouraged Hevesy to ask for "as much [P^{32}] as he can use....We are more than eager to help his important work."[144]

Hevesy received the Nobel prize in chemistry in 1944 for his contributions to the tracer method. The leader of the Rochester

142. Stout and Hoagland, *Am. jl. botany, 26* (1939), 320–4; Arnon, Stout, and Sipos, ibid., *27* (1940), 791–8; J.H. Lawrence, *Nature, 145* (1940), 125–6, and "Summary of biological and medical investigations," Jan 1940 (22/11); Hamilton, *Radiology, 39* (1942), 550–2.

143. Tuve, "Status of the use of radioactive tracers," 10 May 1939 (MAT, 25/"Hamilton Club"), 2–3.

144. Hevesy, *Selected papers,* 111–74, 440–1; Hevesy et al., *Acta biol. exp., 12* (1938), 34–44; Lawrence to Bohr, 17 Feb, quote, and 14 Nov 1938, and Bohr to Lawrence, 25 Jul 1939 (3/3); correspondence with Hevesy, 1937–40 (9/7); Hevesy in Frisch et al., *Trends,* 115.

group, George Whipple, already had a Nobel prize, that of 1934 in physiology and medicine, for work on anemia. Radioiron sent from Berkeley allowed him to wallow more deeply in his favorite subject. His group, in which P.F. Hahn took the lead, showed that anemic dogs accepted iron in any form, collected it rapidly in the bone marrow, and disbursed it rapidly in blood cells. A normal dog would absorb almost no iron at all. It appeared that ingested iron entered the blood stream only if the body's iron store had been depleted; otherwise it passed directly through the intestines. Whipple expected to devise a satisfactory treatment of anemic patients using ordinary iron on the basis of the knowledge labelled iron gave him.[145] The information did not come easily. Most of the time Whipple's group were as deprived of iron as their dogs. The cyclotron could not keep up the supply: the 37-inch made less than 1 $m\mu$Ci of Fe^{59} per μAh, the 60-inch only 0.03 μCi. A radioiron with suitably high specific activity, Fe^{55} (τ = 4 years), can be made by (d,n) on manganese, but it was not available before the war.[146]

The Rochester iron men hoped that the radioisotopes furnished by physicists would turn out to be the "'Rosetta Stone' for the undertaking and study of body metabolism." That was to be very optimistic. The Stone then had nothing to say about the largest part of the bodies of plants and animals: no useful radioactive tracer for hydrogen, carbon, oxygen, or nitrogen had yet been found. No one felt this difficulty more than Kamen and Ruben, who had boldly set forth in 1938 to find the way through photosynthesis armed with C^{11}, which has a half-life of about 21 minutes. An experiment consisted of making the isotope, burning it to carbon dioxide, feeding it to plants, chopping the leaves into a beaker, adding as carrier any substance they guessed might have been labelled with C^{11} by the plant, and examining the various

145. Hahn et al., *Jl. exp. med.*, *69* (1939), 739–53, *70* (1939), 443–51, and *71* (1940), 731–6; Whipple's ambition, recorded by Tuve, "Status of the use of radioactive tracers," 10 May 1939 (MAT, 25/"Hamilton Club"), 7; Hamilton, *Radiology, 39* (1942), 564–6, and *Jl. appl. phys., 12* (1941), 455–6.

146. Whipple to Lawrence, 23 Nov, and reply, 30 Nov 1937, and Lawrence to DuBridge, same date (15/26); letters to Lawrence from Evans, 6 Mar 1939 (7/8), and from Van Voorhis, 5 Feb 1939 (17/14).

carriers. All in an hour or two. The pace forced out their first collaborator, Zev Hassid, who suffered from high blood pressure. They made three of these hectic runs a week for three years.[147]

They made some progress. A plant fed in the dark attached labelled CO_2 to a big molecule RH, R is an unknown radical, making a compound RCOOH. The reaction appeared to be reversible. In the light, however, and the presence of chlorophyll, RCOOH gained water and lost oxygen, irreversibly, to become RCH_2OH. This structure may be considered a molecule R'H, which can fix another CO_2 molecule in the same way to become RCH_2OCH_2OH, and so on. R might then break off, leaving a sugar. The scheme was novel and, in keeping with the requirements of philosophers, had an easily testable consequence that distinguished it from older theories. The consequence: that RH be a very big molecule. For evidence, Kamen and Ruben turned to the ultracentrifuge. The best local setup was at Stanford. Ruben stationed himself there, awaiting with counters ready the delivery of the labelled samples Kamen rushed down from Berkeley. Their lives eased with the discovery that Shell Development Company in Emeryville, adjacent to Berkeley, had a similar centrifuge. The machine showed RH to have a molecular weight between 500 and 1,000.[148] To go farther, to identify the heavy molecule and the intermediates in photosynthesis, Kamen and Ruben needed a longer-lived isotope of carbon. A bout with N^{13}, which has a half-life of 10.5 minutes and which did not quite enable them to decide whether nonleguminous plants can fix nitrogen, further indicated the limits on biochemical research placed by lack of tracers for organic reactions.[149]

Another path lay open. For many years biochemists had been following reactions by tagging compounds with naturally occurring isotopes. They would introduce a substance artificially enriched with deuterium or with C^{13}, which makes up a little over 1 percent of ordinary carbon. They had then only to take the final

147. Hahn et al., *Jl. exp. med., 69* (1939), 739; Kamen, *Radiant science,* 84–7.
148. Ruben, Hassid, and Kamen, *JACS, 61* (1939), 661–3, *62* (1940), 3443–55; Kamen and Ruben, *Jl. appl. phys., 12* (1941), 326; Kamen, *Radiant science,* 105–9; Hamilton, *Radiology, 39* (1942), 567–9, and *Jl. appl. phys., 12* (1941), 457–8.
149. Ruben, Hassid, and Kamen, *Science, 91* (14 June 1940), 578–9.

products of interest and assay them for the relative abundance of the rare isotope in a mass spectrograph. Urey's colleague at Columbia, Rudolf Schoenheimer, the great master of the technique, routinely detected changes of 1 percent in isotopic abundance, which, in the case of carbon, meant one part in ten thousand. With some encouragement from the Research Corporation, Urey had perfected a method to enrich the concentration of C^{13}, which he turned over to Columbia University to patent at the end of 1939.[150] At about the same time, the Eastman Corporation sought advice about the likely market for the stable isotope N^{15} as a tracer for nitrogen. Urey formed and did not hide the opinion that any foundation truly wishing to advance biochemistry and biology should assist in making rare natural isotopes available and not throw its money into cyclotrons.

Lawrence was then at the beginning of his negotiations with the Rockefeller Foundation for what became a request for a million dollars. Around October 1, 1939, Lawrence summoned Kamen and ordered him to find a radioactive carbon, nitrogen, and/or oxygen to silence Urey. "He said I could have both the 37-inch and the 60-inch cyclotrons and all the time I needed, as well as help from whomever I requested—Segrè, Seaborg, anyone!" Naturally Kamen chose to center his all-out search on carbon, and all the more after he had demonstrated with Segrè's help that no useful oxygen or nitrogen could be made by irradiating oxygen with alpha particles or deuterons. He pinned his hopes on an internal target of graphite, which he baked with deuterons for 5,700 μAh during the first six weeks of 1940. On February 13, Kamen terminated his exposure and left the hot target for Ruben to analyze.[151]

Kamen recalled that he had turned to deuterons on graphite in "desperation and resignation." The only likely candidate for a useful radiocarbon was C^{14} made by (d,p) on the rare isotope C^{13}.

150. Poillon to Lawrence, 21 Sep, and reply, 28 Sep 1937 (15/17A); Urey to Douglas C. Gibbs, secretary of University Patents, Inc., 22 Nov 1939, 1 Mar 1940 (Urey P, 8). The patent was filed 3 Aug 1940. The agreement, as usual with University Patents, gave 7 percent of gross receipts to the inventor; letters to Urey from Gibbs, 8 Aug, and from Howard S. Neiman, 13 Aug 1940 (Urey P, 8).
151. Kamen, *Radiant science*, 127–30, and *Science, 140* (1963), 588–9.

But for reasons similar to those that inhibited the discovery of the activity of H^3, C^{14} did not appear to possess the characteristics desired by radiobiologists. For C^{14} had already been discovered at the Laboratory. In 1934, in one of his first observations with his cloud chamber, Kurie had seen half a dozen tracks that indicated a new sort of radiation stimulated by neutrons. In addition to the then known (n,α) process, he saw what he interpreted as (n,p) reactions on air, either $N^{14}(n,p)C^{14}$ or $O^{16}(n,p)N^{16}$. His attribution, challenged by the Cavendish, was established by Bonner and Brubaker in 1936.[152] Kurie and Kamen then studied the profusion of (n,p) events made possible by the cyclotron's enlarged neutron flux; they found the recoil tracks of the C^{14} ions and the liberated protons to be plentiful and conspicuous enough to serve as a measure of the stopping power of the air in the cloud chamber.[153]

From general considerations set forth by Bethe and Bacher, not more than one of a set of isobars can be stable. But N^{14} is stable. The beta decay of C^{14} to N^{14} would depend upon their difference in mass, which Bethe and Bacher estimated at 100 or 200 MeV. "Assuming the β-transitions to be allowed, the lifetimes would be between 1/2 and 20 years." In the same bit of beryllium irradiated by deuterons in which he identified Be^{10} (in fact H^3), McMillan had also noted a weak activity of about three months, which he ascribed to C^{14} made via (d,p) on a carbon impurity in the target. He then tried to make C^{14} in quantity by (n,p) by exposing ammonium nitrate to neutrons from the 37-inch cyclotron; the experiment ended with the accidental breakage of the salt's container and was not renewed.[154] Livingston and Bethe accepted McMillan's estimate of a half-life of several months and Oppenheimer's student Phillip Morrison refined and confirmed their calculations. But careful search, notably by Ernest Pollard of Yale, who tried (d,p) on C^{13} and (α,p) on B^{11}, found the product

152. Kurie, *PR, 45* (1934), 904, letter of 15 June, and *PR, 46* (1934), 330; Chadwick and Goldhaber, Camb. Phil. Soc., *Proc., 31* (1935), 612; Bonner and Brubaker, *PR, 49* (1936), 778; Livingston and Bethe, *RMP, 9* (1937), 344.

153. Kamen, *Radiant science*, 123–4, and *Science, 140* (1963), 585.

154. Bethe and Bacher, *RMP, 8* (1936), 101–3, 201; McMillan, *PR, 49* (1936), 875–6; Kamen, *Science, 140* (1963), 586.

protons, and calculated that C^{14} should yield soft beta rays, disclosed no activity ascribable to any carbon isotope.[155] Hence Kamen's doubt that anything interesting would result from bombarding charcoal with deuterons.

The first tests confirmed his pessimism. Carbon removed from a target of calcium carbonate did not excite a thin-walled Geiger counter. When placed inside a screen-walled counter devised by two of Ruben's colleagues, however, it gave some weak signs, about half the usual background count. That equivocal message marked the discovery of the most important radioactive tracer found at the Laboratory. On leap-year day 1940 the good news was sent for publication.[156] The reason that the decay of C^{14} had been so difficult to detect is that its period is very long. Ruben and Kamen first estimated "years;" a larger irradiation (some 13,000 μAh) of a better probe target enriched in C^{13} allowed a much higher and better estimate, thousands of years. That was a puzzle for the theorists, who, however, in the persons of Oppenheimer and L.I. Schiff, helpfully remarked that the nuclei of C^{14} and N^{14} must differ enough in angular momentum to retard the decay by the amount observed.[157] In any case, the long-sought long-lived carbon was in hand, and it remained only to discover how to make it in sufficient amounts to satisfy the expected large demand. Kamen and Ruben returned to Kurie's process, $N^{14}(n,p)C^{14}$, irradiating large carboys of concentrated ammonium nitrate with neutrons from the 60-inch cyclotron. Although the yield was greater and the recovery easier than with deuteron-irradiated graphite, Lawrence ordered the process discontinued. He had heard that ammonium nitrate presented a serious hazard of explosion and no weight of chemical opinion could persuade him of its safety in solution.[158]

155. Livingston and Bethe, *RMP, 9* (1937), 344; Pollard, *PR, 56* (1939), 1168, letter of 14 Nov; Kamen, *Radiant science*, 126. Pollard had irradiated his carbon for twelve hours with 0.5 μA, which, he supposed, would not give a detectable activity if the period of C^{14} were several years or more.

156. Ruben and Kamen (1940), 549, letter of 29 Feb; Libby and Lee, *PR, 55* (1939), 245.

157. Kamen and Ruben, *PR, 58* (1940), 194, late June 1940; Ruben and Kamen, *PR, 59* (1941), 350-1, 354.

158. Ibid., 351-2; Kamen to Urey, May 1940 (10/10); Kamen, *Radiant science*, 139-40, and *Science, 140* (1963), 589-90.

Lawrence lost no time informing Urey of the miraculous appearance of C^{14} and in asking him for a sample of purified C^{13} to serve as a target for Kamen. That, he intimated, would be the true value of Urey's separation process; C^{14} would not depress the market for C^{13}; "quite the contrary." The rules of science are strict. "Needless to say [Urey replied] we will surely send him the material he wants." But then the obvious question arose: why go to the trouble and expense of transmuting C^{13} into C^{14} for tracing if C^{13} will serve au naturel? Calculations by Urey, Kamen, Ruben, and Tuve concurred. As Kamen put it, C^{14} was "an ace in the hole," something for very special applications, such as the study of photosynthesis, where the big molecules involved might dilute the natural isotope past detection. To make a quantity of C^{14} useful for the purpose would require a long time on the machine. "It is quite beyond all probability to make more than one or two such strong samples in your or my lifetime as conditions are at present."[159] Urey gave Kodak his apparatus and encouraged them to proceed. They agreed to make N^{15} and, if that succeeded, to try C^{13}. Kodak did not like Urey's method of separation, which used deadly hydrogen cyanide gas, and they worried that the more prolific route via N^{14} would make C^{13} unnecessary.[160]

Although Lawrence, Kamen, and Ruben joined Urey and others in urging the commercial production of C^{13},[161] Kodak did not move to serve the mass market of biotracers before the war. The frustration of having too little heavy carbon to unlock the secrets of life was eased by war. In 1945 Urey again tried to enlist Kodak. Again the question: if C^{14} can be made plentifully without separated C^{13}, why invest in a heavy-carbon plant? But if not,

159. Lawrence to Urey, 24 Feb 1940; Urey to Lawrence, 29 Feb and 1 Mar 1940; Kamen to Urey, 13 May 1940, quote, all in 17/40; Tuve to Urey, 9 May and 3 June 1940 (MAT, 25/"biophys. 1940").
160. Urey to C.E.K. Mees (Kodak), 27 Apr, 2 May, 12 Aug 1940; Mees to Urey, 1 and 29 May 1940, all in Urey P, 3/M; Urey to Hevesy, 25 Oct 1940 (Urey P, 2), and to Lawrence, 27 Oct 1940 (17/40).
161. Lawrence to Urey, 3 Oct 1940 (17/40); Urey to Mees, 6 Nov 1940 (Urey P, 3/M).

Kodak should come to the rescue, lest the country "lose years in applying the tracer technique to chemical and biological problems."[162] Kodak wisely remained out of the picture. The tremendous neutron fluxes in the piles built during the war created C^{14} in plenty. Thus the sciences of life gained "the most important tracer available among the artificial radioactive elements."[163]

162. Urey to Otto Beeck, Shell Development Co., 21 Mar, and to Paul H. Emmett, Mellon Institute of Industrial Research, 12 Apr and 9 Jul, and to A. Keith Brewer, National Bureau of Standards, 16 Jul 1945; letters to Urey from W.O. Kenyan, Kodak, 12 Sep, Mees, 11 Jul, and A.O.C. Nier, 20 Aug 1945, quote, all in Urey P, 1/"misc. corresp."
163. Kamen in Wisconsin, Univ., *Symposium*, 150.

Little-Team Research
with Big-Time Consequences

Almost all the Laboratory's work in physical science centered on or developed from the two main concerns with which it began: machine design and nuclear transformations. Few people stayed long enough or became independent enough to use the Laboratory's facilities to pursue significant topics outside this framework. The most successful in making room for himself was Luis Alvarez, whose discovery of K-electron capture won the approval of that otherwise merciless critic of Laboratory life, Maurice Nahmias. "I am very happy for him because he is the most likeable [of the staff] and does not waste his time with 'new periods.'"[1] But the study of K-electron capture lay so close to the Laboratory's established activities that it was immediately assimilated into period hunting. Alvarez's greatest departures from general practice concerned properties of the neutron, involved collaboration with Stanford, and did not open continuing research lines at Berkeley.

Like Alvarez, McMillan had the freedom and the obligation of the young professor to develop his own research lines. He tended to extend, systematize, or make more precise research domains at the center of the Laboratory's work: for precision, his early close studies of the energies of gamma rays emitted in certain nuclear transitions; for systematization, his up-to-date files on nuclear transformations; for extension, his theory of cyclotron focusing.[2]

1. Nahmias to Joliot, 27 June 1937 (JP, F25); cf. Kamen, *Radiant science*, 308.
2. McMillan, *PT, 12:10* (1959), in Weart and Phillips, *History*, 265–6, mentions cyclotron theory, Alvarez's neutronics, and Segrè's technetium as examples of growing sophistication of experiments at the Laboratory in the late 1930s.

McMillan's mastery of all aspects of the Laboratory's nuclear physics and much of its chemistry gave him a decisive advantage in following up the discovery of nuclear fission. It was he who secured the first beachhead in the new territory of transurania. He was accompanied by several accomplished investigators, including Segrè and Seaborg, who annexed the field for the Laboratrory.

1. NUCLEONICS

The neutron was a frustrating object in the laboratory. It went everywhere, refusing obedience to the electric and magnetic fields to which other particles submitted; it provoked a wide—indeed, too wide—range of nuclear transformations; and it was dangerous to living things. The danger, as we know, inspired the Laboratory to invent neutron therapy and caused all large-cyclotron laboratories to surround their machines with a protective barrier, usually of water. Most of the neutrons entering the water quickly slowed to thermal velocities by sharing their energy in collisions with hydrogen atoms and either died in their bath or emerged relatively harmless into the experimental space. The promiscuity of neutrons—their easy union with most nuclei—usually increases as their velocity diminishes, as Fermi had found; in order to explore nuclear responses to their wanton behavior, it was very desirable to work with neutrons of uniform speed. Interest in making homogeneous or "monochromatic" beams of neutrons grew strongly in 1936, when Bohr offered his liquid-drop model of the nucleus in explanation of the anomalously high absorption by certain substances of bombarding particles of particular energies. To study his idea further—to find the energies of particle beams with which various nuclei "resonated"—required control of the beam velocity. And just here the chargeless neutron, the most promising tool for probing resonances, eluded the will of the experimenter.

Although neither the cyclotron nor its keepers at Berkeley were temperamentally adapted to the careful study of resonances, control of neutron beams was of central interest there. Shortly after Fermi's discovery of the power of slow neutrons, several members of the Laboratory discussed production of a roughly monochromatic neutron flux by interrupting the deuteron beam to the cyclotron's beryllium target. The general idea, according to Alvarez's reference to it in 1938, was to shut off the radio frequency potential intermittently and to turn on an amplifier attached to a neutron detector when the accelerating potential cut out. The only neutrons counted by the chamber would be those created in the beryllium during the deuteron bursts (radio frequency on) with such velocities that they could reach the detector during its live time (radio frequency off). The idea was temporarily dropped when the Laboratory learned of a rough-and-ready adaptation of an old technique to the new purpose. Physicists at Columbia ran the output from a Rn-Be source through paraffin and then into a chopper consisting of two disks, each made in alternate sections of duraluminum and cadmium, which are, respectively, transparent and opaque to slow neutrons. When the disks spin around the same axis, only slow neutrons that passed through a transparent section in the first disk with a speed that brought them to a similar section in the second could reach the detector.[3]

The urgency of the study of resonance absorption—"the point of greatest interest today in nuclear physics," to quote British opinion in the fall of 1938—reopened the matter. The neutron chopper did not give monochromatic rays, among other reasons because fast neutrons could pass right through the disks. Two new solutions to the problem were offered that same fall, one by Alvarez, who developed the scheme discussed in the Laboratory in 1935, and the other by a group around G.P. Thomson at Imperial College, London. By then both parties had been working at their projects for some time.[4] A team at the University of Utrecht may

3. Alvarez, *PR, 54* (15 Oct 1938), 609; Dunning et al., *PR, 48* (1935), 704, letter of 7 Oct; Lawrence to Pegram (Columbia), 24 Oct 1935 (4/8), reporting a trial by Stern, then visiting Berkeley, to measure velocities of slow neutrons. The final report of the Columbia experiments by the junior member chiefly involved is Fink, *PR, 50* (1936), 738–47, rec'd 13 Aug.

4. *Nature, 142* (17 Sep 1938), suppl., 520, report of a discussion on nuclear physics at the British Association for the Advancement of Science, 18 Aug 1938, led by Bohr, quote. Alvarez first presented his method in public at the APS meeting in San Diego, 22–24 June 1938, abstract in *PR, 54* (1938), 235; the British group, in the persons of P.B. Moon and C.D. Ellis, gave first notice at the BAAS meeting just mentioned and first results on absorption in Fertel et al., *Nature, 142* (5 Nov 1938), 829, letter of 10 Oct.

also have hit on a solution independently, although their first published description of their method was prompted by a preliminary report of Alvarez's.[5]

The machines built by Thomson and by Alvarez highlight typical features of British and California approaches to instrumentation. The Imperial College group used as source Oliphant's version of the Cockcroft-Walton machine, capable of 150 to 250 kV; Alvarez used the 37-inch cyclotron, then delivering deuterons at 8 MeV. Alvarez interrupted his deuteron beam by suppressing the current at the plates of the rf tubes supplying the accelerating potential; the method required considerable electrical power and some electronics and yielded long deuteron bursts (4 msec) that were not very well defined. The British modulated the current through the discharge tube that created the deuterium ions by amplifying the current from a photocell activated by a light beam interrupted by a tuning fork or rotating shutter; it required little power and gave bursts of about 0.5 msec.

Both parties sent the neutrons from the beryllium target through a standard paraffin "howitzer" (fig. 9.1) and into a cadmium-lined pipe to remove stray particles (cadmium has an extraordinary appetite for slow neutrons); and both employed as primary detector the by then standard BF_3 ionization chamber activated by alpha particles from the reaction $B^{10}(n,\alpha)Li^7$. Alvarez fed the output from his chamber into an amplifier regulated by an elaborate electronic timing circuit to process pulses only during a short time after each deuteron burst; by setting the time in accordance with the distance from the beryllium to the detector, he could arrange that only neutrons of (or, rather, around) a selected velocity would be counted. The British continuously applied the output of their chamber and fed the result to an oscilloscope, which indicated the times of the deuteron bursts as well as the times of the associated chamber pulses. A motion picture of the oscilloscope traces preserved them for analysis. The Dutch competition planned to work much in the British manner, obtaining their neutrons from a d-d reacton in a HT tube, modulating the accelerating potential, and synchronizing the detecting amplifier

5. Milatz and Ter Horst, *Physica*, 5 (Aug 1938), 796.

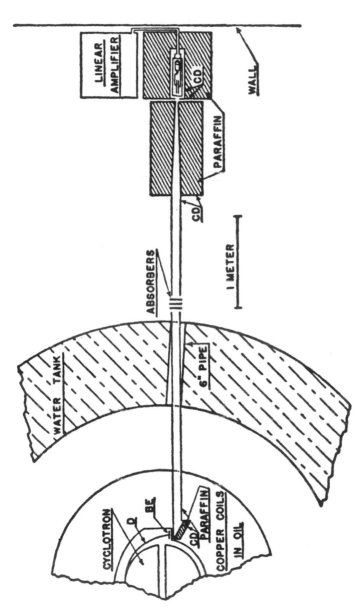

FIG. 9.1 Alvarez's setup for collimating and detecting fast neutron beams. The beam arising from deuterons on beryllium runs through a cadmium-lined pipe in the cyclotron's water shield. The cadmium takes out slow neutrons. Alvarez, *PR, 54* (1938), 613.

and oscilloscope with the modulator. They criticized Alvarez's method for permitting the detection of only one neutron velocity at a time.[6]

This property made Alvarez's instrument what might be called a monochromator: although its chamber received an inhomogeneous beam, the amplifier insured that only neutrons of the selected velocity were counted. Did this mean that the experimenter could consider that he had a homogeneous beam? A deep question that, or perhaps only a matter of words: it called forth expression of a philosophy of science, a very rare, and perhaps unique utterance from the Laboratory in the 1930s. "On the operational viewpoint," Alvarez wrote, in reference to the then widely accepted teachings of Percy Bridgman, "one is justified in asserting that the beam is composed solely of thermal neutrons." The British had no need for Bridgman's philosophy. Their instrument was a velocity spectrometer: the intervals between the chamber pulses and the associated deuteron bursts showed the velocity distribution among the slow neutrons reaching the detector. Their cumbersome method of oscilloscope recording and analysis worked only for small fluxes, however. They ordinarily counted about 25 neutrons a minute. Alvarez's Berkeley rate was 1,000/min.[7]

The definitive form of the neutron velocity spectrometer was made by R.F. Bacher and his students at Cornell. It followed the British by modulating the deuteron source and Alvarez by measuring timed neutrons through a system that split the signal from the BF_3 detector, routing one half to a continuously active amplifier and the other to one alive only when neutrons of the velocity of interest arrived. They obtained very nice records of resonance that confirmed consequences of Bohr's nuclear theory as calculated by Bethe. And—a matter that proved more important to the common man—they measured neutron absorption in the isotopes of uranium. Their design was further perfected at Illinois and at

6. Alvarez, *PR, 54* (1938), 235, and ibid., 609–17, rec'd 20 Aug 1938; Fertel et al., *Nature, 142* (5 Nov 1938), 829, and *PRS, A175* (1940), 316–31; Milatz and Ter Horst, *Physica, 5* (1938), 796.

7. Alvarez, *PR, 54* (1938), 609–10; Fertel et al., *PRS, A175* (1940), 322. On the popularity of Bridgman, Schweber, *HSPS, 17:1* (1986), 61–7.

Stanford in 1941. By then the British team had long since retired into the scientific war against Germany.[8]

Alvarez continued with neutronics in collaboration with a refugee from Germany's war against science, Felix Bloch, who had joined the physics department at Stanford in 1934. Their experiment—the exact determination of the magnetic moment of the neutron—was a perfect matching of skills, interests, and ambitions. Bloch had given its theory in 1936, after a visit to Heisenberg that convinced him that Stanford's deepest and quickest route to a place in nuclear physics was via an experimental program in neutronics. He discussed the measurement at the periodic joint Stanford-Berkeley physics seminar and interested Laslett and Alvarez, who mentioned it as a possible application of his monochromator.[9] The business did not seem overly promising, however, since earlier attempts to detect magnetic behavior in the neutron had failed and the very concept of a magnetic moment of an uncharged particle smacked of oxymoron. There seemed no other explanation, however, of the results obtained by Otto Stern and his associates O.R. Frisch and Immanuel Estermann with their hydrogen beams in Hamburg.

According to classical theory, the ratio between the magnetic moment μ and the angular momentum p of a spinning sphere of charge e and mass M is $\mu/p = e/2Mc$. One might expect therefore that in quantum theory the magnetic moment of an elementary charged particle with angular momentum $sh/2\pi$, where s is its spin quantum number, would be $\mu = (eh/4\pi mc)s$. But in the case of the electron, spectra required both $\mu_e = eh/4\pi Mc$ and $s = \frac{1}{2}$; a contradiction supressed by introducing a numerical factor g into the relationship, $\mu/p = g_e e/2mc$, and setting $g_e = 2$. The Dirac theory of the relativistic spinning electron produced spontaneously, among many other marvels, the value $g_e = 2$. It was therefore assumed that if the theory applied to protons, $s_p = \frac{1}{2}$, $g_p = 2$, and the relevant magneton, let us call it μ_0, would be smaller than μ_e in the ratio of the masses of the electron and proton, m/M. But in 1933 Stern and his associates announced that the proton

8. Baker and Bacher, *PR*, *59* (1941), 332–48, rec'd 30 Dec 1940; Bethe, "History" [1960], 3; Haworth and Gilette, *PR*, *69* (1946), 254; Fryer, *PR*, *70* (1946), 235–44.

9. Chodorow et al, *Bloch*, vii; Alvarez, *PR*, *54* (1938), 609.

moment, μ_p, amounted to around $5\mu_0$. A quick measurement on the heavy water sent by Lewis showed that the moment of the deuteron was smaller than the proton's, although its spin was twice as great. By 1936, owing to further work by Stern in Pittsburgh, where he set up after the Nazis closed his institute, and by Rabi, who greatly improved the technique, the magnetic moments of proton and deuteron were established at between 2.5 and 3.0 μ_0 and around 0.8 μ_0 respectively.[10]

The obvious way to square the numbers was to assume that their difference measured the moment of the neutron. Since the deuteron was known to have a spin of 1 and since indirect evidence made the neutron's spin ½, it appeared that the spin, and hence the moment, of the deuteron's constituents added together, whence $\mu_n \approx -2\mu_0$, the minus sign signifying the relationship between spin and moment characteristic of the electron. But how to explain that μ_n does not equal zero? Following a suggestion made by G.C. Wick, who drew on Fermi's theory of beta decay, Bloch and other theorists supposed that the neutron spends some of its time dissolved into an electron and a proton and that in this state "it" can interact with a magnet and so show a moment. Since $\mu_e \approx 2000\mu_0$, the neutron need not spend much of its time in pieces in order to show an average moment of $-2\mu_0$. Since Fermi's theory treats protons and neutrons on the same footing, the excess moment of the proton was supposed to arise from its temporary disaggregation into a positron and a neutron. Since by the symmetry of the theory, this excess should be equal and opposite to the apparent neutron moment, $\mu_p + \mu_n \approx \mu_0$, which came close to the deuteron moment.

All this was, of course, only inference, an effete proceeding necessary, perhaps, in astrophysics, but unmanly with objects produced by the billions in the laboratory. Bethe supposed that knowledge of the neutron moment would continue to come exclusively from such indirect arguments: "the magnetic moment of the neutron is hardly accessible to direct measurement." Bloch thought otherwise and offered a calculation of the effects of μ_n on

10. Rigden, *HSPS, 13:2* (1983), 339–53; Bethe and Bacher, *RMP, 8* (Apr 1936), 91–2, give $\mu_p = 2.9\mu_0$ and $\mu_d = 0.85\mu_0$ from the latest experiments of Rabi's group.

the scattering of very slow neutrons. The calculation suggested that μ_n might be deduced by scattering slow neutrons from magnetic atoms and by passing them through thin magnetized plates. (Slow neutrons are required in order that their wavelength have about the same size as the atoms scattering them.) Since the total transmission through the plates should depend upon their relative magnetization, Bloch hoped that with such a setup he could obtain a direct indication of the existence of the neutron's suppositious moment.[11]

Bloch set something going. Bethe and Livingston organized an experiment at Cornell that showed a 2 percent difference in transmission through plates with parallel and antiparallel magnetism and calculated that the result was not incompatible with Bloch's theory and $\mu_n = -2\mu_0$. Julian Schwinger then published a long calculation treating neutron scattering by Dirac theory that predicted a larger effect in the transmission experiment than Bloch had computed. In commenting on a draft of Schwinger's paper, Bethe, fresh from his try at μ_n, reaffirmed his view that direct measurements were not likely soon to improve upon subtraction of μ_p from μ_d: "It will be a long time before the direct determination will give the neutron moment to anywhere near this accuracy," which he reckoned at 0.15 μ_0.[12] Next, Rabi proposed a refinement in his method of spin flipping that J.R. Dunning and his students at Columbia immediately adapted to Bloch's transmission experiment.[13] They showed that the transmission of neutrons through a thin magnetized plate *increases* with the magnetization and thickness of the plate and with the slowness of the neutrons and that Rabi's method could partially unpolarize the partially polarized beam emerging from the first plate (fig. 9.2b).

11. Bloch, *PR, 50* (1936), 259–60, and Inst. Henri Poincaré, *Ann., 8:1* (1938), 70; Bethe and Bacher, *RMP, 8* (Apr 1936), 91, 205–6; Rigden, *HSPS, 13:2* (1983), 253–4.

12. Hoffmann, Livington, and Bethe, *PR, 51* (1937), 214–5, rec'd 3 Dec 1936; Schwinger, *PR, 51* (1937), 544–52; Bethe, "Comments to Schwinger," 15 Jan 37 (HAB, 14/22/976).

13. Rabi, *PR, 51* (1937), 652–4, rec'd 1 Mar. Frisch, von Halban, and Koch, *Nature, 139* (1937), 756, and *PR, 53* (1938), 720, and Bloch, in a letter to Frisch, 12 Jan 39 (Frisch P), and in Alvarez and Bloch, *PR, 57* (1940), 112, claimed to have invented a similar apparatus independent of Rabi.

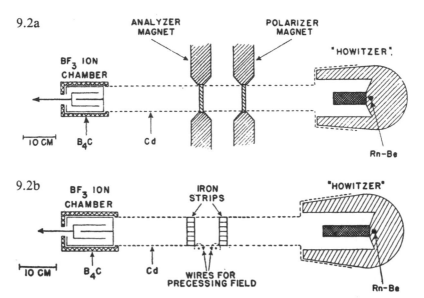

FIG. 9.2   Methods for detecting the nuclear moment of the neutron: 9.2a, Bloch's; 9.2b, Rabi's. In both cases fast neutrons from alpha particles on beryllium shoot from a paraffin howitzer and into a cadmium-coated pipe. In the first case, the polarization of the neutrons is changed by passing through magnetized iron; in the second case, by traversing a precessing magnetic field. Powers, *PR, 54* (1938), 834, 836.

Then, with the help of further calculations by Schwinger, they deduced from the amount of transmission with and without spin flipping that μ_n is negative and lies between 1 and 3 μ_0. No revelation that; and, because they could not determine with any accuracy the strength and variation of the magnetic field causing the flips, they could say no more. "Further refinements are in progress."[14]

While Dunning's associates were flipping neutrons Rabi's way, a group at Copenhagen led by Frisch, who had settled into Bohr's institute in 1934, independently suggested the use of a depolarized field in an experiment of Bloch's type. They worked with small fields and smaller effects; they could count only about 100 transmitted neutrons/minute from their weak Rn-Be source; they

14. Powers, Carroll, and Dunning, *PR, 51* (1937), 1112–3, letter of 18 May; Powers et al., *PR, 52* (1937), 38–9, letter of 19 June.

learned nothing more than that a value of $\mu_n = -2\mu_0$ was not incompatible with their experiments, and they gave up. "It would be hopeless to discuss these results any further and to try and enclose the magnetic moment of the neutron between definite limits."[15] Meanwhile Bloch, who did not agree that Schwinger's calculation came closer to reality than his, had proposed to Laslett that he try to find effects of μ_n by scattering off nickel and iron. Laslett went to work at about the same time the Cornell group did; unlike them, he found no positive results, nothing that would permit any useful quantitative statement about μ_n. Bloch himself tried to detect something useful in the scattering of neutrons (obtained from d-d synthesis) on cobalt. But in the fall of 1937 many, perhaps most, nuclear physicists shared the "pessimism about the moment of the neutron"—the belief that experimental difficulties of direct determination of μ_n would not soon be overcome— expressed by Bloch's friend Egon Bretscher.[16]

In June 1938 Bloch and Rabi, then visiting Stanford, went to Berkeley to watch Alvarez demonstrate his neutron monochromator (plate 9.1). Bloch returned to Stanford, to have a try at μ_n with a Rn-Be source; but he did no better than others, and came to think that useful quantitative results could only be obtained "when an intense neutron source is available that will make it possible to use monochromatic neutrons."[17] In September, he and another émigré, Hans Staub, decided to build a high-tension apparatus for the purpose by exploiting a disused 170 kV x-ray outfit and the d-d reaction; but before they had finished, as Staub tells the tale, Bloch "quite unexpectedly [!] got the apparatus" to work with Alvarez and the Berkeley cyclotron. Was the surprise

15. Frisch, von Halban, and Koch, *Nature, 139* (1937), 756, letter of 7 Apr, and ibid., 1021, letter of 12 May; *PR, 53* (1938), 721–3. Frisch et al. used the remanent field in their magnetized plates, some 10 kG; the Columbia group (Powers, *PR, 54* (1938), 832–3) required fields almost twice as large to produce any noticeable magnetic scattering, as did Alvarez and Bloch, *PR, 57* (1940), 112, who accordingly doubted that the results of Frisch et al. were "significant."

16. Bloch, *PR, 51* (1937), 994; Livingston to Lawrence, 2 Dec 1936, and Lawrence to Livingston, 15 Dec 1936 (12/12); Laslett and Hurst, *PR, 52* (15 Nov 37), 1035–9, rec'd 24 Aug; Bloch to Bretscher, 1 Sep 37 (Mrs. Bretscher).

17. Cooksey to Lawrence, 17 June 38 (4/21); Alvarez and Bloch, *PR, 57* (1940), 112, and Bloch to Frisch, 12 Jan 39 (Frisch P); Bloch, Inst. Henri Poincaré, *Ann., 8:1* (1938), 78.

Lawrence's willingness to allow a lengthy bit of physics to tie up the machine? Alvarez and Bloch began to work together in the late fall of 1938. The final report of the Columbia measurements (μ_n is "probably" around $-2 \pm 0.5\mu_0$), published about that time, gave them an easy mark to better. It is doubtful that the news that Frisch planned another go, with the enthusiastic support of Bohr, would have caused Alvarez and Bloch any anxiety.[18] They had the barren field of exact neutron magnetonics to themselves.

Their great advantage over earlier investigators was the neutron flux from the 37-inch cyclotron. Whereas the Columbia and Copenhagen groups had counted a few million neutrons, Alvarez and Bloch counted two hundred million in their year of experimenting. The hundredfold increase in beam allowed them to detect and correct many subtle instrumental effects that menaced their measurement; although for their definitive value of μ_n they counted only four million neutrons, a number readily obtainable from a good Rn-Be source, they needed the previous two hundred million, which could have been obtained only from an accelerator. The final experimental design is shown schematically in figure 9.3.

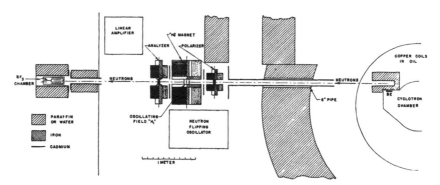

FIG. 9.3   Alvarez and Bloch's definitive experimental setup. Compare fig. 9.1. Alvarez and Bloch, *PR, 57* (1940), 116.

Neutrons from a beryllium target, struck by deuterons and slowed down by passage through paraffin, ran down a cadmium-lined tube stuck through the water shielding around the cyclotron. The

18. Staub in Chodorow et al., *Bloch*, 194–5; Powers, *PR, 54* (1938), 827–8, rec'd 22 Sep; Frisch to Meitner, 19 Sep 1938 (Frisch P).

paraffin so successfully removed unwanted fast neutrons that it proved unnecessary to use Alvarez's modulated beam as originally intended; and the cadmium swallowed slow neutrons so effectively that the entire arrangement introduced a strong collimated beam consisting primarily of slow neutrons to the thin piece of iron that, when magnetized, acted as a polarizer. The design profited perhaps from the contemporaneous work of Aebersold on the collimation of neutron beams for therapy.

The partially polarized neutrons entered the flip space, where they felt two magnetic fields: a constant one, H_0, to provide direction for the orientation of the magnetic moments, and an oscillating field, H_1, to incite the flips. The analyzer consisted of a second piece of iron in a second powerful electromagnet. The detector was the standard BF_3 chamber. The electromagnets at either end were borrowed, one from a colleague who used it for Zeeman spectroscopy, the other from Shell Development Company; the coil energizing the field H_0 had been cannibalized from the 11-inch cyclotron and the solenoid that made H_1 was wound from flat copper strips through which the neutrons passed in entering and leaving the flip region. The currents energizing H_0 and H_1 were kept regular by tapping them from the cyclotron's automatically stabilized supply.[19]

A measurement consisted of reading the number of neutrons registering in the detector as H_0 swept through the value H_n that maximized the number of flips. (This occurs when f_n, the frequency of the oscillating field, equals the frequency with which the neutron moments process around H_0.) The underlying concept: with H_1 off and polarizer and analyzer magnetized in parallel, a certain flux of neutrons will be counted; with H_1 on, a fraction of the neutrons will have their moments reversed, somewhat fewer than before will pass the analyzer, and a smaller flux will register. When this reaches a minimum, resonance obtains and μ_n comes immediately from the value of the precessional frequency: $\mu_n = f_n h / 2H_n$. Or, rather, it comes immediately after the values of f_n and H_n have been determined. Alvarez and Bloch invented a method of measuring these quantities that allowed greater

19. Alvarez and Bloch, *PR*, *57* (1940), 116–8, rec'd 30 Oct 39.

accuracy than standard methods would have given. Write $\mu_n = g_n\mu_0$, g_n being the value of the neutron moment in nuclear magnetons. From the resonance condition of the experiment, $f_n = 2H_n g_n \mu_0/h$. But from the resonance of protons in the cyclotron, $f_p = eH_p/2\pi Mc$, we have $f_p = 2H_p\mu_0/h$; hence $g_n = (f_n/f_p)(H_p/H_n)$. The measurement reduces to finding the ratios of frequencies and fields, which was not difficult, and requires no knowledge of the values of the absolute constants. In effect, the cyclotron supplies the values of e/Mc.[20]

The counting itself was done electronically via a clever circuitry that monitored the unsteady neutron output of the cyclotron. (At best the cyclotron ion beam could be held constant to 1 percent; Alvarez and Bloch worked to an accuracy of a tenth of a percent.) The circuit divided the counting time into intervals of a few seconds during which the flipping fields were alternately on and off. When off, the amplified output of the BF_3 chamber was routed to one of a pair of counters; when on, to the other; comparison of the two allowed correction for fluctuations in the initial beam strength. After making this correction, Alvarez and Bloch got the nice sharp dip of resonance indicated by figure 9.4. With the values of frequency and field thus implied, they obtained $g_n = -1.935$, which they judged to be good to about 1 percent.[21] At the time the latest Columbia values for μ_p and μ_d were $2.785\mu_0$ and $0.855\mu_0$ respectively, both to an accuracy of 0.7 percent. Consequently, to within experimental error the old relation, $\mu_d = \mu_p + \mu_n$, still held, although it seemed very unlikely that the intrinsic moments of the constituents should not be altered to some extent by their combination. It was primarily to achieve these measures and amplify their effects that Bloch and his co-workers undertook to build a cyclotron at Stanford. His subsequent

20. Cf. Norman Ramsey in Trower, *Discovering Alvarez* (1987), 30–1.

21. Alvarez and Bloch, *PR, 57* (1940), 120–1, quoting an error of $\pm0.02\mu_0$, which they later raised to $\pm0.03\mu_0$ without explanation (*PR, 57* (15 Nov 1940), 352). This did better than Bloch hoped for in January 1939, when, after Bohr told him about Frisch's latest try, Bloch responded with his and Alvarez's first results (between -1.9 and -2.3 μ_0), and anticipated an eventual accuracy of 2 percent. Bloch to Frisch, 12 Jan 39 (Frisch P). Six months later they had fixed on the average, -2.1 μ_0; Lawrence to Smyth, 12 June 1939 (16/32).

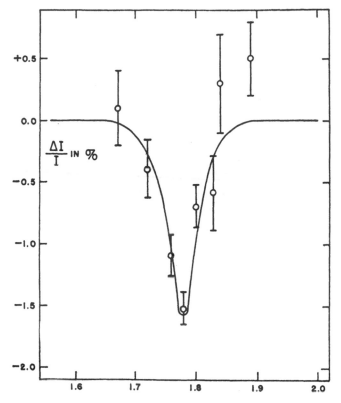

FIG. 9.4   Alvarez and Bloch's definitive measurement. The abscissa is the current energizing H_0 in arbitrary units; the ordinate, the fractional change (in percentage) of the intensity of the neutron beam under the influence of the field. Alvarez and Bloch, *PR, 57* (1940), 120.

development of a technique for precise measurement of nuclear magnetic moments, and the work he did with it, brought Bloch the Nobel prize for physics for 1952, a handsome payoff for his switch to experimental neutronics sixteen years earlier.[22]

22. Alvarez and Bloch, *PR, 57* (1940), 121–2; Rigden, *HSPS, 13:2* (1983), 367; Bloch, Nicodemus, and Staub, *PR, 74* (1948), 1025–45. Alvarez and Bloch's result was high, and Rabi's low, but both within the estimated error; with more precise measurements after the war, the additivity of the moments failed to hold by $0.0223\mu_0$.

2. EUROPEAN TRANSURANICS

In 1938 Fermi received the Nobel prize in physics "for his demonstrations of the existence of new radioactive elements produced by neutron irradiation, and for his related discovery of nuclear reactions brought about by slow neutrons." What was meant by "new elements" appears from a letter of nomination from G.P. Thomson, who specified besides the method of slow neutrons the "proof of the existence of radio-active elements of atomic number greater than 92 produced by the action of neutrons on uranium."[23] Fermi had advanced his claim cautiously; but like Lawrence, he had supporters who snatched at any novelty to justify their confidence. He was swept forward by a speech by his protector, O.M. Corbino, who, to Fermi's dismay, told the Accademia dei lincei on June 3, 1934, that his protégé had made element 93. In the style of Millikan and Lawrence, Corbino added that "the power to produce such transformations in sufficient quantity would give mankind not only immediate possession of the rarest elements, but also command of an almost limitless supply of energy."[24] Although warned that the chemical procedures that appeared to show that the "transuranics" could not be known elements might be faulty, the Rome group acquired confidence by repetition, especially when the experienced team of Otto Hahn and Lise Meitner confirmed their findings. In his Nobel lecture in Stockhom, Fermi unreservedly claimed the discoveries of elements 93 and 94, and, as further proof of their chemical individuality he revealed the euphonic names, "ausonium" and "hesperium," by which they were known in Rome.[25]

23. Nobel Foundation, *Directory,* 56; Thomson to Nobel Prize Committee, 24 Jan 1938 (GPT, J31). Fermi was first nominated in 1935, for his work on neutron activation; Crawford, Heilbron, and Ullrich, *Nobel population,* 142.

24. Dragoni, *Physis, 15* (1973), 352–63; Corbino, *Ric. sci., 5:1* (1934), 615, quoted by Dragoni; Segrè, *Fermi,* 76–7.

25. The strengthening of the claims of the Rome group can be followed in Fermi, *CP, 1,* 750 (June 1934), 743 (text of July 1934), 667–8, 791 (texts of early Feb 1935), and 1040 (Nobel lecture of Nov 1938). Doubts about the transuranic attribution were voiced immediately after the initial claim by Ida Noddack, *Angew. Chem., 47* (1934), 453f., and by von Grosse and Agruss, *Nature, 134* (1934), 773. Cf. Krafft, *Strassmann* (1981), 314–21.

The world's center of transuranic research from 1935 to 1938 was the Kaiser-Wilhelm-Institut für Chemie in Berlin. There the old team of Meitner and Hahn had been reestablished at Meitner's initiative to follow up the discoveries of the Rome group.[26] By 1937 Hahn, Meitner, and their chemist colleague Fritz Strassmann had found and systematized no fewer than nine activities arising from the irradiation of uranium by neutrons. Three of these activities they ascribed to uranium itself, to a complex U^{239} capable of decaying in three different modes, each producing a different activity. It was as if U^{238} could catch three distinct lethal diseases by swallowing a neutron; that the offspring inherited the form of death of its parent; and that the curse could be followed to the fifth generation. The members of this highly degenerate family, all linked by beta decay, are named in figure 9.5. The first two chains were known to be initiated by both fast and thermal neutrons, the third only by fast neutrons of a narrow range of energies, which "resonated" with the uranium nucleus.[27]

This scheme, which dates from 1937, evidently requires U^{239} to exist in three isomeric states. When Meitner first suggested the possibility in 1936, she could appeal only to the isolated example of Hahn's double isomer in protoactinium; but by 1938 several other examples, among them the isomers of zinc, manganese, and tellurium discovered at the Laboratory, supplied indirect support for the possibility of a multiple U^{239}. Even the successive decays had a partial analogy around the middle of the periodic table in what Lawrence called an "induced chain reaction," for example, $Cd \rightarrow In \rightarrow Sn$. The apparent inheritability of the decay modes of uranium, however, and the lengths of the chains in two of them, had no analogue among known radioactive processes.[28] Another problem, which haunted all experiments with "transuranics," was their chemical identity. The teams of Rome and Berlin, as well as Irène Curie's in Paris, whose work was to topple the multiple isomers, all worked with natural sources of neutrons. The

26. Meitner, *Naturw. Rundschau, 16* (1963), 167.

27. Meitner, Hahn, and Strassman, *Zs. f. Phys., 106* (1937), 249–70, table on 249; Walke, *Rep. prog. phys., 6* (1939), 33; Turner, *RMP, 12:1* (Jan 1940), 4–5.

28. Weart in Shea, *Otto Hahn,* 98–101; Lawrence to Cork, 24 Feb 1937 (5/1); Meitner, in Bretscher, *Kernphysik,* 27, 34–41; supra, §8.2.

I. $_{92}U + {}_0n \rightarrow ({}_{92}U + n) \xrightarrow[\text{10 sec}]{\beta} {}_{93}EkaRe \xrightarrow[\text{2.2 min}]{\beta} {}_{94}EkaOs \xrightarrow[\text{59 min}]{\beta}$

   $_{95}EkaIr \xrightarrow[\text{66 hr}]{\beta} {}_{96}EkaPt \xrightarrow[\text{2.5 hr}]{\beta} {}_{97}EkaAu$

II. $_{92}U + {}_0n \rightarrow ({}_{92}U + n) \xrightarrow[\text{40 sec}]{\beta} {}_{93}EkaRe \xrightarrow[\text{16 min}]{\beta} {}_{94}EkaOs \xrightarrow[\text{5.7 hr}]{\beta}$

   $_{95}EkaIr \xrightarrow[\text{60 days}]{\beta} ?$

III. $_{92}U + {}_0n \rightarrow ({}_{92}U + n) \xrightarrow[\text{23 min}]{\beta} {}_{93}EkaRe.$

FIG. 9.5   The standard misunderstanding of the relations among the "transuranics" before 1939. After Meitner, Hahn, and Strassman, *Zs. f. Phys., 106* (1937), 249.

consequent weakness of their sources, as well as the radiochemical complexity of the material they studied, which included the natural descendents of uranium as well as transuranics and fission products, made it very difficult and tedious to determine the genetic relations of the fleeting products and their likely places in the unknown periods beyond uranium in the table of the elements. Despite the many uncertainties, however, most radiochemists and nuclear physicists throughout the world accepted the transuranics at face value.[29] The common consequence of neutron irradiation of the nucleus was capture followed by beta decay; the most cataclysmic consequence, the release of an alpha particle. Cleavage, the division of a heavy nucleus into approximately equal parts, did not come into serious consideration, even though Bohr's analogy of a nucleus to a liquid drop, which readily accommodated the concept of fission in 1939, was available from 1936 and widely advertised by Bohr on a lecture trip to the United States in the spring of 1937 (fig. 9.6). Whence the blind spot? "Because," as the theoretical physicists understood the matter two years before the discovery of fission, "because the disintegration of an element into two approximately equal halves is so very improbable a process."[30]

29. Weart in Shea, *Otto Hahn,* 102–4, 113–25; Quill, *Chem. rev., 23* (1938), 119–20, 138; Segrè in Acc. dei XL, *Memorie, 8:2* (1984), 165–70.

30. Bohr, *Science, 86* (1937), 161–5; Wigner to Bethe, 10 Dec 1937 (HAB, 3), quote.

FIG. 9.6 Bohr's playful analogy to the capture of a particle by a nucleus. The marbles in the dish represent the bound nuclear constituents, that on the edge the incoming particle. The newcomer's energy will quickly be shared by collision among the constituents. Bohr, *Science, 86* (1937), 161.

The Paris group began to fish in the muddy waters of transuranics in 1935. By 1937 they–the fisherfolk were Irène Curie and Paul Savitch—had caught a big one, a 3.5-hour activity formed from uranium by neutron capture; since this $R_{3.5h}$, as they dubbed it, did not have the chemistry of uranium or any "transuranium," they made it (temporarily) an isotope of thorium, according to $U^{238}(n,\alpha)Th^{235}$. This was to produce still another isomer of U^{239} with its own special plan of decay. To complicate matters further, Curie and Savitch observed that nothing seemed to prevent cross decays between members of the first two chains of figure 9.5.[31] Hahn and Meitner could not find a radioactive Th^{235} in their irradiated uranium; Curie and Savitch tried some chemistry on theirs and decided that $R_{3.5h}$ behaved like a rare earth. The only possibilities, they thought, were actinium, which does resemble the rare earths, or a new transuranic quite distinct in chemical properties from the substances investigated by Hahn and Meitner. Either assignment, as they said, faced serious difficulties. In the late spring of 1938, they eliminated one possibility by showing that

31. Curie and Savitch. *Jl. phys. rad., 8* (1937), 385, rec'd 1 Aug 1937, in Joliot and Curie, *Oeuvres,* 621–2; Quill, *Chem. rev., 23* (1938), 147–9.

$R_{3.5h}$ followed lanthanum rather than actinium. "It seems therefore that this body can only be a transuranic with properties very different from those of other known transuranics."[32]

These properties had been established largely on the supposition that the elements beyond uranium would be homologues of the elements beneath them in the periodic table: 93 was expected to behave like rhenium, 94 like osmium, and so on. This was the scheme generally preferred by chemists. To physicists, however, another alternative lay open: transuranics have the properties of uranium or actinium and become more like rare earths with increasing atomic number. The possibility of a second rare-earth series beyond uranium fitted Bohr's version of the periodic table (fig. 9.7), which had in its favor the prediction, confirmed in 1922 against the opinion of chemists that the then unknown element 72 was not a rare earth. By following the chemists instead of Bohr, Curie and Savitch could find no place in the periodic table near uranium suited to receive $R_{3.5h}$.[33] No more could Hahn, Meitner, and Strassmann, who, having convinced themselves by chemistry that element 93 did not resemble either of the alternatives suggested by Bohr, actinium or uranium, accepted by default the common opinion of chemists and assimilated it to rhenium.[34]

The analytic chemist of the Berlin team, Strassmann proposed to derive $R_{3.5h}$ from uranium in three steps: two successive alpha emissions to produce radium followed by a beta decay to the nearest rare-earth homologue, actinium. He and Hahn had a look; Meitner had fled Germany in July 1938, just after Curie and Savitch grudgingly made $R_{3.5h}$ transuranic. Hahn and Strassmann found more than they wanted. With barium as a carrier, they identified three sorts of isomeric radium, giving rise to three different heritable decay sequences of the type indicated in figure 9.5. The first product in each sequence had a period of a few days. No doubt, according to Hahn and Strassmann, the incompetent French team had confused the several actinium isotopes in

32. Curie and Savitch, *CR, 206* (1938), in Joliot and Curie, *Oeuvres,* 625–7, séances of 21 Mar and 30 May 1938; Weart in Shea, *Otto Hahn,* 10–48.
33. Curie and Savitch, *Jl. phys. rad., 9* (1938), in Joliot and Curie, *Oeuvres,* 634–6, text of July 1938; Quill, *Chem. rev., 23* (1938), 101–5; von Grosse, *JACS, 57* (1935), 440–1; Heilbron, *Moseley,* 133–7.
34. Hahn, *Naturw. Rundschau, 15* (1962), 44.

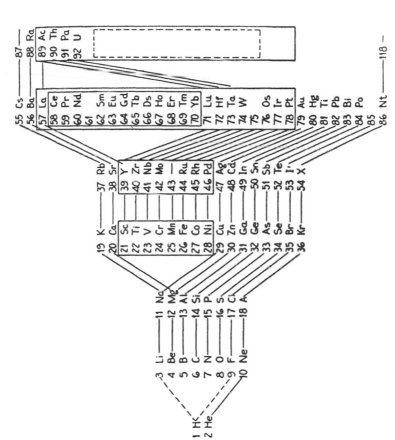

FIG. 9.7 Two methods of visualizing the chemical allegiance of element 93. 9.7a: as a homologue of the rare earths, as in the now accepted "actinide" series. 9.7b: as a homologue of the heavy metals, the view most widely approved in the late 1930s. Quill, *Chem. rev., 23* (1938), 101.

their $R_{3.5h}$ activity. But the basis of the Berlin scheme, (n,2α), seemed scarcely plausible after experiments proposed by Meitner showed that even thermal neutrons of vanishingly small energy provoked it.[35] Several physicists, including Bohr, "expressed their astonishment that slow neutrons should initiate two successive alpha-processes in uranium." Hahn and Strassmann looked again, found another radium and actinium isomer, and struggled to separate their many radiums from barium.[36]

After several weeks of painstaking effort, they decided to quit. We are indebted to the Nazis for the opportunity to follow the dawning discovery of the cause of their failure in detail, in letters from Hahn to his absent partner. On December 19, 1938, he wrote Meitner: "There is in fact something so remarkable about the 'radium-isotopes' that we are now only telling you....The fractionization did not work. Our Ra-isotopes behave like Ba." Their brilliant chemistry had landed them in a situation Hahn had considered desparate. "Perhaps you can suggest some fantastic explanation. We know ourselves that it [uranium] really cannot fall apart into Ba....So, think." It was too much for Meitner. "The assumption of so wide-reaching a disintegration seems very difficult to me at the moment, but there have been so many surprises in nuclear physics that one can not reject anything immediately as impossible." This letter crossed another appeal from Hahn: "We cannot keep silent about our results, even if they are perhaps physically absurd. You see, you would do a good piece of work if you could find a way out."[37] When Hahn wrote these lines, his secretary was typing his and Strassmann's epoch-making announcement of their dilemma. "As chemists, we must rename [our] scheme and insert the symbols Ba, La, Ce in place of Ra, Ac, Th. As nuclear chemists closely associated with physics, we cannot yet convince ourselves to make this leap, which

35. Krafft in Shea, *Otto Hahn*, 152–3, following Strassmann's account, in Krafft, *Strassmann*, 208, 234. Hahn and Strassmann began to pursue $R_{3.5h}$ in mid October; Hahn to Meitner, 25 Oct and 2 Nov 1938, ibid., 239.

36. Hahn to Feather, 2 June 1939, quoted by Weart in Shea, *Otto Hahn*, 110; Hahn, *Vom Radiothor*, 116–31; Krafft, *Strassmann*, 239–53; Hahn, *Naturw. Rundschau*, *15* (1962), 45–6.

37. Hahn to Meitner, 19 and 21 Dec 1938, and Meitner to Hahn, 21 Dec 1938, in Krafft, *Strassmann*, 263–5.

contradicts all previous experience in nuclear physics."[38]

This cry of schizophrenia contrasts with the smooth interdisciplinarity of Lawrence's Laboratory. Lawrence had asked indifferently, "Shall we call [our new science] nuclear physics or shall we call it nuclear chemistry?" It did not require an answer in Berkeley, where interdisciplinarity was attained by addition—at first by the physicists' acquiring a few techniques of the analytical chemist, subsequently by acquiring chemists, then physicians and physiologists. In Berlin, each member of the team had to adjust to the professional constraints of the others. The peculiar intensity and quality of the collaboration of Hahn, Meitner, and Strassmann, which was more than the sum of its parts, may explain why the discovery of fission occurred in Berlin, and not in Rome, Paris, or Berkeley.[39]

On the proofs of the paper disclosing their conflict of allegiance, Hahn inserted what, in a letter to Meitner, he dubbed "a new fantasy." "We have not proved that the transuranics are *not* Ma [Tc], Ru, Rh, Pd....Would it be possible for uranium 239 to disintegrate into a Ba and a Ma? A Ba 138 and a Ma 101 would give 239....If there is anything in this, the transuranics, including 'ausonium' and 'hesperium' would be dead. I do not know whether or not that would make me very unhappy."[40] The geographical separation of the group decomposed it into a chemical division (Hahn and Strassmann) and a physical one (Meitner and her nephew Frisch, who came from Bohr's institute to spend the Christmas holidays with her in Sweden). Each side approached the problem from its disciplinary viewpoint; no doubt the menacing and unstable social and political circumstances made freshly attractive the relative security of firm professional indentifications. As a physicist, Meitner could conceive that a splitting of the uranium nucleus might be possible, although not into Ba and Ma; as a former full participant in the team, she could not disbelieve in the sequences (2) and (3) of figure 9.5, and hence in transuranics; and, as a refugee, she feared that "it would be no good

38. Hahn and Strassmann, *Nwn*, *27*, (1939), 15, rec'd 22 Dec 1938.
39. Lawrence, "Artificial radioactivity," 17 (40/17); Krafft in Shea, *Otto Hahn*, 135–44, following Strassmann in Krafft, *Strassmann*, 210, emphasizes the importance of the intellectual interdisciplinarity of the team.
40. Hahn to Meitner, 28 Dec 1938, in Krafft, *Strassmann*, 267–8.

recommendation for my new beginning to have to repudiate three years of work." In any case, she said, and quite rightly, that she had good reasons to suppose that the 23-minute uranium of sequence (3), the one excited by resonance capture, should give rise to a transuranic.[41]

It was Meitner who first raised the problem of the relation between the transuranics and the apparent production of barium from radium. While Hahn and Strassmann worked urgently to confirm by every chemical means that their "radiums" were barium, Meitner and Frisch, who were the only physicists informed of their progress, worked out that "fission," as they called it, could easily [!] be interpreted in a purely classical, nonquantum manner as the sundering of a distended droplet-nucleus. On this assumption, Hahn and Strassmann's bariums (atomic number 56) and their lanthanum descendents should be accompanied by radiokrypton (atomic number 36 = 92 – 56) and its relatives. Frisch had confirmed the general process of fission by irradiating a thin layer of uranium with neutrons from a Rn-Be source and catching the heavily ionizing fission fragments, which he estimated to have an atomic weight of at least 70, in an ionization chamber. The demonstration further menaced the transuranics. Meitner and Frisch put forward as "rather plausible" that the long sequences (1) and (2) of figure 9.5, which "always puzzled us," should be assigned to isotopes of technetium (atomic number 43) decaying through a chain terminating with cadmium (atomic number 48).[42] This interpretation was very materially strengthened in March 1939 when Meitner and Frisch, together using the high-tension set at Bohr's institute, collected enough of the recoiling fission products to follow radioactive decays. They obtained curves very similar to those the united Berlin trio had for sequences (1) and (2) of the transuranics: "the 'transuranium'

41. Meitner to Hahn, 29 Dec 1938 and 1 and 3 Jan 1939, ibid., 268, 271; Frisch to Hahn, 1 Jan 1939, ibid., 271, explaining Meitner's anxiety that it might appear that her former colleagues had awaited her departure to correct their joint work.

42. Hahn to Meitner, 2 and 10 Jan, 2 Feb 1939, ibid., 269, 276, 295 (is keeping his results secret despite local pressure); Meitner and Frisch, *Nature, 143* (11 Feb 1939), 239–40, dated 16 Jan; Frisch, ibid., suppl., 276, also 16 Jan; Meitner to Hahn, 18 Jan 1939, in Krafft, *Strassmann*, 282.

periods, too, will have to be ascribed to elements considerably lighter than uranium."[43]

Meitner delayed conveying this news to Hahn in order not to upset his birthday celebration. That was because Hahn and Strassmann, having been forced to recognize fission as chemists, had, on the same professional qualification, declined to relinquish the transuranics. Contrary to their mop-up of the "radium" decays, in which they proved conclusively that they dealt with barium and also found decisive evidence of the presence of the associated radiokrypton, they could not find any sequences of light elements chemically identical with the long transuranic lines.[44] Here the poverty of their means, the limitation of their methods, the complexity of the situation, and the desire to preserve what they could of their earlier results made an impossible barrier. Lacking a strong artificial source—had they only had a cyclotron![45] —they could not produce fission products in sufficient quantities to unscramble their decays or recognize the extent of the problem before them. Looking back in 1960, Hahn and Strassmann could identify with certainty only the 10-sec and 40-sec "uraniums" as mixtures of Xe and Kr; the 66-hr element "95" as 78-hr Te; and the 2.5-hr element "96" as 2.26-hr I. The others, according to a repetition of the experiments of 1938 made in 1971, were quite complicated mixtures, entirely unresolvable by the chemical means earlier available.[46]

By insisting upon ascribing each of their "transuranic" activities to a distinct isotope and clinging as long as they could to the sequence of decays they had worked out in 1938, Hahn and

43. Meitner and Frisch, *Nature, 143* (18 Mar 1939), 471–2, dated 6 Mar; they found the 16- and 25-minute, and the 5.7-, 66-, and 2.5-hour "bodies;" cf. Meitner to Hahn, 10 Mar 1939, in Krafft, *Strassmann,* 303. Joliot, *CR, 208* (1939), 341, séance of 30 Jan (*Oeuvres,* 650–2), anticipated Meitner and Frisch's demonstration, and Anderson et al., *PR, 55* (1 Mar 1939), 512, letter of 16 Feb, independently proposed it.

44. Meitner to Hahn, 10 Mar 1939, in Krafft, *Strassmann,* 302; Hahn and Strassman's struggles and their retention of the transuranics appear from their laboratory notes and letters to Meitner in ibid., 270, 276–87, 293–301; Hahn and Strassmann, *Nwn, 27* (1939), 89–95, rec'd 28 Jan 1939, confirm the Ba, find the Kr, and affirm the transuranics and their previous work with Meitner.

45. Hahn to Meitner, 16 Jan 1939, in Krafft, *Strassmann,* 279.

46. Krafft, *Strassmann,* 228–33; Hahn, *Vom Radiothor,* 144–8; Menke and Herrmann, *Radioch. acta, 16* (1971), 119–23.

Strassmann could not identify their "transuranics" with any lighter elements by chemical procedures. And they refused to return to the schizophrenia of interdisciplinarity: "We have in no way touched on physics in the entire business, but have only done chemical separations over and over again. We know our limitations and also of course that in this particular case it makes good sense to do only chemistry."[47] They accordingly continued to hold to the existence of the three isomeric heavy uraniums and the long transuranic sequences for at least four months after their discovery of fission had made all the earlier attributions doubtful. Although they recognized the force of the demonstration by Meitner and Frisch, they preferred not to believe it: "We can find no holes in your interpretation, and on the basis of your results must really kill the transuranics. But for us—Strassmann and myself—this result is completely incomprehensible. For we have not been able to say what these transuranics can be....In any case you have the first result, based on experiments, not on vague conjectures, that will be entirely clear for a physicist....Are you really sure that you have *our* transuranics in the [material you tested]?"[48]

On January 3 Frisch told Bohr about the "small bomb" he and his aunt were giving physicists as a New Year's gift. "Today I was able to speak to Bohr about the exploding uranium. The conversation had lasted only five minutes when Bohr agreed with us in everything. The only thing he thought remarkable was that he had not thought of the possibility earlier, since it follows so directly from current ideas about the structure of the nucleus."[49] Nine days later Frisch started his search for fission fragments. In four days he had his "conclusive" physicist's proof; in nine days, printer's proofs. Meanwhile Bohr had set sail for Princeton, where he was to stay for several months. Wishing to preserve the confidence of Frisch and Meitner until their papers were published, he did not report either the concept of fission or the detection of the fragments, of which he was informed by cable; but his assistant Léon Rosenfeld, unaware of the confidentiality, disclosed

47. Hahn to Meitner, 7 Feb 1939, in Krafft, *Strassmann*, 295.
48. Hahn to Meitner, 13 Mar 1939, and, for the last sentence, 30 Mar 1939, ibid., 304, 306.
49. Frisch to Meitner, 3 and 20 Jan 1939 (Frisch P).

both. Bohr then made the results public on January 26, at the opening session of the fifth Washington Conference on Theoretical Physics. The newspapers picked up the story and accelerator laboratories from Princeton to Berkeley moved into the new fields at American speed.[50]

Fermi, established at Columbia, turned to "the uranium split business with which half the world seems to be occupied...as soon as the cyclotron gave a beam." His search for the fragments found by Frisch was rewarded on January 25. The following day two sets of experimenters, one from the Carnegie Institution and the other from Johns Hopkins, rushed to their laboratories immediately after leaving Bohr and saw the fragments before the Washington Conference ended on January 28. Both used d-d neutrons from high-tension accelerators and both confirmed, by liberal application of paraffin and cadmium, that uranium fission occurs with fast and with thermal neutrons.[51] The Columbia group took more time to publish, employed a Rn-Be source of known intensity in the manner of Amaldi and Fermi, and measured the relative probability of fission by thermal and fast neutrons as 20 to 1.[52]

3. FOLLOWING UP FISSION

Back in Berkeley

When news of fission reached the Laboratory through the daily press about January 29, the staff could not be kept at the assembly of the 60-inch cyclotron, which had been its preoccupation. "All of us couldn't resist our curiosity to look into the matter a bit."

50. Frisch to Margaret Hope, 22 Jan 1939, and to Bohr (unsent draft), 15 Mar 1939 (Frisch P); Frisch, *Nature, 143* (25 Feb 1939), suppl., 276, dated 16 Jan; Frisch, *What little,* 116–17, and in Rozental, *Bohr,* 145–7; Wheeler in Stuewer, *Nuclear physics,* 272–3; Badash et al., APS, *Proc., 130* (1986), 205–11.

51. Paxton to Nahmias, 12 Feb 1939 (JP, 28), quote; Roberts, Mayer, and Hafstad, *PR, 55* (15 Feb 1939), 417 (Carnegie Institution), letter of 4 Feb; Fowler and Dodson, ibid., 418 (Johns Hopkins), letter of 3 Feb.

52. Anderson et al., *PR, 55* (1 Mar 1939), 511–2, letter of 16 Feb; Amaldi and Fermi, *PR, 50* (1936), 899. The two cross-sections, estimated on the assumption that U^{238} is responsible for both, was $2 \cdot 10^{-24}$ and 10^{-25}.

There were various approaches. Oppenheimer proved that fission is impossible. Alvarez wired East for further information, learned that Tuve's group had detected fission fragments, and immediately did the same, for the benefit of Oppenheimer and others who would not accept "so revolutionary [an effect]...until it had been confirmed in several laboratories."[53] Alvarez and his collaborator, G.K. Green, in Berkeley as a National Research Fellow, saw the effects of thermal and fast neutrons and went beyond other hasty confirmers by estimating the time delay between irradiation and fission. They obtained an upper limit of 0.003 sec using a pulsed neutron beam modulated in Alvarez's manner; the limit was soon lowered to 0.001 sec by the intermittent neutron generator at Imperial College.[54] They all missed the so-called "delayed neutrons," emitted a few seconds after cessation of the irradiaton in a small percentage of fissions. These dilatory particles, ejected during the rearrangement of fission fragments, give a handle for the control of the fission process.[55]

Green and Alvarez used Alvarez's preferred detector, a fast electronic affair, a thin-walled ionization chamber (on one of whose plates the uranium sat) connected to a linear amplifier and oscilloscope. Dale Corson, then in charge of the Laboratory's cloud chamber, teamed with Thornton to see fission fragments in a slower, but more evocative, way. They put uranium oxide on a collodion foil in a chamber filled with air, water vapor, and alcohol; irradiating the whole with neutrons procured from the 37-inch cyclotron, they obtained the nice result shown in plate 9.2. Assuming the short side tracks to belong to recoiling ions from the chamber gases, they inferred that the lower fission fragment must have had an atomic weight of at least 75 to have continued its course without apparent deviation. Two weeks after Corson and Thornton sent off the best of their picutures to the *Physical Review*

53. Lawrence to A.J. Allen, 7 Feb 1939 (1/14), quote, to Cockcroft, 9 Feb 1939 (4/5), and to Cork, 7 Feb 1939 (5/1); Alvarez to Darrow, 26 Mar 1939 (6/9), quote, and *Adventures*, 72–7.

54. Green and Alvarez, *PR, 55* (15 Feb 1939), 417, letter of 31 Jan; Gibbs and Thomson, *Nature, 144* (1939), 202, letter of 17 Jul.

55. Roberts et al., *PR, 55* (1939), 510–1, 664, letters of 18 Feb and 10 Mar; Booth, Dunning, and Slack, ibid., 876 (17 Apr). The existence of delayed neutrons was at first contested; Flügge, *Nwn, 27* (1939), 403.

(they could choose among 25 obtained in 885 exposures to cyclotron neutrons), Joliot, a past master of cloud chamber technique, exhibited at the Paris Academy of Sciences the single photograph of a recoiling fission product that he had got in 902 exposures to Rn-Be neutrons. To physicists, the several quick observations of strongly charged, heavy particles constituted "conclusive evidence" of the phenomenon brought to light by Hahn and Strassmann after months of tedious, masterful analytical chemistry.[56]

Seaborg heard about fission at the Journal Club meeting on January 30. "The news excited me very much. After the seminar, I spent hours walking the streets in Berkeley, chagrined that I had not recognized that the 'transuranic' elements...were not really 'transuranic' elements; I felt stupid." (He had judged the results of Curie and Savitch to be "rather strange.") His follow-up, or, rather, that of Joseph Kennedy, to whom he assigned the task, was not a chemical but a physical investigation, a search for the fast beta particles that they thought might be released in the successive transformations of the fission products. Kennedy spent some weeks at the quest, but to no avail. Nor could fast beta particles be detected concurrent with fission. They gave up and notified their colleagues of their failure.[57]

In March, reinspired by the "elegant chemical separations" and "startling conclusions" of Hahn and Strassmann's definitive demonstration that uranium can give birth to barium, Kennedy and Seaborg tried another way to obtain information about the fission fragments. Once the chemical nature of one partner in a fission is known, the nature of the other follows from subtraction of atomic numbers; but a similar inference from an isotope of the one to an isotope of the other is not possible if free neutrons come out from a bursting uranium nucleus along with the fission partners. Kennedy and Seaborg looked for evidence of these neutrons by placing a Ra-Be source on the axis of a cylinder,

56. Corson and Thornton, *PR*, *55* (1 Mar 1939), 509, letter of 15 Feb; Walke, *Rep. prog. phys.*, *6* (1939), 35, quote; Joliot, *CR*, *208* (1939), 647, séance of 27 Feb, in *Oeuvres*, 653–4; Turner, *RMP*, *12:1* (1940), 7–10.

57. Seaborg, *Jl.*, *1*, 362 (11 Jul 1938), 436 (30 Jan 1939), 447 (18 Feb); Kennedy's trials lasted from 9 Feb to 17 Apr (ibid., 441 ff.); Kennedy and Seaborg, *PR*, *55* (1 May 1939), 877, letter of 17 Apr.

surrounding it with two concentric shells, the inner of uranium oxide and the outer of water, and counting the neutrons reaching the water's surface. They could not confirm their expectation that more neutrons should reach their detectors with the uranium in place than without it.[58]

There was another and more urgent reason for desiring to know not only whether, but also how many, free neutrons may be emitted during fission. Should the number be large enough, the process of fission, once instituted by the energy of the physicist and his machines, might continue on its own, perhaps with explosive violence. That had been plain to the Columbia group the moment they had confirmed Frisch's experiment. "Here is real atomic energy!" Dunning wrote in his laboratory notebook on confirmation day. And on the morrow, in the same exclamatory style: "Secondary neutrons are highly important! If emitted would give possibility of self perpetuating neutron reaction." Many others saw the same: the omniscient Fermi; the busy Szilard and Szilard's sour partner Arno Brasch, who twitted him about the now worthless patents on chain reactions with which he had hoped to direct the course of events.[59] At Berkeley they also understood. "We are trying to find out whether neutrons are generally given off in the splitting of uranium; and if so, prospects for useful nuclear energy become very real!" "It may be that the day of useful nuclear energy is not so far distant after all."[60] Alvarez spearheaded the hunt, though not with his customary thoroughness. He directed his beam of slow neutrons into a bottle of uranium oxide and looked for fast neutrons arising from fission. He saw none.[61] He also tried, with the help of a chemist, Kenneth Pitzer, to detect an increase in the temperature of uranium irradiated by neutrons. Again the experiments were inconclusive. They were resumed by Malcolm Henderson, back in Berkeley for the

58. Seaborg, Jl., 1, 460–1 (15–16 Mar).
59. Dunning "Notebook," 25–6 Jan 1939, in Badash et al., APS, Proc., 130 (1986), 210; Szilard to Strauss, 22 Feb 1939 (Sz P, 17/198); Brasch to Szilard, 5 Mar 1939 (Sz P, 29/306).
60. Resp., Lawrence to Cockcroft, 9 Feb 1939 (4/5), and to A.J. Allen, 7 Feb 1939 (1/14); a similar message is in Lawrence to Cork, 7 Feb 1939 (5/1), to Fermi, same date (7/15), and to Van Voorhis, 9 Feb 1939 (17/43).
61. Alvarez, "Adventures," 22.

summer, who detected heat enough to indicate that the fission fragments recoiled with an average energy of around 175 MeV, in agreement with other estimates made in other ways.[62] No further efforts to detect fission neutrons at the Laboratory have come to light, although Lawrence said publicly that the possibility of obtaining energy from uranium depended upon the number of neutrons released in fission.[63]

His lethargy in prosecuting this possibility, with which he had been flirting for several years, may be accounted for as follows. For a change, time was more precious than money. Lawrence preferred not to delay completion of the 60-inch cyclotron for the chance that the Laboratory would be the first to find and advertise fission neutrons. Nor was a cyclotron the best tool for the search. The groups who first established the multiplication of free neutrons during fission worked with Ra-Be sources. The earliest in the field was Joliot's. They compared the neutron intensity at various distances from their source of photoneutrons (Raγ-Be) when it sat in tanks containing plain water and water strewn with uranium oxide (hence they measured primarily the effects of *slow* neutrons). Fermi and two of his associates at Columbia used the same technique, which was adapted from one the Rome group had invented to measure neutron absorption; while another pair at Columbia, Szilard and Walter Zinn, observed fast neutrons from uranium struck by slow ones. In Paris they overestimated the average number of neutrons emitted during uranium fission as 3.5; in New York they made it 2, and then, in a more careful measurement by a combination of the two Columbia groups, who pooled their radium and uranium, 1.5.[64] One preferred the higher or

62. Lawrence to Fermi, 7 Feb 1939 (7/15); Henderson to Lawrence, 21 June and 13 Sep 1939 (9/6); Henderson, *PR, 56* (1939), 703, and *PR, 58* (1940), 774–80.

63. Interview with University Explorer, 23 Feb 1939 ("Cyclotron releases atomic energy"), and with CBS, 15 Apr 1939 (40/15). The relation betweeen the Berkeley cyclotron and atomic energy from uranium was hinted at in Februrary by Birge, in a speech in honor of Lawrence (U.C., *1939 prize*, 22–3), and again by the University Explorer in September ("World's largest atom smasher") (40/15).

64. von Halban, Joliot, and Kowarski, *Nature, 143* (1939), 470–1, letter of 8 Mar, and ibid., 680, letter of 7 Apr, in Joliot and Curie, *Oeuvres,* 559–61, 664–6; Anderson, Fermi, and Hanstein, *PR, 55* (15 Apr 1939), 797–8, dated 16 Mar; Szilard and Zinn, ibid., 799–800, also 16 Mar; Anderson, Fermi, and Szilard, *PR, 56* (1939), 284–6, rec'd 3 July; Anderson, *BAS, 29:4* (1973), 10–1.

lower figure according to whether one hoped for nuclear power or feared a nuclear bomb.[65]

The only important immediate contribution of the Laboratory to the unravelling of the fission process was made by a junior member, Philip Abelson, who, by combining an undergraduate chemistry major with graduate training in physics, constituted an untried one-man interdisciplinary team. In 1937, in the course of the customary search for new activities, he had confirmed the general scheme of transuranics then recently put forward by Meitner, Hahn, and Strassmann, and added a new activity, of 17 hours, whose place he could not find. Then, under the inspiration of Alvarez's work on K-electron capture, he thought to identify "transuranics" by their characteristic x rays.[66] In March 1938, in "transuranics" prepared in quantity at the cyclotron, Abelson isolated an activity of 77 hours, which he identified with the 66-hour body that the Berlin group had fixed as element 95, and whose characteristic x rays he managed to detect with a cheap spectrometer he had constructed. By extrapolating beyond uranium the increase with atomic number of the penetrating power of L rays—characteristic radiation involving the second or L shell of the atom—Abelson convinced himself that the x rays he detected belonged to an atom of nuclear charge 95. The revelation of fission dumbfounded him.

After a day in the dumps, Abelson looked again and learned that the "L rays" of 77-hr element 95 were in fact K rays of 2.4-hr iodine (atomic number 53). The connection, as Abelson worked it out: the iodine, Meitner et al.'s 2.5-hr element 96, descends from the 77-hr body, which must accordingly be tellurium; tellurium's beta decay can knock out a K electron, setting up the production of iodine's K ray. The process, taken all together, constituted "an unambiguous and independent proof of Hahn's hypothesis of the cleavage of the uranium nucleus."[67] Abelson's rays and Dale Corson's pictures caused Oppenheimer to adjust his evaluation of

65. Weart, *PT, 29:2* (1976), 23–30, and Weart, *Scientists in power,* 75–92.

66. Abelson, *PR, 53* (1938), 211–2, APS meeting, Stanford, 17–18 Oct 1937.

67. Abelson, *PR, 55* (15 Feb 1939), 418, letter of 3 Feb, *PR, 56* (1 Jul 1939), 4–5, and *BAS, 30:4* (1974), 48–51. The Cavendish independently fingerprinted via x rays: Feather and Bretscher, *Nature, 143* (1939), 516, used Li-d neutrons to provoke fission and identified fragments by the emergent x rays's absorption.

fission from "impossible" to "unbelievable." And if the unbelievable should give rise to some extra neutrons? "A ten cm cube of uranium deuteride (one should have something to slow the neutrons without capturing them) might very well blow itself to Hell."[68]

Abelson carried on. In three months of hard work, reprieved from crew duty, he found in the products of uranium irradiated by neutrons no fewer than five activities ascribable to antimony, seven to tellurium, and four to iodine. He thereby not only showed the complexity of the German transuranics, but also demonstrated that fission fragments were ordinary isotopes. Two of his telluriums and one of his iodines had the same periods and beta-ray spectra as activities then recently identified in the Laboratory by Seaborg, Livingood, and Kennedy. Abelson could therefore specify decay sequence, half-life, element, and, in some cases, the isotope of the products he studied. In accomplishing his work he mobilized the great experience of the Laboratory in nuclear chemistry, as represented by Segrè and Seaborg, and, of course, the Laboratory's prop and pride: Abelson attributed his success "in large measure to the huge neutron intensities of the cyclotron."[69] The cyclotron deserved the credit, and more. Its neutrons made fission products strong enough to swamp the activities of the natural descendents of uranium, which regrew during chemical analysis of the irradiated samples. With their "enormously weak preparations," Hahn and Strassmann had great difficulty finding iodine among the uranium products and succeeded only after Abelson's demonstration. Their confirmation resulted in what they called the "definitive suppression" of their long transuranic chains.[70]

And what about the Curie-Savitch body that precipitated the studies that led to fission? As Meitner and Frisch observed, a barium isotope arising from uranium should be accompanied by

68. Resp., Oppenheimer to W.A. Fowler, 28? Jan 1939, and to G. Uhlenbeck, 5 Feb 1939, in Smith and Weiner, *Oppenheimer*, 207, 209.
69. Abelson, *PR, 55* (1939), 670, 876, and *PR, 56* (1939), 1–2, 6–9 (quote); cf. Turner, *RMP, 12:1* (1940), 1, and Walke, *Rep. prog. phys., 6* (1939), 38–9.
70. Hahn to Meitner, 13 and 20 Mar 1939, in Krafft, *Strassmann,* 304–5; quotes from, respectively, Hahn to Meitner, 18 June 1939, ibid., 310, and Hahn and Strassmann, *Nwn, 27* (1939), 451, rec'd 19 June.

an isotope of krypton that might decay through rubidium, strontium, and yttrium to a stable zirconium. Hahn and Strassmann
immediately started to look for these elements, of which they soon
reported sightings.[71] Now yttrium has chemical properties similar
to those of lanthanum, from which Curie and Savitch had not
been able to separate $R_{3.5h}$. The exact investigation of the
sequence from strontium on was left to one of Hahn's assistants,
who found an yttrium of the necessary period, but declined to pro-
nounce definitively that it decayed to zirconium and constituted
the inspirational French "lanthanide." It did, and does.[72]

Utopia or Armageddon

There remain the connections among the processes that the Berlin group had tied up in the problematic isomers of U^{239}. Recall
that the long chains in figure 9.5 arise from fast and thermal neutrons respectively, and the short chain from resonance capture.
Meitner and Hahn kept the third heavy uranium, of half-life 23
minutes, as a true U^{239} and tacitly referred both the long chains to
the fission of U^{238}. They clung to this interpretation despite its
odd consequence, that U^{238} would be fissionable by both fast and
thermal neutrons but not by ones with moderate speeds, because
the two chains seemed to be present in about equal strengths.
They knew that a lighter isotope of uranium existed, but since it
makes up less than 1 percent of natural uranium, it did not appear
to them to be the ancestor of either long chain. Bohr's highly
skeptical associate, George Placzek, adduced the ambiguity of the
role ascribed to U^{238} as a strong argument against the possibility
of fission. Thus inspired, Bohr found a way to father thermal-
neutron fission on U^{235}. It was only necessary to detach the
fissions from the old decay chains: experiment could not distinguish which uranium isotope gave rise to which fragment; the

71. Krafft, *Strassmann,* 281–4; Meitner and Frisch, *Nature, 143* (1939), 239–40;
Hahn and Strassmann, *Nwn, 27* (1939), 94.
72. Lieber, *Nwn, 27* (1939), 421–3; Hahn to Meitner, 1 and 18 June, and
Meitner to Hahn, 16 June 1939, in Krafft, *Strassmann,* 309, 310. The relationship
between Y-La-Ac, on the one hand, and Ba-Ra on the other, gave rise to a priority
dispute between the French and the Germans over the discoverer of fission; Krafft,
ibid., 322–7.

genetic connections in the first steps of the long chains were illusory.

Bohr argued from the excitation energies of the compound nuclei U^{239} and U^{236}. In the former case, the impacting neutrons must bring in the explosive energy, because, as an unpaired nucleon in the odd-numbered isotope 239, it will not excite the compound nucleus very much by its binding energy; in the latter case, the pairing of the neutron in the even isotope 236 contributes enough to the general excitation to create a chance of fission. The slower the neutron, the longer it takes to pass a nucleus and the greater the chance of its absorption.[73] Hence the final interpretation of the "transuranic" chains: one comes from fast-neutron fission of U^{238}, the other from thermal-neutron fission of U^{235}. Placzek did not think that Bohr's reasons were very compelling.[74] Others thought them strong enough to undertake the very difficult task of separating enough U^{235} from natural uranium to investigate its fissile properties directly.

Among the bold were the enterprising pair Seaborg and Kennedy. In the summer of 1939 they began construction of a tube, eventually over twenty feet in length, which they fixed to the outside of the chemistry building. Their scheme adapted a recent serendipitous invention made by Klaus Clusius and Gerhard Dickel, chemists at the University of Munich, who subjected a mixed gas to thermal gradient between a hot wire running down the tube's axis and the cooler tube walls. The resultant motion of the gas is hard to calculate but easy to describe: lighter molecules move radially inward under the thermal stress and axially upward by convection, heavier ones radially outward and axially downward. In the circulation, light molecules concentrate at the top and heavy ones at the bottom of the tube. Clusius and Dickel reported an almost complete separation of the major isotopes of chlorine.[75] Kennedy and Seaborg planned to work with uranium hexafluoride gas, or "hex," as it came to be known from its

73. Bohr, *PR, 55* (15 Feb 1939), 418–9, letter of 7 Feb.

74. On Placzek, see Frisch to Meitner, 8 Jan 1939, and Placzek to Frisch, 2 Mar 1939 (Frisch P); Wheeler in Stuewer, *Nuclear physics,* 276–8.

75. Clusius and Dickel, *Nwn, 26* (1938), 546; Welch, *Ann. rep. prog. chem., 36* (1940), 153–8.

unsavory character, and they set up a fluorine generator to cast the spell. A little hex eventuated; a graduate student, Arthur Wahl, joined the project; Clusius's column was installed and Clusius written, in the best and most naive tradition of open science, for advice about uranium separation. Then, on January 12, 1940, the fluorine generator exploded while the three would-be hex splitters were working on it. The next day Kennedy was ill. Seaborg consulted a chemistry book: "Uranium is an extremely powerful, slow-acting poison." Kennedy stayed sick.[76] Although his problem turned out to be mononucleosis, the incident destroyed the group's enthusiasm for hex. They put their columns—eventually they had three in operation—to separating the uncursed isotopes of hydrogen, carbon, and chlorine.[77]

Another reason that Seaborg's group did not pursue uranium splitting was that others were doing it with greater success. By early 1940, the mass spectrograph run by Alfred Nier at the University of Michigan had collected enough U^{235} for the Columbia group to make possible a positive test of Bohr's conjecture about its fission by thermal neutrons. By the summer of 1940, when Kennedy and Seaborg made a tour of laboratories including Nier's, they learned about several attempts to acquire U^{235} in bulk: Urey and Tuve by diffusion methods, Beams by centrifugation, all starting with hex. The Columbia group had the help of Aristid von Grosse, a former collaborator of Hahn's and an expert in hex manfacture.[78] That did not exhaust the competition. There were also Frisch, then relocated in England, where he and Otto Blüh, a refugee from Prague, set up a Clusius tube late in 1939; Clusius himself, who had declined to supply the additional information that Seaborg had requested; and Wilhelm Krasny-Ergen, who had established a Clusius distillery in Stockholm, from which, "had his activities not been suspended by the political situation," he expected a sevenfold enrichment in eighty days.[79] Although

76. Seaborg, *Jl.*, *2*, 553–68 (15 Nov 1939–23 Jan 1940).
77. Ibid., 571–99 (4 Feb–10 May); Kennedy and Seaborg, *PR*, *57* (May 1 1940), 843; Seaborg, Wahl, and Kennedy, *Jl. chem. phys.*, *8* (1940), 639.
78. Seaborg, *Jl.*, *2*, 557 (2 Oct 1939), 616–8 (1–6 Jul 1940); Nier, Booth, Dunning, and von Grosse, *PR*, *57* (1940), 546, letter of 3 Mar, and ibid., 748; Urey to L.C. Bigelow (hex chemist at Duke), 22 Dec 1939 (Urey P, 1/8).
79. Oliphant to Bretscher, 3 June 1939 (Mrs. Bretscher); Frisch to Ebbe Rasmussen, 28 Oct 1939, and to J. Koch, 28 Dec 1939 (Frisch P). Cf. Krasny-Ergen, *Nature, 145* (1940), 742.

Tuve volunteered to send enough hex for preliminary experiments, Kennedy and Seaborg wisely withdrew from uranium separation. Perhaps Lawrence's belief that centrifugation held the greatest promise for large-scale separation of heavy isotopes influenced their decision.[80]

The confirmation that U^{235} has a large cross-section for fission by slow neutrons gave physicists greater confidence and worry that an explosive chain reaction could be achieved. "Physicists are anxious that there be no public alarm over the possibility of the world being blown to bits by their experiments," *Science Service* reported in melodramatic ignorance just after the confirmation of fission at the end of January 1939. The world was not alarmed, despite a revelation in the *New York Times* that a little U^{235} could wipe out New York City and leave a hole halfway to Philadelphia.[81] During 1939 physicists calculated what might be possible, but in ignorance of the relevant cross-sections and reactions they could only conjecture. An effort to keep pertinent data secret assisted the progress of ignorance and speculation. Although Szilard's novel notion that physicists should censor themselves failed to persuade Joliot and so failed internationally, enough was withheld in the United States that physicists as well placed as Tuve and Frisch were "hard pressed to get some data on uranium fission."[82] The most sanguine discussion of the future of fission came from a colleague of Hahn's at the Kaiser-Wilhelm-Institut für Chemie, Siegfried Flügge, who offered a path to a "uranium machine."

Flügge started from the obvious: for a chain to succeed, enough slow neutrons must be obtained and caused to provoke fissions before they are lost or captured ineffectually. To slow the fast

80. Seaborg, *Jl.*, *2*, 605, 623 (1 June, 5–6 Aug 1940); Lawrence to Dunning, 14 Mar 1940 (6/19), re Nier's success with U^{235}; infra, §10.2.

81. *Science service*, 30 Jan 1939, quoted by Anderson, *BAS*, 29:4 (1973), 9; W.L. Laurence, *New York Times*, 5 May 1939, 25, in Badash et al., APS, *Proc.*, *130* (1986), 213–6, who also quote *Newsweek*, *13* (27 Mar 1939), 32, to the same effect.

82. Tuve to Breit, 2 Aug 1939, quoted by Barraca, unpublished ms. (1985), 12; Frisch to Bloch, 15 Jul 1939 (Frisch P); Weart, *PT*, *29:2* (1976), 23–30, *AJP*, *45* (1977), 1049–60, and *Scientists in power*, 75–92.

fission neutrons, it is necessary only to pass them through water, which, to be sure, also captures them; but, according to Flügge's calculations, based on the probability that a slow neutron will cause a fission as measured at Columbia and on his own estimate of the liability of a fast neutron to loss in water while slowing down, a mixture of fifteen kilograms of uranium oxide per liter of water will sustain a chain reaction. That assumed the conservative estimate that an average fission sets free two fast neutrons. It also assumed that the reaction once started could be controlled. Flügge planned a power plant, not a bomb. A method of control had been published by two members of Joliot's team. It rests on the principle that the faster the neutron, the lower its probability of provoking a fission in uranium, and on the fact that the appetite of cadmium nuclei for slow neutrons is almost independent of the temperature. Therefore sprinkle a little cadmium dust in the water along with the uranium oxide. The chain begins, the mixture heats up, the neutrons move faster, the number of fissions goes down; equilibrium will be reached at a temperature fixed by the amount of neutron-removing cadmium present. According to Flügge, with 0.2 gram of cadmium per liter as seasoning, his uranium stew would boil along at a safe and steady 350°. The reactor vessel would require a diameter of a little over a meter. From it heat could be removed and used to make steam to drive a turbine; a cubic meter of uranium thus exploited could generate electricity for eleven years at a rate equivalent to Germany's consumption on the eve of the Depression.[83]

By the time Flügge's design was published, Fermi and his colleagues had decided that water absorbs too many neutrons to serve as decelerator, or moderator, in a uranium machine. Hopes of capitalizing quickly on fission faded; by August 1939, when Tuve managed to learn what had been learned, he concluded that "all indications are that no chain can occur but it is pretty close." At the same time, Bohr was calculating the amount of hydrogen moderator needed to slow the neutrons. He arrived at a ratio of ten atoms of hydrogen to one of uranium, which he thought would prohibit a fast chain or strong explosion. Recalculation reduced

83. Flügge, *Nwn, 27* (June 1939), 406–10; Anderson et al., *PR, 55* (1939), 511–2, letter of 16 Feb; Adler and von Halban, *Nature, 143,* (1939), 793.

the ratio to one to one, which allowed a possibility. The Paris group was more negative. Their experiments with a homogeneous mixture of three atoms of hydrogen to one of uranium gave evidence of fissions caused by secondary and tertiary neutrons, but not of a divergent multiplication of fissions; and by the end of October 1939, in a paper they withheld from publication, they concluded that "it is almost certainly impossible" to promote a divergent chain reaction in a homogeneous blend of naturally occurring uranium, oxygen, and hydrogen.[84] Of course, a moderator other than ordinary water, or a heterogeneous mixture of uranium and water, might perform better. The published consensus of physicists, as expressed in the several reviews of the year's fission research composed toward the end of 1939, was that exploitation of nuclear energy would not occur in the near future and might not be possible at all.[85]

The announcement on May 1, 1940, of the details of Columbia's fissioning of Nier's latest sample of U^{235} transformed the discussion. William L. Laurence, a science writer for the *New York Times*, informed its readers on May 5 that five or ten pounds of U^{235} could drive an ocean liner or submarine indefinitely. So much light uranium might not be hard to procure. Nier had been able to enlarge his sample two hundred times in a matter of weeks. "It is not impossible that a few months or a year hence may see the realization of this quest." Nothing would be simpler than exploiting the new fuel. "All that is needed to put it to work running motors and steamships is to place it in a tank of water....The water would be turned into steam....New water supplied would keep the process going indefinitely." Furthermore, the business would be automatic and self-regulating, since the

84. Anderson, Fermi, and Szilard, *PR, 56* (1939), 284–6, rec'd 3 Jul; Fermi to A.H. Compton, 31 Aug 1943 (Fermi P); Tuve to Breit, 2 Aug 1939, quoted by Baracca, unpublished ms. (1985); Meitner to Frisch, 17 Aug and 18 Sep 1939, re Bohr (Frisch P); von Halban, Joliot, Kowarski, and Perrin, *Jl. phys. rad., 10* (1939), 428, rec'd 19 Sep 1939, in Joliot and Curie, *Oeuvres, 669–72*; von Halban, Joliot, and Kowarski, *CR, 229* (1949), 909, 30 Oct 1939, in Joliot and Curie, *Oeuvres, 676.* Turner, *PR, 57* (15 Feb 1940), 334, letter of 25 Jan, reached the same conclusion as the French group.

85. Walke, *Rep. prog. phys., 6* (1939), 43; Turner, *RMP, 12:1* (1939), 20–1; Frisch, *Ann. rep. prog. chem., 35* (1939), 15–7.

heating of the water speeds up the neutrons and slows or stops the fission process until more water, "the colder the better," is added. Thus utopia. But evil men were trying to suborn the natural good behavior of U^{235}. "Every German scientist in this field, physicists, chemists, and engineers, it was learned, have been ordered to drop all other researches and devote themselves to this work alone." Fortunately they lacked the essential machine for further investigation, the cyclotron. A race had started, a race with stakes that were incalculable, or almost so. According to Laurence, a pound of U^{235} improperly treated would have the same explosive power as 15,000 tons of TNT. An effective separation plant, therefore, was "a secret to be given only to the United States government."[86]

The physicists were not pleased with Laurence's mixture of fact and fiction. Nier declared that his handiwork had little present commercial or military value, the amount of U^{235} so far isolated being "hardly enough to spring a mousetrap." S.K. Allison rated producing utilizable atomic energy as "just as feasible as getting gold out of the ocean." George Pegram, who, as chairman of Columbia's physics department, was constantly pressed by the press, urged his colleagues to "stress what seems to be the fact, namely, that energy from uranium, even if it became available, would apparently not be cheap energy by any means and would not be very explosive energy." Pegram sold this point of view to the informed and responsible chief of the *Times*'s science section, Waldemar Kaempfert, who squared accounts as follows. To make a pound of U^{235} would cost more than the expenses of the federal government for an entire year; to make a gram by Nier's improved technique would take over a century. "The prospect of using U^{235} in the present war is zero."[87]

While Kaempfert calculated, Krasny-Ergen's plan was published in *Nature*. It set Laurence off again. According to him, 10,000 of Krasny-Ergen's units could make a pound of U^{235} in forty days; at $100 per unit, a uranium factory would cost only $10 million. "It may be expected, therefore, that Germany will take measures at

86. Laurence, *New York Times*, 5 May 1940, 1, 51.
87. *New York Times*, 6 May 1940, 19; Pegram to Bethe, 14 May 1940 (HAB, 3); Kaempfert, *New York Times*, 7 May 1940, 24, and 12 May 1940, D7.

once to install such a plant." Kaempfert again flew to the rescue: Krasny-Ergen's method of thermal diffusion would require 17 million kWh to separate a gram of U^{235}; to make a kilogram, 34 million tons of coal would have to be burnt, at a cost of $68 million. "The more we think this over the more we are convinced that we would not invest ten cents in a uranium public utility company....We doubt if the Germans have the time or the stupidity to bother much about isolating uranium-235 in large quantities."[88]

Meanwhile Lawrence was examining a plan for a German power plant driven by light uranium. He had this information from Clifford Williams of Shell Development, who had it from Peter Debye, who had recently left the directorship of the Kaiser-Wilhelm-Institut für Physik in Berlin. According to Debye, all his former staff were engaged in developing U^{235} as a power source (fig. 9.8); Germany was "frantically mining uranium in Czechoslovakia;" and the native metal was to be separated by diffusion. It was perhaps indirectly from this disclosure that the author of the article on Lawrence in *Scientific American* obtained the information that "the Nazis are trying to lay hands on all the uranium they can find." The least secret bit of science in the United States in the summer of 1940 was that (as the *San Francisco Chronicle* put it) atomic power "will transform the face of the earth the moment the production of the magic element U-235 can be cheapened."[89]

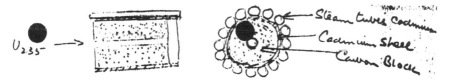

FIG. 9.8   Design for a German U^{235} plant. Williams to Lawrence, 23 May 1940 (18/26).

88. Laurence, *New York Times*, 30 May 1940, 1, 18 (the *Times* reported Krasny-Ergen's method on 26 May, §II.8, with the estimate that it would be 11,000 times faster than Nier's); Kaempfert, ibid., 9 June 1940, §II.5.

89. Williams to Lawrence, 23 May 1940 (18/26); Schuler, *Sci. Am.*, *163:2* (Aug 1940), 71; J.D. Ratcliffe, in "This week" magazine, *San Francisco Chronicle, 25 Aug 1940, quoted by Seaborg, Jl., 2*, who observed that Ratcliffe disclosed more than "we discuss in public gatherings." Yet Segrè, *Encyclopedia Americana*, 1941 (reprinted in 1946), s.v. "uranium," and Lawrence, *Science, 94* (5 Sep 1941), 223, say much the same thing as Ratcliffe.

4. PIONEERING IN TRANSURANIA

According to Meitner and Frisch, the third of the uranium "isomers" of figure 9.5, the 23-minute activity, was indeed an isotope of uranium. According to accepted theory, it should have reduced its surplus of neutrons by emitting beta particles. Among the first to seek the product of such a transformation, a nucleus heavier than any previously detected on earth, was Edwin McMillan (plate 9.3). He had an advantage over others in having at his disposal the large activating neutron flux from the 37-inch cyclotron and a highly cultivated technique for the study of the radiations from nuclear disintegrations. Although his time had been taken up with cyclotron problems, he retained the independence he had shown in studying gamma rays while most of the rest of the Laboratory were building machines or exploding deuterons.[90]

In his follow up of fission, McMillan characteristically examined the penetration of the recoiling fragments through the standard absorber, aluminum. He obtained a maximum range of about 2.2 cm, in rough agreement with slightly earlier experiments by Joliot. McMillan also examined the decay of the products left with the irradiated uranium, products unable to propel themselves through a single sheet of cigarette paper. He found an activity of 25 minutes, which he conjectured might be the same as the 23-minute body of the Berlin group, and he detected a strong activity of about 2 days' duration. He did not suggest a source for this long period, but promised further absorption measurements to fix the ranges of the recoiling fragments.[91] Segrè took up the study of the nonrecoiling product and refined the periods to 23 minutes and 2.3 days. The first he identified with the heavy uranium of Berlin. Was the second a fission fragment or a transuranic? Segrè did his chemistry, made the 2.3-day body a rare earth (which it was), identified it as a fission fragment (which it was not), and declared it to be a light element. He searched for the alpha emitter that might reasonably be expected to terminate the beta transformations beginning with U^{239}. No luck. His conclusion, after discussing the matter with McMillan and Seaborg: the 23-

90. Supra, §4.2.
91. McMillan, *PR*, *55* (1 Mar 1939), 510, letter of 17 Feb.

minute U^{239} becomes a long-lived, undetected element 93. And, underlined: *"Transuranic elements have not yet been observed."*[92] Once bitten, twice shy. As a member of Fermi's group, Segrè had erred in claiming a transuranic; as a cautious and institutionally insecure member of the Laboratory, he did not wish to repeat his mistake, and thus missed the transuranic he had. He bent his efforts to straightening out the decay chains of fission fragments.[93]

Segrè's erroneous negative finding was immediately confirmed by John Irvine, a chemist at MIT, who concentrated U^{239} by taking advantage of the effect on chemical bonding of excited nuclear states. He was unable to detect any activity in a precipitate made by treating his enriched material with a compound of rhenium, which, following the old opinion of the nondiscoverers of the transuranics, element 93 should resemble. In his summary of the state of research on fission, completed in December 1939, Louis Turner of Princeton accepted the straightforward conclusion from the experiments of Segrè and Irvine. The elusiveness of element 93 and the hypothetical terminating alpha emitter nonetheless bothered him. He reduced his bother by the good guess that 94^{239} is the alpha emitter and U^{235} the great grandson of U^{238}. That at least kept the scandal in the uranium family.[94]

Meanwhile Abelson, who had finished up at Berkeley with measurements of the wavelengths of K rays from radioactive substances near the middle of the periodic table,[95] began to doubt the assumption on which Segrè had based his denial of transuranic status to McMillan's 2.3-day activity. The old alternative to likening transuranics to the elements beginning with rhenium and osmium remained: 93 and 94 could well resemble rare earths, and these "actinides," as they were later christened, should have a chemistry like uranium's. In his spare time at Tuve's laboratory, where he went in September 1939 to help with its 60-inch

92. Segrè, *PR*, *55* (1939), 1104–5, letter of 10 May; confirmed by Hahn and Strassmann, *Nwn*, *27* (1939), 451, and endorsed by Walke, *Rep. prog. phys.*, *6* (1939), 40.

93. Cf. Abelson, *BAS*, *30:4* (1974), 51–2; Langdorf, *PR*, *56* (1939), 205, letter of 15 Jul; Segrè and Wu, *PR*, *57* (1940), 552, letter of 29 Feb.

94. Irvine, *PR*, *55* (1939), 1105, letter of 16 May; Turner, *RMP*, *12:1* (1940), 27, and *PR*, *57* (15 Jan 1940), 157, letter of 31 Dec.

95. Abelson, *PR*, *56* (15 Oct 1939), 753–7, rec'd 8 Aug.

cyclotron, Abelson showed that the 2.3-day body did not behave consistently like a rare earth. McMillan, too, had his doubts about Segrè's diagnosis. His further tests had shown that the 2.3-day activity remained firmly with the irradiated uranium and that its intensity exceeded that of all the long-lived fission fragments collected by recoil. Furthermore, when cadmium guarded the uranium target from assault by slow neutrons, the intensity of the fission products fell dramatically; whereas the intensities of the 23-minute and 2.3-day activities not only did not change appreciably, but remained in the same ratio, "suggesting a genetic relation between them," and the consequent identification of the longer period with element 93.[96]

Abelson visited Berkeley in May, with orders from Tuve to "make every effort to settle the identity of the 2.3-day substance." Should it come from U^{235}, the possibility of a chain reaction even in separated uranium appeared doubtful. Abelson and McMillan joined forces and soon found a distinct chemical difference between the 2.3-day activity and rare earths. (The difference, the effect of the presence of an oxidizing agent on certain reactions, explained the erratic results of previous investigators, who had not controlled the oxidizing power of their solutions.) In respect of these reactions, it resembled uranium, and, since it had nothing in common with rhenium, McMillan and Abelson referred it to a possible "second 'rare earth' group of similar elements starting with uranium." It remained to show that the 2.3-day activity grew from the 23-minute U^{239}. Samples collected from the parent at 20-minute intervals all decayed with a period of 2.3 days. The decay, by beta emission, produced element 94. McMillan and Abelson supposed, with Turner, that 94^{239} transformed into U^{235} by alpha emission, which, in fact, it does. Their search for the telltale alpha particles did not succeed, however, and they inferred that, if 94^{239} were unstable against alpha emission, it must have a half-life of a million years. They overestimated by a factor of 400.[97]

96. McMillan and Abelson, *PR, 57* (15 June 1940), 1185–6, letter of 27 May; Abelson, *BAS, 30:4* (1974), 51–2; McMillan, in *Les prix Nobel en 1951*, 168–9.
97. McMillan and Abelson, *PR, 57* (1940), 1185–6; McMillan to W.B. Reynolds, 10 Apr 1951 (12/31); Tuve, "Report for June," 5 Jul 1940 (MAT, 21/"extra copies").

Seaborg knew about the work on element 93 as it progressed, and it made him "eager to work in this exciting field." He assigned Arthur Wahl the task of satisfying his eagerness. While Seaborg and Segrè sought new fission products in uranium struck by very fast neutrons, Wahl perfected chemical means for concentrating 93^{239}. By the middle of October, both projects were well advanced: Seaborg and Segrè found two chains of decays from Pd through Ag to Cd, and Wahl, following the procedure devised by McMillan and Abelson (oxidation by bromate ion), had isolated the 2.3-day activity from several samples of uranium irradiated with neutrons.[98] But neither he nor his senior colleagues could find the suppositious alpha-emitting 94. They decided to try another route. During the summer of 1940, McMillan had invoked the traditional Berkeley bombardment and sent deuterons against natural uranium. He caught a glimpse of a second isotope of element 93 with a beta activity slightly more energetic than 93^{239}'s; and he also saw a sign of its alpha-emitting descendent. After Kennedy had made a special thin-walled counter to follow this descendent, Seaborg wrote McMillan, who had left Berkeley for war work at MIT, that he, Kennedy, and Wahl would be "very glad to collaborate with you on the isolation of the new isotopes of element 93 and uranium found in the bombardment of uranium with deuterons." McMillan replied that he would be very pleased to have Seaborg continue the work.[99]

McMillan had more than a curiosity about the secrets of nature in the continuation of the search for element 94. His and Abelson's discovery had been announced to the press with the usual fanfare: "A development that may bring man a step closer to the release of atomic force and energy which brought the universe into existence;" "something which may quite conceivably prove more influential in the destiny of the world than any single battle of the current World War."[100] The publicity prompted a

98. Seaborg, *Jl.*, *2*, 605–6 (2 June 1940), quote, 622, 631; ibid., 635–41, and Segrè and Seaborg, *PR*, *59* (15 Jan 1941), 212–13, letter of 12 Dec 1940; Segrè et al., UCRL-2791 (22 Dec 42) (12/31).

99. Seaborg, *Jl.*, *2*, 641–2, 687 (10 and 29 Oct 1940); ibid., 657, 661 (28 Nov and 10 Dec); Seaborg to McMillan, 28 Nov, and reply, 8 Dec 1940 (12/30).

100. Resp., *Oakland Tribune,* 9 June 1940, 4A, and *Oakland Post Enquirer* (12/30).

reprimand from Lyman Briggs, director of the National Bureau of Standards, who headed a committee that tried to keep potentially useful information about uranium secret from foreign powers. The potential usefulness lay in the possibility that 94^{239} might be fissionable like U^{235}. Although Alvarez had discussed "fishing" 94^{239} with McMillan in January 1940 and Louis Turner had written Lawrence in July proposing that the 60-inch be put to making enough 94 to test the strong likelihood that 94 could be fished, Abelson and McMillan "did not see any possible connection of our work on element 93 with the fission problem."[101]

Neither the mistake nor the publicity would recur, McMillan wrote Briggs; henceforth all discoveries about 93 and 94 would be submitted to his committee for determination of their sensitivity. As an example of his good behavior and his findings, McMillan disclosed his provisional results about the products of uranium bombarded by deuterons: the unknown isotope 93^7 produces an alpha-emitting body, presumably an isotope of element 94, with the chemical properties of thorium. Briggs replied that the "most important contribution you could make at this time" was to discover whether the alpha emitter fissioned with slow neutrons. "Even rough data will be valuable, and facilities do not exist elsewhere which will permit attempting the work." Only the 60-inch cyclotron could produce a strong enough sample of 93 to decay into enough 94 to offer the possibility of detecting its fission. Briggs's committee—which probably meant Fermi—estimated that with a sample as strong as the one McMillan and Abelson had used, McMillan should be able to see one fission every ten seconds, provided that 94 was about as fissionable as U^{235}.[102] That sample had been extremely strong, about 10 mCi of 93, almost certainly, as Alvarez wrote Turner, "the most heavily bombarded substance in history." Its manufacture had so strained the 60-inch that Alvarez doubted that its like would be seen again before the 184-inch started up. "I don't think that we shall be able to fish 94 for some time."[103]

101. Briggs to McMillan, 13 Aug 1940; Turner to Lawrence, 11 Jul 1940; Alvarez to Turner, 19 Aug 1940; and McMillan to Briggs, 31 Aug 1940 (all in 12/30); Turner, PR, 69 (1946), 366, letter of 29 May 1940; infra, §10.2.
102. McMillan to Briggs, 31 Aug 1940, and reply, 5 Oct 1940 (12/30).
103. Alvarez to Turner, 19 Aug 1940 (12/30).

At first fishing may not have been on the agenda of Seaborg's crew. On December 14, 1940, they prepared a modest sample of $93^?$ by irradiating uranium oxide with 175 μAh of deuterons in the 60-inch cyclotron. Wahl took on the purification of 93 as part of his doctoral work. His excellent preparation when interrogated by counters showed a beta and gamma emission sufficiently distinct from those of 93^{239} to point to the presence of a new isotope. An alpha emitter, apparently the descendent of $93^?$, also showed itself. Unfortunately, the half-life of $93^?$ fell out too close for comfort to that of 93^{239} (2.3 days), which was also produced to some extent in the deuteron bombardment. From another sample prepared in January, the half-life of $93^?$ appeared to be 2.1 days, consistent with the increase of the alpha activity that grew from it. On this evidence and some shaky chemistry of the alpha emitter, "Things look[ed] good for element 94," Seaborg wrote McMillan on January 20, 1941. He added: "*no one* else knows about the most important of these results (the element 94) except Wahl and Kennedy....The Committee [Briggs's] will want us to keep the results VERY SECRET!" On January 28, 1941, Seaborg, McMillan, Kennedy, and Wahl announced the discovery of $94^?$ in a letter to the *Physical Review* withheld from publication by Briggs's committee. For the "?", they offered the choice of 235, 236, and 238; it was, in fact, 238, made by (d,2n) on U^{238}, and so identified by Kennedy, Segrè, Wahl, and Perlman in the fall of 1942.[104]

Confirmation that the alpha emitter was an isotope of element 94 required its chemical separation from 93. Wahl tried many methods before he took up with the powerful oxidizing agent, persulfate ion, on the suggestion of a Berkeley chemist, Wendell Latimer, who was not authorized by the Briggs committee to know anything about the matter. (McMillan regarded the introduction of this agent, which promotes the 94 ion to a state soluble in hydrofluoric acid, as the most important contribution of Seaborg, Kennedy, and Wahl to the discovery of element 94.) By the last

104. Seaborg, *Jl.*, 2, 662–7 (14–31 Dec 1940); Seaborg to McMillan, 11 and 20 Jan 1941 (12/30), and to Fermi, 11 Jan 1941 (Fermi P); Seaborg et al., *PR, 69* (1946), 366–7, letter of 28 Jan 1941; Segrè et al., UCRL-2791 (22 Dec 1942) (12/31).

week in February, all the 93 had decayed into element 94. Wahl hit what remained with persulfate, dissolved it in the presence of fluoride ion, and purified it. "These experiments," wrote Seaborg, Wahl, and Kennedy, in an understated celebration of Wahl's work, "make it extremely probably that this alpha-radioactivity is due to an isotope of element 94."[105]

Meanwhile another game with 94 was in play at the Laboratory. In December 1940, Segrè, then in New York, talked with Fermi about the possibility that 94 would fission with slow neutrons, a matter then also receiving the attention of Bohr and Wheeler. Lawrence, in New York as usual during the giving season, was persuaded by Fermi and Segrè to order enough 94 from the 60-inch to test the hypothesis. On January 9, Segrè and Seaborg made a little 93^{239} by neutron bombardment and, by comparing yields, deduced that more 94 could be made by neutrons than by deuterons for the same cyclotron time.[106] Fermi calculated that a kilogram of uranyl nitrate would be target enough; Seaborg and Segrè calculated that the amount of 93 they would have to make to give the 1 μg of 94 they hoped to have would be so radioactive—some 250 mCi—that it would have to be manipulated at long distance or by remote control. After practice runs with uranyl nitrate sent by Fermi—Cooksey had not been very generous with Laboratory funds for the project—Segrè and Seaborg irradiated 1.2 kg of the stuff with neutrons from 3,368 μAh of deuterons from the beryllium target of the 60-inch. For the next four days, from March 3 to March 6 inclusive, they extracted the 93, wearing goggles and lead-impregnated gloves, working with remote controls, carting their improving sample from one piece of apparatus to another in a lead bucket suspended from long poles. The precious product was sealed in a shallow platinum dish under a layer

105. Seaborg, Wahl, and Kennedy PR, 69 (1946), 367, letter of 7 Mar 1941; Seaborg, Jl., 2, 681, 685, 691–3 (23–4 Feb, the night of the isolation); McMillan to Reynolds, 10 Apr 1951 (12/31). McMillan had suggested to Seaborg, letter of 8 Dec 1940 (12/30), that 94 might behave like tetravalent uranium, "very hard to oxydize to a higher state." Cf. Seaborg, Transuranic elements, 2–5, in Frisch et al., Trends, 106, and in New York Times, 16 Jul 1985.

106. Segrè et al., UCRL-2791 (22 Dec 1942) (12/31), 4; Fermi to A.H. Compton, 31 Aug 1943 (Compton P, Chicago); Seaborg and Wahl, JACS, 70 (1948), 1133–4 (text of 21 Mar 1942), on relative yields.

of duco cement. Kennedy joined the team to monitor the decay of the 93^{239} into 94^{239}.[107]

At this moment, early in March 1941, Seaborg's enterprises came together. Wahl's technique for the separation of 94^{238}, which Seaborg had kept secret from his partner, the alien Segrè, who had not been formally authorized to receive the information by the Briggs committee, was, of course, applicable to the big sample of 94^{239}. "These results came just in time," Seaborg wrote McMillan, "to be of great help to me [!] in the 94^{239} project which I [!] am doing for the Uranium Committee."[108] By the end of March, the 93 had decayed to negligibility. Kennedy, Seaborg, and Segrè brought the sample—about 0.5 μg of 94^{239} mixed with rare earths and other dross—near the beryllium target of the 37-inch cyclotron, irradiated it with neutrons, and had the satisfaction of detecting about one fission per minute per μA of deuterons, from which they guessed that 94^{239} has a cross-section for slow neutrons about one-fifth that of U^{235}. The thick sample did not lend itself to such measurements, however: most fission fragments stopped within it or its cement topping. Wahl used the newly found chemistry of 94 to thin it. On May 17, 1941, it again went under the 37-inch. Almost twice as many fission fragments were counted as would have escaped from an equivalent amount of U^{235}. There was no longer a doubt: 94^{239}, which the Berkeley team estimated to have a half life of 10,000 years when left to itself, can be fished by slow neutrons.[109] During the summer of 1941, a large sample of 94^{239} was prepared at the cyclotron and purified by Wahl. Studies of its chemistry by Wahl and Seaborg, of its fissionability under fast neutrons by Segrè and Seaborg, and of its low rate of spontaneous fission by Kennedy and Wahl

107. Segrè to Fermi, 10 Jan 1941 (Fermi P), and in *Ann. rev. nucl. sci.*, *31* (1981), 11; Wheeler in Stuewer, *Nuclear physics*, 276; Seaborg, *Jl.*, *2*, 668, 673, 677, 679–80 (3, 11, 15, 21, 24 Jan 1941), 683–4, 687 (3, 5, 12 Feb), 694–9 (3–6 Mar); Seaborg to McMillan, 11 Jan 1941 (12/30).

108. Seaborg to McMillan, 8 Mar, and reply, 26 Mar 1941 (12/30): "Your 94 results are really something!" Cf. Segrè et al., UCRL-2791 (22 Dec 1942) (12/31).

109. Seaborg, *Jl.*, *2*, 708, 717, 723–6 (28 Mar, 17 Apr, 12, 17–18 May); Kennedy, Seaborg, Segrè, and Wahl, *PR*, *70* (1946), 555–6, rec'd 29 May 1941; Seaborg, *Transuranic elements*, 8–10, and in Frisch et al., *Trends*, 107.

confirmed the suspicions that it belonged to a heavy rare-earth group and might make a fine explosive.[110] That proved decisive for the career of Glenn Seaborg and may prove so for the rest of the human race.

110. Seaborg and Wahl, *JACS,* *70* (1948), 1130–3; Segrè et al., UCRL-2791 (22 Dec 1942) (12/31); Kennedy and Wahl, *PR,* *69* (1946), 367–8, letter of 4 Dec 1941.

X
Between Peace and War

The academic year 1939/40, which was not a good time for most of the world, was one of great achievement and even greater promise at the Laboratory. As the Nazis prepared to invade Poland, Alvarez and Cornog cleared up the isobars of mass three; as the Nazis overran Belgium, Kamen and Ruben established the radioactivity of C^{14}; on the weekend of December 22, while Finnish and Russian troops bled one another in the snow, the Laboratory presented a sample of its wide-ranging activities to the American Physical Society, meeting in Berkeley. Alvarez and Cornog disclosed their discovery of tritium; Helmholz described gamma-ray conversions in technetium and other metals; Kruger held out hope for radiological cure of cancer; Corson and Mackenzie discussed activities in bismuth and polonium irradiated by alpha particles from the 60-inch; and, to round out the spectrum, Cornog presented an engineering accomplishment, the pumping system of the mighty cyclotron.[1] The machine hummed along, "almost a push-button cyclotron." Its breakdown in January 1940 measured the extent of troubles in the land of milk and honey.[2]

The most promising enterprises of the year were the expansion of the medical program owing to the completion of the 60-inch, the pursuit of transuranic elements, and, above all, the planning of

1. *PR, 57* (1 Feb 1940), 248–50. Much the same range of activities was reported at the summer meeting in Seattle, 18–21 June, *PR, 58* (15 Jul 1940), 192–4, 197: Alvarez and Cornog on the period of tritium; Kamen and Ruben on the same of C^{14}; Alvarez on stripped carbon atoms in the 60-inch, and two papers on apparatus.

2. Lawrence to Thornton, 6 Mar 1940 (17/14), and to DuBridge, 14 Mar 1940 (6/17); Aebersold to R.D. Evans, 9 Feb 1940 (7/8), and supra, § 6.1, on breakdown.

what would be the largest cyclotron ever made. The macho character of cyclotroneering found clear expression as Lawrence's "boys" prepared to create their "he-man cyclotron," "the father of all cyclotrons," on a hill dominating the Berkeley campus. Oliphant's Birmingham bomber, though bigger than the 60-inch, would be—so Lawrence put it in a letter to his cousin—but a "toy," a thing for children, in comparison with the manly machine on the drawing boards in Berkeley.[3] The requirements of this machine were to bring the war closer to the Laboratory; and the machine itself was to become the chief instrument of the Laboratory's eventual mobilization.

1. THE HE-MAN CYCLOTRON

Skinning Cats

It took a man of singular determination and self-confidence to propose a cyclotron capable of accelerating particles to 100 MeV. The substantial cost—projected at perhaps a million dollars—was not at first the major impediment. Nature, not money, seemed to set a limit to the size of cyclotrons. The difficulty, that the increase of mass with speed claimed by the theory of relativity would destroy the synchronism expressed in the cyclotron equation (2.1), had been noticed in 1931, by Livingston and by Feenberg; but the limit, whatever it might be, evidently did not affect the performance of the first cyclotrons, and the menace faded from view. When presenting Lawrence with the Comstock prize in November 1937, W.D. Coolidge saw no obstacles: "The limit to the particle energies which can be generated in this way is not yet in sight."[4] Precisely at that moment, however, Bethe and his student Morris E. Rose declared that in their calculations relativity limited the maximum energies obtainable in a cyclotron to about those achievable with the 37-inch machine. They observed that to

3. Quotes from, resp., G.K. Green to Cooksey, 21 Dec 1939 (8/4); Oliphant to Lawrence, 20 Nov 1939 (14/6); and Lawrence to L.T. Haugen, 20 Dec 1939 (9/1).

4. Lawrence and Livingston, *PR, 40* (1932), 34; Feenberg, reported in Slater to Lawrence, 4 Sep 1931 (35/3); Coolidge, *Science, 86* (5 Nov 1937), 406.

compensate for the continually rising mass, the magnetic field must *increase* toward the periphery of the orbit to keep the circulating particles in phase with the oscillator. But to focus the particles in the median plane, the field must *decrease* from the center outward. The cyclotroneer wants both resonance and focusing; nature requires a choice.

According to Bethe and Rose, the best that can be done is to sacrifice exact resonance; but even so, and with the best field design they could contrive, maximum energy would be 5.5 MeV for protons, 8 MeV for deuterons, and 16 MeV for alpha particles. This estimate supposed 50 kV on the dees; with 100 kV something more could be done, since the accelerated particles would acquire energy more quickly and so have more of it when they finally fell out of phase with the accelerating voltage. Still the outlook was grim. With a magnetic field of 18 kilogauss and a final orbit 37 cm in radius, deuterons of 11 MeV would emerge, "the highest obtainable with as much as 100 kV dee voltage." For such a cyclotron, pole faces 34 inches in diameter would suffice. "Therefore it seems useless to build cyclotrons of larger proportions than the existing ones."[5]

When the Laboratory received this news, it was engaged in what, according to the Cassandras of Cornell, was a wasteful and useless task. But its experience with the 37-inch gave it confidence that the 60-inch could go beyond 10 MeV despite the most refined calculations to the contrary. Lawrence wrote Bethe that relativity had not yet begun to inconvenience cyclotroneers; the existing inhomogeneities in the magnetic field of the 37-inch defocused more menacingly than the mass increase and indicated considerable room for maneuver in the 60-inch. And if shimming were to fail, other possibilities existed, for example, placing wire mesh across the mouths of the dees so as to obtain by electrical force the focusing that would be lost by adjustment of the magnetic field to secure resonance. "We have learned from repeated experience that there are many ways of skinning a cat."[6]

5. Bethe and Rose, *PR, 52* (15 Dec 1937), 1254–5, letter of 24 Nov.

6. Lawrence to Bethe, 4 Dec 1937 (HAB, 3), and, in a similar vein, to cyclotroneers J.R. Dunning, 11 Dec 1937 (6/19), and J. Cork, 13 Dec 1937 (5/6).

This response was not bluff. For a year or so Robert Wilson had been poking around inside the cyclotron tank, determining empirically the strength of the vertical component of the electric field near the dee mouths and of the radial component of the fringing magnetic field, which drives ions toward the meridian plane. The investigation of the circulating current, which eventuated in the internal target for isotope production, was part of his study. Wilson painstakingly worked out the trajectories of ions beginning their courses at any distance from the median plane and reaching the center of the gap in phase with the maximum field there. His numerical integrations showed that from about 10 cm out, where the focusing effect of the electric field becomes negligible (it decreases with the particles' energy), the magnetic field swiftly reduced the vertical amplitude of the beam from a spread of some 5 cm near the ion source to about 1 cm at the exit slit. Probe measurements confirmed Wilson's semi-empirical deduction of beam width as a function of orbital radius. He therefore felt confident in recommending that the aperture of the dees also be made to decrease with radius, thereby reducing their capacitance and easing the performance requirements for power oscillators to accelerate protons.[7]

Wilson presented his results in a seminar about the time that Bethe and Rose's letter of November 24 was circulating in Berkeley. The circumstances inspired McMillan to estimate the defocusing effect of relativity. It was he who found that for the 37-inch defocusing arising from inhomogeneities in the magnetic field exceeded that from relativity by a factor of four. Also, McMillan calculated from experience at Berkeley that the beam could fall out of phase with the electric field by more than 60° and still get through the cyclotron; and on this basis he calculated that the maximum energy of deuterons achievable without altering basic cyclotron design was perhaps $20\sqrt{2}$ MeV with 100 kV on the dees. With Lawrence's grids, he thought, any amount of electrostatic focusing could be attained.[8] On this last point McMillan had a short and victorious duel by mail with Bethe, who had thrust forward the opinion that "no change of the shape of the dees, no

7. Wilson, *PR, 53* (1 Mar 1938), 408–20, rec'd 27 Dec 1937.
8. McMillan to Bethe, 14 Dec 1937 and 14 Jan 1938 (HAB, 3).

insertion of grids at the dee openings, etc., can have any appreciable effect on the electric focusing." McMillan parried that Bethe had mistakenly assimilated a dee with grid to an open dee of smaller aperture; Bethe concurred, and allowed the possibility of doubling the energy limit.[9]

Meanwhile Rose, who had also been working for a long time on cyclotron focusing, sweated to get his theory ready for the press. Whereas Wilson and McMillan relied on their experience with a single machine, Rose began with general equations of motion in changing electric and magnetic fields and deduced, by clever substitutions, a differential equation for the excursion of an ion from the median plane as a function of the phase of the radio frequency voltage it met as it crossed between the dees. His treatment of the general case—which "had been considered much too complicated for solution by many"—agreed with the conclusions about electric and magnetic focusing reached in Berkeley.[10] Rose could do more: from his differential equation he could deduce the maximum energy obtainable without defocusing the beam when the gradient of the magnetic field compensates for relativity. He ended more generous than he and Bethe had begun. They allowed, in a note added to their initial announcement on December 4, that a field giving an angular velocity too large for resonance at the start and too small at the finish could deliver deuterons of 17 MeV with V = 50 kV, a number Rose raised to 21.1 MeV. These numbers would be multiplied by $\sqrt{2}$ if 100 kV were placed across the dees. Rose thought that no greater potential could be reached without severe difficulty and that grids would not have the power Lawrence supposed. "It seems very possible that the energies mentioned [21.1 MeV deuterons] represent the natural upper limit for the cyclotron with the given dee voltage of 50 kV, at least without very radical changes in design."[11]

9. McMillan to Bethe, 3 Jan 1938, and reply, 7 Jan (McMillan P).

10. Bethe to Condon, 17 Feb 1938, and to C.A. Corcoran, 3 Apr 1937 (HAB, 3).

11. Bethe and Rose, *PR, 52* (15 Dec 1937), 1255; Rose, *PR, 53* (1 Mar 1938), 392–408, esp. 404–7, and ibid. (15 Apr 1938), 675, text of 25–6 Feb; Cooksey to M. White, 21 Jan 1938 (18/21), supposing that "Bethe would admit the possibility of small deuteron currents at 30 or 40 MeV."

As Bethe conceded to McMillan, after explaining that he and Rose had published hurriedly because "we considered the *existence* of a relativistic limit so important that we thought we should communicate it to cyclotronists as quickly as possible, without endeavoring to give accurate figures," "it makes all the difference in the world whether the limit is 8 MV or 20."[12] Or 100. The question came before that high tribunal of science, *Time*, whose investigative science editor, Walter Stackley, drew from Lawrence a firm rejection of the 20 MeV limit. There was new work under way at Berkeley, Lawrence said, "which may increase the energy maxima materially." "We believe that there are experimental possibilities of improving focusing conditions which remove the limitation on energy to some unknown point."[13] And, just at this point, L.H. Thomas, known to physicists as the discoverer of a relativistic effect important in atomic theory (the "Thomas precession"), described a novel way to achieve both focusing and resonance by a magnetic field that had notably different strengths in several pie-shaped sections into which he divided the median plane. Thomas's ingenious suggestion received some attention at Berkeley and more at Stanford, where Oppenheimer's former student Leonard Schiff continued the calculations. Although the scheme, which is difficult to put into practice, was not exploited at the time, it gave ample evidence that nature did allow for several methods of cat-skinning.[14]

The opinion of the experienced cyclotroneer about Bethe's limit is nicely reflected in notes by the Rockefeller Foundation's Tisdale. After recording that "Joliot's cyclotron, by a lucky chance, is designed just to the limit of the theoretical voltage," which would have been at once a triumph and an end to the Foundation's investments in cyclotrons, Tisdale reported that Paxton would have none of it. "P[axton] considers that the mass effect is not very important."[15] The cyclotroneer did not doubt that so refined a thing as a relativity effect could be beaten by brute force.

12. Bethe to McMillan, ca. 20 Nov 1937, draft (HAB, 3).
13. Resp., Stackley to Bethe, 16 Feb 1938 (HAB, 3), and Lawrence to Paxton, 9 May 1938 (14/18).
14. Thomas, *PR, 54* (1938), 580–98; Schiff, ibid., 1114–5; Van Voorhis to Cooksey, 20 Nov 1938 (17/43). Thomas cyclotrons were built after the war.
15. Tisdale, notes on a lunch with Joliot et al., 14 Jan 1938 (RF).

Thornton: "Difficulties in reaching high voltages seem to me quite real....But of course there are a number of ways [by] which one may get around [Bethe's] objections." Wells: "It can probably be compensated by applied magnetic inhomogeneities of the field or by properly chosen electrostatic fields." Compton: "[It] can be passed (theoretically) by altering pole pieces and electrostatic focusing. Thus no limit is now assignable." Oliphant: "I am not deterred by papers which have been written on the maximum energy obtainable from a cyclotron."[16] Gentner looked to Thomas's method. Cockcroft preferred to follow Lawrence, who thought azimuthally changing magnetic fields impractical and no longer favored fitting the dees with wires. Instead, he bruited a solution in the style of the Old West: put a million or two million volts on the dees and drive the beam home before it knows that it has been defocused.[17]

Skinning Fat Cats

Lawrence was planning to build far beyond the Bethe-Rose limit even before the 60-inch machine, which itself crossed the suppositious threshold, came on line. It was not relativity, but money, he said in a radio broadcast in the spring of 1939, that stood in his way. "Right now we are considering the possible financial difficulties of constructing a cyclotron to weigh 2,000 tons and to produce 100 million volt particles....It would require more than half a million dollars."[18] Both the size and the price were to grow during the next year with the help of the University of Texas and the Nobel Foundation, and with the encouragement of big-thinking colleagues. "I hope your new apparatus is really big," Chadwick wrote, with 60 or 70 MeV in mind. "Best wishes for the beam to end all beams. The best is none too good for the Berkeley boys," wrote I.I. Rabi of Columbia, who would later try

16. Resp., Thornton to Lawrence, 13 Dec 1937 (13/5); Wells, *Jl. appl. phys., 9* (Nov 1938), 681; A.H. Compton, notes after a visit to the Laboratory, 16 Sep 1938 (4/10); Oliphant to Lawrence, 19 Jul 1938 (14/6).

17. Gentner, *Ergebn., 19* (1940), 150; Cockcroft, *Jl. sci. instr., 16* (1939), 40; Lawrence, foreword to Mann, *The cyclotron,* [vii–viii], 5 Oct 1939.

18. Lawrence on CBS broadcast, "Adventures in science," 15 Apr 1939, repeated in similar terms on NBC, "World's largest atom smasher," 10 Sep 1939 ("University Explorer" no. 457), both in (14/15).

to snatch the best from Berkeley in the interests of East Coast physics.[19]

Lawrence faced difficulties beyond the relativistic and the financial. For one, there was no uncontested space for a 2,000-ton cyclotron on the Campus. An engineering annex had been needed to house the 27-inch; a special building had been erected for the 60-inch; real estate as large as the Campus would be reserved for the new machine. Then there was a taint of overreaching, of imprudent haste, of gluttony, in the plan. "In some quarters it might be considered no less than shocking that we should be looking towards a larger cyclotron almost before the 60 inch is in operation."[20] And finally, there was the disagreeable fact that no major discovery had yet been made in any cyclotron laboratory. As Arthur Compton and his colleague A.J. Dempster pointed out to the Rockefeller Foundation, cosmic-ray physicists had made several of the most spectacular discoveries in physics during the 1930s, in particular the positron and the mesotron, and cosmic rays come gratis.[21]

To this objection Lawrence replied with a claim about the might-have-been and a statement of the what-should-be. The claim: cyclotron physicists had missed the discoveries through a compulsion to perfect their machines; in due course they would have found what others detected earlier with more primitive means. The statement: a discovery has little value unless it can be turned to practical use. "It means a great deal more to civilization, let us say, to find a new radiation or a new substance that will cure disease than it would to discover a super nova." On this reasoning, Joliot and Curie's find would have been barren had it not been for the Berkeley cyclotron. And, Q.E.D., "the discovery of mesotrons in cosmic rays will be of little value in the course of time unless there is developed a way of producing them, and learning of their manifold properties—ultimately to be put in the service of mankind."[22]

19. Chadwick to Lawrence, 16 Apr 1938 (3/34), and Rabi to Lawrence, 17 June 1939 (14/4).

20. Lawrence to Weaver, 16 Sep 1939 and 13 Jan 1940 (15/30).

21. Compton's attitude as inferred from Cooksey to Poillon, 25 Jul 1940 (13/29A), and Compton to Weaver, 29 Jan 1940, enclosing letter of the same date by A.J. Dempster (RF, 1.1/205).

22. Lawrence to Weaver, 21 Feb 1940, answering Weaver's letter of 13 Feb

The invocation of mesotrons and the hint that the projected cyclotron might make them came to the fore only after the University of Texas had set going a mechanism that would provide more money than Lawrence thought possible. He and Sproul turned, indeed spun, to one prospective donor or influential intermediary after another. For a time Walcott, the former senator with the leukemic son and a trustee of the Carnegie Institution, looked like an especially valuable contact. Walcott had contacts in big steel; funding would be easy, Lawrence said, if the steel were donated. The son, a physician, came to Berkeley to work with, and receive radiophosphorus from, John Lawrence. A very strong affection developed between the Lawrences and Cooksey and the Walcotts; but it did not bring steel for nothing or save Walcott's son.[23] Other possibilities: Lewis Strauss, proposed by Oppenheimer and approached through Coolidge; Spencer Penrose, the dying benefactor of the Penrose Foundation, approached through Frank Jewett of Bell Labs; Edsel Ford and General Electric, approached through Dave Morris.[24]

To his own considerable surprise, Weaver turned out to be the route to the pot of gold. We know his attitude on Rockefeller Foundation support of research cyclotrons. During the negotiations over Paxton and Laslett's foreign missions, he had formed the notion that Lawrence was "a happy-go-lucky sort of individual," a good scientist, but indecisive and not overly solicitous about the inconvenience his changes of plan caused others.[25] And in his dealings with Lawrence in the spring of 1939, Weaver had not been pleased by the escalation of the Laboratory's request between discussion and submission.[26] It is doubtful that he

(15/30), which mentioned the neutron, positron, artificial radioactivity, mesotron, and fission as matters missed by cyclotroneers.

23. Lawrence to J.H. Lawrence, 11 Sep 1939, and to F. Walcott, 14 Oct and 7 Nov 1939; letters to J.H. Lawrence from A. Walcott, 10 Oct 1939 and 29 May 1940, all in (18/6).

24. Lawrence to Coolidge, 6 Oct 1939, to Strauss, 14 Oct 1939, to Jewett, 7 Nov 1939 (14/23); to Morris, 8 Nov 1939 (13/13); Coolidge to Strauss, 10 Oct 1939, and Jewett to Lawrence, 27 Oct 1939 (19/23); Morris to E. Ford, 14 Nov 1939 (13/13), and to Fosdick, same date (RF, 1.1/205).

25. Weaver to Tisdale, 17 May 1937 (RF, 5000D).

26. Lawrence to Weaver, 11 Mar 1939, and Weaver's reply, 16 Mar 1939 (15/29).

received with much enthusiasm the news that Lawrence was coming East to look for donors of the $750,000 he reckoned as the amount yet to be raised for an instrument of 1,500 to 2,000 tons to crack the region above 100 MeV. The $750,000 arose by subtraction of the $250,000 Sproul promised to raise from the million that Lawrence, who liked round numbers, thought necessary. On the advice of Poillon, who judged that a request for a cool million would put off donors, Lawrence set the total at a lukewarm $900,000 and the balance at $650,000. This was the amount Morris requested of Edsel Ford, with an overheated inducement: "As this is an instrument which will enlarge the frontier of science almost beyond belief it should be something epoch-making and will link the names of those connected with it alongside of Newton and Einstein."[27]

The justification for this instrument, as outlined to Sproul early in October, when it had grown to 2,000 tons, had no more substance than the rationale Morris offered Ford. There was a more definite and practical reason, however. The success of the cyclotron had inspired competitors, including two clones of the 60-inch; if the Laboratory wished to stay ahead, it must cross the new frontier, where, as cosmic ray studies indicated, "strikingly new and important things" were to be found. Sproul wanted to keep Berkeley ahead. He promised (so Lawrence relayed to Weaver) not only to raise part of the capital outlay but also to finance the operation of the he-man machine. Still, Weaver did not expect that his trustees would take much interest in the proposal, or in any costly esoteric project, in the state of the world in the fall of 1939. Here Lawrence guessed more accurately than Weaver. "I personally am banking on the trustees' taking the view that it is in just such times as these that the Rockefeller Foundation should undertake such important projects, thereby demonstrating a stability and confidence in the progress of civilization."[28]

Lawrence opened negotiations with the Rockefeller Foundation in New York on October 27. Weaver accepted the desirability of

27. Lawrence to Weaver, 16 Sep 1939 (15/29); Poillon to Lawrence, 18 Nov 1939 (15/18), and Morris to Ford, 14 Nov 1939 (13/13).
28. Lawrence to Sproul, 10 Oct 1939, draft (16/42); to Weaver, 14 Oct 1939 (15/29); to Walcott, 7 Nov 1939 (18/6).

a cyclotron that endowed particles with 100 or 200 million electron volts; he encouraged Lawrence to think big, to beware of "initial presentation [of the project] on too small a scale;" and he insisted that the plan make clear that the cyclotron would be a national, even international, facility, "located at the University of California...[but] built for all science." Weaver assimilated Lawrence's project to what he called the Foundation's "national laboratory," the 200-inch telescope and its facilities abuilding on Mount Palomar; and he estimated its costs correspondingly, at $1.5 million including operating expenses for a decade.[29]

The announcement on November 9 that Lawrence had received the Nobel prize for physics in 1939 (about which we shall say much more in a moment) strengthened Weaver's commitment to the Palomar of the vanishing small. With the ardor that the higher administration of the Foundation had once censored as excessive, he celebrated Lawrence's prize in a confidential bulletin sent to the Rockefeller trustees and invited the prize winner to put forth a detailed plan for presentation to the next trustees' meeting, in April 1940, a plan so complete that it would kill any fear that similar or competing requests would arise.[30] "This is the sort of thing which should be done superbly—or not at all. And done superbly it is of compelling attractiveness." Lawrence responded that it should be superb, and raised the weight of the magnet to 3,000 tons, or perhaps (indecisively or flexibly) a little more, and he promised to have full plans for a 180-inch and a 205-inch cyclotron ready for discussion with Weaver in Berkeley in January. Lawrence naturally favored the larger version, as offering a chance of delivering 400 MeV alpha particles.[31]

When Weaver arrived on January 7, he was hit with a plan for a magnet weighing over 4,000 tons. He had come with Lawrence's estimate of $750,000 in mind and the notion that it,

29. Weaver, "Diary," 27 Oct 1939 (RF, 1.1/205).

30. Jewett to Lawrence, 13 Nov and 26 Dec 1939 (14/33); RF, "Trustees' confidential bulletin," Dec 1939 (RF, 1.1/205); Kohler in Reingold, 267–9. A favorable internal review of Weaver's policies in 1938 had strengthened his position.

31. Weaver to Lawrence, 14 Nov 1939, and replies, 22 and 30 Nov 1939 (15/29); Lawrence to F.B. Jewett, 12 Jan 1940 (13/25). The 3,000-ton magnet is also mentioned in Lawrence to Spears, 14 Nov 1939 (16/41).

and perhaps as much again in operating expenses and auxiliary equipment, could be raised in equal shares by the Rockefeller Foundation, the University of California, and industry. "The size to which I found the project had grown, when I arrived at Berkeley [in January]," Weaver sighed, "carried me so far beyond any figures which I had ever discussed." Lawrence wanted $1.5 million of an estimated $2 million from the Foundation. In Lawrence's upbeat report to Poillon, Weaver did not "seem to be unduly distressed...[and] went away far more eager to consummate the project than when he came." Weaver had fallen under the spell of the California sunshine man and of the 60-inch cyclotron, then treating cancer patients and, on demand, charring plywood with a directed energy beam of deuterons released into the air. Weaver suggested to Sproul "the bare possibility" that the Foundation might give as much as a million dollars.[32] Back in the cold East, Weaver discovered that the Foundation's president, Raymond Fosdick, who had been enthusiastic about the project in December, had lost his conviction, and doubted that the trustees would give even $500,000. The January plan was dead. Or so Weaver wrote Sproul, whose recent appointment as a Rockefeller trustee closed the funding loop. "It does not seem to me a desperately serious matter if this project is delayed somewhat. Professor Lawrence is fortunately still young, there is a great deal of rich experience which can be gained with the 60-inch cyclotron, and there is a negligible danger that anyone else will run away with the ball."[33] This was to ignore Texas, still out in left field awaiting its fly, and Lawrence's flexibility.

During January and early February, friends of the Laboratory brought pressure on Fosdick and their acquaintances among the Rockefeller trustees. Among the friends were old supporters like Poillon, who hoped, perhaps, that something might be realized at last from the Research Corporation's cyclotron patents; Karl Compton, a Foundation trustee; former ambassador Morris, of the

32. Weaver, "Diary," 7–9 Jan 1940 (RF, 1.1/205).
33. Lawrence to Weaver, 13 Jan 1940, and Weaver to Sproul and to Lawrence, 23 Jan 1940 (15/30); Weaver to Fosdick, 25 Jan 1940 (RF); Lawrence to Poillon, 31 Jan 1940 (15/18). The 4,000-ton magnet is also mentioned in M. White to Cooksey, 8 Jan, Salisbury to Weaver, 12 Jan 1940 (both in 18/21), and Lawrence, "Report to the Research Corporation...[for] 1939," 2 (15/18).

Macy Foundation; and Alfred L. Loomis, who spent the riches he amassed as an investment banker on a private laboratory and the encouragement of physical research. An expert instrument designer himself, Loomis was much taken with the Laboratory on his first visit there late in 1939; his wide influence among officials of corporations and foundations made his support of the project, which he pledged in December, a most valuable acquisition.[34]

Weaver also made a play among the trustees. He pointed out to Karl Compton the happy parallel between Lawrence's project and the 200-inch telescope. "Such a cyclotron would, I think, be correctly and generally viewed as the definitive instrument for the investigation of the nucleus—the infinitesimally small—just as the 200" telescope is viewed as the definitive instrument for the investigation of the universe—the infinitely great." Compton visited the Laboratory and returned "radiant over all the wonderful things he saw in Berkeley" and convinced that the new machine "should be built adequately large to reach the range of energy above 160 million volts in order to attack the problem of mesotron forces."[35] After a visit from Weaver, another Foundation trustee, George Whipple, a frequent recipient of hot iron from Berkeley, declared himself keen on the project, and certain that funds would be forthcoming from somewhere; he spoke "with an enthusiasm which is very unusual for him concerning Lawrence and his group, saying that the way they do things out there is 'just right.' "[36]

Weaver also collected professional evaluations. He asked Bohr, Bush, both Comptons, W.D. Coolidge, Jewett, Joliot, and Oliphant whether "expert opinion of the world of science is reasonably unanimous in viewing [the giant cyclotron] as one of the most interesting, the most potentially important, and the most promising projects in the whole present field of natural science."

34. Lawrence to Poillon, 31 Jan and 12 Feb 1940 (15/18); Lawrence to Mrs. Sproul, 4 Dec 1939 (37/24), and to Walcott, 4 Apr 1940 (18/6), re Loomis; Alvarez, *PT, 36:1* (1983), in Weart and Phillips, *History*, 204.

35. Weaver to K.T. Compton, 25 Jan 1940, and to Lawrence, 6 Mar 1940 (15/30); R.D. Evans (MIT) to Lawrence, 10 Apr 1940, re Compton's radiance (7/8); Lawrence to Poillon, 12 Feb 1940, reporting Compton's proposal (15/18).

36. Weaver, "Report of visit to U. of Rochester," 20 Feb 1940 (RF 1.1/205), quote; DuBridge to Lawrence, 19 Mar 1940 (6/17).

The replies might have made Lawrence blush. Bohr: "It would be greeted with utmost pleasure by all physicists." Bush: "This opportunity is the most interesting, the most potentially important, the most promising project of large magnitude in the whole field of natural science." A.H. Compton: "If anyone can make a success of a 2000-ton cyclotron, Lawrence can....On the whole, the investment would be a nice one." K.T. Compton: "I would definitely place it in the number one position by a large margin." Coolidge: "Now is the time to do it while the exceptional combination of enthusiasm, intelligence, experience and skill of Dr. Lawrence and his group are available." Jewett: "[Its] value...is of course beyond question." Joliot: "The realization of such an apparatus is likely to bring important results....Lawrence is, without any doubt, the most qualified man to undertake its construction." Oliphant: "It is essential that the construction of the cyclotron should be carried to the limit by Professor Lawrence."[37]

All this lobbying cancelled Fosdick's timidity. At a meeting in mid February, which Poillon attended, "the Rockefeller [administrative] group distinctly favor[ed] the larger [cyclotron] because of the certainty of its performance within and above the 160,000,000-volt range."[38] The 160 MeV referred to one of several designs that Lawrence had supplied when he realized there was no chance of $1.5 million from the Foundation. The 205-inch, perhaps so chosen to beat Palomar, fell to 184 inches, the largest size of commercially available steel plate. And the 150-inch stayed in the running. On February 20, 1940, Lawrence provided Weaver with four options: (a) 184 inches, $1.5 million, handsomely housed and fully equipped, operating at 2,500 kW to kick ions to 200 MeV before relativity could take its toll, the "conservatively ideal in exploiting the limit of the cyclotron method;" (b) 184 inches, $1 million, cheaply housed and partially equipped, operating at 700 kW and perhaps yielding 100 MeV deuterons, easily stepped up to (a); (c) 184 inches, $875,000, a skeleton, deficient in copper and steel, producing 75 MeV

37. Weaver's letter requesting evaluations is dated 25 Jan 1940; replies from Coolidge, 26 Jan, A.H. Compton, 29 Jan, K.T. Compton, 29 Jan, Jewett, 16 Feb, Bush, 17 Feb, Oliphant, 19 Feb, and Bohr, 28 Feb (RF, 1.1/205).

38. Poillon to Lawrence, 16 Feb 1940 (15/18); Weaver to Lawrence, 18 Feb 1940 (15/30).

deuterons, capable of upgrading to (a); and (d) 150 inches, $750,000, able to reach 100 MeV with an oscillator more powerful than (c)'s, but not easily refashioned into (a). Lawrence took his stand between the most desirable and the least expensive: "It seems to me that attention should be concentrated on projects 'b' or 'c', of course very much preferably 'b'." The 160 MeV probably referred to option (b) and alpha particles, since, with his ear for audience, Lawrence had advised Weaver to couch his statements in terms of alpha energies, which are twice those of deuterons for the same cyclotron parameters and somewhat less afflicted by relativistic mass increase.[39]

The fundamental alternative—184 inches versus 150 inches—represented a hedged bet. On the one hand, option (a) and its upgradable lower forms would quite possibly be able to materialize mesotrons. DuBridge and Karl Compton emphasized the desirability of building the machine that, as Compton put it, allowed a "reasonable expectation of producing mesotrons." The reasonableness depended on estimates of the mesotron's mass. Karl Compton thought 160 MeV might do; DuBridge, "energies of the order of 100 million electron volts."[40] Oppenheimer and Fermi, who happened to be in Berkeley, put the mass of the mesotron between 70 and 120 MeV, gave it a 90 percent chance of falling under 100 MeV, and advised that the higher the bombarding energy—the closer to option (a), "which exploits the full practical potentialities of the cyclotron method"—the greater the chance of making mesotrons in the cyclotron.[41] Lawrence rated the materialization of mesotrons "the most fundamental experimental problem that one can formulate at the present time," and thought he could succeed with 150 MeV. But he did not promise. Although mesotrons might fail to materialize, the energy region above 100 MeV was nonetheless certain to be rich: "we cannot help but entertain the possibility of nuclear chain reactions by starting

39. Lawrence to Weaver, 13 Jan, 20 Feb, and 12 Mar 1940 (15/30).

40. DuBridge to Weaver, 27 Jan 1940, and K. Compton to Weaver, 29 Jan 1940 (RF, 1.1/205).

41. Lawrence to Weaver, 20 and 21 Feb 1940 (15/30); Lawrence, "Case for the 184" cyclotron" [1940], 2–4 (26/1), quote. Wheeler and Ladenburg, *PR, 60* (1 Dec 1941), 754–61, give 90 MeV as the best value of the mesotron mass derivable from existing experiments.

them off with sufficiently energetic particles and that maybe a hundred million volt particles will do the trick....Should this prove to be true, we will have a discovery of great immediate practical importance. On the one hand, we will have a practical philosopher's stone transmuting elements on a large scale; and, as a corollary thereto, we will have tapped, on a practical scale, a vast store of nuclear energy."[42]

Despite these formidable arguments, Lawrence retained option (d). He thus deprived the trustees of the Rockefeller Foundation of the option of arguing that if they could not put up enough for mesotrons, they should put up nothing at all. Lawrence had opened his mind on the matter to Weaver during a telephone conversation at the end of January: "The point is that it is far more important to get into the new territory now. We would rather build a, say, haywire outfit and actually have been up there than to take a chance on going up later and maybe not getting there at all." Occasionally he thought to go for the 150-inch and not risk its loss by groping for mesotrons, and he so advised Weaver by telegram. As he explained his position to Poillon, who had heard similar arguments from him before, the most important thing was "to accomplish the original and primary objective of attacking the energy range in the atom above one hundred million volts....We will be in entirely new territory....It is distinctly of secondary importance that we get a little further in by going 50% higher."[43] As in the old days, Lawrence set goals expressible not in terms of progress in physics, but in terms of increase in decimals.

Weaver decided to take two options before the trustees in April: $750,000 for the 150-inch; or $1 million toward the 184-inch, on condition that the University raise at least another $250,000 for it. In either case, the University would have to provide operating costs for a decade.[44] Sproul had already obtained authorization from the Board of Regents for the $250,000 he had promised for construction and either $50,000 or $85,000 a year for maintenance of the smaller or larger machine respectively. Sproul

42. Lawrence, "Case for the 184" cyclotron" (26/1).
43. Log, 29 Jan 1940 (15/30); Lawrence to Weaver, 31 Jan 1940, and to Poillon, 31 Jan and 12 Feb 1940 (15/18).
44. Weaver to Lawrence, 27 Feb 1940 (15/30).

regarded the commitment of such sums by the regents to such a purpose as "pretty overwhelming."[45] It remained only to await the decision of the trustees. They reviewed the opinions of the physicist and engineering consultants from Bush to Oliphant. They had a lesson in nuclear physics and its applications from Karl Compton, who had been coached in Berkeley (plate 10.1), and from Weaver, who drew on inspirational photographs of the Laboratory and its machines supplied by Cooksey. And they heard a heady peroration from their program officer. Weaver compared Lawrence's Laboratory with Bohr's institute; he recommended that the Foundation support the 184-inch project, as an "opportunity to make discontinuous change in [the] rate of progress of science;" and he extolled the "shrewd intelligence, imagination and insight, unselfishness, inspiration for young men, [and the] charm" of the man who would carry the project through.[46] And there would be no trouble carrying it through, as the trustees learned from Jewett, now speaking as head of the National Academy of Sciences: "a matter of engineering calculation [he said] and not one of uncertain speculation."[47]

As a further aid to their deliberations, the Rockefeller trustees felt heavy pressure transmitted through their officers from Lawrence's agents and admirers Poillon and Morris. They were not content with the prospect of a million dollars. "I am making life miserable for Warren Weaver and Raymond Fosdick," Morris had written Lawrence at the end of February. "Confidentially, we are all striving for the million dollar cyclotron under column B, and Howard [Poillon] and I are trying to jack up this limit 12 and one half percent. I really feel that all four of us are working for

45. Lawrence to Poillon, 12 Feb 1940 (15/18); Sproul to Lawrence, 9 Mar 1940 (15/30).

46. Weaver, "Cyclotron notes" [Apr] 1940 (RF, 1.1/205). Weaver had lobbied the trustees individually, e.g., letters to Walter Stewart, 8 Mar, and to Herbert Gasser, 21 Mar 1940 (RF, 1.1/205).

47. Jewett to Lawrence, 31 Oct 1939 (14/23), an open letter of recommendation. A.H. Compton, who had not been enthusiastic about the project because of its ascendency over the natural, or cosmic-ray, approach to meson physics, lined up under pressure from Bush, who had become president of the Carnegie Institution, hence an investor in both cosmic-ray research and cyclotrons. Kevles, *Physicists*, 285; Compton to Weaver, 14 Mar 1940 (RF, 1.1/205); Cooksey to Poillon, 25 Jul 1940 (13/29A).

you heart and soul."[48] The quartet missed its pitch by 2.5 percent.

At noon on April 3, 1940, Weaver called Lawrence to announce that the trustees had come down 15 percent above the expected maximum.[49]

> WW: Our trustees voted $1,150,000...
>
> EL: Really, Warren, $1,150,000...
>
> WW: And with the $250,000 that makes $1,400,000, which you see is the full original budget.
>
> EL: The full original budget....Its hard to tell you how I feel. This is the most wonderful thing that has ever happened....I'm coming to New York, and it will give me a chance really to explain my feelings to you. This is the most wonderful thing that one can think of in the world.

Lawrence had the feeling he was "walking on air." So did his successful agents Morris and Poillon. "You can scarcely overestimate the joyous feeling that resulted from the news," Poillon said, and indeed he had earned the right many times over to share in this tribute to the machine and Laboratory he had backed from the beginning. The munificent grant represented many things: dollars, to be sure, but also the affection, respect, and confidence in which Lawrence's fellow physicists and prominent men of business held him. As Dave Morris wrote: "This really great triumph should mean much to you in more ways than one. There was no disagreement anywhere along the whole line. Great and small, technical and lay, they all backed the PLAN and YOU. Do get full emotional satisfaction from such rare unanimity: you deserve it."[50] According to the formal agreement, the Rockefeller Foundation and the University would put up money as Lawrence needed it in the proportion of 23:5 until June 30, 1944, when, barring "unforeseen difficulties," the machine was to have been completed.[51]

48. Barton to Weaver, 12 Apr 1940 (15/30); Morris to Lawrence, 27 Feb 1940 (13/13).

49. Notes of call (15/30).

50. Quotes from, resp., Lawrence to Poillon, 4 Apr 1940, and Poillon to Lawrence, 3 Apr 1940 (15/18); Morris to Lawrence, 4 Apr 1940 (13/13).

51. Weaver to Sproul, 8 Apr 1940, and reply, 30 Apr 1940 (15/30).

The University immediately obtained a fifth of its commitment of $250,000 from the Research Corporation. Lawrence asked the Markle Foundation for the balance, toward which it gave $50,000 at Weaver's urging, and tried to get Westinghouse to underbid General Electric's generous offer to make the 184-inch's power supply at cost, which Westinghouse declined to do. He spent two weeks touring Wall Street with Loomis, asking for help in knocking down the price of steel and other material and equipment. Despite the pressure of war orders, which left little incentive for price concessions, Loomis and Lawrence did very well on Wall Street. The balance of the University's share of the capital costs eventually came from the federal government, in consequence of those unforeseen but foreseeable difficulties that prevented completion of the machine before June 1944. And the war also made good the shortfall in Lawrence's ideas of eluding relativity; the machine when finished in 1946 operated on a principle invented by McMillan in 1945, perhaps as a result of his wartime experience with radar.[52]

In a public explanation of the gift, and before the unforeseen difficulties interrupted the building of the 184-inch cyclotron, Fosdick wrote: "With so much creative human talent employed in devising increasingly powerful engines of destruction it is at least some comfort to know that today in the United States work is proceeding on two of the mightiest instruments the world has ever seen for the peaceful exploration of the Universe." The 200-inch telescope and the 184-inch cyclotron would respectively open up the infinitely great and the infinitely small, alleviating "the insatiable curiosity which is the mark of civilized man." To be sure, the cyclotron would do something practical: it would produce specialized radioactive isotopes, perhaps beams of therapeutic value, perhaps even clues to the exploitation of atomic energy. But above all, "like the 200-inch telescope, it is a mighty symbol, a token of man's hunger for knowledge, an emblem of the undiscourageable search for truth which is the noblest expression

52. Lawrence to Archie Wood, Markle Foundation, 18 May 1940, and Kenneth Priestley to Cooksey, 7 Apr 1944 (re a gift for $25,000), both in (12/35); Weaver, "Diary," 22–24 Apr, 28 May 1940 (RF, 1.1/205); Lawrence to Condon, 13 Apr 1940 (4/15), to M.W. Smith, Westinghouse, 29 Jul, and reply, 31 Jul 1940 (18/19); Lawrence to Sproul, 15 May 1941 (46/9), re travels with Loomis in May 1940.

of the human spirit."[53] This inspired gloss, which was not entirely disingenuous, marks the end of the era of private support for Lawrence's Laboratory.

Lucky Dog

The Rockefeller Foundation had heard three substantive objections to the he-man cyclotron: that relativity would cripple it; that the muse of discovery did not attend cyclotroneers; and that the Foundation would further its program in the applications of physics to biology by favoring the production of natural, rather than artificial, isotopes of the elements of living things. All these objections lost much of their force while the Rockefeller trustees pondered. The second two both collapsed with the detection of H^3 and C^{14}, two solid discoveries that promised to give biologists tracers for the most important ingredients in organic molecules. Lawrence made much of both these products of his Laboratory. He gave the discovery of H^3 pride of place in his report to the Research Corporation for 1939. "Radioactive hydrogen," he wrote, in a gloss he doubtless expected Poillon to pass on to the Rockefeller trustees, "opens up a tremendously wide and fruitful field of investigation in all biology and chemistry."[54]

Toward the end of the dickering, in late February 1940, Kamen and Ruben called on Lawrence, then in bed with one of his colds, to present their first, flimsy evidence of the existence of C^{14}. "He jumped out of bed, heedless of his cold, danced around the room, and gleefully congratulated us." His ecstasy turned to outrage on learning that the report of the discovery in the *Physical Review* bore the names Ruben and Kamen. "He turned on me [Kamen recalled] with ill-concealed anger and demanded to know why my name and the institutional credits placed me and the Rad Lab in a position secondary to the Chemistry Department." The explanation, that Ruben thought he needed all the credit he could amass to gain tenure in a Chemistry Department not free from anti-Semitism, did not placate Lawrence, who thought the Laboratory needed all the credit it could garner to win the Rockefeller

53. Fosdick, *Review for 1940*, 36–41.
54. Lawrence, "Report...for 1939," 1940 (15/18).

sweepstakes. "The best of all," Lawrence wrote Weaver, in an itemization of favorable signs, "is the discovery of carbon 14 by Kamen and Ruben [!] here....All cyclotrons now in existence could be usefully employed in making radio-carbon only."[55]

The third objection—that Lawrence did not know what he was doing, that the maker of the cyclotron knew too little physics, that he had overstepped the limit set by Einstein and nature—was answered emphatically by the certifiers of the world's greatest physicists. On November 9, 1939, Lawrence received the telegram from Stockholm that responded to the recommendations of distinguished physicists throughout the world. The Swedish Academy of Sciences had decided to award him the Nobel prize in physics, "for your having invented and developed the cyclotron and especially for the results attained by means of this device in the production of artificial radioactive elements." He thus fulfilled almost to the volt the prophecy made by Jesse Beams eight years before: "With 10^{-8} amps at 900,000 volts you already have a powerful tool and Boy with 20,000,000 you'll get the Nobel prize."[56]

Lawrence was first proposed formally in 1938, by an American, a Japanese, and an Indian (who proposed a division with Fermi). The prize committee could not decide whether the cyclotron was prizeworthy and chose Fermi for his discovery of radioactive substances and the method of activation by slow neutrons.[57] In 1939 the Compton brothers organized a campaign for Lawrence among former American prizewinners (all prizewinners have a permanent right of nomination); two of them, Clinton Davisson and Irving Langmuir, did propose Lawrence; Carl Anderson preferred Stern; Millikan did not care to exercise his franchise. The usual rationale for the nomination was, as Langmuir put it, Lawrence's "construction of the cyclotron and his studies of radioactivity that have been made possible by its use." Another ground, which

55. Kamen, *Radiant science*, 132, 134; Lawrence to Weaver, 2 Mar 1940 (15/30).

56. Swedish Academy of Sciences to Lawrence, 9 Nov 1939, and Henning Pleijel to Lawrence, 16 Nov 1939 (37/20); Beams to Lawrence, 29 Jul 1931 (2/26).

57. Nobelkommittén för fysik, report, 13 Sep 1938 (Nobel). The nominators: Henry A. Erikson (University of Minnesota), Hantaro Nagaoka, and C.V. Raman. C.D. Anderson named Lawrence as his second choice.

perhaps agrees better with the facts, was expressed by Livingston in a letter forwarded to the Nobel committee by Davisson. Livingston observed that many people before Lawrence had had the idea of the cyclotron. "However the idea developed," Livingston continued, "Lawrence was the first and only one to have enough confidence in it to try it out....His optimistic and inspirational attitude was what convinced me it was worth working on." And thus their division of labor: "Professor Lawrence's ability as a director and organizer and his inspirational leadership amount almost to genius, but the bulk of the development was done by others."

The foregoing enumeration does not exhaust the list of Americans who proposed Lawrence for the physics prize for 1939. Two invited to nominate that year, E.F.W. Alexanderson of General Electric and the eminent surgeon Harvey Cushing, plumped for him; and Bethe and R.C. Gibbs of Cornell decided that if Lawrence did not win, they would use their invitation to nominate for 1940 on his behalf. In a few words, as DuBridge wrote Lawrence after the happy news, "there seems to have been an unanimous feeling for the past two years at least that you were the outstanding candidate among American physicists for the Nobel award."[58] And he had powerful support in places where he was not a favorite son. The Italians—Amaldi, Fermi, and Rasetti—endorsed him unanimously, after Fermi's victory in 1938; the most influential Scandinavian physicists, Bohr and Manne Siegbahn, favored him; and even the British, or anyway those who did not think that Cockcroft and Walton had the prior claim, "would have made the award in the way the [Nobel] Committee did."[59]

There was rejoicing in Berkeley (plate 10.2). Birge worked out that Lawrence was the thirteenth American to win a Nobel prize in science and the first to have won while employed at an American state university. The Physics Department and the administra-

58. K.T. Compton to Lawrence, 4 Oct 1938 (37/20); A.H. Compton to Cooksey, 5 Jan 1939 (4/10); Langmuir to Nobelkommittén, 14 Nov 1938, and Livingston to Davisson, 3 Jan 1939 (Nobel).

59. Gibbs to Lawrence, 13 Nov 1939 (37/21); DuBridge to Lawrence, 20 Nov 1939 (6/17). Rasetti to Lawrence, 16 Nov 1939 (37/24); Bohr to Nobelkommittén, 19 Jan 1939 (Nobel); Siegbahn to Lawrence, 9 Nov 1939 (37/20); E.V. Appleton to Lawrence, 30 Nov 1939 (27/21).

tion of the University, which had reduced Lawrence's teaching load and given him money, therefore deserved a share of the honor, Birge wrote Sproul. "I think if the outside world realized more fully the handicaps under which we work, in getting and retaining men of real eminence, it also would consider the whole collection of events leading up to this award as a seeming miracle." The regents proclaimed that the prize ranked Lawrence "with the greatest scientists of the world." Lawrence knew better and graciously associated his collaborators with the honor.[60] In turn they gave him a high-spirited party—a "jubilation"—at their favorite Italian restaurant. Aebersold provided an apt text, which the celebrants sang to the tune of "A Ramblin' Reck from Georgia Tech:"[61]

> The prexy came around to see the gadget put to test
> Of course the young professor wished to show it at its best
> "You may fire the thing when ready, boy," the eager prexy cried
> So Lawrence pushed the switches in and quickly stepped aside.
>
> He aimed it at the window pane and smashed out all the glass
> It hit a poor old alley cat right square upon his—face
> He turned it on some students and it swept them off their feet
> He bombed the Campanile and he moved it down the street.
>
> And then he bombed some common lead and turned it into gold
> The prexy jumped around with joy and loudly shouted, "Hold
> I am convinced the thing is good—no more I'll have to go
> To the Solons up in Sacrament' to ask them for some dough.

The publicity surrounding the prize—which made Lawrence the subject and victim of the advertisements he had courted—was its most important and useful feature.[62] As Weaver observed in his telegram of congratulations: "Some of us think they were a year or two late. But this definitive recognition is nonetheless particularly useful just now." What Weaver had in mind appears more clearly in Lawrence's letter of thanks to Siegbahn. "It was already clear

60. Birge to Sproul, 24 Nov 1939 (2/32); Underhill to Lawrence, 17 Nov 1939 (37/20); Sproul in U.C., *1939 prize,* 5; Lawrence in ibid., 33.

61. Aebersold to Snell, 10 Nov 1939 (1/9); the party took place on Nov 17.

62. Cooksey to Roger Hickman, 12 Jan 1940 (8/28); letters in (37/21–24). Cf. Zuckerman, *Scientific elite,* 221–36.

that the difficulties in the problem [of attaining 100 MeV] were no longer technical but purely financial. The added prestige to the work of our laboratory which the Nobel award brings...will make it possible for us to raise the large financial support for this great project."[63] The prize, like the successful operation of the 60-inch cyclotron, would undercut those who doubted the feasibility and/or desirability of a machine rated at five times the Bethe-Rose limit and encourage foundations willing to take risks endorsed by the Nobel authorities. Lawrence understood the power of the prize to confer legitimacy on new fields or doubtful adventures; in a report written in 1938, he had pointed out the useful advertisement provided by Nobel laureate Joliot's decision to build a cyclotron.[64] Several successful fund-raisers made the same observation as Weaver, among them Walter Alvarez, Cottrell, and Poillon.[65] And Lawrence spread the message widely in his answers to the several hundred letters and telegrams he received; to more than fifty of these correspondents he excused the honor, and opened his thoughts, by observing that the cyclotron would be the beneficiary of his placement among the demigods of science.

The size of the projected cyclotron grew along with its prospects in the nourishing light of the Nobel prize. The plan for a 2,000-ton machine with 120-inch pole pieces, considered, at a cost of $500,000, to be at the edge of the attainable in October 1939, swelled to a dream of a 5,000-ton atom smasher with poles of 205 inches. Answering Bohr's congratulations of November 14, Lawrence referred to a machine of 3,000 tons; answering G.W.C. Kaye's on December 30, he mentioned his Christmas wish of 5,000.[66] It is very likely that without the Nobel prize Lawrence would not have had the boldness to have doubled his design and his costs. The 184-inch machine owes its existence as much to the prize givers of Stockholm as to the exertions of Warren Weaver.

63. Weaver to Lawrence, 10 Nov 1939 (37/24); Lawrence to Siegbahn, 4 Dec 1939 (37/20).

64. Lawrence, "The work," 1937, 12, and *Science, 90* (3 Nov 1939), 407–8.

65. Letters from W. Alvarez, 20 Nov 1939, Cottrell, 11 Nov 1939 (37/21), and Poillon, 10 Nov 1939 (37/23).

66. Letter to Bohr (37/21) and to Kaye (37/22); F.W. Loomis et al. to Lawrence, ? Nov 1939 (37/23), mention the profession's expectation of 120 inches.

The award to Lawrence honored not only the invention of an instrument but also the creation of the environment necessary to exploit it. What the Nobel committee had in mind appears best from the award letter quoted earlier: they were impressed by the invention and development of the machine and "especially" by its application to the production of radioisotopes. The same emphasis appears in the committee's report to the Swedish Academy of Sciences, which dwells on output figures: the cyclotron easily makes artificial sources of gamma rays equivalent to a hundred grams of radium and gives a hundred times the neutrons from a kilogram of radium mixed with beryllium. Reference to production does not occur, however, in the official citation ("for the invention and development of the cyclotron and for results obtained with it, especially with regard to artificial radioactive elements") or in Bohr's statement of Lawrence's achievement ("for the extraordinarily great contribution to the study of the reactions of atomic nuclei that he has made by construction of...the so-called cyclotron").[67] The official citation and Bohr's recommendation suggest that Lawrence himself made some notable contribution to radiochemistry or nuclear physics with the help of the cyclotron. Indeed he did: the cyclotron laboratory. But he himself had not uncovered much new about the nucleus. Had the Nobel committee wished to distinguish inventors of an accelerator who had made a fundamental discovery with it, they did not have far to look. Cockcroft and Walton fit the description and had the additional advantage over Lawrence of priority. And their work was considered prizeworthy. Otto Schumann of Munich nominated them for 1935; Rutherford and Fowler did so in 1937; Chadwick took up their cause in 1938 and 1939; and the Nobel committee agreed that both their "pioneering" splitting of the nucleus and their exact determination of atomic masses had "a special importance." In 1951 they did share the prize "for their pioneer work on the transmutation of atomic nuclei by artificially accelerated atomic particles."[68]

67. Nobelkommittén för fysik, report, 12 Sep 1939, 12–4 (Nobel); Nobel Foundation, *Directory,* 56; Bohr to Nobelkommittén, 19 Jan 1939 (Nobel).

68. Nobelkommittén, 25 Sep 1935, re Schumann's proposal, and 7 Sep 1937, 11–2 (Nobel); Crawford, Heilbron, and Ullrich, *Nobel population,* s.v. "Cockcroft," "Walton;" Nobel Foundation, *Directory,* 58. Cf. Lawrence to Cockcroft, 1 Dec 1939 (37/21): "I hope also to be sending you a congratulatory letter on the Nobel prize next year or later, as I feel very strongly that you well deserve it."

In preferring Lawrence to Cockcroft and Walton, the Nobel committee on physics went against its own arguments and precedents, although in a direction of which Nobel would have approved. Lawrence's was the first award for the development of an instrument for physics if we leave C.T.R. Wilson's prize of 1927 out of the reckoning. Efforts to give prizes for hardware alone had subsequently failed. In 1935 Walther Nernst proposed Hans Geiger. The Nobel committee thus evaluated his candidacy: "One can say that the Geiger counter together with the prizewinning Wilson chamber are the experimental instruments that have made possible the brilliant discoveries in nuclear physics....But Geiger himself has not taken any noteworthy part in the work that led to these important discoveries."[69] A campaign of many years on behalf of Aimé Cotton, who built a very large electromagnet (100 tons, 75 cm pole tips, 64 kG) for the Paris Academy of Sciences and used it for important spectroscopic studies, likewise did not convince the committee. Pieter Zeeman (prize of 1902) might compare Cotton's magnet with Aston's mass spectrograph or the Rowland grating, and insist that progress in physics comes equally from ideas and from machines; C.E. Guillaume (prize of 1920) might declare Cotton's magnet precious and its research potential prizeworthy; Pierre Sève of Marseilles might demand a reward for "the construction of an instrument unique in the world..., the Laboratoire du Gros Electroaimant, where many workers under [Cotton's] direction have already obtained extremely important results in all branches of physics;" but the committee rejected it all, on the ground that Cotton had not made any discovery with his magnet important enough to deserve a Nobel prize.[70]

The relaxation of this condition in Lawrence's favor owed much to the progress and popularity of nuclear physics and to the scale of the machines and laboratories it required. One of Cotton's last nominators, C.E. Guye of Geneva, realized that, if the Nobel committee went for machines, they would probably prefer those of nuclear to those of atomic and molecular physics.

69. Nobelkommittén, report for 1935, 6 (Nobel).
70. Letters from Zeeman, 18 Jan 1931, from Guillaume, 25 Jan 1933, and from Sève, 21 Jan 1934. Nobelkommittén, report for 1931, 9–10 (Nobel).

The only hope of the old school, he thought, was to slip in before the committee could decide which of the inventors of accelerators or discoverers of new particles to reward first. It was in connection with the confusion and profusion of claimants that the eventual need to pick an accelerator physicist was first expressed in the correspondence of the Nobel committee. Dick Coster of the University of Groningen, writing in December 1933, suggested that the "artificial disintegration of nuclei by fast protons" might prove deserving; two years later Caltech's Richard Tolman took advantage of the throng—Lawrence, Lauritsen, Van de Graaff, and Cockcroft and Walton—to dismiss all in favor of Caltech's Anderson.[71] By 1938 Lawrence had outdistanced the rest in the building not only of machines but also of laboratories.

An instructive evaluation of the situation as it appeared to three nominators that year is preserved in the correspondence of O.W. Richardson (prize of 1928), who liked neither big nor nuclear physics. Bohr (prize of 1922) talked over options with him in December 1938. "After discussing a number of distinguished names, to several of which I would have been prepared to offer a measure of support, he finally decided on the combination of Lawrence and Kapitsa [another big machine, big laboratory man]. Well, what has Lawrence done? invented an instrument which would have been more or less obvious to anybody unfamiliar with the difficulties of experimental technique, made it to work, and done nothing with it, except to incite a large number of very able experimental physicists all over the world, unsuccessfully, to emulate his efforts. The wiser of them seem to have handed this trouble over to their students, but it is doubtful if that will help their generation! As for Kapitsa!!!" The addressee of this blast, G.P. Thomson (prize of 1937), who did some nuclear physics, returned the suggestion of Cockcroft without Walton. Bohr had also considered a prize for Cockcroft alone, but all recognized its unacceptability. In the end Bohr dropped Kapitsa, and Richardson and Thomson nominated E.V. Appleton, whose investigations concerned the upper atmosphere.[72]

71. Letters to Nobelkommittén för Fysik from Guye, 19 Jan 1934; Coster, 14 Dec 1933; and Tolman, 2 Jan 1935 (Nobel).
72. Richardson to Thomson, 12 and 24 Dec 1938 (Richardson P, 5).

Appleton had to wait until 1947. The Nobel committee was infatuated with nuclear physics and its machinery and willing to reward discoverers of the one and inventors of the other. Oliphant read Lawrence the significance of the committee's decision in 1939: "It is extremely encouraging to find that the Nobel Prize Committee, in common with many other authorities, is now recognizing the tremendous importance of technique in scientific investigations....The technical side of the subject is now recognized as equally important with advances that follow from the use of these techniques, and more important, I hope, than the theories that endeavor to explain them....It is certain that you have no difficulty now in raising funds for your 'father of all cyclotrons.' "[73]

German submarines kept Lawrence from the prize giving in Stockholm. The citation and medal were sent to the Swedish consul in San Francisco and presented at a ceremony on the Berkeley campus presided over by Sproul. Birge gave a speech reciting the accomplishments of Lawrence and the Laboratory. It was the evening of February 29, 1940. At the end of his prepared remarks, Birge announced the discovery of C^{14} to the crowd that filled the hall. "On the basis of its potential usefulness," Birge said, with exaggeration appropriate to the hour, "this is certainly much the most important radioactive substance that has yet been created."[74] Lawrence's reply carried two examples of his most effective technique. For one, he let Birge say what he himself wished to say without incurring any obligations: "As [Birge] has indicated, there are substantial prospects that [the next cyclotron] will be the instrument for finding the key to the almost limitless reservoir of energy in the heart of the atom." For another, he did not miss the opportunity for fund-raising. "It goes without saying that such a great recognition at this time will aid tremendously our efforts to find the necessary large funds for the next voyage of exploration into the depths of the atom." "[This] very considerable financial problem...we must now hand over to President Sproul."[75]

73. Oliphant to Lawrence, 20 Nov 1939 (14/6).
74. According to Kamen, *Radiant science*, 132–3.
75. Quotes from, resp., Lawrence in U.C., *1939 prize*, 36; draft of his acceptance talk (40/23); and *1939 prize*, 35.

As we know, description of the cyclotron had always been a parade ground for military metaphors. The additional possibilities offered by the source of the Nobel prize carried the "University Explorer" to heights truly and doubly inspired. "Ernest Lawrence," he declared, "has discovered a blasting technique far more potent than anything Alfred Nobel ever dreamed of." R.W. Wood, an elder statesman of physics, improved the metaphor into prophecy. He wrote Lawrence: "As you are laying the foundations for the cataclysmic explosion of uranium (if anyone accomplishes the chain reaction) I'm sure old Nobel would approve."[76]

2. NEW JOBS FOR CYCLOTRONEERS

On June 27, 1940, four men representing together the nation's best research universities and technical schools, its largest private foundations and most advanced applied science, and its most prestigious scientists obtained from Roosevelt a commission to set up a National Defense Research Committee (NDRC) under the authority of the forgotten Council of National Defense of World War I. These pooh-bahs of science and technology were Vannevar Bush, accomplished electrical engineer, former vice president of MIT, head of the Carnegie Institution, head of the National Advisory Committee for Aeronautics; Karl T. Compton, physicist, president of MIT, trustee of the Rockefeller Foundation; James B. Conant, chemist, president of Harvard; and Frank B. Jewett, electrical engineer, head of Bell Labs, president of the National Academy of Sciences.[77] All four have appeared in our pages as promoters of cyclotrons. They were to recruit from the corps of cyclotroneers many of the men who would direct the major wartime laboratories. They knew that cyclotroneers understood how to work at the borders and edges of science and technology, and how to work in teams: cyclotroneers and their fellow travelers did not fear big projects, did not disdain to scrounge when necessary, did not insist on perfection or protocol. They were ideal people for crash programs.

76. "Atom smasher wins highest scientific honor," University Explorer, [Nov 1939] (40/15); R.W. Wood to Lawrence, 13 Nov 1939 (37/24).
77. Hewlett and Anderson, *New world*, 24–5; Baxter, *Scientists against time*, 14–16.

In the Ether

In October 1940 Alfred Loomis, who had been appointed head of microwave work under Karl Compton's division of NDRC, called members of his committee to meetings at his home and laboratory in Tuxedo Park, New York. Lawrence was among them. They agreed that the technology made possible, and the emergency recommended, the establishment of a central national laboratory to do "anything and everything that was needed to make microwaves work."[78] To confuse the enemy, and in honorific obfuscation, the center, located at MIT, took the name "Radiation Laboratory." It invented or perfected many sorts of radars operating at wavelengths around and below 10 cm. In devising these most important aids to detection and navigation of ships and planes, blind landings, gun laying, and so on, the MIT Radiation Laboratory drew twice over on the experience of cyclotroneers. For one, it acquired appropriately socialized staff with relevant technical knowledge. ("If all the energy which has gone into nuclear physics since 1932 were turned onto this problem [the air menace]," Cockcroft had predicted, "it would be solved.") To head the new Rad Lab, the Microwave Committee chose Lee DuBridge of Rochester, whom Lawrence rated as the most desirable laboratory leader of his generation, an excellent physicist, administrator, and team player, the perfect cyclotroneer. As DuBridge later recalled, with an exaggeration that illustrates the strength of the brotherhood he represented so well, "our whole initial group at the MIT laboratory were the cyclotroneers—all had been associated with Ernest [Lawrence] either remotely or intimately." A meeting on applied nuclear physics in Cambridge around November 1, 1940, arranged by Livingston and attended by 600 physicists from all across the country, provided a perfect recruiting ground.[79]

78. Loomis to Lawrence, 1 Oct, and reply, 3 Oct 1940 (46/8); MIT, *Five years,* 12, quote.
79. Cockcroft to Lawrence, 24 Oct 1938 (4/5); Lawrence to Zeleny, 5 Mar 1940 (18/44); Guerlac, *Radar,* 260–1; DuBridge, interview by James Culp, Oct 1981, 11, quote; Condon, *BAS, 1:11* (15 May 1946), 8–9.

Table 10.1 lists the ten cyclotroneers engaged during the first three months of the MIT Rad Lab's operations. One came and stayed as director; six entered as or quickly became group leaders; five left as division leaders (there were seven R&D divisions in all); two ended on the laboratory's steering committee. Lawrence's devotion to Loomis and conviction of the importance of radar caused him to send his very best men, Alvarez and McMillan, and his expert on radio frequency systems, Salisbury. McMillan, who left MIT soonest, worked on field tests of an interception system against aircraft and on airborne radar for detecting and homing on ships. Salisbury directed efforts in his specialty. Alvarez, who stayed the longest of the Berkeley group, moved furthest. After helping to develop radar for attack planes, he invented a system to guide approaching aircraft from the ground and directed work on other aids to detection, navigation, and identification.[80] Another man from Berkeley, Lauriston Marshall of the Department of Electrical Engineering, had a career at the MIT laboratory similar to a cyclotroneer's: he came to work on magnetrons, rose immediately to group leader, then chairman of the Ship Committee and director of the British Branch of the Radiation Laboratory, and closed the war as head of the Laboratory's Operational Research Section attached to the Headquarters of the U.S. Air Force in the Pacific.

Marshall represents the second way in which accelerator laboratories contributed essentially to radar. In the late 1930s, he helped transform Sloan's latest piece of cyclotronics into a generator of radio waves at a frequency and power suitable to radar. This was a tube designed to remove the obstacle encountered in the early 1930s to the development of the Wideröe linac into a useful tool in nuclear physics. To recall the old difficulty, a Wideröe machine for protons of reasonable size and ambitious energy would require a power oscillator working at a wavelength of two or three meters, about an order of magnitude shorter than commercial tubes could easily handle. As he remembered his

---

80. Lawrence to Dunning, 9 Nov 1940 (6/19); MIT, Rad. Lab., "Staff;" Alvarez, *Adventures,* 86–101. We omit mention of radar work at Bell Labs, not because it lacked importance, but because it lacked cyclotroneers; Fisk, Hagstrum, and Hartman, *Bell System techn. jl., 25:2* (1946), 1–189.

Table 10.1

Cyclotroneers Recruited to the MIT
Radiation Laboratory in 1940/1

Name	Institution	At Rad Lab	First assignment	Last post(s)	Destination
Alvarez, L.	Berkeley	12/40–9/43	airborne radar, then gp ldr, attack plane radar	div hd, Beacons	MED
Bacher, R.	Cornell	2/41–6/43	gp ldr, indicators	div hd, Receivers	MED
Bainbridge, K.	Harvard	11/40–8/43	gp ldr, modulators	div hd, Transmitters	MED
DuBridge, L.	Rochester	11/40–12/45	director	director	stayed
McMillan, E.	Berkeley	11/40–8/41	field testing	field testing	sonar lab
Pollard, E.	Yale	1/41–11/45	mbr, indicator design; proj. eng., coastal surveillance	ass div hd, Ground and Ships; mbr steering com'tee	stayed
Ramsey, N.	Columbia	11/40–9/43	gp ldr, magnetrons	liaison with Army Air Force	MED
Salisbury, W.	Berkeley	1/41–3/42	gp ldr, rf components	gp ldr, rf components	
Van Voorhis, S.	Rochester	12/40–1045	gp ldr, roof systems	gp ldr, x-band receivers	stayed
White, M.	Princeton	11/40–12/45	mbr, pulser group	div hd, Airborne Systems; mbr, steering com'tee	stayed

Source: Radiation Laboratory, *Staff* (1946).

tinkering to improve their performance, Sloan "just pushed triodes and tetrodes to high power, high frequency, beyond anything the cyclotron needed." By May 1940, he and Marshall had a tube that oscillated at 50 cm and with great power—some 2,500 watts. Cooksey reported this news to Loomis, who had not waited for the NDRC to begin to push work on microwave electronics. Loomis was elated.[81] The cause of elation: the size of radar sets diminished, while the detail of the objects they could see increased, with decline of wavelength down to about 1 cm, where atmospheric absorption begins to make trouble. The Sloan-Marshall tube, or "resnatron," held promise as the fast-paced heart of a powerful, centimetric microwave transmitter for airborne use.[82]

As soon as he became head of the NDRC's microwave work, Loomis asked Lawrence to take responsibility for the further development of the resnatron "in a big way," as "the major war research of the University [of California]." The Research Corporation, which had patented the resnatron, gave $4,500; Loomis provided $1,500 from his deep pocket and a promise of $20,000 of NDRC funds. Lawrence agreed to sail "full speed ahead" with the help, if necessary, of moneys diverted from the 184-inch cyclotron.[83] The good ship resnatron was then not the only centimetric pulser at sea in the Bay Area. Loomis had asked Lawrence to send Sloan and Marshall to San Carlos, California, to join their work with efforts under way there to perfect something called the klystron.[84] Like the resnatron, the klystron resulted from efforts at a university—in its case Stanford—to overcome the frequency limit of commercial oscillator tubes. Stanford wanted a very powerful x-ray tube, no less than 3 MV, but did not want to pay for it. Considering strategies in 1934, William Hansen, an instructor in Stanford's physics department, thought to set up oscillations in a

81. Sloan, interview by A. Norberg, 10 Dec 1974, 52–3, quote (TBL); Cooksey to Loomis, 14 May 1940, and other correspondence of Loomis, Cooksey, and Lawrence, May 1940 (46/8).

82. Terman, *Elect. radio eng.*, 810, 854, 1018–9.

83. Loomis to Lawrence, 9 Jul and 12 Aug 1940, and Lawrence to Loomis, 26 Jul and 6 and 29 Aug, 28 Sep 1940 (46/8).

84. Loomis to Lawrence, 26 June, and reply, 1 Jul 1940 (46/8); Poillon to P.R. Bassett, Sperry, 26 June, and to Lawrence, 2 Aug 1940, and Lawrence to Poillon, 6 and 24 Aug 1940 (15/18).

cavity, a transmission line, as it were, with no inner electrode. The thing itself is simple enough in principle and, moreover, of convenient size: a cube 10 cm on a side resonates at a wavelength of 14 cm. It may be driven by a transmission line that creates an appropriate magnetic field within a loop coupled to the cavity. A struck bell is a crude, but serviceable, analogy. In 1937 Hansen had a visionary plan to drive electrons to 100 MeV within his "rhumbatron," as he called his cavity resonator. He sent off a report, in the usual way, to a professional journal; but Stanford's administrators, sensing something big, made him hold it back for a year while they made sure of the commercial possibilities.[85]

The rhumbatron transformed into the klystron when Hansen's former roommate, Russell Varian, discovered a novel way to control electrons. That was in the summer of 1937. Varian and his brother Sigurd, a former commercial pilot, then worked with Hansen as unpaid research associates on the design of a microwave device for navigating and detecting airplanes. This purpose had seized Sigurd Varian, who knew the dangers of commercial flying and could imagine those of enemy action. The klystron (fig. 10.1) consists of two reentrant cavities C_1 and C_2, separated by a drift space RS. A beam of electrons from the tubular cathode at the top accelerates under the dc voltage between M and the grid P, whence they drift through the field-free region PQ and into the neck of the "buncher" cavity C_1. Between Q and R they suffer an oscillating field created through the loop F. This experience causes them to collect into bunches in the drift space RS: an electron that crosses just before the oscillating field rises to zero will be slowed; one that crosses when the field goes positive will speed up. As they drift, the retarded early electron, the on-time electron, and the hurrying late electron will congregate. The neck of the "catcher" cavity C_2 stands where the density is greatest and the induced field has the proper phase to oppose the motion of the bunches (and so derive energy from them). The congregations pass at the frequency of the buncher, thereby exciting very strong

85. *Science news,* in *Science, 84* (16 Oct 1936), suppl., 9, reported that Hansen had electrons of 5 MeV; Nahmias to Joliot, 28 Apr 1937 (JP, F25), mentioned the plan for 100 MeV; Hansen, *Jl. appl. phys., 9* (1938), 654–63, rec'd 17 Jul 1937; Ginzton, IEEE, *Spectrum* (Feb 1975), 33.

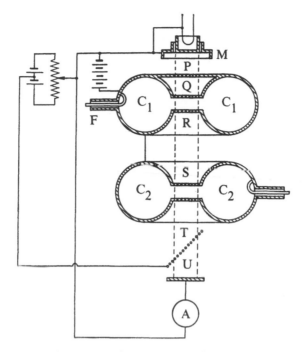

FIG. 10.1 The klystron as first described by the Varian brothers. R. Varian and S. Varian, *Jl appl. phys.*, 10 (1939), 324.

oscillations in the tuned catcher. The deceleration of the bunches in the catcher prevents them from overcoming the opposing voltage between T and U, a fact recorded by the meter at A. The whole business rests in a vacuum. By August 1937 a cardboard version coated with copper foil was working at λ = 13 cm. The Sperry Corporation undertook to develop it for aircraft detection.[86] This was the line of research to which Loomis wished to couple development of the Sloan-Marshall tube. As it happened, the Varians and Hansen went to Sperry's research laboratory in New York, where the klystron developed into a versatile circuit element. Hansen spent much of his time in the East lecturing about microwaves at the MIT Radiation Library.

86. Ginzton, IEEE, *Spectrum* (Feb 1975), 34–9; Varian and Varian, *Jl. appl. phys.*, 10 (1939), 321–7; Webster, ibid., 501–8, 864–72. The business can be followed in detail in the Hansen Papers, Stanford Univ. Library.

Neither Berkeley's nor Stanford's entry into what had become a worldwide race for a transmitter of centimetric radar could generate the necessary power in 1940. The device on which the MIT Rad Lab was raised later in the year had much in common with them, however; it came from a university laboratory much concerned with particle accelerators and exploited the bunching principle introduced by the Varian brothers. This "cavity magnetron" came from Oliphant's institute at the University of Birmingham. Its inventors, J.T. Randall and H.A.H. Boot, arranged an anode consisting of a thick ring scalloped by cavities around a central cylindrical cathode (fig. 10.2). Electrons move toward the anode under a dc potential and a magnetic field strong enough to bend them all back to the cathode. When the cavities resonate under an external impulse, their radio frequency fields, leaking into the anode space, slow some electrons enough that the magnetic force on them (which is proportional to their velocity) no longer suffices to return them to the cathode. Electrons so affected bunch together and add energy to the resonant cavities as they pass en route to the anode. In effect, the cavity magnetron is to the klystron what the cyclotron is to the linac.[87]

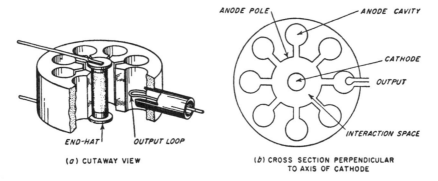

FIG. 10.2   The cavity magnetron. Modulation of the radial electrostatic field between the cylindrical cathode and the concentric anode-block by a high-frequency field leaking from the cavities causes some of the electrons to bunch. Terman, *Elec. rad. eng.*, 689.

87. Terman, *Elect. radio eng.*, 689–95; Hagstrum, IRE, *Proc.*, *35* (June 1947), 548–64.

The cavity magnetron was the centerpiece of many ingenious devices that a British technical mission under Henry Tizard, which included Cockcroft and Fowler, showed American military and scientific men in September 1940 in the hope of returns in kind. The magnetron made an impression. The Naval Research Laboratory then had a klystron transmitter operating at 10 watts. The British cavity magnetron gave a thousand times as much power at the same wavelengths. Cockcroft and E.J. Bowen explained its operation in detail at a gathering at Loomis's estate at the end of September. Two weeks later, on October 12–13, the Microwave Committee met with Bowen and Cockcroft, again at Tuxedo Park. They decided to copy Britain in entrusting development of radar to interdisciplinary teams of academic scientists and engineers. Lawrence ran to Loomis's telephone. "During the next few weeks [as Bowen recalled] he was to telephone every physicist of consequence in the United States." A month or so later the MIT Rad Lab, "a central laboratory built on the British lines, was in operation."[88]

The magnetron did not end the war for the resnatron. After Marshall threw in his lot with the MIT group and Sloan at last won his Ph.D. in Berkeley, the new doctor took his tube for treatment at Westinghouse's Research Laboratory in Pittsburgh. There it waxed exceedingly robust. Westinghouse built forty-two of the final design, each of which weighed 500 pounds. These fat resnatrons knocked out everything on the air. They were of first importance in jamming German radar on D-Day.[89]

Elsewhere

In August 1941 Lawrence pulled McMillan from the ether at MIT and dropped him in the water at San Diego. A new laboratory, run jointly by the navy and Jewett's division of NDRC, was being organized there under a contract with the University of California. Its director, Vern Knudsen, professor of physics at UCLA, had built up a small, strong group in applied acoustics with support from the movie industry, for which he had built

88. Clark, *Tizard,* 264–70; MIT, *Five years,* 12, quote; Bowen, *Radar days,* 157–9, 168–78, quotes; Guerlac, *Radar,* 248–50.
89. Anon., *Westinghouse eng., 6:2* (Mar 1946), 47.

sound stages. But neither Hollywood nor UCLA had given him work on the scale on which he was now to perform: to study the physics of underwater sound, especially means to measure its spead precisely; to improve or design new methods of underwater detection and evasion appropriate to conditions in the Pacific; and to develop training manuals and devices for operators with very little technical knowledge. Knudsen turned for help to the physicist at the University of California most experienced in big operations. Lawrence "came down [to San Diego], spent time with us..., and participated in formulating our research program."[90]

Perhaps most usefully, Lawrence furnished McMillan, much to the irritation of the leaders of the MIT Rad Lab. With the breadth of view and ingenuity that had made him so valuable a member of Lawrence's Laboratory, McMillan contributed to all phases of the work at San Diego. He devised an echo repeater, "Beeping Tom," the first contribution from Knudsen's shop accepted by the navy, which simultaneously freed submarines from service as training targets and sonar operators from the need for practice at sea. He was particularly effective, according to Knudsen's successor, G.P. Harnwell, "in criticizing and directing the program of the laboratory in the fundamental investigations assigned to it."[91]

The first group of cyclotroneers entirely mobilized by NDRC was Tuve's force in the Carnegie Institution. In September 1940 they put aside their Crocker clone for "nights and days with defense work." They had taken on the task of knocking enemy planes from the sky. At the time, conventional wisdom rated very highly an antiaircraft system that could hit one plane in 2,500 shots. Under this mild inhibition, the Luftwaffe could bomb and strafe without much worry about guns on the ground. Following conversations between Lauritsen and Tuve and the navy's Bureau of Ordnance in August, the NDRC contracted with the Carnegie Institution for "preliminary experimental studies on new ordnance

90. Baxter, *Scientists against time,* 172–4, 180; Hackman, *Seek and strike,* 251–2; Knudsen, interview with L. Delsasso and W.J. King, 18 May 1964 (AIP), 24–8, 46–7, quote.

91. McMillan to Lawrence, 18 June, 4 Jul, and 22 Aug 1941, and Lawrence to McMillan, 8 Jul 1941 (12/30); McMillan to Lawrence, 18 June 1942 (12/31); Harnwell to Bush, 5 Oct 1942 (McMillan P).

devices." Tuve learned about what the British had done from Cockcroft and Fowler and set out to make a fuse activated by radio that would detonate near its target. Everyone working on the Carnegie cyclotron—Tuve himself, L.R. Hafstad, R.B. Roberts, G.K. Green, and Philip Abelson—went to work to make a radio sufficiently small and tough to fit into the space of an ice cream cone and withstand the inspiring forces—some tens of thousands of times greater than the force of gravity—exerted during firing on a five-inch shell. The Carnegie's administration, however, preferred to see its expensive cyclotron brought to completion; and, in a gambit we shall see repeated, requested Tuve's men to return to their machine as a measure of national defense. He rejected the request as selfish and the aim as ineffectual. "Representatives of every Cyclotron Laboratory in the country have individually asked us what they could use their cyclotrons for in defense work and no valid ideas have been forthcoming....It is easy for an enthusiastic entrepreneur to make a casual remark that a cyclotron can be classified as a defense project. If the Institution staff had no other defense work of clearly greater urgency, this would be our position [too] as it was previous to August."[92]

Tuve did find a little war work for his new three-story cyclotron laboratory. He and his associates dropped miniature radio tubes from its roof onto the concrete driveway below as a test of fragility. Enough survived to prompt contracting with their makers. When he declared the worthlessness of cyclotrons for national defense, Tuve had three sorts of tiny tubes that could withstand firing in a five-inch shell. By May 1941, a basic design for a fuse triggered by radio was in hand; but premature firings and duds troubled its tests during the summer. The bombing of Pearl Harbor brought new urgency and manpower to the project and its relocation to a large garage in Maryland. This new facility, dubbed the Applied Physics Laboratory of Johns Hopkins University, which contracted for its operation, improved reliability and invented an ingenious mechanism to prevent unintended explosions. Late in 1942, 4,500 shells, perfectly safe to their users,

92. Tuve, "Report for September 1940" (MAT, 21/"extra copies"); quotes from, resp., Tuve to Condon, 23 Oct 1940 (MAT, 24/"MIT conf."), and Tuve to Fleming, 25 Jan 1941 (MAT, 25/"cycl. 1940").

reached the Pacific Fleet. In their first engagement they brought down a Japanese bomber in four shots. The project compared in importance, success, and expense with the making of the atomic bomb.[93]

One of Bush's first prizes as chairman of NDRC was the Advisory Committee on Uranium. This body, chaired by Lyman J. Briggs, director of the National Bureau of Standards, had resulted from the famous letter alerting Roosevelt to the possibility of nuclear weapons, signed by Einstein but composed by Szilard and his fellow Hungarian refugees Eugene Wigner and Edward Teller. Briggs was no cyclotroneer. His committee, which had the frequent optimistic advice of Szilard, had not accomplished much by May 1940, after seven months of existence. The delay and the apparent indifference of the armed services to the opportunities opened by fission made the refugees impatient and, perhaps, self-important. "We ought not to try to save the country for the Americans [Wigner wrote Bethe], but to push them to save themselves."[94] Without their pushing, however, a typically American instrument—a body of self-moving private citizens appointed by the president to mobilize scientists within and outside government—had come into being that would vitalize the uranium project.

The citizens, the founders of NDRC, had been drawn into the business of the uranium committee a few months before the creation of their organization. The news from Columbia in March 1940 that, as predicted by Bohr, U^{235} was the party guilty of fission by slow neutrons, directed attention to the importance of the separation of uranium isotopes. In April, at the meeting of the American Physical Society in Washington, Beams, Fermi, Nier, Tuve, and Urey decided that Beams's ultracentrifuge offered the best hope for separation in kilogram amounts. In May, Beams, Cooksey, Karl Compton, and Lawrence reached a similar conclusion. Compton notified Bush (as head of the rich Carnegie

93. Baxter, *Scientists against time,* 221–33, 241–2. Over 130 million proximity fuses were made at an average unit cost of $20; their manufacture eventually monopolized 25 percent of the American electronics industry and 75 percent of the facilities for molding plastic.

94. Wigner to Bethe, 21 May 1940 (HAB, 14/22/976); Hewlett and Anderson, *New world,* 19–24.

Institution), who already knew about Beams's work from Tuve, whose department had deliberated purchasing a centrifuge for biological work. Tuve had rated Beams's model, which cost $5,000, as the best available for separating isotopes (he had tracers in mind) and biological materials. Tuve now recommended to Bush that the Carnegie Institution give Beams $10,000 to determine whether U^{235} could be spun free from U^{238}. In Tuve's opinion, centrifugation offered "the only hope of separating the isotopes of any but the light elements in quantity." Neither thermal diffusion nor the mass spectrograph (electromagnetic separation) seemed competitive to him.[95]

Bush agreed to provide money and call meetings. The Naval Research Laboratory had been helping Beams with supplies and apparatus at a level estimated by Tuve at $2,000 a year. After Bush's intervention, the army and navy put up $100,000 to study the separation of isotopes, primarily by centrifugation, but also by thermal liquid diffusion, as proposed by Carnegie's Abelson, then recently returned from Berkeley and work on element 93.[96] A possibility that Tuve had not considered explicitly, diffusion of uranium hexafluoride gas through tiny holes in a "barrier," which would slightly enrich the lighter isotope, appealed to Urey and others at Columbia, who obtained money from NDRC in the winter of 1940/41 to follow it up. The runaway favorite in July 1941, as judged by a budget then proposed by Briggs's committee, was Beams's centrifuge ($95,000); Columbia's gaseous diffusion ($25,000) came a poor second. At just this moment, however, the British intervened as decisively as they had in the fall of 1940.[97]

For a year and a half, Chadwick, Cockcroft, Oliphant, Thomson and other leading British physicists knew that a bomb might be made from 10 kg or less of separated U^{235}. The relevant

95. Compton to Bush, 9 May 1940 (KTC); Hewlett and Anderson, New world, 23–4; Abelson and Tuve, "The current status of the ultracentrifuge as a research tool," 17 Jan 1940 (MAT, 25/"biophys. 1940"); Tuve to Bush, 13 Apr 1940 (MAT, 19/"Beams"); Beams, RMP, 10 (1938), 248–51; Loofbourow, RMP, 12 (1940), 324–9.

96. Bush to Compton, 14 May, and Loomis to Compton, 17 May 1940 (KTC); Tuve to Bush, 13 Apr 1940 (MAT, 19/"Beams"); Hewlett and Anderson, New world, 27, 32.

97. Hewlett and Anderson, New world, 40–3.

considerations had been put forward in February 1940 by the émigrés O.R. Frisch (Cambridge) and Rudolph Peierls (Birmingham), who assumed, among much else, that fast neutrons as well as slow ones could cause fission in U^{235}. "From rather simple theoretical arguments," they wrote, without arguing, "it can be concluded that almost every collision produces fission and that neutrons of any energy are effective." This was a capital point: a slow neutron bomb would be more likely to fizzle than to devastate. And where procure the U^{235}? Frisch and Peierls suggested gaseous thermal diffusion. Another émigré, Franz Simon (Oxford), showed that diffusion through a barrier could do much better. In his optimistic calculations, completed in December 1940, a plant covering forty acres and employing 1,200 people could turn out 1 kg of 99 percent pure U^{235} in a day. During the first six months of 1941, these prophecies drew strength from rough measurements by Tuve's group, which confirmed the fundamental hypothesis of fast-neutron fission on a sample of U^{235} provided by Nier. Peierls exulted: "There is [now] no doubt that the whole scheme is feasible (provided the technical problems of isotope separation are satisfactorily solved)." The official report of the British uranium committee (called the MAUD Committee) of July 1941 endorsed and refined the original Frisch-Peierls memorandum: twenty-five pounds of active material, a gaseous diffusion plant costing £5 million, a bomb deliverable at the end of 1943 equivalent in destructive power to 1,800 tons of TNT.[98]

The MAUD report changed American thinking. Although everybody had known that a chain reaction, if achieved, might make possible a nuclear explosive, Briggs's uranium committee did not have a bomb as its goal. Looking back with the greater wisdom of 1943, Fermi recalled that he knew of no one working with either fission or element 94 in the United States who appreciated their potential as explosives until the spring (or, better, the early summer) of 1941. Ignorance of British thinking was not the reason for this devaluation. Oliphant had written Lawrence in May 1939 that the British defense authorities insisted on looking

98. Gowing, *Britain and atomic energy,* 58, 67–8 (quote), 77, 390, 392, 394–8; Smyth, *Atomic energy,* 66.

into the possibilities of a bomb, however remote, "as there are rumours that great developments have taken place recently along these lines in Germany." Lawrence, conceding the possibility, had asked "Segrè and some of the other boys" to see whether they could fission lead or bismuth. Typically, he saw neither a danger nor the likelihood of imminent success, but an entrepreneurial opportunity. "This sort of thing is another reason why the British government should come forward with generous support of nuclear physics."[99] Nor did disclosure of the Frisch-Peierls report by the Tizard mission inspire Lawrence or other leaders of American nuclear physics to set a high priority on making bombs. Cockcroft thought them overly skeptical about military applications and overly fascinated with the possibilities of nuclear power.[100] But then Britain, not the United States, was at war.

The pace of the Briggs committee exasperated some of its members, particularly Urey, and busy outsiders like Lawrence. In March 1941 Lawrence managed through Karl Compton to have himself assigned by a reluctant Bush—who did not like being pressured—to the post of temporary consultant to Briggs. Lawrence obtained a modest increase ($2,000) in support of the Laboratory's work on elements 93 and 94 and a contract to Nier for 5 mg of U^{235}. His role was that of gadfly; he did not urge a change of program but greater vigor and less secrecy in pursuing the self-sustaining pile.[101] At the instigation of Briggs, Bush asked Jewett to convene a committee of the National Academy of Sciences to evaluate the uranium program. Jewett appointed A.H. Compton, Coolidge, Lawrence, John Slater of MIT, and John Van Vleck of Harvard. Their report, finished in May, did not emphasize a bomb; it was vague and uplifting, in the style of Lawrence's requests for major funding. The uranium project should be supported for the general significance of achieving a chain reaction and for the importance of even a moderate separation of uranium isotopes; if successful, the project might produce,

99. Fermi to A.H. Compton, 31 Aug 1943 (Fermi P); Oliphant to Lawrence, 30 May, and reply, 15 June 1939 (14/6).

100. Hewlett and Anderson, *New world*, 28–9; Hartcup and Allibone, *Cockcroft*, 124.

101. Hewlett and Anderson, *New world*, 35–9; Bush, *Pieces*, 60; Compton, *Atomic quest*, 47.

in order of military importance, radioelements in sufficient quantities to poison enemy territory, a power plant for submarines, and a bomb, the last unlikely before 1945.[102]

Bush and Conant then knew about British hopes for a bomb, which Conant had learned of during a sojourn in London in the early spring as liaison between NDRC and British defense authorities. He and Bush dismissed the NAS report as too vague on bombs and too fanciful on power. They asked for another. Jewett added two engineers to the committee. Its report, of 11 July, did not differ significantly from its predecessor's. Conant inclined to squelch the project. No one seemed to know much of the MAUD report or to take it seriously. Lawrence heard about it in September 1941, not from Briggs, who appears to have kept the report secret even from himself, but from Oliphant, on tour of American laboratories engaged in radar and other war work. What had been missed in the United States, judging from a letter from Coolidge, who had chaired the NAS committees, was the connection between fast-neutron fission, ten-kilogram explosives, and practical gaseous diffusion. Oliphant raised Lawrence's enthusiasm for nuclear bombs. Oliphant thought that 10 kg of pure U^{235} might be within reach, perhaps by cyclotronics. And there was also element 94, which the Laboratory had shown to be fissionable and MAUD had mentioned as an alternative, if unlikely, explosive.[103] At the end of September 1941, just after meeting with Oliphant, Lawrence attended the fiftieth anniversary celebrations of the University of Chicago. There he met with Compton and Conant and urged that a new NAS committee be empanelled to consider the uranium project in the light of the MAUD report.[104]

102. Smyth, *Atomic energy,* 50–2; Hewlett and Anderson, *New world,* 36.

103. Cockburn and Ellyard, *Oliphant,* 102–6; Coolidge to Jewett, 11 Sep 1941, ibid., 106; Gowing, *Britain and atomic energy,* 433. According to Cockburn and Ellyard, *Oliphant,* 104, Fermi doubted the possibility of a fast-neutron explosive.

104. Compton, *Atomic quest,* 8; Smyth, *Atomic energy,* 51–2; Lawrence, "Historical notes," 26 Mar 1945. Compton misdates the meeting in Chicago to early or mid September, before Oliphant reached Berkeley; Lawrence arrived in Chicago on 25 Sep, according to Lawrence to Compton, 5 Sep 1941 (4/10).

The illumination from London came just after Bush had reorganized and extended his empire. On June 28, 1941, he took over the directorship of a new agency, the Office of Scientific Research and Development (OSRD), which gave him responsibility for development of instruments of war as well as for research of military interest, authority to coordinate the efforts of various agencies, and immediate access to the president. Conant took the chairmanship of NDRC and Briggs remained as chairman of the uranium committee, which Bush raised to an independent unit, "S-1," of OSRD. He and Conant followed Lawrence's and others' promptings and returned to the NAS to ask for a new review with emphasis on the U^{235} bomb and the gaseous-diffusion plant. While the new committee deliberated under the chairmanship of A.H. Compton, Bush conferred with the president. He left the White House on October 9 with authority to expedite research on nuclear weapons in every way possible short of the construction of production plants. Compton's committee, which included Lawrence and Oppenheimer, endorsed the British findings, with some qualifications that proved wrong. Oppenheimer expected that 100 kg of U^{235} would be needed for a bomb. The committee thought that centrifugation might work. They raised the cost of separation to $100 million.[105]

On December 18, less than two weeks after Pearl Harbor, the complete S-1 section met at the National Bureau of Standards. It was time for a crash program. And to let contracts. Lawrence spoke up, immediately and eloquently, for study of electromagnetic separation on a large scale. The committee immediately recommended a sum of $400,000. It was easier than dealing with the Rockefeller Foundation.[106]

105. Hewlett and Anderson, *New world*, 44–50.
106. Ibid., 52.

3. NEW JOBS FOR CYCLOTRONS

Machine Work

As Tuve observed when shelving the almost finished Carnegie cyclotron in favor of defense work, many other directors of cyclotron laboratories were imagining how their instruments might be commissioned in the war effort. Those with half-built machines worried that otherwise they would not be able to acquire needed material; those with functioning production machines wanted to keep their staffs together and their clients supplied. But what purpose central to the national defense could a cyclotron serve? In the fall of 1939, J. Stuart Foster, casting about frantically for arguments to go forward with long-laid plans for a cyclotron at Toronto, proposed radioactive labelling of foods, metals, and strategic materials, to discover what was being shipped to enemy territory, and of documents, to detect them when stolen; radioactive beacons, to demarcate combat zones, batteries, and so on. Lawrence's reply to Foster's fancies suggests that he had not thought about military uses of his invention. "It is difficult for me to suggest concrete practical applications of the cyclotron in warfare, but it seems to me there are possibilities along the lines you have suggested."[107]

As late as June 1941, directors of cyclotron laboratories not privy to the doings of the uranium committee saw only mundane uses for their equipment. A.L. Hughes, of Washington University in Saint Louis, whose cyclotron neared completion, to cyclotron headquarters: "I suppose that Defense application of the cyclotron falls into two classes, the production of radioactive yttrium as a substitute for radium in the examination of castings and the production of tracer elements to help chemists solve their problems." Lawrence replied encouragingly. Hughes's machine got its first beam on December 10, 1941, just after Bush and Conant had decided to pursue bombs made of U^{235} and perhaps also bombs of element 94. Washington's deuteron beam was very large, almost 0.5 mA. It went to work producing samples of plutonium, in

107. Foster to Lawrence, 22 Sep, and to Cooksey, 27 Sep 1939, and Lawrence to Foster, 28 Sep and 6 Oct (quote) 1939 (7/18).

which it outdid the Crocker cracker. News gets around. At the end of December, A.C.G. Mitchell of Indiana, who had inquired over a year earlier for defense work on his cyclotron to keep Kurie and Laslett, asked Lawrence whether there might now be a "possibility that we could get a project at Indiana to use our magnet for a defense job?"[108]

Karl Compton, a man very much in the know, tried to solve the financial problems of the MIT and Harvard cyclotrons and to further the purposes of OSRD by having them declared "stand by defense group[s]." He arrived at this strategy in November 1941, after discussion with Lawrence. Compton's reasons: both laboratories needed high-priority items for supplies; both faced the breakup of their groups unless their members could "feel that they are engaged on recognized national defense work;" and lack of money. In fact, the Harvard cyclotron had received a top priority rating in August 1941, along with a contract from the NDRC to make radioisotopes for other NDRC projects; but no requests had come for over seven months.[109] Harvard got a little business—some radiosilver, some radioarsenic—in February 1942, following a directive from NDRC to use Harvard rather than Berkeley unless "very active material is required." Insufficient demand closed it down, all set up; in which condition it was later carted to Los Alamos, where it saw important service in fast-fission work under the command of Robert Wilson.[110]

The Berkeley cyclotrons early entered into business relations with the NDRC. A contract for production of miscellaneous isotopes existed by January 1941. It supported five research assistants. The Laboratory charged the government $25.25 an hour for

108. Hughes to Lawrence, 24 June and 10 Dec 1941, and Lawrence to Hughes, 10 Jul 1941 (18/12); Thornton, interview, 28 Aug 1975 (TBL), and Segrè et al., UCRL-2791 (22 Dec 1942) (12/31), on productivity of the Washington cyclotron; Mitchell to Lawrence, 28 Oct 1940, and 27 Dec 1941 (13/9).

109. K.T. Compton to M.C. Winternitz, 24 Nov 1941 (4/12); Irvin Stewart (NDRC) to Hickman, 19 Aug 1941; Hickman to F.T. Gucker, 29 Jan 1942 (all in UAV, 691.60/3).

110. Gucker to contractors of NDRC, 15 Feb 1942; letters to Hickman from George L. Clark, Illinois, 12 Feb, requesting radiosilver, and from E.O. Wiig, Rochester, 17 Feb 1942, ordering radioarsenic (all in UAV, 690/60.3); Oppenheimer to A.C. Compton, 29 Sep 1942 (12/31); Wilson in Holton, *Twentieth cent.*, 476–7.

overhead computed at 50 percent of direct costs. For a time in the spring, the machine ran 24 hours a day (it then dropped back to a more normal 12) to supply other NDRC contractors and the usual internal and external users. These orders could mount up. In the fall of 1941, the Laboratory billed $2,700 for 104 hours of cyclotron time to make 500 μCi of radioiodine and 30 mCi of radioarsenic for experiments at Caltech.[111] The contracts also provided new equipment for chemical separations, mechanisms for remote handling of hot isotopes, and so on, permanent gains from evanescent products. Kamen oversaw most of the production, under "terrific tension."[112]

The NDRC also picked up the cost of the work on element 94. Early in January 1941, Seaborg prepared at Lawrence's request an account of bombardment time and materials expended. He understood the purpose: "I believe that he is considering the possibility of having the government finance this work as an official project." In accounts rendered on March 11, 1941, $835 was charged to making element 94 by deuterons and $8,310 to the test of its fissionability by slow neutrons, exclusive of salaries. By far the largest item in both cases was cyclotron time, 315 hours completed or estimated, most of it at the 60-inch.[113] As we know, Lawrence acquired further support through the uranium committee when he advised Briggs later that March. The demonstration in May by Kennedy, Seaborg, Segrè, and Wahl that 94^{239} fissions with slow neutrons and the follow-up in July by Segrè and Seaborg, which showed that it does so with fast ones as well, and much more readily than U^{235}, brought a good deal more from the NDRC, some $21,500.[114]

111. "NDRC project..., Production of radioactive materials," Jan 1941, and "Cost of 60-inch operation," [1941] (22/3); Cornog to Volkoff, 19 Mar 1941 (5/2); Brobeck to Salisbury, 29 May 1941 (3/11); W.J. Norton to Lawrence, 30 June 1941 (20/13), reserving overhead to a special fund; Cooksey to Lawrence, 25 Sep 1941 (4/25); Griggs to O. Lundberg, 5 Nov 1941 (5/7).

112. "NDRC, production, orders," 11 Mar 1941 (22/3); Lawrence to Dunning, 6 May 1941 (6/19); Kamen to McMillan, [Feb] 1941 (10/10); cf. Gleason, *Pop. sci. monthly, 136:5* (1940), 225.

113. Seaborg to McMillan, 12 Jan 1941 (121/30); "Deuterons on uranium problem" and "Fission problem," 11 March 1941 (22/3).

114. Seaborg, *Transuranium elements,* 51.

The demonstrated fissionability of 94^{235} with fast neutrons confirmed its candidacy as an explosive. It might therefore appear odd that the third, and definitive NAS report commissioned by Bush, the report of November 1941 promoted by Lawrence, did not mention 94. But that was to follow the lead of Bush and the British, who had considered 94 chiefly in relation to a power generator. Joseph Rotblat of Chadwick's group at Liverpool had guessed at the fissionability of 94 in June or July 1940 and Bretscher had done the same by November, reasoning from the theory of Bohr and Wheeler. To go further, Bretscher told the MAUD Committee, he would need a sample procurable only at Berkeley; and he agitated "whether we should ask for facilities to enable us to work there."[115] When Oliphant visited the Laboratory in September 1941, he saw what could be accomplished with good funding, practiced investigators, and a big cyclotron. "When I saw at Berkeley the work going on there on 93 and 94 and saw activities of more than one curie of 93 I felt that you [Bretscher] were struggling against very difficult circumstances!" On this testimonial, Bretscher asked again whether he might not have samples of 93 and 94, "for which one requires neutron sources of the strength available in California."[116] But by then the new intensity of the S-1 program had put a premium on every atom of 94^{239} made in the Berkeley cyclotrons.

By the time of Pearl Harbor, the 37-inch and 60-inch machines had spent much of their time for almost a year on work that the NDRC deemed to be in the national interest. By then the 184-inch had also come under government protection, if not into government service. The commissioning occurred during the summer of 1941. Lawrence had been prompt in ordering the steel and copper, at the discounted prices that he and Loomis had negotiated with U.S. Steel (a savings of $30,000) and Phelps-Dodge; and also the power supply from GE, at its generous price without overhead charge. The contracts for the metal were let in July, and

115. Chadwick to Bretscher, 5 Nov 1940 and 26 Feb 1946; Bretscher to Chadwick, 20 Feb 1946; Bretscher, "Summary of the report given at the meeting of the MAUD Committee on 9th April [1941]," 3, quote; all in Mrs. Bretscher's possession.
116. Oliphant to Bretscher, 17 Oct 1941, and Bretscher to Chadwick, 5 Jan 1942 (Bretscher P).

for the construction of the magnet in September; the first installment of steel, some twenty-eight tons, a good chunk to be sure but
less than 1 percent of the whole, went up the newly built road to
the hilltop site on the last day of October 1940.[117] Problems with
the unions in December occasioned delays in building, but design
work went forward expeditiously. Around New Year the parameters had been set, or almost so, to obtain deuterons of 100 MeV:
the minimum gap between the poles would be 40 inches, or even 4
feet, to allow sufficient clearance between the dees and the
chamber walls to permit peak voltages of a million across the
accelerating gap. By the middle of February 1941, Lawrence had
succeeded in spending $534,550 on the he-man cyclotron.[118]

In the spring, however, just as workers were to start raising the
cyclotron around the 30-foot tall magnet frame, construction met
resistance that money alone could not overcome. To assure its
supplies, the Laboratory tried to buy up all the copper and brass
available locally before the government restricted or acquired the
stock. In the summer, some contractors refused to ship equipment they had already made unless the 184-inch project could
acquire a priority rating. In the fall, Salisbury, visiting GE's plant
to enjoy the sight of the power transformers advertised as partially
completed, learned that the completion referred only to the blueprints. "I was told that nothing would be done for at least a year
unless we could get a priority." The same message came from
Phelps-Dodge. "This unexpected turn of events made it necessary
for us to appeal to the National Defense Research Council,"
Lawrence wrote the Rockefeller Foundation, in explanation of the
admission of a new partner to their agreement. "Washington has
been most cooperative....[The] construction program is now going
full steam ahead."[119] Lawrence viewed the intervention as a favor,

117. Lawrence to A.H. Compton, 22 Jul 1940 (4/10); to M. Smith, Westingouse,
29 Jul 1940 (18/19); to Weaver, 31 Jul and 24 Sep 1940 (15/30); Cooksey to Wilson, 31 Oct 1940 (18/29).

118. Brobeck to Cooksey, 19 Dec 1940 (26/1); Cooksey to Hafstad, 22 Jan 1941
(3/32); Lawrence to Weaver, 4 Mar 1940 (15/30), on projected dee voltage; expenditures as of 17 Feb 1941 (22/3).

119. Brobeck to Snell, 29 May 1941, and "The cyclotron," Oct 1941, 10 (3/11).
Cooksey to L.G. Baker, 7 May 1941 (22/3); Salisbury to Cooksey, 23 Sep 1941
(16/6); Cooksey to Lawrence, 27 Sep 1941 (4/25); Lawrence to George W. Gray,
17 Oct 1941 (15/31).

not as a measure for defense; most of the materials needed had already been shipped to Berkeley before priorities obtruded; the "blessing of the OSRD," Lawrence told the Research Corporation, had fallen on the project "in line with a wise and foresighted policy of encouraging steady scientific progress in the midst of the stress of war activities." He hoped to have the cyclotron ready for trials in 1943.[120]

If this encouragement were truly the purpose of the OSRD, virtue reaped its reward. The big magnet proved a fateful instrument of war. But no one saw it in the spring or summer of 1941. The uranium committee had paid for the electromagnetic separation of microgram samples of U^{235} only for determination of its nuclear parameters. No more than its British counterpart did the uranium committee or the first two NAS committees consider electromagnetism to be an option for large-scale separation. The third NAS committee—the one that reported on November 6—mentioned, without specifying, "other methods" than centrifugation and gaseous diffusion then under or needing investigation.[121]

As an expert on large magnets, Lawrence took on responsibility within S-1 for electromagnetic separation of small samples of light uranium for experimental purposes. Nier came to Berkeley in November to help Brobeck convert the 37-inch cyclotron into a mass spectrograph. The principle of operation is represented in figure 10.3: gaseous uranium ions from the source traverse circular paths under the old Poulsen magnet, the lighter isotope following the tighter circle and ending (if all went well) in a collecting cup. The first rough test at the end of November gave encouraging results; the mechanism, which became known as the "calutron" after the institution that gave it birth, appeared likely to contribute samples of U^{235} for experimental purposes.[122]

This performance suggested that the chief technical difficulties in large-scale separation—the skimpiness of the beam and the menace of space charge that deflects the ions from their circles—

120. Lawrence, "Report to the Research Corporation [for 1941]" (15/19); Lawrence to Gray, 17 Oct 1941 (15/31).

121. Smyth, *Atomic energy*, 66, 71–2, 165; Hewlett and Anderson, *New world*, 36.

122. Brobeck, "Reminiscences," 5 Nov 1982, 2 (TBL); Hewlett and Anderson, *New world*, 50; Smyth, *Atomic energy*, 188–9.

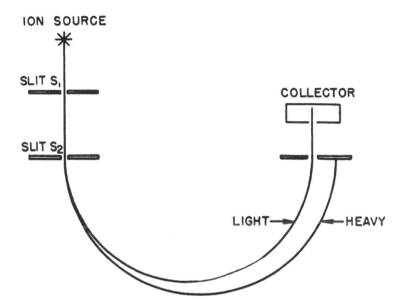

FIG. 10.3 Principle of the calutron. It is not easy to obtain the clean and copious separation indicated. Smyth, *Atomic energy*, 164.

might be overcome. Already in October, Lawrence and Henry Smyth of Princeton had discussed the difficulties and persuaded each other to be optimistic. They had quite different corrections in mind. Lawrence thought to apply cyclotron technology and to build bigger; Smyth, who had some experience with Princeton's mass spectrometer, sought a new approach that would do without the collimating slits and so make use of an extended source. The ever-resourceful Robert Wilson found a way to help his colleague Smyth and challenge his teacher Lawrence. Wilson proposed to do without slits and magnetic fields by adapting the principle of the klystron: a wide beam of uranium ions would pass through a cavity oscillating at radio frequencies and emerge in two sets of bunches, one of each uranium isotope; at the cross section of the drift space where the bunches are best defined, a transverse rf field would work, phased to draw aside the heavier isotopes while allowing the lighter to pass on to a collector. The "isotron" (so named for no reason at all) as well as the calutron received funding at the meeting of the uranium committee on December 18, 1941.[123]

123. Smyth, *Atomic energy,* 66, 197–9.

In January 1942, promising experimental versions of both machines existed. But the isotron scarcely had a chance. Lawrence's means were not limited to what the government chose to give him—Brobeck had converted the 37-inch to a calutron on money from the Research Corporation—or to local or unpracticed help. A call went out, and cyclotroneers came home—James Cork, J.R. Richardson, and Robert Thornton, among others—to join or rejoin men who had not left. (Tables 10.2 and 10.3 tell who ended where.) These men knew their business. They put poles on the frame of the 184-inch magnet and went to work squashing the bugs in calutron prototypes tested under the Rockefeller Foundation's investment in the higher aspirations of the human species. They squashed the isotron soon enough. "Ernest wanted to cannibalize our group," Wilson remembered. "We resorted [in vain] to every device of politics and rhetoric to forestall the take-over."[124] It took most of the rest of the war, most of the efforts of the Laboratory, and most of a billion dollars to make successful calutrons.

Glimpses of a New Era

Government work had its bright side. The salaries of research assistants and associates rose sharply, some tripling, when the civil service classified them. Nonacademic staff also prospered, although less dramatically. Griggs, who received no adjustment to salary in 1940/41 despite the large increase in her duties, got 9 percent more in 1941/42; Harvie, much underpaid as shop foreman in 1940/41 at $2,000, received three times the increase under the NDRC than he had from Sproul. As for the academic staff, Lawrence paid a monthly salary of a tenth the annual rate plus a subsistence allowance of $150 a month to people he wanted to attract.[125] Then there were deferments. Seaborg pulled a low draft number. It did not worry him: "I expect to be involved in some sort of scientific work for the war effort instead of being drafted."

124. Smyth, *Atomic energy*, 189–90, 198–9; Wilson in Holton, *Twentieth cent.*, 474; Cork to Lawrence, 1 June 1942 (5/1); Richardson to Lawrence, 20 Mar 1942 (15/22).

125. Kamen, *Radiant science*, 141; Lawrence to Sproul, 2 Sep, and reply, 20 Sep 1940 (23/17); to J.H. Loughbridge, 12 June 1942 (12/40); printed budget, 1940/41–1942/43 (23/17).

Table 10.2 Subsequent Careers of Radiation Laboratory Regulars, 1932/40 (Berkeley Ph.D.s)

Name	Ph.D	Departure	Destination(s)	Type of job	War service[a]	Postwar career[b]
Abelson, P.H.	1939	1939	Carn. Inst.	cyclotron	Nav. Res. Lab.	nonacademic
Aebersold, P.C.	1939		UCRL	biophysics	MED	AEC
Backus, J.G.	1940		UCRL	cyclotron	UCRL	prof.
Coates, W.M.	1933	1933	Columbia U.	Sloan X-ray	died 1937	died 1937
Condit, R.I.	1942	1942	MED	war work	MED	industry
Cornog, R.A.	1940	1941	Navy Ordnance	war work	Navy; MED	prof.
Corson, D.R.	1938	1940	U. Miss.	physics	MIT	prof., admin.
Helmholtz, A.C.	1940	1942	MED	war work	MED	prof.
Kalbfell, D.C.	1939	1939	Standard Oil	physics	OSRD	industry
Kennedy, J.W.	1939, chem.			phys. chem.	UCRL	prof.
Laslett, L.J.	1937	1937	Bohr Inst., U. Ind	cyclotron	MIT	prof.; UCRL
Linford, L.H.	1935	1935	Utah St. Ag. Coll.	math. & physics		prof.
Livingston, M.S.	1931	1934	Cornell; MIT	cyclotron		BNL
Livingston, R.S.	1941	1943	Tenn. Eastman	war work	MED	AEC
Lofgren, E.J.	1946	1944	MED	war work	MED	UCRL
Lyman, E.M.	1938	1938	U. Illinois	cyclotron	MIT	prof.
MacKenzie, K.R.	1940	1946	UCLA	physics	UCRL	prof.
Nag, B.C.	1940	1940	U. Calcutta	physics, cyclotron		prof.
Paxton, H.C.	1937	1937	Paris; Columbia U.	cyclotron	MED	industry
Raymond, R.C.	1941	1941	Am. Cyanamid	physics	MIT	prof.; industry
Richardson, J.G.	1937	1937	U. Mich; U. Ill.	cyclotron	MED	prof.
Simmons, S.J.	1939	1939	U. Pittsburgh	cyclotron		died 1946
Sloan, D.H.	1941	1942	Westinghouse	engineer	Westinghouse	prof., UCB
Tuttle, L.W.	1948	1940	NDRC	war work	MED	AEC; prof.
Wahl, A.C.	1942, chem	1943	MED	war work	MED	prof.
White, H.J.	1933	1935	Res. Corp.	physics	MIT	nonacademic
White, M.G.	1935	1935	Princeton	cyclotron	MIT	prof.
Wilson, R.R.	1940	1940	Princeton	physics	MED	prof.; Fermilab
Wright, B.T.	1941	1941	NDRC	war work	MED	UCLA
Wu, C.S.	1940	1942				prof.
Yockey, H.P.	1942	1944	Tenn. Eastman	war work	MED	AEC; Aberdeen

Note: 'Regulars' are those who worked in the Laboratory to prepare for a career in science; they do not include visitors on sabbatical or fellowship leave, or volunteers without career ambitions, like Emo and Lucci.

a. MED, Manhattan Project; MIT, Radiation Laboratory, MIT; OSRD, Office of Scientific Research and Development; UCRL, University of California Radiation Laboratory.

b. AEC, Atomic Energy Commission; BNL, Brookhaven National Laboratory; UCB, University of California, Berkeley.

Table 10.3   Subsequent Careers of Radiation Laboratory Regulars, 1932/40 (Postdocs and Undergraduates)

Name	Date of arrival	Date of departure	Destination(s)	Type of job	War service[a]	Postwar career[b]
Alvarez, L.	1936	1940	MIT	war work	MIT→MED	prof., UCB
Brobeck, W.M.	1937			engineer	UCRL/MED	engineering
Cooksey, D.	1935			assoc. dir.	UCRL/MED	admin., UCRL
Cowie, D.B.	1935	1938	Swarthmore	student		
Farley, W.W.	1938	1941	MIT	war work	MIT	industry
Green, G.K.	1938	1939	Carn. Inst.	cyclotron		BNL
Hamilton, J.G.	1938			med. phys.	UCRL/MED	UCRL
Henderson, M.C.	1933	1935	Princeton	cyclotron		UCRL
Hurst, D.G.	1936	1937	Cambridge	cyclotron	AEC/Canada	AEC
Kamen, M.D.	1937	1941		cyclotron		AEC/Canada
Kinsey, B.B.	1933	1936	Liverpool	cyclotron		prof.
Kurie, F.N.D.	1933	1938	Indiana	cyclotron		Navy Res. Lab.
Langsdorf, A.	1938	1939	Wash. U.	cyclotron		AEC
Lawrence, J.H.	1936				UCRL	prof., UCRL
Lewis, M.N.	1938	1942	Vassar	physics		
Livingood, J.J.	1932	1938	Harvard	cyclotron	OSRD	AEC
McMillan, E.M.	1932	1940	MIT	war work	MIT→MED	prof., UCB
Mann, W.B.	1936	1938	London			NBS
Salisbury, W.W.	1937	1941	MIT	war work	MIT	industry
Scott, K.G.	1938	1941	med school	student		
Seaborg, G.T.	1936	1942	Chicago	war work	MED	prof., UCB
Snell, A.H.	1934	1938	Chicago	cyclotron	MED	AEC
Thornton, R.L.	1933	1936	Mich.; Wash. U.	cyclotron	MED	prof., UCRL
Van Voorhis, S.N.	1935	1938	Rochester	cyclotron	MIT	Lincoln Lab
Walke, H.	1935	1937	Liverpool	cyclotron	died 1939	died 1939

Note: "Regulars" are those who worked in the Laboratory to prepare for a career in science; they do not include visitors on sabbatical or fellowship leave, or volunteers without career ambitions, like Emo and Lucci.

a. MED, Manhattan Project; MIT, Radiation Laboratory, MIT; OSRD, Office of Scientific Research and Development; UCRL, University of California Radiation Laboratory.

b. AEC, Atomic Energy Commission; BNL, Brookhaven National Laboratory; UCB, University of California, Berkeley.

In April 1941 Lawrence wrote that the Laboratory had no trouble procuring deferments for all its graduate students and research fellows. In August, Griggs reported to the American Institute of Physics that only one member of the staff had been called to duty, a research assistant who was also an ensign.[126] No one faced unemployment. "There aren't any men available here." Lawrence raided the movie studios for engineers, junior colleges for physicists, high schools for shop boys and storeroom clerks.[127] And, when the OSRD began to outfit the 184-inch magnet for calutron work, he reached the end, or rather, the beginning, of the supply of trained manpower. "We have government defense jobs up here for almost any number of undergraduate majors in physics."[128]

And now the downside. All these nouveaux arrivés destroyed the spirit of the Laboratory. Ninety people, most of them employees of a few months' seniority, attended the Christmas party in 1940. "It wasn't a cozy gang....We all missed you like hell." Thus insider still inside, Kamen, to insider then outside, McMillan. Some had ceased to be insiders. When Wolfgang Gentner visited Berkeley in 1939, travelling under the patronage of the German government, he had been given the freedom of the Laboratory.[129] For several years Lawrence and Cooksey had extended themselves to help Sagane and others copy the 60-inch. None of that was possible in the fall of 1940. The regents ordered that the Laboratory be closed to foreign scientists. The reasons, as explained by Lawrence to Nishina: overcrowding and "a certain amount of work in progress of a confidential character." The boycott extended to information about cyclotrons, or so Cooksey understood it; and he evaded requests from the Japanese for blueprints of the Crocker cyclotron. The connection was pointed out by Brobeck to Kurie, who had proposed a symposium on cyclotronics. "The cyclotron is now [March 1941] working on national

126. Seaborg, *Jl.*, *2*, 654 (22 Nov 1940); Lawrence to Hughes, 7 Apr 1941 (1/30); Griggs to Barton, 14 Aug 1941 (2/2).

127. Lawrence to Cork, 4 June 1941 (13/5), quote; correspondence with Paramount studios (14/16); Kenneth Simpson to Lawrence, 9 May 1942 (16/26); Cooksey to Norton, 26 Jan 1942 (20/13), re the boys, who worked a 54-hour week.

128. Lawrence to Curtis Haupt, 8 May 1942 (9/2).

129. Kamen to McMillan, 25 Dec 1940 (10/10); Gentner, "Bericht über die Reise nach Nordamerika...zum Studien der Cyklotronapparaturen," attached to Bothe to Generalverwaltung der KWG, 25 May 1939 (MPG, Akten/1063,5).

defense and may soon become an official defense project, so that it might not be wise or even possible to publish information about it."[130]

Among the disagreeable features of war is falling toward the level of one's enemies. A few years earlier Rutherford and Lawrence had joked about restriction of access to German laboratories. Rutherford: "This state of affairs in Nazi-land is rather amusing, and when some of our men from the Cavendish wished to visit Berlin to see Debye's laboratory [at the new Kaiser-Wilhelm-Institut für Physik], he wrote to Cockcroft that official permission would have to be granted by the Government before he could admit them!" Lawrence: "Your account of the state of affairs in Germany is almost unbelievable. One would think that with such a scientific tradition the German people could not adopt such an absurd course of action in scientific affairs."[131] In the fall of 1940 the reciprocal action seemed not absurd but prudent.

The regents did not stop with closing the Laboratory to touring potential enemies. They further ordered the firing of aliens paid from state funds. The comptroller made plain to Lawrence how deep the edict cut. "We regret that under the regulation...Mr Segrè is not eligible for employment by the University. Immediate steps should be taken to dismiss this employee from your staff." When this ukase came, Segrè and Seaborg were engaged in establishing the fissionability of element 94 by fast neutrons. Their collaboration had been inhibited earlier by the secrecy of the uranium commitee, which asked that Seaborg direct his sensitive results directly to Briggs and not confide in Segrè, who had a tendency to inform another alien, Fermi. Lawrence had done something to regularize the situation so that (as McMillan wrote Seaborg) "You will be able to talk to Segrè after all." Lawrence saw to it that the conversations continued. He obtained

130. Lawrence to Nishina, 22 Aug 1940 (9/38), and to R.E. Worley, 4 Nov 1940 (3/32); Cooksey to Griggs, 17 Dec 1940, and Takamina to Cooksey, 2 and 10 Jan 1941 (9/42); Brobeck to Kurie, 12 Mar 1941 (3/11). Cooksey urged Livingston not to publish a treatise on cyclotrons he circulated in manuscript, lest it enable the "Tokyo group" to get their machine in good form; draft letter, 18 Nov 1943 (12/12).

131. Rutherford to Lawrence, 21 Dec 1936, and reply, 11 Feb 1937 (ER), quoted in Oliphant, PT, 19:10 (1966), 48, and in Weart and Phillips, *History*, 189–90.

permission from the provost to have the Physics Department hire Segrè as a part-time lecturer (then much in demand) paid for not by state funds but from the Rockefeller money, the balance of his time to be spent as a research fellow on the same fund in the Laboratory. Similar arrangements kept Kenneth MacKenzie (a Canadian) and C.S. Wu.[132]

Throughout this time of partial mobilization—from the award of the Rockefeller grant in April 1940 to Pearl Harbor—Lawrence tried to keep the Laboratory at peace work as well as at war work. The beginning of the period coincided with the end of the phony war and the beginning of the end for France. "The war has taken an almost incredible turn for the worse," he wrote Morris, on May 10. "I hope we will get busy with the problem of arming ourselves....We need to take every advantage of modern science and technology." He did not include himself among those who should busy themselves with the problem. On May 21 he listed "an impossible amount of work immediately ahead," namely, the medical program, nuclear physics, and the building of "the great cyclotron." On May 29, answering Thornton, who had told of the death of his brother, a Canadian, on the battlefield: "With the world situation as it is, it does seem hard to go ahead with our work. However, this is what we must do."[133] And did. According to Bethe, who visited California six weeks later, the hawks of scientific preparedness wheeled around the old campaigner from Caltech, not around Lawrence. "Millikan, and it seems practically everybody at Pasadena is working on defense."[134]

The appointment to Loomis's committee under the NDRC first brought Lawrence significantly into military preparedness. But, as Millikan rightly observed, Lawrence's part was to send others to do the work, not to do it himself. At the Laboratory, too, he played the occasional expeditor in work on the uranium problem

132. J.H. Corley to Lawrence, 29 Jul, and reply, 2 Aug 1941 (16/14); Seaborg, *Jl.*, *2*, 668, 678 (3 and 16 Jan 1941), 694 (1 Mar 1941), reporting disclosing the separation of 94 to Lewis and Latimer ("I am confident they will treat it as a secret"); McMillan to Seaborg, 26 Mar 1941 (12/30).

133. Lawrence to Morris, 10 May 1940 (13/13); to Weaver, 21 May 1940 (15/40); and to Thornton, 29 May 1940 (17/14).

134. Bethe to Lark-Horowitz, n.d., quote, and to C.C. Murdoch, 23 Jul 1940, mentioning especially Tolman and Lauritsen (HAB, 3); Lawrence to A.H. Compton, 22 Jul 1940 (4/10).

and element 94. Unlike Tuve, intensely at work on defense matters from the fall of 1940, Lawrence's main concern until late in 1941 was the construction of cyclotrons. That fall he could still write the Rockefeller Foundation and the Research Corporation that he intended to keep up the program in basic nuclear science. In October he listed pure physics, NDRC projects, medical research, and therapy, in that order, as the chief work of the 37-inch and 60-inch cyclotrons.[135]

Lawrence threw his heart and soul into war work during his meeting with Conant and Arthur Compton at Chicago's golden anniversary at the end of September 1941. Lawrence gave his colleagues a pep talk about the MAUD report and Berkeley's plutonium work. Conant asked whether he would set aside the big cyclotron and devote the next several years of his life to making a bomb. The question "brought up Lawrence with a start. I can still recall the expression in his eyes....He hesitated only a moment. 'If you tell me this is my job, I'll do it.'"[136] Once committed, he pursued the goal with his native optimism and relentlessness. At a meeting with Compton and others late in October, there was some talk of the uncertainties of the undertaking. "This, to my mind, is very dangerous," Lawrence wrote Compton the next day. "It will not be a calamity if when we get the answers to the Uranium problem they turn out negative from the military point of view, but if the answers are fantastically positive and we fail to get them first, the results for our country may well be a tragic disaster. I feel strongly, therefore, that anyone who hesitates on a vigorous all-out effort on Uranium assumes a grave responsibility."[137] Lawrence could be content only with a project through which his creation, his Laboratory, might make a decisive contribution to victory. "Calutron" says it all: the contribution of California, its University, and its Radiation Laboratory to a weapon that might change the course of history.

135. Millikan to Neher, answering Neher's letter of 23 Nov 1940 (Millikan P); Lawrence to Poillon, 22 Sep 1941 (15/19), and to G.M. Gray, 17 Oct 1941 (15/31).
136. Compton, *Atomic quest,* 6; Childs, *Genius,* 317–18, quote.
137. Lawrence to Compton, 22 Oct 1941 (4/10), addressed to "Dear Dr Compton," evidently for wider circulation.

Bibliography

There follows a list of all published items cited in the notes except editorial comments, short anonymous articles, and interventions at scientific meetings. A few unpublished manuscripts are also listed here—rather than described in full where mentioned in the notes, as is our usual practice—if we refer to them several times or if they have some general interest.

Articles from a book are cited in short form—last name of editor, abbreviated title—when several from the same book were used. The full citation can be retrieved by looking up the book in its alphabetical place under its editor, or, if it has no stated editor, under its title.

The following abbreviations are used:

AIP	American Institute of Physics
APS	American Philosophical Society
BAS	*Bulletin of the atomic scientists*
CIW	Carnegie Institution, Washington
CP	*Collected papers*, identified by author
CW	*Collected works*, identified by author
HSPS	*Historical studies in the physical and biological sciences*
IEEE	Institute of Electrical and Electronics Engineers
IRE	Institute of Radio Engineers
ICHS	International Congress of History of Science
JACS	American Chemical Society, *Journal*
JFI	Franklin Institute, *Journal*
NAS	National Academy of Sciences
NRC	National Research Council
Nwn	*Die Naturwissenschaften*
PR	*Physical review*
PRS	Royal Society of London, *Proceedings*
PT	*Physics today*
RMP	*Reviews of modern physics*
RSI	*Review of scientific instruments*
SEBM	Society for Experimental Biology and Medicine
TBL	The Bancroft Library, University of California, Berkeley

Aaserud, Finn. "Niels Bohr as fund raiser." *PT, 38:10* (1985), 38–46.

Abelson, P.H. "Neutron produced activities in uranium." *PR, 53* (1938), 211–2.

———. "Cleavage of the uranium nucleus." *PR, 55* (1939), 418.

———. "Further products of uranium cleavage." *PR, 55* (1939), 670.

———. "The identification of some of the products of uranium cleavage." *PR, 55* (1939), 876–7.

———. "An investigation of the products of the disintegration of uranium by neutrons." *PR, 56* (1939), 1–9.

———. "The identification of characteristic x-rays associated with radioactive decay." *PR, 56* (1939), 753–7.

———. "A sport played by graduate students." *BAS, 30* (1974), 48–52.

———. "Merle A. Tuve." *PT, 35:9* (1982), 90–3.

Adler, F., and Hans von Halban. "Control of the chain reaction involved in fission of the uranium nucleus." *Nature, 143* (1939), 793.

Aebersold, P.C. "The production of a beam of fast neutrons." *PR, 56* (1939), 714–27.

Aitken, H.G.J. *The continuous wave: Technology and American radio, 1900–1932.* Princeton: Princeton University Press, 1985.

Alleluia, Irene B. de, and Cornelius Keller. "Occurrence of technetium." *Gmelins Handbuch der anorganischen Chemie: Technetium. Supplement, 1.* Berlin: Springer, 1982. Pp. 12–25.

Allen, J.S., and L.W. Alvarez. "A thin-walled Geiger-counter." *RSI, 6* (1935), 329.

Allen, R.W., and R.W. Parkinson. *Examination of the botanical references at Nova Albion.* San Francisco: Drake Navigators' Guild, 1971.

Allibone, T.E. "The industrial development of nuclear power." *PRS, A282* (1964), 447–63.

———. "Metropolitan-Vickers Electrical Company and the Cavendish Laboratory." In Hendry, *Cambridge physics,* 150–73.

———. "Dennis Gabor." Royal Society of London. *Biographical memoirs, 26* (1980), 107–47.

Alvarez, L.W. "Nuclear K electron capture." *PR, 52* (1937), 134–5.

———. "Electron capture and internal conversion in gallium 67." *PR, 53* (1938), 606.

———. "Collimated, variable energy beam of pure thermal neutrons." *PR, 54* (1938), 235.

———. "The capture of orbital electrons by nuclei." *PR, 54* (1938), 486–97.

———. "The production of collimated beams of monochromatic neutrons in the temperature range of 300° to 10° K." *PR, 54* (1938), 609–17.

———. "Adventures in nuclear physics." University of California Radiation Laboratory Publication, UCRL-10476 (1962).

———. "The early days of accelerator mass spectrometry." *PT, 35:1* (1982), 25–32.

———. "Alfred Lee Loomis—last great amateur of science." *PT, 36:1* (1983), 25–34, and in Weart and Phillips, *History of physics*, 198–207.

———. *Adventures of a physicist.* New York: Basic Books, 1987.

Alvarez, L.W., and Felix Bloch. "A quantitative determination of the neutron moment in absolute nuclear magnetons." *PR, 57* (1940), 111–22.

———. "The magnetic moment of the neutron." *PR, 57* (1940), 352.

Alvarez, L.W., and Robert Cornog. "He^3 in helium." *PR, 56* (1939), 379.

———. "Helium and hydrogen of mass 3." *PR, 56* (1939), 613.

———. "Radioactive hydrogen." *PR, 57* (1940), 249.

———. "Radioactive hydrogen—a correction." *PR, 58* (1940), 197.

Alvarez, L.W., A.C. Helmholz, and Eldred Nelson. "Isomeric silver and the Weizsäcker theory." *PR, 57* (1940), 660–1.

Amaldi, Edoardo. "Studio dei ciclotroni negli istituti degli Stati Uniti." *Viaggi, 6* (1940), 7–10.

———. "Personal notes on neutron work in Rome in the thirties and post-war European collaboration in high-energy physics." In Weiner, ed., *History of twentieth century physics*, 294–351.

———. "From the discovery of the neutron to the discovery of nuclear fission." *Physics reports, 111* (1984), 1–332.

Amaldi, Edoardo, O. D'Agostino, Enrico Fermi, Bruno Pontecorvo, Franco Rasetti, and Emilio Segrè. "Artificial radioactivity produced by neutron bombardment, II." *PRS, A149* (1935), 522–58, and in Fermi, *CP, 1*, 765–94.

Amaldi, Edoardo, and Enrico Fermi. "On the absorption and the diffusion of slow neutrons." *PR, 50* (1936), 899–928, and in Fermi, *CP, 1*, 892–942 .

———. "Sopra l'assorbimento e la diffusione dei neutroni lenti." *Ricerca scientifica, 7* (1936), 454–503, and in Fermi, *CP, 1*, 841–91.

Amaldi, Edoardo, Enrico Fermi, Franco Rasetti, and Emilio Segrè. "Nuovi radioelementi prodotti con bombardamento di neutroni." *Nuovo cimento, 11* (1934), 442–7, and in Fermi, *CP, 1*, 725–31.

Amaldi, Edoardo, L.R. Hafstad, and M.A. Tuve. "Neutron yields from artificial sources." *PR, 51* (1936), 896–912.

Amaldi, Edoardo, Bruno Pontecorvo, Franco Rasetti, and Emilio Segrè. "Azione di sostanze idrogenate sulle radioattività provocata da neutroni, I." *Ricerca scientifica, 5:2* (1934), 282–3, and in Fermi, *CP, 1*, 757–8.

Ambrosen, Johan. "Atomsønderling og kunstig radioaktivitet." *Fysisk tidsskrift, 32* (1934), 131–57.

American Association for the Advancement of Science. Committee on Patents, Copyrights, and Trademarks. *The protection by patents of scientific discoveries.* New York: Science Press, 1934.

American Association of University Professors. Committee Y. *Depression, recovery, and higher education.* New York: McGraw-Hill, 1937.

American Institute of Chemical Engineers. Patent Committee. "Symposium on patents." American Institute of Chemical Engineers, *Transactions, 28* (1932), 183–234.

Anderson, Elizabeth N. "Hercules and the muses: Public art at the fair." In Benedict, *World's fairs*, 114–33.

Anderson, H.L. "Early days of the chain reaction." *BAS, 29:4* (1973), 8–12.

Anderson, H.L., E.T. Booth, J.R. Fleming, Enrico Fermi, G.N. Glasoe, and F.G. Stack. "The fission of uranium." *PR, 55* (1939), 511–2.

Anderson, H.L., J.R. Dunning, and D.P. Mitchell. "Regulator systems for electromagnets." *RSI, 8* (1937), 497–501.

Anderson, H.L., Enrico Fermi, and H.B. Hanstein. "Production of neutrons in uranium bombarded by neutrons." *PR, 55* (1939), 797–8.

Anderson, H.L., Enrico Fermi, and Leo Szilard. "Neutron production and absorption in uranium." *PR, 56* (1939), 284–6.

Angell, J.R. "The development of research in the United States." NRC. *Reprint*, no. 6 (1919).

Anonymous. "The Lafayette radio station." *Radio review, 2* (1921), 85–93.

———. "The Chemical Foundation." *Chemical industry, 36* (1939), 139–43.

———. "Story of research—the resnatron, world's most powerful microwave tube." *Westinghouse engineer, 6:2* (1946), 47.

Ardenne, Manfred von. *Mein Leben für Forschung und Fortschritt.* 7th ed. Munich: Nymphenburger, 1984.

Arnon, D.I., P.R. Stout, and F. Sipos. "Radioactive phosphorus as an indicator of phosphorus absorption of tomato fruits at different steps of development." *American journal of botany, 27* (1940), 791–8.

Artom, C., G. Sarzana, C. Perrier, M. Santangelo, and Emilio Segrè. "Rate of 'organification' of phosophorus in animal tissues." *Nature, 139* (1937), 836–7.

———. "Phospholipid synthesis during fat absorption." *Nature, 139* (1937), 1105–6.

Aston, F.W. "The atoms of matter: Their size, number, and construction." *Nature, 110* (1922), 702–5.

———. "The constitution of oxygen." *Nature, 123* (1929), 488–9.

Axelrod, Dorothy, P.C. Aebersold, and J.H. Lawrence. "Comparative effects of neutrons and x rays on three tumors irradiated in vitro." SEBM, *Proceedings, 48* (1941), 251–6.

Bacher, R.F. "Elastic scattering of fast neutrons." *PR, 55* (1939), 679–80.

Bacher, R.F., and D.C. Swanson. "Scattering of fast neutrons." *PR, 53* (1938), 676.

———. "On the scattering of fast neutrons." *PR, 53* (1938), 922.

Badash, Lawrence. "Nuclear physics in Rutherford's laboratory before the discovery of the neutron." *AJP, 51:10* (1983), 884–89.

———. *Kapitza, Rutherford, and the Kremlin.* New Haven: Yale University Press, 1985.

Badash, Lawrence, Elizabeth Hodes, and Adolph Tiddeus. "Nuclear fission: Reaction to the discovery in 1939." APS. *Proceedings, 130* (1986), 196–231.

Bainbridge, K.T. "The isotopic weight of H^2." *PR, 41* (1932), 115.

———. "The mass of Be^9 and the atomic weight of beryllium." *PR, 43* (1933), 367–8.

———. "The masses of the lithium isotopes." *PR, 44* (1933), 56–7.

———. "The equivalence of mass and energy." *PR, 44* (1933), 123.

Baker, C.P., and R.F. Bacher. "Experiments with a slow neutron velocity spectrometer." *PR, 59* (1941), 332–48.

Baker, S.W., Jr. *The Rackham funds of the University of Michigan, 1933–1953.* Ann Arbor: University of Michigan Press, 1955.

Baracca, Angelo. "A differentiation between 'big science' vs. 'little science': Lawrence and Tuve, first experiments with deuterons." Unpublished ms., 1985.

Barnes, S.W., L.A. DuBridge, E.O. Wiig, J.H. Buck, and C.V. Strain. "Proton-induced radioactivity in heavy nuclei." *PR, 51* (1937), 775.

Barre, H.A. "New transmission line construction in 1920." *Journal of electricity, 45* (1920), 566–7.

B[arton], H.A. "Notes on physics in industry." *RSI, 5* (1934), 263–4, 309–10; *RSI, 6* (1935), 30.

Barton, H.A., D.W. Mueller, and L.C. van Atta. "A compact high potential electrostatic generator." *PR, 42* (1932), 901.

Baxter, J.P., 3rd. *Scientists against time.* Boston: Little, Brown, 1946.

Beams, Jesse. "High speed centrifuging." *RMP, 10* (1938), 245–63.

Beams, Jesse, and E.O. Lawrence. "On the lag of the Kerr effect." NAS. *Proceedings, 13* (1927), 505–10.

———. "On relaxation of electric fields in Kerr cells and apparent lags of the Kerr effect." *JFI, 206* (1928), 169–79.

Benedict, Burton. *The anthropology of world's fairs. San Francisco's Panama Pacific International Exposition of 1915.* London and Berkeley: Scholar Press, 1983.

Bethe, H.A. "Nuclear physics. B. Nuclear dynamics, theoretical." *RMP, 9* (1937), 69–244.

———. "The Oppenheimer-Phillips process." *PR, 53* (1938), 39–50.

———. "Binding energy of the neutron." *PR, 53* (1938), 313–4.

———. "The history of nuclear physics at Cornell." Unpublished ms., ca. 1960. Bethe Papers, Cornell University Library.

Bethe, H.A., and R.F. Bacher. "Nuclear physics. A. Stationary states of nuclei." *RMP, 8* (1936), 82–229.

Bethe, H.A., and M.E. Rose. "The maximum energy obtainable from the cyclotron." *PR, 52* (1937), 1254–5.

Birge, R.T. "Address." In U.C., *1939 prize,* 11–27.

———. *History of the physics department.* Mimeograph. 5 vols. Berkeley, 1966–?.

Birge, R.T., and D.H. Menzel. "The relative abundance of the oxygen isotopes, and the basis of the atomic weight system." *PR, 37* (1931), 1669–71.

Bjerge, T., K.J. Bronstrøm, J.Koch, and T. Lauritsen. "A high tension apparatus for nuclear research." Danske Videnskabernes Selskab. *Mathematisk-fysiske meddelser, 18:1* (1940).

Blackett, P.M.S. "The ejection of protons from nitrogen nuclei, photographed by the Wilson method." *PRS, A107* (1925), 349–60.

Bleakney, Walker, and A.J. Gould. "The relative abundance of hydrogen isotopes." *PR, 44* (1933), 265–8.

Bleakney, Walker, G.P. Harnwell, W.W. Lozier, P.T. Smith, and H.D. Smyth. "The production and identification of helium of mass three." *PR, 46* (1934), 81–2.

Blewett, J.P., and E.D. Courant. "M. Stanley Livingston." *PT, 40:6* (1987), 88–9, 92.

Bloch, Felix. "On the magnetic scattering of neutrons." *PR, 50* (1936), 259–60.

———. "On the magnetic scattering of neutrons, II." *PR, 51* (1937), 994.

———. "Le moment magnétique du neutron." Institut Henri Poincaré. *Annales, 8:1* (1938), 63–78.

Bloch, Felix, Morton Hamermesch, and Hans Staub. "Neutron polarization and ferromagnetic saturation." *PR, 64* (1943), 47–56.

Bloch, Felix, David Nicodemus, and Hans Staub. "A quantitative determination of the magnetic moment of the neutron in units of the proton moment." *PR, 74* (1948), 1025–45.

Bohr, Niels. "Transmutations of atomic nuclei." *Science, 86* (1937), 161–5.

———. "Resonance in uranium and thorium disintegrations and the phenomenon of nuclear fission." *PR, 55* (1939), 418–9.

Bonner, T.W. "Formation of an excited He^3 in the disintegration of deuterium by deuterons." *PR, 53* (1938), 711–3.

Bonner, T.W., and W.M. Brubaker. "Neutrons from the disintegration of deuterium by deuterons." *PR, 49* (1936), 19–21.

———. "The disintegration of nitrogen by slow neutrons." *PR, 49* (1936), 778.

Booth, E.T., J.R. Dunning, and F.G. Slack. "Delayed neutron emission from uranium." *PR, 55* (1939), 876.

Bothe, Walther, and Wolfgang Gentner. "Kernisomerie beim Brom." *Nwn, 25* (1937), 284.

Bowen, E.G. *Radar days.* Bristol: Adam Hilger, 1987.

Brady, James. "More unpublished physics." *PT, 36:3* (1983), 11–3.

Brasch, Arno. "Erzeugung und Anwendung schneller Korpuskularstrahlen (Atomzertrümmerung)." *Nwn, 21* (1933), 82–6.

Brasch, Arno, and Fritz Lange. "Ein Vakuum-Entladungsrohr für sehr hohe Spannungen." *Nwn, 18* (1930), 16.

———. "Künstliche γ-Strahlung: Ein Vakuum-Entladungsrohr für 2,4 Millionen Volt." *Nwn, 18* (1930), 765–6.

———. "Experimentelltechnische Vorbereitung zur Atomzertrümmerung mittels hoher electrischer Spannung." *Zeitschrift für Physik, 70* (1931), 10–37.

Brechin, Gray. "Sailing to Byzantium: The architecture of the fair." In Benedict, *World's fairs* (1983), 94–113.

Breckenridge, W.A. "Hydroelectric development in California." *Journal of electricity, 44* (1919), 501–5

———. "Southern California Edison announces fifteen-year program." *Journal of electricity, 45* (1920), 347.

Breit, Gregory. "Some recent progress in the understanding of atomic nuclei." *RSI, 9* (1938), 63–74.

Breit, Gregory, and M.A. Tuve. "The production and application of high voltage in the laboratory." *Nature, 121* (1928), 535–6.

Breit, Gregory, M.A. Tuve, and Odd Dahl. "Atomic physics." CIW. *Yearbook, 27* (1927/28), 208.

———. "A laboratory method of producing high potentials." *PR, 35* (1930), 51–65.

Bremer, H.W. "University technology transfer—publish and perish." In Marcy, *Patent policy*, 55–68.

Bromberg, Joan. "The impact of the neutron: Bohr and Heisenberg." *HSPS, 3* (1971), 309–23.

Brooks, John. *Telephone: The first hundred years.* New York: Harper and Row, 1976.

Brown, Laurie, and Lillian Hoddeson, eds. *The birth of particle physics.* Cambridge: Cambridge University Press, 1983.

Brunetti, Rita. "La fisica moderna e i suoi rapporti con la biologia e con la medicina." *Viaggi, 4* (1938), 15–21.

Burrill, E.A. "Van de Graaff, the man and his accelerators." *PT, 20:2* (1967), 49–52.

———. "Robert Jemison Van de Graaff." *Dictionary of scientific biography, 13,* 569–71.

Bush, Vannevar. *Pieces of the action.* New York: Morrow, 1970.

Cabrera, Blas, and H. Fahlenbrach. "Diamagnetism o del agua pesada en los estados liquido y sólido." Sociedad española de fisica y química. *Anales, 32* (1934), 538–42.

Cacciapuoti, B.N., and Emilio Segrè. "Radioactive isotopes of element 43." *PR, 52* (1937), 1252–3.

Cambridge University Reporter, 1930–40.

Cameron, F.G. *Cottrell, samaritan of science.* Garden City, NY: Doubleday, 1952.

Capron, P.C. "Final report [of fellowship in U.S., 1938–39]." Unpublished ms., Louvain, 12 April 1939. Lawrence Papers (3/28).

Carnegie Institute, Washington. *Yearbook,* 1926–32.

Carter, P.A. "Science and the common man." *American scholar, 45* (1975/76), 778–94.

Carty, J.J. "Science and the industries." NRC. *Reprint*, no. 8 (1920).

Catterall, Mary. "The medical uses of fast neutrons." Accademia nazionale delle scienze detta dei XL. *Memorie di scienze fisiche e naturali, 8:2* (1984), 251–4.

Chadwick, James. "The existence of a neutron." *PRS, A136* (1932), 692–708.

———. "Diffusion anomale des particules α. Transmutation des éléments par des particules α. Le neutron." In Solvay, 1933, *Rapports*, 81–112.

———. "The cyclotron and its applications." *Nature, 142* (1938), 630–4.

Chadwick, James, Norman Feather, and Egon Bretscher. "Measurements of range and angle of projection for the protons produced in the photo-disintegration of deuterium." *PRS, A163* (1937), 366–75.

Chadwick, James, and Maurice Goldhaber. "A 'nuclear photo-effect': Disintegration of the diplon by γ-rays. *Nature, 134* (1934), 237–8.

———. "Disintegration by slow neutrons." Cambridge Philosophical Society. *Proceedings, 31* (1935), 612–6.

———. "The nuclear photoelectric effect." *PRS, A151* (1935), 479–93.

Chaikoff, I.L., and A. Taurog. "Application of radioactive iodine to studies in iodine metabolism and thyroid function." In Wisconsin, University, *Symposium* (1948), 292–326.

Changus, G.W., I.L. Chaikoff, and Samuel Ruben. "Radioactive phosphorus as an indicator of phospholipid metabolism, IV: The phospholipid metabolism of the brain." *Journal of biological chemistry, 126* (1938), 493–500.

Charlton, E.E., and W.F. Westendorf. "A 100-million volt induction electron accelerator." *Journal of applied physics, 16* (1945), 581–93.

Chievitz, O., and George Hevesy. "Radioactive indicators in the study of phosphorus metabolism in rats." *Nature, 136* (1935), 754–5, and in Hevesy, *Selected papers*, 60–2.

———. "Studies on the metabolism of phosphorus in animals." Danske Videnskabernes Selskab. *Biologiske meddelser, 13* (1937), and in Hevesy, *Selected papers*, 63–78.

Childs, Herbert. *An American genius: The life of Ernest Orlando Lawrence, father of the cyclotron.* New York: Dutton, 1968.

Chodorow, M., et al., eds. *Felix Bloch and twentieth century physics. Dedicated to Felix Bloch on the occasion of his seventy-fifth birthday.* Houston: Rice University, 1980.

Clark, Austin H. "The press service at the Pittsburgh meeting." *Science, 81* (1935), 315–6.

Clark, R.W. *Tizard.* Cambridge, Mass.: MIT Press, 1965.

Cleland, R.G. *California in our time.* New York: Knopf, 1947.

Clusius, Klaus, and Gerhard Dickel. "Neues Verfahren zur Gasentmischung und Isotopentrennung." *Nwn, 26* (1938), 546.

Coates, W.M. "The production of x-rays by swiftly moving mercury ions." *PR, 46* (1934), 542–8.

Cockburn, Stewart, and David Ellyard. *Oliphant: The life and times of Sir Marcus Oliphant.* Adelaide: Axiom Books, 1981.

Cockcroft, John. "The development of high voltage experiments in Cambridge." Unpublished ms. Cockcroft Papers, Churchill College, Cambridge.

———. "La désintégration des éléments par des protons accélérés." In Solvay, 1933, *Rapports*, 1–56.

———. "The high-voltage laboratory and cyclotron of the Cavendish Laboratory and their application to nuclear physics." British Association for the Advancement of Science. *Report*, 1938, 381–2.

———. "The cyclotron and its applications." *Journal of scientific instruments, 16* (1939), 37–44.

Cockcroft, John, C.W. Gilbert, and E.T.S. Walton. "Production of induced radioactivity by protons." *Nature, 133* (1934), 328.

———. "Experiments with high velocity positive ions, IV: The production of induced radioactivity by high velocity protons and diplons." *PRS, A148* (1935), 225–40.

Cockcroft, John, and E.T.S. Walton. "Experiments with high velocity positive ions." *PRS, A129* (1930), 477–89.

———. "Artificial production of fast neutrons." *Nature, 129* (1932), 242.

———. "Disintegration of lithium by swift protons." *Nature, 129* (1932), 649.

———. "Experiments with high velocity positive ions, I: Further developments in the method of obtaining high velocity positive ions." *PRS, A136* (1932), 619–30.

———. "Experiments with high velocity positive ions, II: The disintegration of elements by high velocity protons." *PRS, A137* (1932), 229–42.

———. "Disintegration of light elements by fast protons." *Nature, 131* (1933), 23.

———. "Experiments with high velocity positive ions, III: The disintegration of lithium, boron, and carbon by heavy hydrogen ions." *PRS, A144* (1934), 704–20.

Coleman, C.M. *P.G. and E. of California: The centennial story of Pacific Gas and Electric Company, 1852-1952.* New York: McGraw-Hill, 1952.

Coles, J.S. "The Cottrell legacy: Research Corporation, a foundation for the advancement of science." In *Cottrell Centennial Symposium: Air pollution and its impact on agriculture, January 13 and 14, 1977.* Stanislaus, Calif.: Stanislaus State University, 1977.

Compton, A.H. *Atomic quest: A personal narrative.* New York: Oxford University Press, 1956.

Compton, K.T. "High voltage." *Science, 78* (1933), 19-24, 48-52.

———. "The battle of the alchemists: Attacks, ancient and modern, on the citadel of the atom." *Technology review, 35* (1933), 165-9, 186-90.

———. "Science makes jobs." *Scientific monthly, 38* (1934), 297-300.

———. "Put science to work! The public welfare demands a national scientific program." *Technology review, 37:4* (1935), 133-5, 152, 154, 156, 158.

Condon, E.U. "Science and international cooperation." *BAS, 1:11* (1946), 8-9.

Connolly, A.G. "Should medical inventions be patented?" *Science, 86* (1937), 383-7.

Constant, E.W. "On the diversity and co-evolution of technological multiples: steam turbine and Pelton water wheels." *Social studies of science, 8* (1978), 183-210.

———. "Scientific theory and technological testability: science, dynamometers, and water turbines in the nineteenth century." *Technology and culture, 24* (1983), 183-98.

Cook, S.F., K.G. Scott, and P.H. Abelson. "The deposition of radio phosphorus in tissues of growing chicks." NAS. *Proceedings, 23* (1937), 528-33.

Cooksey, Donald, and M.C. Henderson. "Threshold counting voltage, dependence on gas, mechanical dimensions and pressure." *PR, 41* (1932), 392.

Cooksey, Donald, F.D.N. Kurie, and E.O. Lawrence. "Design of a six-million volt resonance acceleration unit for the Berkeley magnet." *PR, 49* (1936), 204.

Cooksey, Donald, and E.O. Lawrence. "Six-million volt magnetic resonance accelerator with emergent beam." *PR, 49* (1936), 866.

Coolidge, W.D. "The production of high-voltage cathode rays outside of the generating tube." *JFI, 202* (1926), 693-721.

———. "Presentation of the Comstock Prize." *Science, 86* (1937), 405–7.

Corbino, O.M. "Prospettive e risultati della fisica moderna." *Ricerca scientifica, 5:1* (1934), 609–19.

Cork, J.M., and E.O. Lawrence. "Transmutation of platinum by deuterons: A resonance phenomenon." *PR, 49* (1936), 205.

———. "The transmutation of platinum by deuterons." *PR, 49* (1936), 788–92.

Cork, J.M., J.R. Richardson, and F.D.N. Kurie. "The radiations emitted by radio-aluminum." *PR, 49* (1936), 208.

Cornog, Robert. "Discovery of hydrogen/helium three." In Trower, *Discovering Alvarez*, 26–8.

Corson, D.R., and K.R. MacKenzie. "Artificially produced alpha-particle emitters." *PR, 57* (1940), 250.

Corson, D.R., K.R. MacKenzie, and Emilio Segrè. "Possible production of radioactive isotopes of element 85." *PR, 57* (1940), 459.

———. "Artificially radioactive element 85." *PR, 58* (1940), 672–8.

Corson, D.R., and R.L. Thornton. "Disintegration of uranium." *PR, 55* (1939), 509.

Cottrell, F.G. "Patent experience of the Research Corporation." American Institute of Chemical Engineers. *Transactions, 28* (1932), 222–5.

Crane, H.R., L.A. Delsasso, R.D. Fowler, and C.C. Lauritsen. "High energy γ-rays from lithium and flourine bombarded with protons." *PR, 46* (1934), 531–3.

Crane, H.R., and C.C. Lauritsen. "On the production of neutrons from lithium." *PR, 44* (1933), 783–4.

———. "Disintegration of beryllium by deuterons." *PR, 45* (1934), 226–7.

———. "Radioactivity from carbon and boron oxide bombarded with deutons and the conversion of positrons into radiation." *PR, 45* (1934), 430–2.

Crane, H.R., C.C. Lauritson, and W.W. Harper. "Artificial production of radioactive substances." *Science, 79* (1934), 234–5.

Crane, H.R., C.C. Lauritson, and A. Soltan. "Artificial production of neutrons." *PR, 44* (1933), 514.

———. "Production of neutrons by high speed deutons." *PR, 44* (1933), 692–3.

Crawford, Elisabeth, J.L. Heilbron, and Rebecca Ullrich. *The Nobel population.* Berkeley: Office for the History of Science and Technology; Uppsala: Office for the History of Science, 1987.

Crenshaw, R.S. "The Darien radio station of the U.S. Navy." IRE. *Proceedings, 4* (1916), 35–40.

Crowther, J.G. *The Cavendish Laboratory, 1874–1974.* New York: Science History Publications, 1974.

Curie, Irène, and Frédéric Joliot. "Emission de protons de grande vitesse par les substances hydrogénées sous l'influence des rayons γ très pénétrants." *CR, 194,* (1932), 273–5, and in Joliot and Curie, *Oeuvres,* 359–60.

———. "La complexité du proton et la masse du neutron." *CR, 197* (1933), 237, and in Joliot and Curie, *Oeuvres,* 417–8.

———. "Électrons de matérialisation et de transmutation." *Journal de physique et le radium, 4* (1933), in Joliot and Curie, *Oeuvres,* 444–54.

———. "Un nouveau type de radioactivité." *CR, 198* (1934), 254, and in Joliot and Curie, *Oeuvres,* 515–6.

———. "Séparation chimique des nouveaux radioéléments émetteurs d'électrons positifs." *CR, 198* (1934), 559, and in Joliot and Curie, *Oeuvres,* 517–9.

———. "Artificial production of a new kind of radio-element." *Nature, 133* (1934), 201–2, and in Joliot and Curie, *Oeuvres,* 520–1.

———. "Mass of the neutron." *Nature, 133* (1934), 721.

Curie, Irène, and Pierre Savitch. "Sur les radioéléments formés de l'uranium irradié par les neutrons." *Journal de physique et le radium, 8* (1937), 385–7; *9* (1938), 355–9.

———. "Sur le radioélément de période 3,4 heures formé dans l'uranium irradié par les neutrons." *CR, 206* (1938), 906, and in Joliot and Curie, *Oeuvres,* 624.

Dahl, Odd, L.R. Hafstad, and M.A. Tuve. "On the technique and design of Wilson cloud-chambers." *RSI, 4* (1933), 373–8.

Dancoff, S.M. "Shielding of high energy neutrons by water tanks." *PR, 57* (1940), 251.

Dancoff, S.M., and Philip Morrison. "The calculation of internal conversion coefficients." *PR, 55* (1939), 122–30.

Davis, N.P. *Lawrence and Oppenheimer.* New York: Simon and Schuster, 1968.

Davis, Watson. "Science and the press." *Vital speeches, 2* (1936) 361–5.

Debye, Peter. "Das Kaiser-Wilhelm-Institut für Physik." *Nwn, 25* (1937), 257–60.

Dee, P.I. "Disintegration of the diplon." *Nature, 133* (1934), 564.

DeVault, D.C., and W.F. Libby. "Evidence for gamma-radioactivity of 4.5-hour Br^{80} from radiobromate." *PR, 55* (1939), 322.

Diebner, Kurt, and E. Grassman. "Künstliche Radioaktivität." *Physikalische Zeitschrift, 37* (1936), 359–83; *38* (1937), 406–25; *39* (1938), 469–501; *40* (1939), 297–314; *41* (1940), 157–94.

Dietz, David. "Science, Uncle Sam, and the future." *RSI, 7* (1936), 1–5.

———. "Science and the American press." *Science, 85* (1937), 107–12.

Dobkin, Marjorie M. "A twenty-five million dollar mirage." In Benedict, *World's fairs,* 66–93.

Dragoni, Giorgio. "L'illusoria scoperta del primo elemento transuranico." *Physis, 15* (1973), 351–74.

———. "Un momento della vita scientifica italiana degli anni trenta: La scoperta dei neutroni lenti e la loro introduzione nella sperimentazione fisica." *Physis, 18* (1976), 131–64.

DuBridge, L.A., and S.W. Barnes. "The Rochester cyclotron." *PR, 49* (1936), 865.

Dunning, J.R., and H.L. Anderson. "High frequency filament supply for ion sources." *RSI, 8* (1937), 158–9.

———. "High frequency systems for the cyclotron." *PR, 53* (1938), 334.

Dunning, J.R., G.B. Pegram, G.A. Fink, D.P. Mitchell, and Emilio Segrè. "Velocity of slow neutrons by mechanical velocity selector." *PR, 48* (1935), 704.

Dunnington, F.G. "An optical study of the formation stages of spark breakdown." *PR, 38* (1931), 1535–46.

Durand, W.F. "The Pelton water wheel." *Mechanical engineering, 61* (1939), 447–51.

Elsasser, W.M. *Memoirs of a physicist in the atomic age.* New York: Science History Publications; Bristol: Adam Hilger, 1978.

Entenman, C., Samuel Ruben, I. Perlman, F.W. Lorenz, and I.L. Chaikoff. "Radioactive phosphorus as an indicator of phospholipid metabolism, III. The conversion of phosphate to lipid phosphorus by the tissues of the laying and non-laying bird." *Journal of biological chemistry, 124* (1938), 795–802.

Erf, L.A., and J.H. Lawrence. "Phosophorus metabolism in neoplastic tissue." SEBM. *Proceedings, 46* (1941), 694–5.

———. "Clinical studies with the aid of radioactive phosphorus, I: The absorption and distribution of radio-phosphorus in the blood and in its excretion by normal individuals and patients with leukemia." *Journal of clinical investigations, 20* (1941), 567–76.

Eve, A.S. *Rutherford: Being the life and letters of the Rt. Hon. Lord Rutherford, O.M.* New York: Macmillan, 1939.

F., A. "Atomic transmutation." *Nature, 132* (1933), 432–3.

Feather, Norman, and Egon Bretscher. "Uranium Z and the problem of nuclear isomerism." *PRS, A165* (1938), 530–51.

———. "Atomic numbers of the so-called transuranic elements." *Nature, 143* (1939), 516.

———. "The experimental discovery of the neutron." In Hendry, *Cambridge physics*, 31–41.

Feld, B.T. "The neutron." In Emilio Segrè, ed. *Experimental nuclear physics, 2.* New York: Wiley, 1953. Pp. 209–586.

Fermi, Enrico. "Radioattività prodotta da bombardamento di neutroni." *Nuovo cimento, 11* (1934), 429–41, and in Fermi, *CP, 1,* 715–24.

———. "Radioactività indotta da bombardamento di neutroni, I." *Ricerca scientifica, 5* (1934), 283, and in Fermi, *CP, 1,* 645–6.

———. "Radioactivity induced by neutron bombardment." *Nature, 133* (1934), 757, and in Fermi, *CP, 1,* 702–3.

———. *Collected papers.* Ed. Emilio Segrè et al. 2 vols. Chicago: University of Chicago Press, 1962–65.

Fermi, Enrico, Edoardo Amaldi, O. D'Agostino, Franco Rasetti, and Emilio Segrè. "Artificial radioactivity produced by neutron bombardment." *PRS, A146* (1934), 483–500, and in Fermi, *CP, 1,* 732–47.

Fermi, Laura. *Atoms in the family.* Chicago: University of Chicago Press, 1954.

Fertel, G.E.F., D.F. Gibbs, O.B. Moon, G.P. Thomson, and C.E. Wynn-Williams. "Experiments with a velocity-spectrometer for slow neutrons." *PRS, A175* (1940), 316–31.

Fertel, G.E.F., O.B. Moon, G.P. Thomson, and C.E. Wynn-Williams. "Velocity distribution of thermal neutrons." *Nature, 142* (1938), 829.

Field, S.B. "A historical survey of radiobiology and radiotherapy." *Current topics in radiation research quarterly, 11* (1976), 1–85.

Fink, G.A. "The production and absorption of thermal energy neutrons." *PR, 50* (1936), 738–47.

Fisk, J.B., H.D. Hagstrum, and P.L. Hartman. "The magnetron as a generator of centimeter waves: Developments at the Bell Telephone Laboratories, 1940–1945." *Bell System technical journal, 25:2* (1946), 1–189.

Flammersfeld, Arnold. "Zur Geschichte der Atomkernisomerie." In Frisch et al., *Trends,* 74–7.

Flexner, Abraham. "University patents." *Science, 77* (1933), 325.

Flügge, Siegfried. "Kann der Energieinhalt der Atomkerne technisch nutzbar gemacht werden?" *Nwn, 27* (1939), 402–10.

Foote, P.D. "Industrial physics." *RSI, 5* (1934), 57–66.

Forbes, B.C. *Men who are making the West.* New York: Forbes, 1923.

Forman, Paul, J.L. Heilbron, and Spencer Weart. "Physics *circa* 1900: Personnel, funding, and productivity of the academic establishments." *HSPS, 5* (1975), 1–185.

Fosdick, R.B. *The Rockefeller Foundation: A review for 1940.* New York: Rockefeller Foundation, 1941.

Fowler, R.D., and R.W. Dodson. "Intensity of ionizing particles produced by neutron bombardment of uranium and thorium." *PR, 55* (1939), 417–8.

Fremlin, J.H., and J.S. Gooden. "Cyclic accelerators." *Reports on progress in physics, 13* (1950), 295–350.

Fries, B.A., Samuel Ruben, I. Perlman, I.L. Chaikoff. "Radioactive phosphorus as an indicator of phospholipid metabolism, II: The role of the stomach, small intestine, and large intestine in phospholipid metabolism in the presence and absence of ingested fat." *Journal of biological chemistry, 123* (1938), 587–93.

Frisch, O.R. "Induced radioactivity of flourine and calcium." *Nature, 136* (1935), 220.

———. "Radioactivity and sub-atomic phenomena." *Annual reports of the progress of chemistry, 35* (1939), 7–24.

———. "Physical evidence for the division of heavy nuclei under neutron bombardment." *Nature, 143:suppl.* (1939), 276.

———. "Radioactivity and sub-atomic phenomena." *Annual reports of the progress of chemistry, 36* (1940), 7–22.

———. "The interest is focussing on the atomic nucleus." In Rozental, *Niels Bohr,* 137–48.

———. *What little I remember.* Cambridge: Cambridge University Press, 1979.

Frisch, O.R., Hans von Halban, and Jørgen Koch. "A method of measuring the magnetic moment of free neutrons." *Nature, 139* (1937), 756.

———. "Sign of the magnetic moment of free neutrons." *Nature, 139* (1937), 1021.

———. "Some experiments on the magnetic properties of free neutrons." *PR, 53* (1938), 719–26.

Frisch, O.R., F.A. Paneth, F. Laves, and Paul Rosband, eds. *Trends in atomic physics: Essays dedicated to Lise Meitner, Otto Hahn, and Max von Laue on the occasion of their 80th birthday.* New York: Interscience; Braunschweig: Vieweg, 1959.

Fryer, E.M. "Transmutation of velocity-selected neutrons through magnetized iron." *PR, 70* (1946), 235–44.

Gamow, George. "Zur Quantentheorie der Atomzertrümmerung." *Zeitschrift für Physik, 52* (1928), 510–5.

———. "Negative protons and nuclear structure." *PR, 45* (1934), 728–9.

Geiger, R.L. *To advance knowledge: The growth of American research universities, 1900–1940.* Oxford: Oxford University Press, 1986.

Gentner, Wolfgang. "Die Erzeugung schneller Ionenstrahlen für Kernreaktionen." *Ergebnisse der exakten Naturwissenschaften, 19* (1940), 107–69.

———. "Das Heidelberger Zyklotron." In FIAT Review of German Science, 1939–1946. *Nuclear physics and cosmic rays, 2,* 28–31. Wiesbaden: Dietrich, 1948.

———. "Im besetzten Paris 1940 bis 1942." In Max-Planck-Gesellschaft. *Wolfgang Gentner Gedankfeier.* Munich: Max-Planck-Gesellschaft, 1981. Pp. 41–50.

Giauque, W.F., and H.L. Johnston. "An isotope of oxygen of mass 18." *Nature, 123* (1929), 318.

———. "An isotope of oxygen of mass 17 in the earth's atmosphere." *Nature, 123* (1929), 831.

Gibbs, D.F., and G.P. Thomson. "Possible delay in the emission of neutrons from uranium." *Nature, 144* (1939), 202.

Ginzton, E.L. "The $100 idea: How Russell and Sigurd Varian, with the help of William Hansen and a $100 appropriation, invented the klystron." IEEE. *Spectrum* (1975), 30–9.

Gleason, Sterling. "Giant atom gun to help fight disease." *Popular science monthly, 136:5* (1940), 47–51, 225.

Goldhaber, Maurice. "The nuclear photoelectric effect and remarks on higher multiple tranmissions: A personal history." In Stuewer, *Nuclear physics in retrospect,* 83–110.

———. "Working with Chadwick." In Hendry, *Cambridge physics,* 189–94.

Goldsmith, Maurice. *Frédéric Joliot-Curie: A biography.* London: Lawrence and Wishart, 1976.

Golovin, I.N. *I.V. Kurchatov.* Tr. W.H. Dougherty. Bloomington, Ind.: Selbstverlag, 1969.

Goodchild, Peter. *J. Robert Oppenheimer: Shatterer of worlds.* Boston: Houghton Mifflin, 1981.

Goodspeed, T.H. "The organization and activities of the Committee on Scientific Research of the State Council of Defense of California." NRC. *Bulletin, 5:6* (1923).

Gordon, Elizabeth. *What we saw at Madame World's Fair.* San Francisco: Samuel Levinson, 1945.

Gowing, Margaret. *Britain and atomic energy, 1939–1945.* London: Macmillan, 1964.

Grahame, D.C., and G.T. Seaborg. "The separation of radioactive substances without the use of a carrier." *PR, 54* (1938), 240–1.

Gray, G.W. "Science and profits." *Harper's, 172:4* (1936), 539–49.

Green, G.K., and L.W. Alvarez. "Heavily ionizing particles from uranium." *PR, 55* (1939), 417.

Greenberg, D.M. "Mineral metabolism, calcium, magnesium, phosphorus." *Annual review of biochemistry, 8* (1939), 269–300.

———. "Tracer studies on the metabolism of mineral elements with radioactive isotopes." In Wisconsin, University, *Symposium* (1940), 261–91.

Gregg, Alan. "University patents." *Science, 77* (1933), 257–9.

Grégoire, R. "Noyaux stables et radioactifs." *Journal de physique et le radium, 3* (1938), 419–27.

Grosse, Aristid von. "The mass of the neutron and the constitution of atomic nuclei." *PR, 43* (1933), 143.

———. "The chemical properties of elements 93 and 94." *JACS, 57* (1935), 440–1.

Grosse, Aristid von, and M.S. Agruss. "Fermi's element 93." *Nature, 134* (1934), 773.

Guerlac, H.E. *Radar in World War II.* 2 vols. New York: AIP and Tomash, 1987.

Hacker, B.C. *The dragon's tail: Radiation safety in the Manhattan Project, 1942–1946.* Berkeley: University of California Press, 1987.

Hackman, W.D. *Seek and strike: Sonar, anti-submarine warfare, and the Royal Navy, 1914–1954.* London: HMSO, 1984.

Hafstad, L.R., and M.A. Tuve. "Induced radioactivity using carbon targets." *PR, 47* (1935), 506.

———. "Carbon radioactivity and other resonance transmutations by protons." *PR, 48* (1935), 306–15.

Hagstrum, H.D. "The generation of centimeter waves." IRE. *Proceedings, 35* (1947), 548–64.

Hahn, Otto. "Die 'falschen' Transurane: Zur Geschichte eines wissenschaftlichen Irrtums." *Naturwissenschaftliche Rundschau, 15* (1962), 43–7.

―――. *Vom Radiothor zur Uranspaltung: Eine wissenschaftliche Selbstbiographie.* Braunschweig: Vieweg, 1962.

Hahn, Otto, and Fritz Strassmann. "Über den Nachweis und das Verhalten der bei der Bestrahlung des Urans mittels Neutronen entstehenden Erdalkalimetalle." *Nwn, 27* (1939), 11–5, and in Krafft, *Strassmann,* 255–9, 266–7.

―――. "Nachweis der Entstehung aktiver Barium-Isotopen aus Uran und Thorium durch Neutronenbestrahlung; Nachweis weiterer aktiven Bruchstücke bei der Uranspaltung." *Nwn, 27* (1939), 89–95.

―――. "Zur Frage der Existenz der 'Trans-Urane,' I: Endgültige Streichung von Eka-Platin und Eka-Iridium." *Nwn, 27* (1939), 451–3.

Hahn, P.F., W.F. Bale, R.A. Hettig, M.D. Kamen, and G.H. Whipple. "Radioactive iron and its excretion in urine, bile, and feces." *Journal of experimental medicine, 70* (1939), 443–51.

Hahn, P.F., W.F. Bale, E.O. Lawrence, and G.H. Whipple. "Radioactive iron and its metabolism in anemia: its absorption, transportation, and utilization." *Journal of experimental medicine, 69* (1939), 739–53.

Hahn, P.F., J.F. Ross, W.F. Bale, and G.H. Whipple. "The utilization of iron and the rapidity of hemoglobin formation in anemia due to blood loss." *Journal of experimental medicine, 71* (1940), 731–6.

Halban, Hans von, Frédéric Joliot, and Lew Kowarski. "Liberation of neutrons in the nuclear explosion of uranium." *Nature, 143* (1939), 470–1, and in Joliot and Curie, *Oeuvres,* 559–61.

―――. "Number of neutrons liberated in the nuclear fission of uranium." *Nature, 143* (1939), 680, and in Joliot and Curie, *Oeuvres,* 664–6.

―――. "Sur la possibilité de produire dans un milieu uranifère des réactions nucléaires en chaine illimitée." *CR, 229* (1949), 909, and in Joliot and Curie, *Oeuvres,* 673–7.

Halban, Hans von, Frédéric Joliot, Lew Kowarski, and Francis Perrin. "Mise en évidence d'une réaction nucléaire en chaine au sein d'une masse uranifère." *Journal de physique et le radium, 10* (1939), and in Joliot and Curie, *Oeuvres,* 669–72.

Hale, G.E. "The purpose of the National Research Council." NRC. *Bulletin, 1* (1919), 1–7.

―――. "Science and the wealth of nations." *Harper's, 156* (1928), 243–51.

Hall, B.E. "Therapeutic use of radiophosphorus in polycythemia vera, leukemia, and allied diseases." In Wisconsin, University, *Symposium* (1948), 353–76.

Hamilton, J.G. "Rates of absorption of radio-sodium in normal human subjects." NAS. *Proceedings, 29* (1937), 521–7.

———. "The rates of absorption of the radioactive isotopes of sodium, potassium, chlorine, bromine, and iodine in normal human subjects." *American journal of physiology, 124* (1938), 667–78.

———. "The applications of radioactive tracers to biology and medicine." *Journal of applied physics, 12* (1941), 440–60.

———. "The use of radioactive tracers in biology and medicine." *Radiology, 39* (1942), 541–72.

———. "Medical applications of radioactive tracers." In Wisconsin, University, *Symposium*, 327–52.

Hamilton, J.G., and G.A. Alles. "The physiological action of natural radioactivity." *American journal of physiology, 125:2* (1939), 410–3.

Hamilton, J.G., and J.H. Lawrence. "Recent clinical developments in the therapeutic application of radio-phosphorus and radio-iodine." *Journal of clinical investigations, 21* (1942), 624.

Hamilton, J.G., and M.H. Soley. "Studies in iodine metabolism by the use of a new radioactive isotope of iodine." *American journal of physiology, 127* (1939), 557–72.

———. "Studies in iodine metabolism of the thyroid gland *in situ* by the use of radio-iodine in normal subjects and in patients with various types of goiter." *American journal of physiology, 131* (1940), 135–43.

Hamilton, J.G., M.H. Soley, and K.B. Eichorn. "Deposition of radioactive iodine in human thyroid tissue." University of California. *Publications in pharmacology, 1* (1940), 339–67.

Hamilton, J.G., and R.S. Stone. "Excretion of radio-sodium following intravenous administration in man." SEBM. *Proceedings, 35* (1937), 595–8.

Hansen, W.W. "A type of electrical resonator." *Journal of applied physics, 9* (1938), 654–63.

Harden, Victoria A. *Inventing the NIH: Federal biomedical research policy, 1887–1937.* Baltimore: Johns Hopkins University Press, 1986.

Harnwell, G.P., H.D. Smyth, et al. "The production of H^3 by a cathode-ray discharge in helium." *PR, 45* (1934), 655–6.

Harnwell, G.P., H.D. Smyth, and W.D. Urry. "Purification and spectroscopic evidence for $He_2{}^3$." *PR, 46* (1934), 437.

Hartcup, Guy, and T.E. Allibone. *Cockcroft and the atom.* Bristol: Adam Hilger, 1984.

Harteck, Paul, and H. Streibel. "Versuch zur Trennung der Isotopen des Broms." *Zeitschrift der anorganischen und allgemeinen Chemie, 194* (1930), 299–304.

Haworth, L.J., and F.N. Gilette. "A slow neutron velocity spectrometer." *PR, 69* (1946), 254.

Heilbron, J.L. *H.G.J. Moseley: The life and letters of an English physicist, 1887–1915.* Berkeley: University of California Press, 1974.

———. *"Fin-de-siècle* physics." In C.G. Bernhard, Elisabeth Crawford, and Per Sörbom, eds. *Science, technology, and society in the times of Alfred Nobel.* Oxford: Pergamon Press, 1982. Pp. 51–73.

———. "La radio ed il ciclotrone." *Museoscienza, 22* (1983), 13–24.

———. "The earliest missionaries of the Copenhagen spirit." *Revue d'histoire des sciences, 38* (1985), 194–230.

———. *The dilemmas of an upright man: Max Planck as spokesman for German science.* Berkeley: University of California Press, 1986.

Helmholz, A.C. "The measurement of gamma-ray energies." *PR, 57* (1940), 248.

Henderson, M.C. "The disintegration of lithium by protons of high energy." *PR, 43* (1933), 98–102.

———. "Two radioactive substances from magnesium after deuteron bombardment." *PR, 48* (1935), 855–61.

———. "The heat of fission of uranium." *PR, 56* (1939), 703.

———. "The heat of fission of uranium." *PR, 58* (1940), 774–80.

Henderson, M.C., M.S. Livingston, and E.O. Lawrence. "Artificial radioactivity produced by deuton bombardment." *PR, 45* (1934), 428–9.

———. "The transmutation of flourine by proton bombardment and the mass of flourine 19." *PR, 46* (1934), 38–42.

Henderson, M.C., and M.G. White. "The design and operation of a large cyclotron." *RSI, 9* (1938), 19–30.

Henderson, W.J., L.D.P. King, and J.R. Risser. "The Purdue cyclotron." *PR, 55* (1939), 1110.

Henderson, W.J., L.D.P. King, J.R. Risser, H.J. Yearian, and J.D. Howe. "The Purdue cyclotron." *JFI, 228* (1939), 563–79.

Hendry, John, ed. *Cambridge physics in the thirties.* Bristol: Adam Hilger, 1984.

Herrick, G.W. "Some obligations and opportunities of scientists in the upholding of peace." *Science, 52* (1920), 93–9.

Hertz, Saul. "Treatment of thyroid disease by means of radioactive iodine." In Wisconsin, University, *Symposium,* 377–94.

Hertz, Saul, and A. Roberts. "Application of radioactive iodine in therapy of Graves' disease." *Journal of clinical investigation, 21* (1942), 624.

Hertz, Saul, A. Roberts, and R.D. Evans. "Radioactive iodine as an indicator in the study of thyroid physiology." SEBM. *Proceedings, 38* (1938), 510–13.

Hertz, Saul, A. Roberts, J.H. Means, and R.D. Evans. "Radioactive iodine as an indicator of thyroid physiology: Iodine collected by normal and hyperplastic thyroids in rabbits." *American journal of physiology, 128* (1940), 565–76.

Hevesy, George. "Application of radioactive indicators in biology." *Annual review of biochemistry, 9* (1940), 641–62.

———. "Einige 'Anwendungen' des radioaktiven Eisens." In Frisch et al., *Trends*, 115–20.

———. *Selected papers.* Oxford: Pergamon Press, 1967.

Hevesy, George, et al. "Untersuchungen über die phosphorübertragungen in der Glykolyse und Glykogenlyse." *Acta biologiae expermentalis, 12* (1938), 34–44.

Hewlett, R.G., and O.E. Anderson, Jr. *A history of the United States Atomic Energy Commission, 1. The new world, 1939–46.* University Park, Pa.: Pennsylvania State University Press, 1962.

Highfill, R.R., and B.W. Wieland. "The medical applications of short-lived, cyclotron-produced radionuclides." IEEE. *Transactions in nuclear science, 26* (1979), 2220–3.

Hirosige, Tetu. "Social conditions for prewar Japanese research in nuclear physics." In Nakayama et al., *Science and society*, 202–20.

Hoffmann, J.G., M.S. Livingston, and H.A. Bethe. "Some direct evidence on the magnetic moment of the neutron." *PR, 51* (1937), 214–5.

Holbrouw, C.H. "The giant cancer tube and the Kellog Radiation Laboratory." *PT, 34*:7 (1981), 42–9, and in Weart and Phillips, *History of physics*, 86–93.

Holland, Maurice. "Putting physics to work." *RSI, 6* (1935), 36–9.

Holloway, M.C. "The disintegration of N^{14} by deuterons." *PR, 57* (1940), 347–8.

Holloway, M.C., and B.L. Moore. "The disintegration of C^{12}, C^{13}, and O^{16} by deuterons." *PR, 57* (1940), 1086.

———. "The disintegration of N^{14} and N^{15} by deuterons." *PR, 58* (1940), 847–60.

Holton, Gerald, ed. *The twentieth-century sciences: Studies in the biography of ideas.* New York: Norton, 1972.

Hoover, Herbert. "The vital need for greater financial support of pure science research." NRC. *Reprint*, no. 65 (1925).

Howeth, L.S. *The history of communications-electronics in the United States Navy.* Washington, D.C.: GPO, 1963.

Hughes, T.P. *Networks of power: Electrification in Western society, 1880–1930.* Baltimore: Johns Hopkins University Press, 1983.

Hurst, D.G., and Harold Walke. "Induced radioactivity of potassium." *PR, 51* (1937), 1033–36.

Hull, A.W. "Putting physics to work." *RSI, 6* (1935), 377–80.

Institute of Electrical and Electronics Engineers. "1978 conference. Session on cyclotrons." IEEE. *Transactions, 26* (1979), 1703–32.

Irvine, J.W., Jr. "Concentrating the uranium isotope of twenty-three-minute half-life." *PR, 55* (1939), 1105.

Irving, David. *The German atomic bomb: The history of nuclear research in Nazi Germany.* New York: Simon and Schuster, 1967.

Ising, Gustaf. "Prinzip einer Methode zur Herstellung von Kanalstrahlen hoher Voltzahl." *Arkiv för matematik, astronomi och fysik, 18* (1924), 1–4. Translated in Livingston, *Development*, 88–90.

———. "Högspänningsmetodor för atomsprängning." *Kosmos, 11* (1933), 141–98. (Svenska fysikersamfundet, *Årsbok.*)

Jacobsen, J.C. "Positrons from radio-scandium." *Nature, 139* (1937), 879–80.

———. "Om cyclotronen." *Fysisk tidsskrift, 39* (1941), 33–50.

———. "Construction of a cyclotron. (Institute for Theoretical Physics, Copenhagen.)" Danske Videnskabernes Selskab. *Mathematisk-fysiske meddelser, 19:2* (1941).

Jaeckel, R. "Versuche mit Neutronen aus Aluminium und Beryllium." *Zeitschrift für Physik, 91* (1934), 493–510.

Jane's warships of the world. London: Jane's Publishing Co., 1939.

Jenkin, J.G. "The development of angle-resolved photoelectron spectroscopy, 1900–1960." *Journal of spectroscopy and related phenomena, 23* (1981), 187–273.

Jewett, F.B. "Modern research organization and the American patent system." *Mechanical engineering, 54* (1932), 394–8, 450.

———. "The social effects of modern science." *Science, 76* (1932), 23–6.

Johnston, W.M., R.L. Wolfgang, and W.F. Libby. "Tritium in nature." *Science, 113* (1951), 1–2.

Joliot, Frédéric. "Preuve expérimentale de la rupture explosive des noyaux d'uranium et de thorium sous l'action des neutrons." *CR, 208* (1939), 341, and in Joliot and Curie, *Oeuvres,* 650–2.

———. "Observation par la méthode de Wilson des trajectoires de brouillard des produits de l'explosion des noyaux d'uranium." *CR, 208* (1939), 647, and in Joliot and Curie, *Oeuvres,* 653–4.

———. *Textes choisis.* Paris: Editions sociales, 1959.

Joliot, Frédéric, and Irène Curie. "Rayonnement pénétrant des atomes sous l'action des rayons α." In Solvay, 1933, *Rapports,* 121–56.

———. "Artificial production of a new kind of radio-element." *Nature, 133* (1934), 201–2, and in Joliot and Curie, *Oeuvres,* 520–1.

———. *Oeuvres scientifiques complètes.* Paris: PUF, 1961.

Jones, H.B., I.L. Chaikoff, and J.H. Lawrence. "Radioactive phosphorus as an indicator of phospholipid metabolism, VI: The phospholipid metabolism of neoplastic tissues (mammary carcinoma, lymphoma, lymphosarcoma, sarcoma 180)." *Journal of biological chemistry, 128* (1939), 631–44.

———. "Phosphorus metabolism of the soft tissues of the normal mouse as indicated by radioactive phosphorus." *American journal of cancer, 40* (1940), 235–42.

———. "Phosphorus metabolism of neoplastic tissues (mammary carcinoma, lymphoma, lymphosarcoma) as indicated by radioactive phosphorus." *American journal of cancer, 40* (1940), 243–50.

———. "Radioactive phosphorus as an indicator of phospholipid metabolism, X. The phospholipid turnover of fraternal tumors." *Journal of biological chemistry, 133* (1940), 319–27.

Josiah Macy, Jr., Foundation. *A review by the president of activities for the six years ended December 31, 1936.* New York: The Foundation, 1937.

———. *The Josiah Macy, Jr., Foundation, 1930–1955.* New York: The Foundation, 1955.

Kalbfell, David. "Internal conversion of γ-rays in the 6-hour element 43." *PR, 54* (1938), 543.

Kamen, M.D. "Assay of radioactive isotopes in biological research." In Wisconsin, University, *Symposium* (1948), 141–60.

———. "Early history of carbon-14." *Science, 140* (1963), 584–90.

———. *Radiant science, dark politics.* Berkeley: University of California Press, 1985.

Kamen, M.D., and Samuel Ruben. "Production and properties of carbon 14." *PR, 58* (1940), 194.

———. "Studies in photosynthesis with radiocarbon." *Journal of applied physics, 12* (1941), 326.

Kargon, Robert. *The rise of Robert Millikan: Portrait of a life in American science.* Ithaca: Cornell University Press, 1982.

———. "The evolution of matter: Nuclear physics, cosmic rays, and Robert Millikan's research program." In Shea, *Otto Hahn,* 69–90.

Kellog, Vernon. "The university and research." *Science, 54* (1921), 19–23.

Kennedy, J.W., and G.T. Seaborg. "Search for beta-particles emitted during uranium fission process." *PR, 55* (1939), 877.

———. "Isotopic identification of induced radioactivity by bombardment of separated isotopes: 37-minute Cl^{38}." *PR, 57* (1940), 843–4.

Kennedy, J.W., G.T. Seaborg, Emilio Segrè, and A.C. Wahl. "Properties of 94^{239}." *PR, 70* (1946), 555–6.

Kennedy, J.W., and A.C. Wahl. "Search for spontaneous fission in 94^{239}." *PR, 69* (1946), 367–8.

Kerst, D.W. "The acceleration of electrons by magnetic induction." *PR, 60* (1941), 47–53.

———. "Historical development of the betatron." *Nature, 157* (1946), 90–5.

Kevles, D.J. *The physicists: The history of a scientific community in modern America.* New York: Knopf, 1978.

Kirby, H.W. "History of Technetium." *Gmelins Handbuch der anorganischen Chemie: Technetium. Supplement, 1,* Berlin: Springer, 1982. Pp. 1–11.

Kohler, R.E. "The origin of G.N. Lewis's theory of the shared pair bond." *HSPS, 3* (1971), 343–76.

———. "Warren Weaver and the Rockefeller Foundation in molecular biology: A case study in the management of science." In Reingold, *Science in the American context,* 249–93.

———. *From medical chemistry to biochemistry: The making of a biomedical discipline.* Cambridge: Cambridge University Press, 1982.

Konopinski, E.J., and G.E. Uhlenbeck. "On the Fermi theory of β-radioactivity." *PR, 48* (1935), 7–12.

Krafft, Fritz. *Im Schatten der Sensation: Leben and Wirken von Fritz Strassmann.* Weinheim: Verlag Chemie, 1981.

———. "Internal and external conditions for the discovery of nuclear fission by the Berlin team." In Shea, *Otto Hahn,* 135–65.

Krasny-Ergen, Wilhelm. "Separation of uranium isotopes." *Nature, 145* (1940), 742–3.

Kröger, B. "On the history of the neutron." *Physis, 22* (1980), 175–90.

Kruger, P.G., and G.K. Green. "A million volt cyclotron." *PR, 51* (1937), 57–8.

———. "The construction and operation of a cyclotron to produce one million volt deuterons." *PR, 51* (1937), 699–705.

Kruger, P.G., G.K. Green, and F.W. Stallmann. "Cyclotron operation without filaments." *PR, 51* (1937), 291.

Kurchatov, B., I. Kurchatov, L. Myssovsky, and L. Roussinov. "Sur un cas de radioactivité artificielle provoquée par un bombardment de neutrons, sans capture du neutron." *CR, 200* (1935), 1201–3.

Kurie, F.D.N. "The use of the Wilson cloud chamber for measuring the range of alpha particles from weak sources." *RSI, 3* (1932), 655–7.

———. "A new mode of disintegration induced by neutrons." *PR, 45* (1934), 904–5.

———. "The loss in energy in the disintegration of nitrogen by neutrons." *PR, 46* (1934), 324.

———. "Disintegration with the emission of protons induced by neutrons." *PR, 46* (1934), 330.

———. "The disintegration of nitrogen by neutrons." *PR, 47* (1935), 97–107.

———. "The cyclotron: A new research tool for physics and biology." *General Electric review, 40* (1937), 264–72.

———. "Present-day design and technique of the cyclotron: A description of the methods and application of the cyclotron as developed by Ernest O. Lawrence and his associates at the Radiation Laboratory, Berkeley." *Journal of applied physics, 9* (1938), 691–701.

———. "The technique of high intensity bombardment with fast particles." *RSI, 10* (1939), 199–205.

———. "The growth of nuclear physics." *Naval research reviews, 9:2* (1956), 10–5.

Kurie, F.D.N., J.R. Richardson, and Hugh Paxton. "On the shape of the distribution curves of electrons emitted from artificially produced radioactive substances." *PR, 48* (1935), 167–8.

———. "Further data on the energies of beta-rays emitted from artificially produced radioactive bodies." *PR, 49* (1936), 203.

———. "The radiations emitted from artificially produced radioactive substances, I: The upper limits and shapes of the β-ray spectra from several elements." *PR, 49* (1936), 368–81.

Kuznick, P.J. *Beyond the laboratory: Scientists as political activists in 1930s America.* Chicago: University of Chicago Press, 1987.

Lachman, Arthur. *Borderland of the unknown: The life story of Gilbert Newton Lewis, one of the world's great scientists.* New York: Pageant Press, 1936.

Ladenburg, Rudolph. "The mass of the neutron and the stability of heavy hydrogen." *PR, 45,* (1934), 224–5.

———. "Errata—The mass of the neutron and the stability of heavy hydrogen." *PR, 45,* (1934), 495.

Lamb, W.E., Jr. "The unobservable decay of Na^{22}." *PR, 50* (1936), 388–9.

———. "Five encounters with Felix Bloch." In Chodorow et al., *Felix Bloch,* 133–45.

———. "The fine structure of hydrogen." In Brown and Hoddeson, *Birth of particle physics,* 311–28.

Langdorf, A., Jr. "Fission products of thorium." *PR, 56* (1939), 205.

Langer, R.M., and N. Rosen. "The neutron." *PR, 37* (1931), 1579–82.

Laslett, L.J. "A long period positron activity." *PR, 50* (1936), 388.

———. "A long period positron activity: Na^{22}." *PR, 52* (1937), 529–30.

Laslett, L.J., and D.G. Hurst. "The energy losses of fast electrons." *PR, 52* (1937), 1035–9.

Lassen, N.D. "Lidt af historien om cyclotronen på Niels Bohr Institutet." *Fysisk tidsskrift, 60* (1962), 90–119.

Latimer, Wendell. "A theory of the arrangement of protons and electrons in the atomic nucleus." *JACS, 53* (1931), 981–90.

———. "The existence of neutrons in the atomic nucleus." *JACS, 54* (1932), 2125–6.

Lauritsen, C.C., and R.D. Bennett. "A new high potential x-ray tube." *PR, 32* (1928), 850–7.

Lauritsen, C.C., and H.R. Crane. "Gamma-rays from carbon bombardment with deutons." *PR, 45* (1934), 345–6.

———. "Transmutation of lithium by deutons and its bearing on the mass of the neutron." *PR, 45* (1934), 550–2.

Lauritsen, C.C., H.R. Crane, and W.W. Harper. "Artificial production of radioactive substances." *Science, 79* (1934), 234–5.

Lawrence, E.O. "The role of the Faraday cylinder in the measurement of electron currents." NAS. *Proceedings, 12* (1926), 29–31.

———. "A principle of correspondence." *Science, 64* (1926), 142.

———. "Radioactive sodium produced by deuton bombardment." *PR, 46* (1934), 746.

———. "Transmutation of sodium by deutons." *PR, 47* (1935), 17–27.

———. "Artificial radioactivity." *Ohio journal of science, 35* (1935), 388–405.

———. "Science and technology." *RSI, 8* (1937), 311–3.

———. "The biological action of neutron rays." *Radiology, 29* (1937), 313–22.

———. "The work of the Radiation Laboratory, University of California." August 1937. Lawrence Papers (3/38).

———. "The medical cyclotron of the William H. Crocker Radiation Laboratory." *Science, 90* (1939), 407–8.

———. "Acceptance of the Nobel prize award." In U.C., *1939 prize*, 33–6.

———. "The new frontiers in the atom." *Science, 94* (1941), 221–5.

———. "The evolution of the cyclotron." In *Les prix Nobel en 1951.* Stockholm: Nobelstifting, 1952. Pp. 127–40. Also in Livingston, *Development*, 136–49.

Lawrence, E.O., et al. "Initial performance of the 60-inch cyclotron of the William H. Crocker Radiation Laboratory, University of California." *PR, 56* (1939), 124–6.

Lawrence, E.O., and Donald Cooksey. "On the apparatus for the multiple acceleration of light ions to high speeds." *PR, 50* (1936), 1131–40.

Lawrence, E.O., and F.G. Dunnington. "On the early stages of electric sparks." *PR, 35* (1930), 396–407.

Lawrence, E.O., and N.E. Edlefson. "The ionization of Cs vapor by light of frequency greater than the series limit." *PR, 33* (1929), 265.

———. "The photo-ionization of the vapors of caesium, rubidium, and potassium." *PR, 33* (1929), 1086–7.

———. "The photo-ionization of potassium vapor." *PR, 34* (1929), 1056–60.

———. "On the production of high speed protons." *Science, 72* (1930), 376–7.

Lawrence, E.O., M.C. Henderson, and M.S. Livingston. "The transmutation of fluorine by proton bombardment and the mass of fluorine 19." *PR, 46* (1934), 324–5.

Lawrence, E.O., and M.S. Livingston. "A method for producing high speed hydrogen ions without the use of high voltage." *PR, 37* (1931), 1707.

———. "The production of high speed protons without the use of high voltage." *PR, 38* (1931), 834.

————. "Production of high speed light ions without the use of high voltage." *PR, 40* (1932), 19–35.

————. "The emission of protons and neutrons from various targets bombarded by three million volt deutons." *PR, 45* (1934), 220.

————. "The multiple acceleration of ions to very high speeds." *PR, 45* (1934), 608–12.

Lawrence, E.O., M.S. Livingston, and G.N. Lewis. "The emission of protons from various targets bombarded by deutons of high speed." *PR, 44* (1933), 56.

Lawrence, E.O., M.S. Livingston, and M.G. White. "The disintegration of lithium by swiftly moving protons." *PR, 42* (1932), 150–1.

Lawrence, E.O., E.M. McMillan, and M.C. Henderson. "Transmutations of nitrogen by deutons." *PR, 47* (1935), 273–7.

Lawrence, E.O., E.M. McMillan, and R.L. Thornton. "The transmutation functions for some cases of deuteron induced radioactivity." *PR, 48* (1935), 493–9.

Lawrence, E.O., and David Sloan. "The production of high speed cathode rays without the use of high voltage." NAS. *Proceedings, 17* (1931), 64–70.

Lawrence, J.H. "Some biological applications of neutrons and artificial radioactivity." *Nature, 145* (1940), 125–7.

————. "Early experiences in nuclear medicine." *Journal of nuclear medicine, 20:6* (1979), 561–4.

Lawrence, J.H., P.C. Aebersold, and E.O. Lawrence. "Comparative effects of x-rays and neutrons on normal and tumor tissue." NAS. *Proceedings, 22* (1936), 543–57.

Lawrence, J.H., and E.O. Lawrence. "The biological action of neutron rays." NAS. *Proceedings, 22* (1936), 124–33.

Lawrence, J.H., Bernard Manowitz, and B.S. Loeb. *Radioisotopes and radiation: Recent advances in medicine, agriculture, and industry.* New York: McGraw-Hill, 1964.

Lawrence, J.H., and K.G. Scott. "Comparative metabolism of phosphorus in normal and lymphomatous animals." SEBM. *Proceedings, 40* (1939), 694–6.

Lawrence, J.H., and Robert Tennant. "The comparative effects of neutrons and x rays on the whole body." *Journal of experimental medicine, 66* (1937), 667–87.

Lawrence, J.H., L.W. Tuttle, K.G. Scott, and C.L. Connor. "Studies on neoplasms with the aid of radioactive phosphorus, I: The total phosphorus metabolism of normal and leukemic mice." *Journal of clinical investigation, 19* (1940), 267–71.

Lea, D.E. "Combination of proton and neutron." *Nature, 133* (1934), 24.

Lewis, G.N. *Valence and the structure of atoms and molecules.* New York: Chemical Catalogue Co., 1923

———. "The biochemistry of water containing hydrogen isotopes." *JACS, 55* (1933), 3503–4.

Lewis, G.N., M.S. Livingston, M.C. Henderson, and E.O. Lawrence. "The disintegration of deutons by high speed protons and the instability of the deuton." *PR, 45* (1934), 242–4.

———. "On the hypothesis of the instability of the deuteron." *PR, 45* (1934), 497.

Lewis, G.N., M.S. Livingston, and E.O. Lawrence. "The emission of alpha particles from various targets bombarded by deutons of high speed." *PR, 44* (1933), 55–6.

———. "The disintegration of nuclei by swiftly moving ions of the heavy isotopes of hydrogen." *PR, 44* (1933), 317.

Lewis, G.N., and R.T. Macdonald. "Some properties of pure $H^2 H^2 O$." *JACS, 55* (1933), 3057–9.

———. "The viscosity of $H^2 H^2 O$." *JACS, 55* (1933), 4730–1.

———. "Concentration of H^2 isotope." *Journal of chemical physics, 1* (1933), 341–4.

Lewis, G.N., A.R. Olson, and William Maroney. "The dielectric constant of $H^2 H^2 O$." *JACS, 55* (1933), 4731.

Lewis, G.N., and P.W. Schutz. "Neutron refraction." *PR, 51* (1937), 1105.

Lewis, Oscar. *George Davidson: Pioneer West Coast scientist.* Berkeley: University of California Press, 1954.

Libby, W.F., and D.D. Lee. "Energies of the soft beta-particles of radiation and other bodies: Method for their determination." *PR, 55* (1939), 245–51.

Libby, W.F., M.D. Peterson, and W.M. Latimer. "Alpha radioactivity of argon formed by radiochlorine." *PR, 48* (1935), 571–2.

Liebner, C. "Die Spaltprodukte aus der Bestrahlung des Urans mit Neutronen: Die Strontium-Isotope." *Nwn, 27* (1939), 421–3.

Linford, L.H. "The emission of electrons by swiftly moving mercury ions." *PR, 47* (1935), 279–82.

Livingood, J.J. "Radioactivities of zinc under deuteron bombardment." *PR, 49* (1936), 206.

———. "Deuteron-induced radioactivity in bismuth." *PR, 49* (1936), 876.

———. "Deuteron-induced radioactivities of antimony and tin." *PR, 50* (1936), 385.

———. "Deuteron-induced radioactivities in ruthenium and copper." *PR, 50* (1936), 391.

———. "Deuteron-induced radioactivities." *PR, 50* (1936), 425–34.

Livingood, J.J., F. Fairbrother, and G.T. Seaborg. "Radioactive isotopes of manganese, iron, and cobalt." *PR, 52* (1937), 135.

Livingood, J.J., and G.T. Seaborg. "Deuteron-induced radioactivity in tin." *PR, 50* (1936), 435–9.

———. "Radioactive antimony isotopes." *PR, 52* (1937), 135–6.

———. "Radio isotopes of nickel." *PR, 53* (1938), 765.

———. "Radioactive isotopes of iron." *PR, 54* (1938), 51–5.

———. "Long period of radioactive zinc." *PR, 54* (1938), 239.

———. "Radioactive manganese isotopes." *PR, 54* (1938), 391–7.

———. "Radioactive isotopes of iodine." *PR, 54* (1938), 775–82.

———. "Radioactive antimony from I + n and Sn + D." *PR, 55* (1939), 414–15.

———. "Radioactive isotopes of zinc." *PR, 55* (1939), 457–63.

———. "A table of induced radioactivities." *RMP, 12* (1940), 30–46.

Livingston, M.S.. *The production of high velocity hydrogen ions without the use of high voltages.* Ph.D. thesis, University of California, Berkeley, 14 April 1931.

———. "High speed hydrogen ions." *PR, 42* (1933), 441–2.

———. "The magnetic resonance accelerator." *RSI, 7* (1936), 55–68.

———. "Ion sources for cyclotrons." *RMP, 18* (1946), 293–9.

———. "History of the cyclotron, part I." *PT, 12:10* (1959), 18–23, and in Weart and Phillips, *History of physics,* 255–60.

———. *Particle accelerators: A brief history.* Cambridge, Mass.: Harvard University Press, 1969.

Livingston, M.S., ed. *The development of high-energy accelerators.* New York: Dover, 1966.

Livingston, M.S., and H.A. Bethe. "Nuclear physics: C. Nuclear dynamics, experimental." *RMP, 9* (1937), 245–390.

Livingston, M.S., and J.P. Blewett. *Particle accelerators.* New York: McGraw-Hill, 1962.

Livingston, M.S., J.H. Buck, and R.D. Evans. "The Massachusetts Institute of Technology cyclotron." *PR, 55* (1939), 1110.

Livingston, M.S., M.C. Henderson, and E.O. Lawrence. "Neutrons from deutons and the mass of the neutron." *PR, 44* (1933), 781–2.

———. "Neutrons from beryllium bombarded by deutons." *PR, 44* (1933), 782–3.

———. "Radioactivity artificially induced by neutron bombardment." NAS. *Proceedings, 20* (1934), 470–5.

Livingston, M.S., M.G. Holloway, and C.P. Baker. "A capillary ion source for the cyclotron." *RSI, 10* (1939), 63–7.

Livingston, M.S., and E.O. Lawrence. "The disintegration of aluminum by swiftly moving protons." *PR, 43* (1933), 369.

Livingston, M.S., and E.M. McMillan. "The production of radioactive oxygen." *PR. 46* (1934), 437–8.

Loofbourow, J.R. "Borderland problems in biology and physics." *RMP, 12* (1940), 267–358.

Lozier, W.W., P.T. Smith, and Walker Bleakney. "H^2 in heavy hydrogen." *PR, 45* (1934), 655.

Lyman, E.M. "The shape of the β-ray spectrum of P^{32}." *PR, 50* (1936), 385.

———. "The beta-ray spectra of radium E and radioactive phosphorus." *PR, 51* (1937), 1–7.

Macomber, Ben. *The jewel city: Its planning and achievement.* San Francisco: Williams, 1915.

McGucken, William. "The social relations of science: The British Association for the Advancement of Science." APS. *Proceedings, 123* (1979), 236–64.

———. *Science, society, and state: The social relations of science movement in Great Britain, 1931–1947.* Columbus: Ohio State University Press, 1984.

McMillan, E.M. "Absorption measurements of hard gamma rays from flourine bombarded by protons." *PR, 46* (1934), 325.

———. "Some gamma rays accompanying artificial nuclear disintegrations." *PR, 46* (1934), 868–73.

———. "Artificial radioactivity of very long life." *PR, 49* (1936), 875–6.

———. "Radioactive recoils from uranium activated by neutrons." *PR, 55* (1939), 510.

———. "A brief history of cyclotron development at the University of California." Jan 1940. Lawrence papers (22/11)

———. "The transuranium elements: Early history." In *Les prix Nobel en 1951.* Stockholm: Nobelstifting, 1952. Pp. 165–73.

——. "History of the cyclotron." *PT, 12:10* (1959), 24–34, and in Weart and Phillips, *History of physics,* 261–71.

McMillan, E.M., and P.H. Abelson. "Radioactive element 93." *PR, 57* (1940), 1185–6.

McMillan, E.M., M.D. Kamen, and Samuel Ruben. "Neutron-induced radioactivity of the noble metals." *PR, 52* (1937), 375–7.

McMillan, E.M., and E.O. Lawrence. "Transmutation of aluminum by deutons." *PR, 47* (1935), 343–8.

McMillan, E.M., and M.S. Livingston. "Artificial radioactivity produced by the deuteron bombardment of nitrogen." *PR, 47* (1935), 452–7.

McMillan, E.M., and W.W. Salisbury. "A modified arc source for the cyclotron." *PR, 56* (1939), 836.

Manegold, Karl-Heinz. *Universität, Technische Hochschule, Industrie.* Berlin: Duncker and Humbolt, 1970.

Mann, W.B. "Nuclear transformations produced in zinc by alpha-particle bombardment." *PR, 54* (1938), 649–52.

——. "Recent developments in cyclotron technique." *Nature, 143* (1939), 583–5.

——. "The cyclotron and some of its applications." *Reports on progress in physics, 6* (1939), 125–36.

——. *The cyclotron.* London: Methuen, 1940.

——. "Ernest Orlando Lawrence." Physical Society of London. *Proceedings, 53* (1941), 1–4.

Marcy, Willard, ed. *Patent policy. Government, academic, university concepts.* Washington, D.C.: American Chemical Society, 1978. (American Chemical Society. *Symposium series,* no. 81.)

Maria, Michelangelo de, and Arturo Russo. "'Cosmic-ray romancing': The discovery of the latitude effect and the Compton-Millikan controversy." *HSPS, 19* (1989), in press.

Massachusettes Institute of Technology. Radiation Laboratory Staff Members, 1940–1945. Cambridge, Mass.: MIT Radiation Laboratory, 1946.

——. *Five years at the Radiation Laboratory...1940–1945.* Cambridge, Mass.: MIT, ca. 1947.

Meitner, Lise. "Künstliche Umwandlungsprozesse beim Uran." In Egon Bretscher, ed. *Kernphysik.* Berlin: Springer, 1936. Pp. 24–42.

——. "Wege und Irrwege zur Kernenergie." *Naturwissenschaftliche Rundschau, 16* (1963), 167–9.

Meitner, Lise, and O.R. Frisch. "Disintegration of uranium by neutrons: A new type of nuclear reaction." *Nature, 143* (1939), 239–40.

———. "Products of the fission of the uranium nucleus." *Nature, 143* (1939), 471–2.

Meitner, Lise, Otto Hahn, and Fritz Strassmann. "Über die Umwandlungsreihen des Urans, die durch Neutronenbestrahlung erzeugt werden." *Zeitschrift für Physik, 106* (1937), 249–70.

Mellon, A.W. "The value of individual research." NRC. *Bulletin, 1* (1919), 16–8.

Menke, H., and G. Herrmann. "Was waren die 'Transurane' der dreissiger Jahre in Wirklichkeit?" *Radiochimica acta, 16* (1971), 119–23.

Milatz, J.M.W., and D.T.J. ter Horst. "Spectroscopy of slow neutrons." *Physica, 5* (1938), 796.

Millikan, R.A. "The new opportunity in science." *Science, 50* (1919), 292–7.

———. "Available energy." *Science, 68* (1928), 279–84.

———. "Alleged sins of science." *Scribner's, 87* (1930), 119–29.

Millikan, R.A., and C.C. Lauritsen. "Relation of field currents to thermionic currents." NAS. *Proceedings, 14* (1928), 45–9.

Møller, Christian. "On the capture of orbital electrons by nuclei." *PR, 51* (1937), 84–5.

Morgan, Jane. *Electronics in the West: The first fifty years.* Palo Alto, Calif.: National Press Books, 1967.

Münsterberg, Hugo. "The x rays." *Science, 3* (1896), 161–3.

Murlin, J.R. "Science and culture." *Science, 80* (1934), 81–6.

Nahmias, M.E. *Technique du cyclotron: La désintégration de la matière et la radiobiologie.* Paris: Editions de la Revue d'Optique, 1942.

———. *Le cyclotron: La désintégration de la matière et la radiobiologie.* Paris: Editions de la Revue d'Optique, 1945.

———. *Machines atomiques.* Paris: Editions de la Revue d'Optique, 1950.

Nakayama, Shigeru, D.L. Swain, and Eri Yagi, eds. *Science and society in Japan.* Cambridge, Mass.: MIT Press, 1974.

National Research Council. *National research fellowships, 1919–1938.* Washington, D.C.: NRC, 1938.

———. Physics Survey Committee. *Physics in perspective.* 2 vols. Washington, D.C., 1972.

National Resource Committee. *Research: A national resource.* 3 vols. *1, Relation of the federal government to research. 2, Industrial research. 3, Business research.* Washington, D.C.: GPO, 1938–41.

Nelson, Eldred. "Internal conversion in the L shell." *PR, 57* (1940), 252.

Newson, H.W. "The radioactivity induced in oxygen by deuteron bombardment." *PR, 48* (1935), 790–6.

———. "Transmutation functions of high bombarding energies." *PR, 51* (1937), 620–3.

———. "The radioactivity induced in silicon and phosphorus by deuteron bombardment." *PR, 51* (1937), 624–7.

Nicholls, W.H. "Some economic aspects of university patents." *Journal of farm economics, 21* (1939), 494–8.

Nier, A.O., E.T. Booth, J.R. Dunning, and A. von Grosse. "Nuclear fission of separated uranium isotopes." *PR, 57* (1940), 546.

———. "Further experiments on fission of separated uranium isotopes." *PR, 57* (1940), 748.

Nishina, Yoshio, Tamaichi Yasaki, and Sukeo Watanabe. "The installation of a cyclotron." Institute of physical and chemical research. *Scientific papers, 34:854* (1938), 1658–68.

Nobel Foundation. *Directory.* Stockholm: Sturetryckeriet, 1981.

Noddack, Ida. "Über des Element 93." *Angewandte Chemie, 47* (1934), 653–5.

Norberg, Arthur. *Chemistry in California's history.* Berkeley: Bancroft Library, 1976.

Ofstrosky, M., and Gregory Breit. "The excitation function of lithium under proton bombardment." *PR, 49* (1936), 22–34.

Oldenberg, Otto. "Artificial radioactivity of tantalum." *PR, 53* (1938), 35–9.

Oliphant, M.L.E. "The two Ernests." *PT, 19:9* (1966), 35–49, and *19:10* (1966), 41–54, and in Weart and Phillips, *History of physics,* 173–84, 185–93.

Oliphant, M.L.E., Paul Harteck, and Ernest Rutherford. "Transmutation effects observed with heavy hydrogen." *PRS, A144* (1934), 692–703, and in Rutherford, *CP, 3,* 386–96.

———. "Transmutation effects observed with heavy hydrogen." *Nature, 133* (1934), 413, and in Rutherford, *CP, 3,* 384–5.

Oliphant, M.L.E., A.E. Kempton, and Ernest Rutherford. "The accurate determination of the energy released in certain nuclear transformations." *PRS, A149* (1935), 406–16.

———. "Some nuclear transformations of beryllium and boron, and the masses of the light elements." *PRS, A150* (1935), 241–58.

Oliphant, M.L.E., B.B. Kinsey, and Ernest Rutherford. "The transmutation of lithium by protons and by ions of the heavy isotope of hydrogen." *PRS, A141* (1933), 722–33, and in Rutherford, *CP, 3*, 351–61.

Oliphant, M.L.E., and Ernest Rutherford. "Experiments on the transmutation of elements by protons." *PRS, A141* (1933), 259–81, and in Rutherford, *CP, 3*, 329–50.

O'Neal, R.D., and Maurice Goldhaber. "Radioactive hydrogen from the transmutation of beryllium by deuterons." *PR, 57* (1940), 1086–7.

O'Neill, Edmund. "Work of the Department of Chemistry in war time." *U.C. Chronicle, 20:1* (1918), 82–92.

Oppenheimer, J.R. "On the quantum theory of autoelectric field currents." NAS. *Proceedings, 14* (1928), 363–5.

———. "The disintegration of lithium by protons." *PR, 43* (1933), 380.

Oppenheimer, J.R., and Melba Phillips. "Note on the transmutation function for deuterons." *PR, 48* (1935), 500–2.

Osietzki, Maria. "Kernphysikalische Grossgeräte zwischen naturwissenschaftliche Forschung, Industrie und Politik. Zur Entwicklung der ersten deutschen Teilbeschleuniger bei Siemens, 1943–45." *Technikgeschichte, 55* (1988), 25–46.

Owens, Mark, Jr. "Patent program of the University of California." In Marcy, *Patent policy*, 65–8.

P., A. "Sodio attivato con deutoni." *Nuovo cimento, 12* (1935), 123–4.

Packard, Charles, and C.C. Lauritsen. "The biological effects of high voltage x rays." *Science, 73* (1931), 321–2.

Palmaer, W. "Le prix Nobel de Chimie pour l'année 1935." In *Les prix Nobel en 1935*. Stockholm: Nobelstifting, 1937. Pp. 33–8.

———. [Presentation of Nobel prize to Urey, 1934.] Nobel Foundation. *Nobel lectures...chemistry, 1922–41*. Amsterdam: Elsevier, 1966. Pp. 333–8.

Palmer, A.M., and F.P. Garvan. *Aims and purposes of the Chemical Foundation, Inc., and the reasons for its organization*. New York: Chemical Foundation, 1919.

Palmer, A.M. "University patent policies." Patent Office Society. *Journal, 16:2* (1934), 96–131.

Paneth, F.A., and H. Loleit. "Chemical detection of artificial transmutation of elements." *Nature, 136* (1935), 950.

Paneth, F.A., and G.P. Thomson. "Attempts to produce helium 3 in quantity." *Nature, 136* (1935), 334.

Paul, W. "Early days in the development of accelerators." In *Aesthetics and science: Proceedings of the International Symposium in honor of Robert R. Wilson.* Batavia, Il.: Fermilab, 1979, 25–68.

Pauli, Wolfgang. *Wissenschaftlicher Briefwechsel.* Ed. A. Hermann, K. von Meyenn, and V.F. Weisskopf. 2 vols. Berlin: Springer, 1979–85.

Perlman, I., and I.L. Chaikoff. "Radioactive phosphorus as an indicator of phospholipid metabolism, V: On the mechanism of the action of choline upon the liver of the fat-fed rat." *Journal of biological chemistry, 127* (1939), 211–20.

———. "Radioactive phosphorus as an indicator of phospholipid metabolism, VII: The influence of cholesterol upon phospholipid turnover in the liver." *Journal of biological chemistry, 128* (1939), 735–43.

Perlman, I., Samuel Ruben, and I.L. Chaikoff. "Radioactive phosphorus as an indicator of phospholipid metabolism, I: The rate of formation and destruction of phospholipids in the fasting rat." *Journal of biological chemistry, 122* (1937), 169–82.

Perrier, C., M. Santangelo, and Emilio Segrè. "Radioactive isotopes of zinc and cobalt." *PR, 53* (1938), 104–5.

Perrier, C. and Emilio Segrè. "Some chemical properties of element 43." *Journal of chemical physics, 5* (1937), 712–16, and *7* (1939), 155–6.

———. "Radioactive isotopes of element 43." *Nature, 140* (1937), 193–4.

———. "Technetium: The element of atomic number 43." *Nature, 159* (1947), 24

Pestre, Dominique. *Physique et physiciens en France, 1918–1940.* Paris: Editions des archives contemporaines, 1984.

Pettitt, G.A. "Transmutation." *California monthly, 27* (1931), 18–21.

———. *Twenty-eight years in the life of a university president.* Berkeley: University of California Press, 1966.

Phillips, Melba. "The photoionization of atomic potassium." *PR, 39* (1932), 552.

———. "Photoionization probabilities of atomic phenomena." *PR, 39* (1932), 905–12.

Planck, Max. "Neue Erkenntnisse der Physik." *Die Woche* (1931), 1419–20.

———. "Ein Blick an das Universum: Die Stellung und Bedeutung der heutigen Physik." *Ernte, 13:7* (1932), 31–3.

Pollard, Ernest. "Mass and stability of C^{14}." *PR, 56* (1939), 1168.

Powell, C.F. "Development of nuclear research in Bristol." Typescript. Powell Papers, University of Bristol Library.

Powers, P.N. "The magnetic scattering of neutrons." *PR, 54* (1938), 827–38.

Powers, P.N., H. Carroll, H. Beyer, and J.R. Dunning. "The sign of the magnetic moment of the neutron." *PR, 52* (1937), 38–9.

Powers, P.N., H. Carroll, and J.R. Dunning. "Experiments on the magnetic moment of the neutron." *PR, 51* (1937), 1112–3.

Pupin, Michael. "Romance of the machine." *Scribner's, 87* (1930), 130–7.

Pursell, C.W., Jr. "The anatomy of a failure: The Science Advisory Board, 1933–35." APS. *Proceedings, 109* (1965), 342–51.

———. "The administration of science in the Department of Agriculture." *Agricultural history, 42* (1968), 231–40.

———. "A preface to government support of research and development: Research legislation and the National Bureau of Standards, 1935–1941." *Technology and society, 9* (1968), 145–64.

———. "'A savage struck by lightning': The idea of a research moratorium, 1927–1937." *Lex et scientia, 10* (1974), 146–61.

Quill, L.L. "The transuranic elements." *Chemical reviews, 23* (1938), 87–155.

Rabi, I.I. "Space quantization in a gyrating magnetic field." *PR, 51* (1937), 652–4.

Rabi, I.I., Robert Serber, V.F. Weisskopf, Abraham Pais, and G.T. Seaborg. *Oppenheimer.* New York: Scribner's, 1969.

Raju, M.R. *Heavy particle radiotherapy.* New York: Academic Press, 1980.

Ramsay, N.F. "The neutron magnetic moment." In Trower, *Discovering Alvarez,* 30–2.

Rand, M.J. "The National Research Fellowships." *The scientific monthly, 73:8* (1951), 71–80.

Rasetti, Franco. *Elements of nuclear physics.* New York: Prentice-Hall, 1936.

———. "Sorgenti artificiale di neutroni." *Viaggi, 3* (1936), 77–9.

———. "Risultati moderni della fisica nucleare." *Nuovo cimento, 14* (1937), 376–9.

Rendy, L.S. "The electric power situation in California." *Journal of electricity, 44* (1919), 479–81.

Reich, L.S. *The making of American industrial research: Science and business at GE and Bell, 1876–1926.* Cambridge: Cambridge University Press, 1985.

Reingold, Nathan, ed. *The sciences in the American context: New perspectives.* Washington, D.C.: Smithsonian Institution Press, 1979.

Richter, Steffen. *Forschungsförderung in Deutschland, 1920–1936.* Düsseldorf: VDI, 1972.

Rider, Robin E. "Alarm and opportunity: Emigration of mathematicians and physicists to Britain and the United States, 1933–1945." *HSPS, 15:1* (1984), 107–76.

Riess, Suzanne B. "Ida (Wittschen) Sproul, 1891–1981: The president's wife." Typescript of interviews, 1980–81. TBL.

Rigden, J.S. "Molecular beam experiments on the hydrogens during the 1930's." *HSPS, 13:2* (1983), 335–73.

———. *Rabi. Scientist and citizen.* New York: Basic Books, 1987.

Roberts, R.B., L.R. Hafstad, R.C. Meyer, and P. Wang. "The delayed neutron emission which accompanies fission of uranium and thorium." *PR, 55* (1939), 664.

Roberts, R.B., R.C. Meyer, and L.R. Hafstad. "Droplet fission of uranium and thorium nuclei." *PR, 55* (1939), 416–7.

Roberts, R.B., R.C. Meyer, and P. Wang. "Further observations on the splitting of uranium and thorium." *PR, 55* (1939), 510–1.

Robertson, I.K. *Atomic artillery.* New York: Van Nostrand, 1937.

Robertson, Peter. *The early years: The Niels Bohr Institute, 1921–1930.* Copenhagen: Akademisk Forlag, 1979.

Rollefson, G.K., and W.F. Libby. "Primary photochemical processes in solution." *Journal of chemical physics, 5* (1937), 569–71.

Roosevelt, F.D. "The responsibility of engineering." *Science, 84* (1936), 393–4.

Root, Elihu. "The need for organization in scientific research." NRC. *Bulletin, 1* (1919), 7–10.

Rose, M.E. "Focusing and maximum energy of ions in the cyclotron." *PR, 53* (1938), 392–408.

———. "Maximum energy of ions from the cyclotron." *PR, 53* (1938), 675.

Rosenberg, C.E. "Martin Arrowsmith: The scientist as hero." In Rosenberg, *No other gods: On science and American social thought.* Baltimore: Johns Hopkins University Press, 1976. Pp. 123–31.

Rozental, Stefan, ed. *Niels Bohr: His life and work as seen by his friends and colleagues.* Amsterdam: North Holland, 1968.

Ruben, Samuel, W.Z. Hassid, and M.D. Kamen. "Radioactive carbon in the study of photosynthesis." *JACS, 61* (1939), 661–3.

———. "Photosynthesis with radioactive carbon, II: Chemical properties of the intermediates." *JACS, 62* (1940), 3443–50.

———. "Radioactive nitrogen in the study of N_2 fixation by non-leguminous plants." *Science, 91* (1940), 578–9.

Ruben, Samuel, and M.D. Kamen. "Photosynthesis with radioactive carbon, IV: Molecular weight of the intermediates and a tentative theory of photosynthesis." *JACS, 62* (1940), 3451–5.

———. "Radioactive carbon of long life." *PR, 57* (1940), 549.

———. "Long-lived radioactive carbon: C^{14}." *PR, 59* (1941), 349–54.

Rumbaugh, L.H., R.B. Roberts, and L.R. Hafstad. "Nuclear transformations of the lithium isotopes." *PR, 54* (1938), 657–80.

Russo, Arturo. "Science and industry in Italy between the two world wars." *HSPS, 16:2* (1986), 281–320.

Rutherford, Ernest. "Scientific aspects of intense magnetic fields and high voltage." *Nature, 120* (1927), 809–11, and *PRS, A117* (1928), 307–12.

———. "Address of the president." *PRS, A117* (1928), 300–16.

———. "Heavy hydrogen." *Nature, 132* (1933), 955–6.

———. "The search for isotopes of 'nydrogen and helium of mass 3." *Nature, 140* (1937), 303–5, and in Rutherford, *CP, 3,* 424–8.

———. *Collected papers.* 3 vols. London: George Allen and Unwin, 1962–65.

Rutherford, Ernest, James Chadwick, and C.D. Ellis. *Radiations from radioactive substances.* Cambridge: Cambridge University Press, 1930.

Rutherford, Ernest, and A.E. Kempton. "Bombardment of the heavy isotopes of hydrogen by particles." *PRS, A143* (1934), 724–30, and in Rutherford, *CP, 3* 377–83.

Ryan, W.P. "Discussion on ownership of intellectual property in educational institutions." American Institute of Chemical Engineers. *Transactions, 28* (1932), 209.

Rydell, R.W. "The fan dance of science: American world's fairs in the Great Depression." *Isis, 76* (1985), 525–42.

St. John, C.E. "The Spectroscopic Committee of the Division of Physical Sciences of the National Research Council." *PR, 16* (1920), 372–4.

Salow, H. "Das Miersdorfer Zyklotron." In FIAT Review of German science, 1939–46. *Nuclear physics and cosmic rays, 2.* Wiesbaden: Dietrich, 1948. Pp. 32–38.

Sargent, B.W. "The maximum energy of the rays from uranium X and other bodies." *PRS, A139* (1933), 659–73.

——. "Nuclear physics in Canada in the 1930's." In Shea, *Otto Hahn,* 221–40.

Scherer, J.A.B. *The nation at war.* New York: George A. Doran, [1918].

Schiff, L.I. "On the paths of ions in the cyclotron." *PR, 54* (1938), 1114–5.

Schuler, L.A. "Maestro of the atom." *Scientific American, 163:2* (1940), 68–71.

Schweber, S.S. "The empiricist temper regnant: Theoretical physics in the United States, 1920–1950." *HSPS, 17:1* (1986), 55–98.

Schwinger, J.S. "On the magnetic scattering of neutrons." *PR, 51* (1937), 544–52.

Seaborg, G.T. "Artificial radioactivity." *Chemical reviews, 27* (1940), 199–285.

——. *Journal.* Mimeograph. 2 vols. Lawrence Berkeley Laboratory.

——. *The transuranium elements.* New Haven: Yale University Press, 1958.

——. "Early radiochemical investigations of plutonium." In Frisch et al., *Trends,* 104–14.

Seaborg, G.T., and J.W. Kennedy. "Nuclear isomerism and chemical separation of isomers in tellurium." *PR, 55* (1939), 410.

Seaborg, G.T., and J.J. Livingood. "Artificial radioactivity as a test for minute traces of elements." *JACS, 60* (1938), 1784–6.

Seaborg, G.T., J.J. Livingood, and J.W. Kennedy. "Radioactive tellurium: Further production and separation of isomers." *PR, 55* (1939), 794.

Seaborg, G.T., E.M. McMillan, J.W. Kennedy, and A.C. Wahl. "Radioactive element 94 from deuterons from uranium." *PR, 69* (1946), 366–7.

Seaborg, G.T., and Emilio Segrè. "Nuclear isomerism of element 43." *PR, 55* (1939), 808–14.

Seaborg, G.T., and A.C. Wahl. "The chemical properties of elements 93 and 94." *JACS, 70* (1948), 1128–34.

Seaborg, G.T., A.C. Wahl, and J.W. Kennedy. "Thermal diffusion and separation of radioactive and ordinary hydrogen isotopes." *Journal of chemical physics, 8* (1940), 639–40.

——. "Radioactive element 94 from deuterons on uranium." *PR, 69* (1946), 367.

Segrè, Emilio "An unsuccessful search for transuranic elements." *PR, 55* (1939), 1104–5.

———. "Uranium." *Encyclopedia Americana*, 1946 edn., *22*, 588–9.

———. "Nuclear isomerism." *RMP, 21* (1949), 271–304.

———. "The consequences of the discovery of the neutron." ICHS, X (1962). *Acts, 1*, 149–54. Paris: Hermann, 1964.

———. *Enrico Fermi, physicist.* Chicago: University of Chicago Press, 1970.

———. "Fifty years up and down a strenuous and scenic trail." *Annual review of nuclear and particle science, 31* (1981), 1–18.

———. "A cinquant' anni della radioattività provocata da neutroni." Accademia nazionale delle scienze detta dei XL. *Memorie di scienze fisiche e naturali, 8:2* (1984), 165–75. (Convegno sui neutroni e loro applicazioni nel cinquantenario della scoperta della radioattività indotta da neutroni, 4–5 June 1984.)

———. "The adventurous history of the discovery of technetium." In Mario Nicolini, Giuliano Bandoli, and Ulderico Mazzi, eds. *Technetium in chemistry and nuclear medicine, 2.* New York: Raven; Verona: Cortina, 1986. Pp. 1–10.

———. *Mezzo secolo fra atomi e nuclei.* Milan: Montedison, 1986.

———. "K-electron capture by nuclei." In Trower, *Discovering Alvarez*, 11–2.

Segrè, Emilio, R.S. Halford, and G.T. Seaborg. "Chemical separation of nuclear isomers." *PR, 55* (1939), 321–2.

Segrè, Emilio, and A.C. Helmholz. "Nuclear isomerism." *RMP, 21* (1949), 271–304.

Segrè, Emilio, and G.T. Seaborg. "Nuclear isomerism of element 43." *PR, 54* (1938), 772.

———. "Fission products of uranium and thorium produced by high energy neutrons." *PR, 59* (1941), 212–3.

Segrè, Emilio, and C.S. Wu. "Some fission products of uranium." *PR, 57* (1940), 552.

Seidel, Arnulf. "Technetium in biology and medicine." *Gmelins Handbuch der anorganischen Chemie: Technetium. Supplement, 1*, Berlin: Springer, 1982. Pp. 271–314.

Seidel, R.W. *Physics research in California: The rise of a leading sector in American physics.* Ph.D. thesis, University of California, Berkeley, 1978. (*DAI*, 7904599.)

Serber, Robert. "Electronic orbits in the induction accelerator." *PR, 60* (1941), 53–8.

———. "The early years." In Rabi et al., *Oppenheimer*, 11–20.

———. "Particle physics in the 1930's: A view from Berkeley." In Brown and Hoddeson, *Birth of particle physics*, 206–21.

Shea, W.R., ed. *Otto Hahn and the rise of nuclear physics.* Dordrecht and Boston: Reidel, 1983.

Sheline, G.T., T.L. Phillips, S.B. Field, J.T. Brennan, and Antolin Raventos. "Effects of fast neutrons on human skin." *American journal of roentgenology, 111* (1971), 31–41.

Shenstone, A.G. "An attempt to detect induced radioactivity resulting from ray bombardments." *Philosophical magazine, 43* (1922), 938–43.

Simpson, Lola Jean. "Le Conte Hall, the new physics building." *California monthly, 16:5* (1924), 251–3.

Sinclair, Upton. *Goose-step: A study of American education.* Pasadena, Calif.: The author, 1923.

Six, Jules. "Pourquoi ni Bothe ni les Joliot-Curie n'ont découvert le neutron?" *Revue d'histoire des sciences, 41* (1988), 3–24.

Sloan, D.H. "A radio-frequency high voltage generator." *PR, 47* (1935), 62–71.

Sloan, D.H., and W.M. Coates. "Recent advances in the production of heavy high speed ions without the use of high voltages." *PR, 46* (1934), 539–42.

Sloan, D.H., and E.O. Lawrence. "Production of heavy high speed ions without the use of high voltages." *PR, 38* (1931), 2021–32.

Sloan, D.H., R.L. Thornton, and F.A. Jenkins. "A demountable power-oscillator tube." *RSI, 6* (1935), 75–82.

Smith, Alice Kimball, and Charles Weiner, eds. *Robert Oppenheimer: Letters and recollections.* Cambridge, Mass.: Harvard University Press, 1980.

Smith, Norman. "The origins of the water turbine." *Scientific American, 242:1* (1980), 138–48.

Smyth, H.D. *Atomic energy for military purposes.* Princeton: Princeton University Press, 1945.

Smyth, H.D., G.P. Harnwell, Walker Bleakney, and W.W. Lozier. "The production of helium of mass three." *PR, 47* (1935), 800–1.

Snell, A.H. "The transmutation of argon by deuterons." *PR, 49* (1936), 207.

———. "Radioactive argon." *PR, 49* (1936), 555–60.

———. "The radioactive isotopes of bromine: isomeric forms of bromine 80." *PR, 52* (1937), 1007–22.

Soiland, Albert. "Experimental clinical research work with x-ray voltages above 500,000." *Radiology, 20* (1933), 99–104.

[Solvay, 1933.] Institut international de physique Solvay. *Structure et propriétés des noyaux atomiques: Rapports et discussions du septième conseil de physique....Octobre 1933.* Paris: Gauthier-Villars, 1934.

Sorenson, R.W. "California Institute of Technology's million-volt laboratory." *Journal of electricity, 53* (1924), 242–5.

———. "Vacuum switching experiments at California Institute of Technology." American Institute of Electrical Engineers. *Transactions, 45* (1926), 1102–5.

Spence, R. "John Douglas Cockcroft." *Dictionary of Scientific Biography, 3,* 328–31.

Sproul, R.G. "Opening remarks." In U.C., *1939 prize,* 5–9.

Staub, H.H. "Ten years of neutron physics with Felix Block at Stanford, 1938–1949." In Chodorow et al., *Felix Bloch,* 193–200.

Steenbeck, Max. "Beschleunigung von Elektronen durch elektrische Wirbelfelder." *Nwn, 31* (1943), 234–5.

———. "Persönliche Erinnerungen aus der Betatron-Entwicklung." Jena Universität. *Wissenschaftliche Zeitschrift, Math.-Naturw. Reihe, 13* (1964), 437–8.

———. *Impulse und Wirkungen: Schritte auf meinen Lebensweg.* Berlin: Verlag der Nation, 1977.

Stone, R.S. "Neutron therapy and specific ionization." *American journal of roentgenology, 59* (1948), 771–85.

Stone, R.S., and P.C. Aebersold. "Clinical deductions from physical measurements of 200 and 1000 kilovolt x rays." *Radiology, 29* (1937), 297–304.

Stout, P.R., and D.R. Hoagland. "Upward and lateral movement of salt in certain plants as indicated by radioactive isotopes of potassium, sodium, and phosphorus absorbed by roots." *American journal of botany, 26* (1939), 320–4.

Strauss, Lewis. *Men and decisions.* London: Macmillan, 1963.

Stuewer, R.H. *Nuclear physics in retrospect: Proceedings of a symposium on the 1930's.* Minneapolis: University of Minnesota Press, 1979.

———. "The nuclear electron hypothesis." In Shea, *Otto Hahn,* 19–68.

———. "The naming of the deuteron." *AJP, 54* (1986), 206–18.

Swain, D.C. "The rise of a research empire: NIH, 1930–1950." *Science, 138* (1962), 1233–7.

Szilard, Leo. *The collected works, I. Scientific papers.* Ed. Bernard T. Feld and Gertrude Weiss Szilard. Cambridge, Mass.: MIT Press, 1972.

Szilard, Leo, and T.A. Chalmers. "Chemical separation of the radioactive element from its bombarded isotope in the Fermi effect." *Nature, 134* (1934), 462, and in Szilard, *CW, 1,* 143–4.

———. "Detection of neutrons liberated from beryllium by gamma rays: a new technique for inducing radioactivity." *Nature, 134* (1934), 494–5, and in Szilard, *CW, 1,* 145–6.

Szilard, Leo, and W.H. Zinn. "Instantaneous emission of fast neutrons in the interaction of slow neutrons with uranium." *PR, 55* (1939), 795–800.

Szlvessy, G. "Besondere Fälle von Doppelbrechung." *Handbuch der Physik, 21* (1929), 724–884.

Taylor, F.J., and E.F. Welty. *Black bonanza: How an oil hunt grew into the Union Oil Company of California.* New York: McGraw-Hill, 1950.

Terman, F.E. *Radio-engineering.* 2nd ed. New York: McGraw-Hill, 1937.

———. *Electronic and radio engineering.* New York: McGraw-Hill, 1955.

Texas Industrial Commission. *Texas giants: The new breed.* Austin: The Commission, 1971.

Thackray, Arnold, and Everett Mendelssohn, eds. *Science and values.* New York: Humanities Press, 1974.

Thibaud, Jean. "Production d'ions positifs de vitesse élévée par accélérations multiples." *CR, 194* (1932), 360–2.

———. "Production d'ions positifs de vitesse élévée par accélérations multiples." Congrès international d'électricité, Paris, 1932. Section 1. *Comptes rendus, 2* (1932), 962–7.

———. "Note." In Solvay, 1933, *Rapports,* 72–5.

Thomas, Jerry. "John Stuart Foster, McGill University, and the renascence of nuclear physics in Montreal, 1935–1950." *HSPS, 14:2* (1984), 357–77.

Thomas, L.H. "The paths of ions in the cyclotron, I: Orbits in the magnetic field." *PR, 54* (1938), 580–8.

———. "The paths of ions in the cyclotron, II: Paths in the combined electric and magnetic fields." *PR, 54* (1938), 588–98.

Thornton, R.L. "Artificial radioactivity induced in arsenic, nickel, and cobalt under deuteron bombardment." *PR, 49* (1936), 207, 306.

———. "The radioactive isotopes of zinc." *PR, 53* (1938), 326–7.

Tobey, R.C. *The American ideology of national science, 1919–1930.* Pittsburgh: University of Pittsburgh Press, 1971.

Tobias, C.A. "High energy carbon nuclei." In Trower, *Discovering Alvarez*, 50–3.

Todd, F.M. *The story of the exposition.* 5 vols. New York: Putnam, 1921.

Trower, W.P., ed. *Discovering Alvarez: Selected works of Luis W. Alvarez with commentary by his students and colleagues.* Chicago: University of Chicago Press, 1987.

Turner, L.A. "The non-existence of transuranic elements." *PR, 57* (1940), 157.

———. "Secondary neutrons from uranium." *PR, 57* (1940), 334.

———. "Nuclear fission." *RMP, 12:1* (1940), 1–29.

———. "Atomic energy from U^{238}." *PR, 69* (1946), 366.

Tuttle, L.W., L.A. Erf, and J.H. Lawrence. "Studies on neoplasms with the aid of radioactive phosphorus, II: The phosphorus metabolism of the nucleoprotein, phospholipid, and acid soluble fractions of normal and leukemic mice." *Journal of clinical investigation, 20* (1941), 57–61.

Tuttle, L.W., K.G. Scott, and J.H. Lawrence. "Phosphorus metabolism in leukemic blood." SEBM. *Proceedings, 41* (1939), 20–5.

Tuve, M.A. "The atomic nucleus and high voltage." *JFI, 216* (1933), 1–38.

———. "Depth-dose calculation for super x rays." *Radiology, 20* (1933), 289–95.

———. "Nuclear physics symposium: A correction." *Science, 80* (1934), 161–2.

Tuve, M.A., Odd Dahl, and L.R. Hafstad. "The production and focusing of intense positive beams." *PR, 48* (1935), 241–56.

Tuve, M.A., Odd Dahl, and C.M. van Atta. "A low-power positive-ion source of high intensity." *PR, 46* (1934), 1027–8.

Tuve, M.A., and L.R. Hafstad. "The emission of disintegration-particles from targets bombarded by protons and by deuterium ions at 1200 kilovolts." *PR, 45* (1934), 651–3.

Tuve, M.A., L.R. Hafstad, and Odd Dahl. "High-voltage tubes." *PR, 35* (1930), 1406–7.

———. "High voltage tubes." *PR, 36* (1930), 1261–2.

———. "Experiments on high voltage tubes." *PR, 37* (1931), 469.

———. "High speed protons." *PR, 39* (1932), 384–5.

———. "A stable hydrogen isotope of mass three." *PR. 45* (1934), 840–1.

――――. "High voltage technique for nuclear physics studies." *PR, 48* (1935), 315–37, and in Livingston, *Development*, 28–50.

Twentieth Century Fund. *American foundations and their fields, 3.* Compiled by Dorothy J. Davis. New York: The Fund, 1935.

――――. Ibid., *4.* Compiled by Genera Seybold. New York: Raymond Rich, 1939.

Tyndall, A.M. "A history of the department of physics in Bristol, 1876–1948, with personal reminiscences." Unpublished ms., University of Bristol library.

United States. Atomic Energy Commission. *In the matter of J. Robert Oppenheimer: Transcript of hearing before Personnel Security Branch and texts of principal documents and letters.* Cambridge, Mass.: MIT Press, 1971.

United States. 75th Cong., 1st sess., 1937. *U.S. Statutes at large, 50:1,* 559–62.

University of California. *President's report.* 1916–40.

――――. *The 1939 Nobel prize award in physics to Ernest Orlando Lawrence, February 29, 1940.* Berkeley: University of California, 1939.

Urey, H.C. "The natural system of atomic nuclei." *JACS, 53* (1931), 2872–80.

――――. "Accomplishments and future of chemical physics." *RSI, 8* (1937), 223–7.

Urey, H.C., F.G. Brickwedde, and G.M. Murphy. "A hydrogen isotope of mass 2 and its concentration." *PR, 40* (1932), 1–15.

Van Atta, L.C., D.L. Northrup, C.M. van Atta, and R.J. Van de Graaff. "The design, operation, and performance of the Round Hill Electrostatic Generator." *PR, 49* (1936), 761–76.

Van de Graaff, R.J. "A 1,500,000 volt generator." *PR, 38* (1931), 1919–20.

Van de Graaff, R.J., K.T. Compton, and L.C. van Atta. "The electrostatic production of high voltage for nuclear investigations." *PR, 43* (1933), 149–57.

Van Voorhis, S.N. "The artificial radioactivity of copper, a branch reaction." *PR, 49* (1936), 876.

Varian, R.H., and S.F. Varian. "A high frequency oscillator and amplifier." *Journal of applied physics, 10* (1939), 321–7.

Varney, R.W. "Some physics not in the Physical Review." *PT, 35:10* (1982), 24–9.

Vaughan, F.L. *The United States patent system: Legal and economic conflicts in American patent history.* Norman: University of Oklahoma Press, [1956].

Voge, H.H., and W.F. Libby. "Exchange reactions with radiosulphur." *JACS, 59* (1937), 2474.

Walke, Harold. "Radioactive indicators in the study of phosphorus metabolism in rats." *Nature, 136* (1935), 754–5.

———. "The induced radioactivity of calcium." *PR, 51* (1937), 439–45.

———. "The induced radioactivity of titanium." *PR, 51* (1937), 1011.

———. "A new radioactive isotope of potassium." *PR, 52* (1937), 663.

———. "Induced radioactivity." *Reports of the progress of physics, 6* (1939), 16–47.

Walke, Harold, E.J. Williams, and G.R. Evans. "K-electron capture, nuclear isomerism, and the long-period activities of titanium and scandium." *PRS, A171* (1939), 360–82.

Wallace, H.A. "The social advantages and disadvantages of the engineering-scientific approach to civilization." *Science, 79* (1934), 1–5.

Walton, E.T.S. "The production of high speed electrons by indirect means." Cambridge Philosophical Society. *Proceedings, 25* (1929), 469–81.

———. "Personal recollections of the discovery of fast particles." In Hendry, *Cambridge physics,* 49–55.

Ward, H.B., ed. "The Berkeley meeting of the American Association for the Advancement of Science." *Science, 80* (1934), 43–62.

Washburn, E.W., and H.C. Urey. "Concentration of the H^2 isotope of hydrogen by the fractional electrolysis of water." NAS. *Proceedings, 18* (1932), 496–8.

Weart, Spencer. "Scientists with a secret." *PT, 29* (1976), 23–30.

———. "Secrecy, simultaneous discovery, and the theory of nuclear reactors." *American journal of physics, 45* (1977), 1049–60.

———. "The physics business in America, 1919–1940: A statistical reconnaisance." In Reingold, *Sciences in the American context,* 295–358.

———. *Scientists in power.* Cambridge, Mass.: Harvard University Press, 1979.

———. "The road to Los Alamos." *Journal de physique, 43,* suppl. (1982), 301–13. (Colloque international sur l'histoire de la physique des particules.)

————. "The discovery of fission and a nuclear physics paradigm." In Shea, *Otto Hahn*, 91–133.

Weart, Spencer, and Melba Phillips, eds. *History of physics: Readings from Physics today.* New York: American Institute of Physics, 1985.

Weart, Spencer, and Gertrud Weiss Szilard, eds. *Leo Szilard: His version of the facts. Selected recollections and correspondence.* Cambridge, Mass.: MIT Press, 1978.

Weaver, Warren. *Science of change: A lifetime in American science.* New York: Scribner's, 1970.

Webster, D.L. "Cathode-ray bunching." *Journal of applied physics, 10* (1939), 501–8.

————. "The theory of klystron oscillations." *Journal of applied physics, 10* (1939), 864–72.

Weiner, Charles. "Physics in the Great Depression." *PT, 23* (1970), 31–8.

————. "1932—Moving into the new physics." *PT, 25* (1972), 40–9.

————. "Institutional settings for scientific change: Episodes from the history of nuclear physics." In Thackray and Mendelssohn, *Science and values,* 187–212.

————. "Cyclotrons and internationalism: Japan, Denmark, and the United States, 1935–1945." ICHS, XIV (1974). *Proceedings, 2* (1975), 353–65.

Weiner, Charles, ed. *History of twentieth century physics.* New York: Academic Press, 1977. (Società italiana di fisica. Scuola Internazionale di Fisica "Enrico Fermi." LVII corso. *Rendiconti.*)

Weisskopf, V.F. "Wolfgang Gentner—ein Forscherleben in unserer Zeit." In Max-Planck-Gesellschaft, *Wolfgang Gentner Gedankfeier.* Munich: Max-Planck-Gesellschaft, 1981. Pp. 23–28.

————. *Physics in the twentieth century: Selected essays.* Cambridge, Mass.: MIT Press, 1972.

Weizsäcker, C.F. von. "Metastabile Zustände der Atomkerne." *Nwn, 24* (1936), 813–4.

Welch, A.J.E. "The separation of isotopes by thermal diffusion." *Annual reports of the progress of chemistry, 36* (1940), 153–64.

Weld, L.D. "The college teacher and research." *Science, 52* (1920), 45–8.

Wells, W.H. "Production of high energy particles." *Journal of applied physics, 9* (1938), 677–89.

Wheeler, J.A. "Some men and moments in nuclear physics." In Stuewer, *Nuclear physics in retrospect,* 217–306.

Wheeler, J.A., and Rudolph Ladenburg. "Mass of the meson by the method of momentum loss." *PR, 60* (1941), 754–61.

White, M.G., and E.O. Lawrence. "The disintegration of boron by swiftly moving protons." *PR, 43* (1933), 304–5.

Wick, G.C. "Sugli elementi radioattivi di F. Joliot e I. Curie." Accademia nazionale dei lincei. Classe di scienze fisiche, matematiche e naturali. *Atti, 19* (1934), 319–24.

———. "Istituti di fisica-chimica danesi e svedesi." *Viaggi, 5* (1939), 25–31.

Wideröe, Rolf. "Über ein neues Prinzip zur Herstellung hoher Spannungen." *Archiv für Elektrotechnik, 21* (1928), 387–406. Translated as "A new principle for the generation of high voltages," in Livingston, *Development*, 92–114.

———. "Das Betatron." *Zeitschrift für angewandte Physik, 5* (1953), 187–200.

———. "Die ersten zehn Jahre der Mehrfachbeschleunigung." Jena Universität. *Wissenschaftliche Zeitschrift, Math.-Naturw. Reihe, 13* (1964), 431–6.

———. "Some memories and dreams from the childhood of particle accelerators." *Europhysics news, 15:2* (1984), 9–11.

Wilson, C.G. *California Yankee: William R. Staats, business pioneer.* Claremont, Calif.: Saunders, 1946.

Wilson of High Wray, Lord. "Natural sources of power." In Trevor I. Williams, ed. *A history of technology, 6:1. The twentieth century.* Oxford: Clarendon Press, 1978. Pp. 195–222.

Wilson, R.R. "Magnetic and electrostatic focusing in the cyclotron." *PR, 53* (1938), 408–20.

———. "Magnitude of accelerated current in the cyclotron." *PR, 54* (1938), 240.

———. "My fight against team research." In Holton, *Twentieth century sciences*, 468–79.

Wilson, R.R., and M.D. Kamen. "Internal targets in the cyclotron." *PR, 54* (1938), 1031–6.

Wisconsin, University. *A symposium on the use of isotopes in biology and medicine.* Ed. P.P. Wilson. Madison: University of Wisconsin Press, 1948.

Withrow, J.R. "Ownership of intellectual property in educational institutions." American Institute of Chemical Engineers. *Transactions, 28* (1932), 203–9.

Wynn-Williams, C.E. "An automatic magnetic field stabilizer of high sensitivity." *PRS, A145* (1934), 250–7.

———. "The scale-of-two counter." In Hendry, *Cambridge physics*, 141–9.

Yukawa, Hideki, and Shoichi Sakata. "On the theory of the β-disintegration and the allied phenomena." Physical-Mathematical Society of Japan. *Proceedings, 17* (1935), 467–79, and *18* (1936), 128–30.

Zinnser, Hans. "The next twenty years." *Science, 74* (1931), 397–404.

Zirkle, R.E., and P.C. Aebersold. "Relative effectiveness of x rays and fast neutrons in retarding growth." NAS. *Proceedings, 22* (1936), 134–8.

Zirkle, R.E., P.C. Aebersold, and E.P. Dempster. "The relative biological effectiveness of fast neutrons and x rays upon different organisms." *American journal of cancer, 29* (1937), 556–62.

Zuckerman, Harriet. *Scientific elite: Nobel laureates in the United States.* New York: Free Press, 1977.

Index